The Skeptic's Book of Lists

Karl Mamer

The HiveMind Group

Print Edition 1.19 October 2021

ISBN: 978-1-7778611-1-7

http://skepticsbookoflists.com/

Disclaimer

All trademarks used in this book are the property of their respective owners. All list items in this book have been checked by the author using at least two independent sources. However, the author assumes no responsibility for errors, inaccuracies, or omissions regarding people, products, and organizations mentioned herein. The author neither claims nor implies any authority or expertise on any topic, person, or organization covered in this book. No list item writeup should be construed as the author claiming or asserting a fact. This book is a work of opinion. List items represent the author's personal assessment of current research and what constitutes consensus opinion on people, products, and organizations mentioned in this book. Your opinion on the history, nature, conduct, and motivations of people and organizations mentioned herein may vary with that of the author and/or his sources.

Thank You

For my wife: Without you, I would not have had the time to do this.

Uncle Sam: For introducing me to the original *The Book of Lists*.

Terry Brown: For his hot takes on UFO culture.

Sean Stone: For confirming the coffee mug superstition and supplying the photographic evidence.

Neil Croll: For post-release editing.

Contents

A Skeptical Book of What?

If you were a precocious tween or teen in the early 1980s, given more to digging through the books on your relatives' shelves than spending actual time with your relatives, you probably encountered a curious book called *The Book of Lists*, or more fully *The People's Almanac Presents The Book of Lists*. It was a compendium of, well, lists. Authors David Wallenchinsky, Irving Wallace (died 1990), and Amy Wallace (died 2013) wrote and compiled hundreds of lists in the style of radio top ten lists. Not all lists were exactly ten items. You had lists like "Twenty famous high school dropouts", "Fifteen famous events that happened in a bathtub", "Nine outstanding winners of the Congressional Medal of Honor", "Seven famous men who died virgins", and "Twelve Nazi war criminals still at large".

Now if a book of lists sounds boring to you, I should add it wasn't just a dry listing of names, dates, and tabulated figures. Each list tended to flesh out items on that list with fascinating descriptions that usually focused on the more unusual aspects of the people, items, or incidents on the list. To a precocious tween or teen, *The Book of Lists* offered a fascinating pre-Internet portal into a world of stuff they didn't teach you in school about history, politics, true crime, and, yeah, sex.

I mean how many 13-year-olds in North America in 1981, if not for stumbling on *The Book of Lists* on your uncle's bookshelf, got a chance to learn about Rasputin; his purported invulnerability to poison, bullets, and stabbing implements; and his reputed talents in the bed chambers of Russian nobility? Like I say, to a curious young person of that era, the book was a springboard into whole domains of knowledge and esoteric minutiae. It expanded your vocabulary, teaching you words like "libertine", "phlegmatic", "braised trake", and "antidisestablishmentarianism".

Looking back, I've come to realize *The Book of Lists* was my generation's version of web surfing. The list format, like web surfing, requires a low investment of the reader's time and commitment. There's no particular order you have to read the book. You can flip to a random page and scan for something that piques your interest then flip to another page.

This book, *The Skeptic's Book of Lists*, seeks to recapture that joy of discovering new and fascinating tidbits of information but from a skeptical and rational perspective. Lists come from the world of skepticism, science, pseudoscience, culture, and more. List items are summarized with the intent to give the reader a good overview and hopefully stoke the reader's curiosity to head out onto the net and dig much, much deeper.

— Karl Mamer, October 4, 2021

P.S. Are You New Around Here?

Now I know I've suggested you can read this book in any order, but for the gentle reader who may be fairly new to the culture and jargon of the skeptical movement, I would invite you to flip to the last chapter's list "Twenty Skeptical Neologisms and Phrases" for a 101 on some of the terms and phrases used throughout this book. Though I trust the gentle reader can pick up what phrases like "unsinkable rubber ducky" mean in context, my wife (who has doubled as my editor) isn't as sure. So, please. Go. Now.

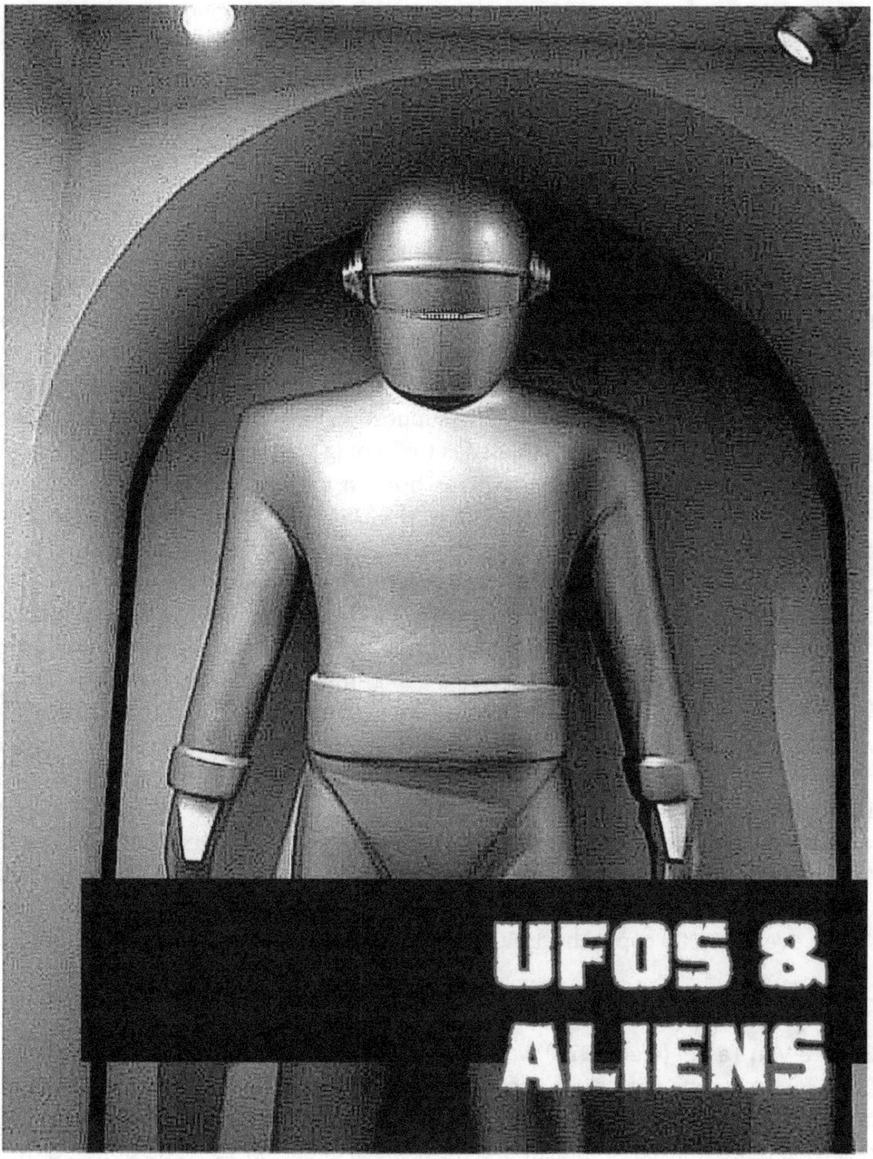
UFOS &
ALIENS

Four UFO Contactees Named George

The Contactee movement of the late 1940s and early 1950s was characterized by people claiming some form of contact with space aliens that went beyond merely seeing anomalous lights in the sky. This didn't necessarily imply physical contact with aliens. Many of the Contactees of that era claimed space aliens were contacting them via telepathic or other supernatural means. Typically, these aliens were described as benevolent beings, trying to relay messages of universal peace and the dangers of nuclear technology.

GEORGE ADAMSKI

George Adamski (died 1965) is probably the most famous of the Contactees of the 1950s. He is best known for two things. 1) His woefully incorrect predictions about what the surface of Venus would be like. 2) His creation of over 500 hoaxed UFO photos. Many of his photos became iconic imagery of what UFOs were supposed to look like. Of note was his photo of a supposed alien "scout craft", which looked a lot like a metallic lamp shade that uses three ping pong balls glued to the bottom as a means of propulsion. Adamski was born in what was then the Kingdom of Prussia and emigrated with this family to the USA in 1893. In the 1920s Adamski became interested in Theosophy and founded an organization called the Royal Order of Tibet. In 1940 Adamski moved to a ranch near Mount Palomar with a few of his followers. He ran a cafe and campground for local tourists and gave lectures on Eastern philosophy. Adamski went by the title of Professor and told his followers he grew up in a Tibetan monastery as a child. In the late 1940s, Adamski and his followers started seeing UFOs in the skies over their ranch. His claims of receiving wisdom from ancient Eastern ascendant masters transmogrified into claims of receiving wisdom from space aliens living on various planets in our solar system. In 1952, Adamski told his followers he was visited in person by a "Nordic" Venusian named Orthon. Orthon preached friendship and the dangers of nuclear war. In 1954 Adamski co-wrote *Flying Saucers Have Landed,* a book about his conversations with Orthon as well as other so-called Nordic aliens. The book sold remarkably well, selling over 100,000 copies by 1960.

GEORGE VAN TASSEL

George Van Tassel (died 1978) was an aircraft mechanic and inspector for various California aerospace firms like Lockheed and Hughes. In 1947, he moved his family out to a desert near Joshua Tree, California. There he claimed to have started receiving telepathic communications from space aliens. In 1952, he bundled these communications together in book format and released them under the title *I Rode a Flying Saucer.* It is considered the first Contactee book ever published. The title was misleading as nowhere in the book does Van Tassel describe being aboard a flying saucer. It is purely a compendium of conversations with space aliens named

Totalmon, Clatu (do note *The Day the Earth Stood Still* was released in 1951), and Ashtar. These aliens mostly warned Van Tassel about the hydrogen bomb. Out in the desert, he built his own airstrip and opened a cafe. He started hosting the annual Interplanetary Spacecraft Convention, which at its peak attracted over 10,000 attendees. Although he didn't ride a flying saucer in 1952, the following year the space aliens seemed to have granted his wish. In 1953 a Venusian named Solgonda supposedly visited him and took him for a ride on a spaceship. The alien also gave him blueprints for a dome-shaped building called an "Integratron". Van Tassel ended up building the structure in the aptly named town of Landers, California. He believed people who entered the Integratron and stood under its dome would be bathed in electromagnetic energy with healing properties. In addition, Van Tassel alleged the Integratron was a time machine. Upon his death, there was a plan to turn the Integratron into a disco but those plans fell through. Today it's a registered National Historic Site.

GEORGE KING

George King (died 1997) was a London cab driver who was interested in both Theosophy and yoga. In 1954 he allegedly achieved a yogic state that allowed space aliens from Venus and Mars to contact him. One of their leaders, a Venusian by the name of Aetherius, told him to prepare himself to be the leader of an interplanetary parliament. So, King formed the Aetherius Society and taught his followers to chant into psychic batteries that, to the non-discerning eye, appear to be normal car batteries.

Aetherius Society UK HQ in London
Photo Credit: Karl Mamer

As is the custom among leaders in the Contactee movement, he wrote his own book. King published *You Are Responsible!* in 1961. Much like the books by Adamski and Van Tassel, it was a compendium of his conversations with space aliens warning him about the dangers of nuclear energy. King set himself apart from most of the other Contactees by a curious mania for buying himself numerous diploma mill degrees and junk honors and royal titles from hucksters.

GEORGE HUNT WILLIAMSON

George Williamson (died 1967) was an archaeology student at the University of Arizona until he was expelled in 1951. Although he never attained a degree, he did prefer to be called "Doctor". Williamson was a follower of Nazi sympathizer William Dudleym Pelley. Pelley was an antisemitic cult leader and sometimes presidential candidate for the so-called Christian Party. Williamson eventually became a follower of Adamski in the early 1950s and was touted as a key eyewitness to Adamski's Orthon visitations. Not content to merely receive space brother wisdom at the feet of Adamski, Williamson started to insist he too was receiving revelations from Adamski's aliens via a Ouija board. This led to a falling out with Adamski. Williamson started his own Contactee group called the Brotherhood of the Seven Rays. Using his training as an archaeologist, Williamson helped seed in the public consciousness the notion of ancient aliens. In a 1958 book called *Secret Places of the Lion*, Williamson claimed humans had been visited throughout the ages by space aliens. They helped humankind and influenced our evolution. They also left clues to their visits in the designs of ancient temples and catacombs throughout the world. He was also one of the first in the Contactee movement to suggest not all the aliens were benevolent space brothers.

Twelve Contactee Aliens

Contactees of the 1950s and early 1960s usually claimed they were chosen by an alien representative to spread some message of importance to the good people of Earth. Many of these Contactee aliens were given names that sounded very otherworldly to 1950s sensibilities but today sound quaint or comical.

AKON

Akon is a male alien from a planet called Menton. Menton orbits Proxima Centauri. In the late 1950s, Contactee Elizabeth Klarer (died 1994) claimed to have had a romantic relationship with Akon that resulted in a space baby. Klarer was required to deliver the baby on Menton and was not allowed to return to Earth with her child.

ALAN

Alan (which is supposed to be pronounced like "a-lawn") is an alien that instructed Daniel Fry (died 1992) on physics and the history of Atlantis and Lemuria. Fry wrote about his encounter in his 1954 book *The White Sands Incident*. He went on to publish a newsletter for several decades called *Understanding* to expound upon the science and history Alan revealed. *Saucer Smear* once described the newsletter as "mind-numbing".

BUCKEY AND BOB

Contactee Buck Nelson (died 1982) claimed he was given the "Twelve Laws of God" from a pair of Venusians who took the Earth names Buckey and Bob. Nelson may be unique among Contactees as he also claimed to have met an alien dog named Bo. Bo was said to be a 385-pound Venusian dog with black fur. At UFO conventions, Nelson would sell envelopes of the dog's hair.

COSMIC AWARENESS

In 1962, former US Army officer William Ralph Duby (died 1967) believed he was receiving messages from something he called the Cosmic Awareness. The Cosmic Awareness shared with Duby its "Law of Love" which offers the appealing, if unrealistic, suggestion "Love IS the path of least resistance". Duby started to publish his communications in a newsletter and formed a publishing company called the Organization of Awareness. He located it in Olympia, Washington. He attracted a small following. Cosmic Awareness messages followed a standard form. An adherent would ask a question and Duby channeling the Cosmic Awareness would answer. Even the simplest question generated a response that ran on for several pages and generally made little to no sense. It was pure stream of consciousness writing. After Duby died in 1967 the group splintered. Groups like the Servants of Awareness were formed by adherents who each claimed they were the new "interpreter" of the Cosmic Awareness. Most splinter groups died off within a few years. Paul Shockley (died 2006) was able to retain control of the original group and renamed it Cosmic Awareness Communications. Shockley continued as the official interpreter until his death. The group attracted a wider following when the group started publishing its question/answer newsletters on Usenet groups like alt.alien.visitors in the early 1990s.

JW

JW was a Venusian that Gloria Lee (died 1962) claimed to be in contact with telepathically. According to Lee, Venusians had long ago lost the use of their vocal cords and communicated psychically. JW apparently provided Lee with plans for world peace that involved building a space station. Lee went to Washington to try and get politicians to fund the construction of her space station. No one in Congress took her seriously. In hopes she could get politicians to listen, she checked herself into a hotel room and staged a hunger strike. That attracted no publicity. After a bit more than a month into her hunger strike, she slipped into a coma and died.

LIN-ERRI

Lin-Erri is a female alien from a planet called Korendor. Lin-Erri, along with fellow Korendorians Orii, Gery-Sol, ArKay, and Master Kalen-Li, started contacting a teenage ham radio operator named Bob Renaud in 1961. Renaud would write up their communications and pass them along to publisher Gabriel Green (died 2001). Green would then publish them in his *Flying Saucers International* zine. What began with radio transmissions evolved into visitations by Lin-Erri and Orii. They took Renaud for rides on their flying saucer and showed him the location of underground bases. After Alexei Leonov (died 2019) became the first Earth man to walk in space in 1965, Renaud began to assert he had beaten Leonov's EVA by some 13 months.

NEPTUNE

In his 1955 book *The Secret of the Saucers*, Orfeo Angelucci (died 1993) wrote about being visited by an alien that called itself Neptune. A few years before the Neptune encounter, Angelucci had written a bizarre crank book on his theories of physics. It didn't win over any scientists. Amazingly, Neptune arrived on the scene a few years later to assure Angelucci he was 100% right.

ORTHON

Orthon was George Adamski's alien liaison. Orthon was a blond-haired man from Venus who communicated with Adamski telepathically. Orthon took Adamski on a trip to Venus and showed him advanced cities on the surface. The alien also revealed two additional planets in the solar system astronomers had yet to discover. The legend of Orthon managed to outlive Adamski following his death in 1965. A group of Danish Ufologists made Orthon the center of a doomsday cult, predicting a worldwide catastrophe in 1967. The cult went as far as to build their own bomb shelter in preparation for the looming disaster.

RENTAN

UFO zine publisher Gabriel Green promoted the Bob Renaud transmissions. However, he himself claimed to be in contact with an alien from Alpha Centauri named Rentan. Rentan convinced Green to run for president in 1960. Green's campaign slogan, printed on buttons, was "Abe in 1860. Gabe in 1960". Not quite "I like Ike" but not bad, all the same.

CAPTAIN AURA RHANES

Truman Bethurum (died 1969) claimed he had multiple contacts with a space vixen named Captain Aura Rhanes. She was apparently from a planet called Clarion. Bethurum became so romantically obsessed with Captain Aura Rhanes that he hired a secretary who he believed resembled the alien. His second wife eventually divorced him, citing Bethurum's infatuation with his space vixen.

VALIANT THOR

Valiant Thor was a Venusian space brother who was supposedly advising the US military about the dangers of nuclear warfare. Frank E. Stranges (died 2008) wrote about him in his 1967 book *Stranger at the Pentagon*. Valiant Thor wasn't just advising the Pentagon, he also took Stranges under his wing and taught him much about humanity. Stranges was an evangelical Christian who also happened to fervently believe in UFOs. Stranges managed to merge his two interests by turning the space brothers into ambassadors from his god. Valiant Thor wasn't just here to warn about nuclear technology. The alien's other mission was to help "mankind return to the Lord." Stranges managed something no other Contactee had ever done. He offered a crystal-clear photo of his alien. The photo, however, wasn't that impressive. It appears to be a picture of people at a UFO conference and one particularly clean-cut middle-aged man in the crowd was identified by Stranges as Valiant Thor. During his time, Stranges managed to parlay his fame into a cameo on a 1971 episode of Gerry Anderson's *UFO* series (a precursor to *Space: 1999*). He also attracted two unusually high-profile followers. Fans of Stranges' Valiant Thor tale included someone claiming to be the nephew of Admiral Byrd and the great granddaughter of President Eisenhower.

MR. X

Mr. X is a space alien from Saturn that appeared to Reinhold Schmidt (died 1974). In 1957, Schmidt wrote a book titled *Edge of Tomorrow* which detailed his encounters with Mr. X. Mr. X, along with his fetching female crew, appeared as humans although they supposedly came from Saturn. Mr. X took Schmidt aboard his mothership, which for some reason had propellers. Mr. X revealed Jesus was actually a Venusian and he had helped Jesus travel to Earth. By way of proof, Mr. X showed Schmidt Jesus' scout UFO which was parked under the pyramids. Schmidt joined the Contactee circuit, speaking at various conventions around the USA. He used that opportunity to target elderly people in attendance. He spun a tale that aliens had shown him a mine with a seam of magical quartz that could cure cancer and he needed funds to set up a mine. He was able to use that story to bilk the elderly out of about $30,000 before he was arrested.

Ten Thoroughly Unconvincing UFO Cases

THE ROSWELL SLIDES

You know Ufologists are desperate for some concrete evidence of Roswell when they turn to pictures of dead children to rally the faithful. UFO researcher Tony Braglia announced he had the smoking gun evidence: Kodachrome slides of the alien body found at Roswell! Images from the so-called Roswell slides would be revealed at a, umm, ticketed event. Braglia would reveal the mummified corpse of a dead alien under glass. To skeptics who got a sneak peek at the slides, it sure looked like a museum exhibit. It was hard to miss there was a placard on the glass case with some blurry writing. Using photo imaging software, a French skeptic managed to unblur the writing and revealed it was, yup, an old vacation slide of a museum exhibit. The placard described the non-alien under glass as the mummified body of a two-year-old boy that was displayed for a time in a Mesa Verde National Park museum.

THE MAURY ISLAND INCIDENT

Fred Crisman (died 1975) was a big fan of Richard Sharpe Shaver's *Shaver Mysteries* which were published in *Amazing Stories*. The *Shaver Mysteries* was a pet project of the magazine's editor Ray Palmer. Shaver told of an underground world controlled by evil "detrimental robots" known as "Deros". To sum up the *Shaver Mysteries* in a single sentence is to do a disservice to the story's scope and influence on modern crank mythology. But suffice it to say, even though *Amazing Stories* never came right out and claimed the stories were factual, there were true believers. Editor Palmer skirted the line as close as he could. He published a letter to the editor by Fred Crisman. In it Crisman affirmed Shaver's stories were real as he himself once shot his way out of a Deros lair with a submachine gun. Crisman must have impressed Palmer because Palmer bothered to read a second letter by Crisman wherein he alleged he had an actual piece of a UFO. Palmer, being in Chicago, wasn't willing to travel out to Crisman's home in Washington state's Puget Sound area. Palmer did pass on Crisman's story to Kenneth Arnold. Yeah, the guy who started the whole UFO flap. Arnold met with Crisman and a Crisman associated name Harold Dahl. Crisman and Dahl untruthfully identified themselves as harbor patrolmen. While on patrol, Dahl had witnessed a UFO disposing of some metallic waste on Maury Island. Crisman and Dahl showed the metal to Arnold. Arnold was super impressed and called some friends in Army intelligence to investigate. Two officers flew up from California. When they took a look at Crisman's metal, it was apparent it was just aluminum. The officers caught a B-25 plane back to their base. Tragically, the plane caught fire and crashed en route, killing all aboard. Crisman, not the least remorseful this hoax resulted in multiple deaths, claimed their plane was shot down by a mysterious energy weapon. Ufologists today continue to perpetuate this made up story.

THE BOB WHITE ARTIFACT

In 1985 Bob White and his friend were driving along the Colorado-Utah border. They claimed to see a strange light in the sky which they believed was a UFO. A piece of the UFO allegedly broke off and White managed to retrieve it. Until his death in 2009, White had been trying to convince people this piece of dark grey metal (which looks like either a petrified pinecone or something that may have broken off an H.R. Geiger sculpture) was a piece of a genuine UFO. White had the object tested a number of times (notably by Robert Bigelow's NIDS group) but he never seemed satisfied with the results. Testers kept telling him there was nothing extraterrestrial about his artifact. Those familiar with metal work have identified it as a kind of metallic stalagmite that builds up during industrial grinding processes. It can look pretty alien to someone unfamiliar with the process by which the object forms.

THE UMMO AFFAIR

In the 1960s UFO organizations in Europe began receiving letters claiming to be written by aliens from the planet UMMO (which was always written in uppercase). The UMMO writing campaign started in 1966 and petered out in the 1970s. Over three hundred letters were sent, amounting to about 1,300 pages of text. The letters seemed to pick up where the Contactee movement left off in the 1950s, with the exception the science was a bit more updated. By the 1960s we well understood Venus and Mars weren't very hospitable to advanced life, so the UMMOites were described as being from the star Wolf 424. Like the aliens of the US Contactee movement, they were of the blond Nordic variety. The letters also had a more European bent: less American style proselytizing and more café society riffing on life, the universe, and everything. True believers pointed out astounding scientific knowledge contained within the letters. People with their heads firmly screwed on quickly saw the "astounding science" was science anyone could pull out of a set of encyclopedias. Oddly the only person to ever see an alien from UMMO was José Peña, a Spanish telecommunications technician. New letters always seemed to get sent after Peña reported a sighting. As well, the groups that received those letters had Peña as a member. It was pretty obvious to most that Peña was the author of the UMMO letters. Peña himself confirmed this in 1993 when he confessed to perpetrating the hoax.

ED WALTERS GULF BREEZE

One might be lucky to see one UFO. And before the advent of high-quality cell phone cameras, one might be lucky to get one good photo of said UFO. Ed Walters of Gulf Breeze, Florida was unusually blessed when, starting in 1987, he was supposedly visited multiple times by UFOs and managed to snap eerily detailed night shots of a circular, brightly lit flying saucer. MUFON pronounced the Gulf Breeze incident as one of the most incredible cases in the history of UFOs. To many photo experts, his photos were indistinguishable from a double exposure of a night background and a model UFO. Believers shot back Walters' photos were taken with a Polaroid instant camera which everyone knew couldn't be used to take a double exposure. It's not difficult to find instructions online today on how to do double exposures with such a camera. People clearly were able to figure this out in the 1980s too. Evidence of the double exposure trick was solidified when Walters moved homes and the new owners stumbled on a foam model in the attic. It was an exact match for the UFO in Walters' photos. Walters asserts the model was planted to discredit him. One wonders why anyone would hatch a plan to discredit Walters that involves waiting for him to eventually move out of his house.

THE SCHIRMER ABDUCTION

Warren Smith (died 2003) wrote on a wide variety of woo topics, from cryptids to hollow Earth. When evidence was lacking, he simply made up facts to fit. Smith himself admitted he was a hoaxer and a close associate of his wrote a whole book on Smith's hoaxing. So, anything Smith documented about the Schirmer abduction gets slotted into the highly suspect category. Following the widely reported abduction of Betty and Barney Hill, Smith seems to have wanted to get on the abduction bandwagon. He turned an otherwise unremarkable 1967 UFO sighting by a police officer named Herb Schirmer into a full-fledged abduction experience by human-looking aliens dressed up like the aliens from the B-movie *Mars Needs Women*.

ALIEN AUTOPSY

This one is a real mess. The Fox channel aired a show in 1995 built around a film of an alleged autopsy of one of the Roswell aliens. Supposedly shot in 1947, the film showed conveniently masked pathologists dissecting a little grey alien. Obvious tells the film was a fake were the fact the autopsy techniques in the film were woefully out of line with long established practices and the alien's organs were without structure. The organs appeared like random bits of meat from a supermarket packed into a rubber doll. The effects artist who created the doll indicated the viscera was made from sheep brains, chicken, and pig knuckle, all mixed with raspberry jam. The film supposedly came to Fox by way of Ray Santilli, who up until then was a producer of music and music-themed documentaries. Santilli, while living and working in London, got a copy from a conveniently anonymous retired American military man. Santilli later claimed the film was indeed faked but he actually had a real alien autopsy film that was of such poor quality, Santilli was forced to remake the film.

MCMINNVILLE UFO PHOTOGRAPHS

In 1950, Oregon farmers Paul and Evelyn Trent claimed to have taken two photos of a UFO that was moving slowly over their farm. The photos turned out to be quite a sensation and were featured in the pages of *Life* magazine. The so-called UFO appears to be an upside-down pie plate hovering (or dangling) under a powerline. Apparently, space aliens also design their UFOs with a small knob on top where you could tie a string. The couple asserted the photos were taken near the day's end but the lighting suggests the photos were taken in the morning.

BOO VIDEO

Ufologist Stan Romanek (who was convicted in 2017 for possession of child pornography) somehow managed to talk his way onto Larry King's CNN show. On King's show he exhibited a grainy video of an alien he shot peeking in his window. The alien is a classic big-eyed grey. It bobs its head around a bit and does appear to blink but the video is so low def it's hard to tell. Many in the world of UFOs found his video cheap and laughable, dubbing it the "Boo Video". A so-called expert countered if this was a hoax it would be an expensive one, costing upwards of $50,000. The expert later turned out to be a friend of Romanek. A Colorado skeptics group knocked off an even better version for $90.

THE BILLY MEIER CASE

Billy Meier is a Swiss citizen who alleges he's had multiple contacts with a race of space aliens called Plejarens. As evidence, he has offered over the decades a wide range of photos, videos, and predictions about future events. One of his classic UFO photos (he refers to UFOs as "beamships") has been dubbed the "wedding cake UFO" for its resemblance to, well, a wedding cake. It's really quite ridiculous looking. UK skeptic Phil Langdon was able to demonstrate how one can easily create the wedding cake UFO from a round shipping container lid and various household objects. Because Meier lost part of one arm in an accident, Langdon's documentation of his reconstruction also shows how you can build it with only one full arm. Langdon has gone on to duplicate, with stunning exactitude, most of Meier's UFO photos and videos. Meier at some point claimed his Plejarens friends took him time traveling back to the age of the dinosaurs. As proof he offered an actual photo of a *Pteranodon* in full flight! Oddly, the photo has a certain blobsquatch fuzziness to it. Skeptical researcher Derek Bartholomaus on his site billymeierufocase.com shows the photo is beyond a shadow of a doubt a photograph of a *Pteranodon* illustration from a 1972 book called *Life Before Man*.

Eight Alleged Saucer Crashes Besides Roswell

In the minds of UFO believers, anything seen in the sky and on fire is a crashing saucer until proven otherwise. It beggars belief that aliens could engineer saucers that can travel faster than the speed of light but can't avoid light aircraft or shield their electronics from military radar. Lists of alleged saucer crashes published on the Internet suggest these things are coming down on average every other year. Below are some of the more famous crashes UFO believers keep returning to, despite excellent evidence for more prosaic explanations. All share a few common features: they lack physical evidence, they're based primarily if not wholly on the unreliable memories of eyewitnesses, and they are usually preceded by initial, sensational news reports that immediately reach for the ET hypothesis.

AURORA, TEXAS

Near the end of the 19th century, looming dirigible technology, like the Zeppelin, was becoming a popular sensation. They would be the next big thing. Although it wouldn't be well into the 20th century that dirigibles were actually being flown in North America, people in the USA started reporting sightings of these craft. Newspapers of the day gladly ran stories of "mystery airship" sightings. The stories grew increasingly embellished. Mystery airships were seen to sport huge wings and piloted by strange looking people wearing unfamiliar uniforms. The ultimate mystery airship story was the supposed crash of one into an Aurora, Texas judge's windmill on April 17, 1897. News of the crash was sent to the *Dallas Morning News* by a local correspondent named S.E. Haydon. His story told of people finding the body of a Martian and pieces of metal of unknown origin on the judge's land. The Martian was buried in the local cemetery, but not before being stripped of its other worldly goods. On the body of the Martian pilot was a flight log written in some strange hieroglyphic characters. The similarities to the Roswell myth are pretty deep. Haydon's story is a known fabrication. Haydon was a friend of the judge and he had a history of submitting satirical articles to various publications. This was not unusual. Newspapers of the day had a habit of running fantastical stories alongside serious news. Everyone's assumption was that readers were clever enough to discern fact from fiction. Aurora, Texas itself was literally a dying town in 1897. The town's cash crop was wiped out by a boll weevil infestation and locals were dying of yellow fever. The airship story might have been a Hail Mary attempt by Haydon to put his community back on the map. Haydon's jape was forgotten for decades until in 1973 when Hayden Hewes, founder of the International UFO Bureau, stumbled on the story. He thought he had located the grave and requested permission to exhume it. He was turned down. This didn't stop a few aged locals from stepping forward with stories of having seen the wreckage. In 1998, KDFW, a Dallas TV station, thought they had an "X-file" in their own backyard and did a multi-part story on the Aurora "UFO crash". They failed to turn up any physical evidence. This rubber ducky popped up again in 2005 on an episode of History channel's *UFO Files*. Three years later History took another kick at the can by featuring the legend in a 2008 episode of *UFO Hunters*.

AZTEC, NEW MEXICO

This alleged saucer crash in Aztec, New Mexico is considered the other Roswell by some. The crash supposedly happened eight months after the events at Roswell. Most, however, considered it a made-up story created by two scammers named Silas Newton (died 1972) and Leo Gebauer. Newton and Gebauer were trying to sell fake gadgets they claimed could find gold and oil. Of course, this was all based on alien technology recovered from a saucer that crashed near Aztec. The story was popularized in the early 1950s in *Variety* magazine by columnist Frank Scully and later in a book he wrote called *Behind the Flying Saucers*. According to Scully's account the saucer came from Venus. It was flown by three-foot-tall aliens that drank heavy water to keep hydrated. The story fell apart when Newton and Gebauer were exposed as con men by a piece of investigative journalism by J.P. Cahn. Cahn, a reporter for the *San Francisco Chronicle*, published his expose in *True* magazine. *True* magazine was roughly the 1950s version of *Maxim* magazine. Most UFO investigators for the next several decades recognized the story as a hoax. However, rubber duckies always pop back up again. The story has found new life in recent years. In 1987 a UFO researcher named William Steinman tried to revive the story by publishing a book called *UFO Crash at Aztec*. The book was not well received by the UFO community. It was described by some as an exercise in paranoia. Even Stanton Friedman (died 2019) dissed it. However, *The Aztec Incident* published by Scott and Suzanne Ramsey in 2016 gave the old hoax serious new life. *The Aztec Incident* is marked by not only a high level of credulity on the part of the authors but features some Obama "birtherism" thrown in for good measure. Despite the book's racist jab at Obama, Friedman happily wrote a forward to it. He was apparently now on the Aztec crash bandwagon. Since 2015, the legend has reached ever wider audiences, heavily promoted to true believers by *Coast to Coast AM* mainstay Linda Moulton Howe.

CHIHUAHUA, MEXICO

Why should New Mexico (a la Roswell) get all the glory? Why not regular old Mexico? Well, strap in, because the Mexican state of Chihuahua is home to its own Roswell. According to legend, in August of 1974, US authorities were tracking a saucer coming over the Gulf of Mexico and heading straight for Texas. For some reason it changed course for the bordering Mexican state of Chihuahua. There the saucer collided mid-air with a small Cessna plane and went down in flames. It's troubling that a spaceship that should be strong enough to shrug off colliding with interstellar matter at the speed of light should be so easily downed by a tin can flying at a few hundreds of miles per hour. Contemporary reports indicate the Cessna was running drugs. Witnesses reported finding bags of cash and pot in the wreckage. The state of Chihuahua is just south of Texas. It is not uncommon for small planes of questionable airworthiness to crash in the area while trying to make drug runs. No one gave much thought to the 1974 crash until 1992 when someone started mailing various UFO researchers what was claimed to be a retyped report about the crash. The report was written by a mysterious organization calling itself the Deneb Group. In its retelling of the 1974 Cessna crash, a CIA recovery team was sent into Mexico to aid the Mexican government's recovery efforts. The CIA helped the Mexican recovery team load a silver UFO onto the back of a flatbed truck. Mysteriously the Mexican recovery team was murdered on the way back to their base. The CIA team figured it now had no choice but to help themselves to the saucer and bring it back to the USA. Remember, this whole story is based on credulous UFO researchers accepting the reality of a retyped document (i.e., not even something pretending to be an original document on government letterhead).

KECKSBURG, PENNSYLVANIA

In the late afternoon of December 9, 1965, a spectacular meteor was seen burning through the skies of Ontario, Michigan, Ohio, and Pennsylvania. This meteor (dubbed by some as "the Great Lakes Fireball") was well covered by scientific journals of the day. Given the meteor's brightness and size of its fireball, witnesses assumed the meteor was at a significantly lower altitude than it actually was. When a witness saw it disappear behind a house or a tree line, it was easy to assume the meteor had crashed. There were reports of it crashing down in Michigan and Ohio. Witnesses in Kecksburg, Pennsylvania (which is south of Pittsburgh) likewise filed reports of something crashing in a nearby forest. Pennsylvania state police officials assumed there was a crash and sealed off the area to investigate. On the possibility it was some errant Soviet space technology, Army engineers were sent in. The local paper, the *Greensburg Tribune Review*, played up the UFO angle and the legend was born. Local scuttlebutt soon established the "official" story: Despite military denials nothing was found, a strange acorn-shaped spacecraft was recovered and hauled away on a flatbed truck. People who claimed to have witnessed the recovery reported the UFO's exterior featured hieroglyphs. The Kecksburg legend gained a huge boost from an early 1990s episode of *Unsolved Mysteries*. As part of the episode's recreation of the alleged events, they constructed a large acorn-shaped prop. The production crew left the prop with the town which was then erected on a raised platform as a tourist attraction.

LONG ISLAND, NEW YORK

If one were to credulously accept the claims of the Long Island UFO Network (LIUFON), you would believe there were no less than three saucer crashes on or near Long Island, New York in the late 1980s and 1990s. The first was supposedly in 1989. LIUFON's founder John Ford asserted orange lights seen over Long Island's Moriches Bay were not flares (as some observers suggested) but a byproduct of an air battle between a UFO and a squadron of government helicopters. Somehow the helicopters (craft not particularly suited for air-to-air combat) were able to blow the UFO out of the sky and kill seventeen space aliens. LIUFON alleged their bodies were taken to the Brookhaven National Laboratory. Brookhaven denies they were ever sent bodies of space aliens. In November 1992, Long Island's Southaven Park was closed to visitors for four days to accommodate duck hunting. LIUFON wasn't buying. The UFO group believed the area was cordoned off after a saucer exploded over the park. Government agents were sent in to retrieve six escape pods jettisoned by the UFO. But, wait, the loss of two saucers was still not enough to keep those pesky aliens away from Long Island. In the summer of 1995 eastern Long Island was beset by forest fires. LIUFON, yet again, knew the real story. The fires were caused when government particle beam weapons bagged a third saucer that went down in flames and started the forest fires. There would probably be no end to these increasingly improbable tales of UFOs crashing into Long Island if it weren't for the fact LIUFON's founder John Ford was arrested for the attempted murder of three Republican politicians. He was determined unfit to stand trial in the late 1990s and was held in a psychiatric hospital until 2019. Naturally, UFO buffs don't accept the possibility the dude could have been straight up bonkers. He was clearly railroaded on false charges because he was getting too close to the truth!

PADCAYA, BOLIVIA

Citizens Against UFO Secrecy (CAUS) is a UFO society formed in the late 1970s. Its mission is to get the US government to release super-secret UFO files via Freedom of Information requests and lawsuits. In 1978, the group's newsletter *Just Cause* published a series of articles about a supposed UFO crash near Padcaya, Bolivia in the spring of 1978. According to a UPI wire story from May 1978, Bolivians reported seeing something crash into a mountain range near the village of Padcaya. The UPI story suggested it might have been some kind of object returned to Earth from a space launch and NASA was investigating. *Just Cause* picked up on this story and launched a three-issue investigation. They found the claim of a NASA investigation erroneous. The USA did, however, send two Air Force officers to Bolivia to aid in a search. In 1978, the US government was taking reports of crashed space objects seriously. A few months before, in January of 1978, a nuclear-powered Soviet spy satellite called Cosmos 954 had crashed in Northern Canada. Radioactive wreckage covered hundreds of square miles of Canada's Northwest Territories and resulted in an incredibly expensive cleanup operation. The Bolivian Air Force overflew the alleged crash area a number of times but found no evidence of a downed object. *Just Cause* tapped Bob Pratt (died 2005), a reporter for the *National Enquirer*, as their man on the ground. Relying on the *National Enquirer* for accurate, factual reporting is always a mistake. Pratt reported wild tales offered by locals of seeing a UFO execute a series of desperate maneuvers before plowing into the side of a mountain. Pratt claimed he himself overflew the crash site and saw a mysterious landslide that appeared to be scorched. He also claimed the Bolivian army sent a mounted detachment to investigate the landslide and found ten monolith-like stones cut with precision. Pratt's own story about the UFO crash he was writing for the *National Enquirer* never saw the light of day. It was mysteriously killed by the editors.

SHAG HARBOUR, NOVA SCOTIA

Shortly before midnight on October 4, 1967, several residents reported seeing a low-flying object plunge into the waters of Shag Harbour. Witnesses reported a whistling sound followed by a loud bang. The local RCMP was notified and quickly investigated, assuming it was a downed aircraft. Canadian coast guard rescue boats were sent out but no bodies, survivors, or wreckage was found. It quickly became apparent no aircraft in the area were reported missing. The Canadian coast guard logged it with the Canadian air force as an "unidentified flying object" (which does not mean spaceship). The Canadian navy dispatched a dive team to the harbor but found nothing. The local paper played up the UFO angle for maximum effect and ran a rumor that the Canadian navy actually pulled something dark and mysterious out of the water. Today, the Shag Harbour incident is usually described as Canada's Roswell. Shag Harbour has capitalized on the legend to some degree. There's an annual Shag Harbour UFO festival and in 2007 the Shag Harbour UFO Incident Museum was opened. In 2019 the Canadian mint released a special glow-in-the-dark coin commemorating the incident.

ALDGATE, AUSTRALIA

In 1980, two ten-year-old boys in Aldgate, Australia claimed to have witnessed an air battle between two UFOs. The larger UFO was reported to have zapped the smaller UFO with some kind of red beam. Another independent witness later reported seeing a speedboat-like object crash land into a tree around the same time. The police were called to investigate but the crashed UFO disappeared by the time officers arrived. The only trace evidence was a broken branch. One of the ten-year-old witnesses described the air battle as looking like two blowflies flitting about. There's not a lot to really hang one's hat on here.

Ten Defunct UFO Organizations

NATIONAL INVESTIGATIONS COMMITTEE ON AERIAL PHENOMENA (1956-1990)

For a time during the late 1950s and early 1960s, NICAP was America's most influential UFO organization. Its board of governors included several prominent military men, the most famous being former Marine Corp aviator Donald Keyhoe (died 1988). Membership peaked out at about 14,000 dues-paying members. The group was characterized by a more sober "nuts and bolts" approach to the subject of UFOs. Members viewed the Contactee movement with a certain distaste. The group jumped the shark when it backed the Condon Committee's investigation into UFOs. When the Condon Committee concluded they could find no compelling reason to continue studying the UFO phenomenon, NICAP pulled its support. People were not impressed and membership dropped. Although the group was never financially stable, even during its heyday, it managed to limp along until 1990 when it officially pulled the plug. It sold its archive of UFO reports and research to the Center for UFO Studies.

AERIAL PHENOMENA RESEARCH ORGANIZATION (1952-1988)

Although not nearly as big as NICAP in terms of membership (peaking at about 1,400 members), many considered APRO as influential and authoritative as NICAP. It was staffed with far fewer ex-military which may explain why it tended to avoid NICAP's obsession with government UFO cover ups. Like NICAP, it avoided links to the Contactee movement. Whereas NICAP was characterized by a more investigative approach to UFOs, APRO tended to be more credulous and more apt to pass on stories without researching them. APRO might be best known for its involvement in promoting Travis Walton's claims of being abducted by aliens. APRO lost a large number of members to the upstart Mutual UFO Network (MUFON) in 1969. It limped along until the death of its founders in the late 1980s.

SAUCER AND UNEXPLAINED CELESTIAL EVENTS RESEARCH SOCIETY (1954-2012)

Saucer and Unexplained Celestial Events Research Society (which cleverly gets shortened to SAUCERS) was founded in 1954 by James W. Moseley (died 2012). The group might have had a membership of one. Moseley mostly used it as a front organization to publish one of the first successful UFO ezines called *Saucer News*. It's probably most famous for revealing George Adamski's methods for creating his UFO pictures (basically using a Chrysler hubcap and some ping pong balls). Moseley sold *Saucer News* to Gray Barker in 1968.

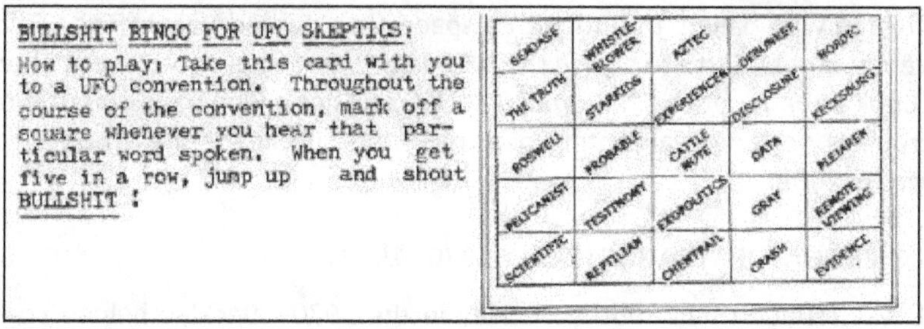

Cover art from the March 15th, 2008 issue of Saucer Smear

In 1976, Moseley got back into the publishing game with *Saucer Smear*. *Saucer Smear* was a departure from *Saucer News*. It offered a biting, sarcastic, and gossipy take on the world of UFOs. The zine's motto was "Dedicated to the highest principles of UFOlogical journalism" which was a bit of a swipe at the standards in the field. *Saucer Smear* folded after the death of Moseley in 2012 from cancer of the esophagus.

GROUND SAUCER WATCH (1957-LATE 1980s)

Ground Saucer Watch (GSW) was another group similar to NICAP. It brought together people of a scientific bent to analyze UFO reports. It was founded by William Spaulding. Spaulding was an engineer (drink!) by profession. It was based out of Phoenix, Arizona. The group was probably best known for launching several Freedom of Information Act requests for the release of UFO materials. As well, the group sued the CIA in 1981 when it felt the CIA had not released *all* its UFO files. The group lost that suit. The GSW also had a reputation for being the go-to group for photo analysis of alleged UFOs. The date of their demise is not clear but the group seems to have run into financial troubles and dropped off the UFO scene radar by the late 1980s.

INTERNATIONAL UFO BUREAU (1958-2017)

The so-called International UFO Bureau was founded in Oklahoma in the late 1950s by UFO hunter Hayden Hewes when he was 13. At its peak the IUFOB seemed to have only a few hundred members. The group's biggest moments in the sun both came in 1973. In that year, Hewes revived the long dead 19th century legend of an "airship" crashing in Aurora, Texas. His group made news when they unsuccessfully petitioned the town to let them exhume a grave in a local cemetery on the suspicion it contained the body of a space alien. In 1973, President Jimmy Carter (then the governor of Georgia) sent the group a UFO sighting report.

NATIONAL INVESTIGATIONS COMMITTEE ON UNIDENTIFIED FLYING OBJECTS (1967-2008)

The National Investigations Committee on Unidentified Flying Objects (NICUFO) was founded by UFO preacher Frank E. Stranges in 1967 in California. Stranges was a Contactee who palled around with a space alien named Valiant Thor. Thor was apparently sent to Earth to advise the US military. Stranges wrote about Thor in his 1967 book *Stranger at the Pentagon*. NICUFO still has a web page up but it reports the group is no longer accepting new members. It's likely the group died with Stranges in 2008.

UFO INVESTIGATORS LEAGUE (EARLY 1970s-1990s)

UFOIL was founded by Timothy Beckley in the 1970s. Beckley helped co-author *Shaver Mysteries* material for *Fate* magazine editor Raymond Palmer. Beckley seemed particularly interested in publishing about the Men in Black non-phenomenon. A dues-paying member of UFOIL was given a membership badge that ostensibly gave the member "authority to officially investigate UFO cases and report your findings to headquarters". Members also got an official investigators certificate and a copy of Beckley's *UFO Investigators League Field Manual*. His field manual contained useful tips on how to "best interrogate witnesses." UFOIL didn't seem to do a lot on its own but members, with their official badges and interrogation manuals, spun off a few active state chapters including Kentucky's KUFOIL and Ohio's OUFOIL. The group today seems to mostly exist as a PO Box in New Brunswick, New Jersey.

FUND FOR UFO RESEARCH (1979-2006)

The Fund for UFO Research (FUFOR) was founded in 1979 with UFOlogist and physicist Bruce Maccabee as its head. Its stated purpose was to use its funds to "support all reasonable and scientific efforts" in the investigation of the UFO phenomenon. FUFOR's fund wasn't a particularly large pile of cash. In the 1980s, the organization was taking in between $2,000 and $6,000 a year and spending about 25% of that funding investigations. The group seems to have ceased any appreciable activity by 2006.

LONG ISLAND UFO NETWORK (1988-1997)

The Long Island UFO Network was established in 1988 by John Ford, a retired bailiff. Ford's group peaked at north of 400 members. The UFO group seems to have been operated out of Ford's home until he was arrested in 1997 for allegedly plotting the murder of three local government officials that he believed were covering up local UFO crashes. Ford's alleged plan was to slowly poison the officials with radium. Ford's attempted murder case never made it to court as court psychiatrists diagnosed him as delusional and not fit to stand trial. He remained in a psychiatric care facility until a court order released him in 2019.

NATIONAL INSTITUTE FOR DISCOVERY SCIENCE (1995-2004)

NIDS was founded in 1995 by billionaire woo-meister Robert Bigelow. Although NIDS did not claim to be just a UFO organization, they are best known for their so-called research into cattle mutilations and black triangle UFOs. The group was probably most famous for buying Skinwalker Ranch. Reputed to be home to almost anything your crazy mind can cook up, Skinwalker Ranch turned out to be just a ranch. For a time, the FAA was suggesting instead of sending them reports of UFOs, people should send them to Bigelow's NIDS group. Bigelow seems to have disbanded NIDS in 2004 when he concluded black triangle UFOs were probably military aircraft.

Six Excellent UFO Skeptics

DEREK BARTHOLOMAUS

Derek Bartholomaus has been a long-time thorn in the side of UFO Contactee Billy Meier. Bartholomaus' site (billymeierufocase.com) doggedly deconstructs how Meier's UFO evidence can be easily and exactly duplicated with common household objects. For example, he's demonstrated how to make an exact duplicate of Meier's ridiculous looking "Wedding Cake" UFO from a container lid, a flowerpot tray, and a dozen other items you can pick up from the corner hardware store. He did it all while showing how you can do it with one hand. Meier lost an arm in an accident and it's sometimes suggested Meier is not physically capable of carrying off alleged hoaxes. Bartholomaus' dogged research has also turned up the terrestrial sources of various paranormal photos Meier has offered as proof of his contact with aliens. In one case, Meier produced a picture of two women space aliens named Asket and Nera. Bartholomaus discovered the women in the photo were actually a pair of dancers who regularly performed on *The Dean Martin Show* in the 1960s.

JACK BREWER

Jack Brewer runs the *UFO Trail* blog. There he takes a hard-nosed look at UFO stories in the headlines. He seems to specialize in alien abductions and government cover up claims. No one seems to be able to read a fuzzy, heavily redacted 200-page FBI document like Brewer. In 2016 Brewer published *The Greys Have Been Framed: Exploitation in the UFO Community.* In that book, he surveys how charlatans have exploited true believers.

DAVID CLARKE

David Clarke is a professor who teaches and researches folklore at Sheffield Hallam University in the UK. He was a main driver behind getting the UK's Ministry of Defense to release its UFO files. Clarke's blog lacks a snappy name (drdavidclarke.co.uk) but it offers serious yet readable analyses of various UK-related UFO legends.

PHILIP J. KLASS

Philip J. Klass (died 2005) was a senior editor for the venerated *Aviation Week* magazine for more than three decades. He became interested in UFOs in the mid-1960s. Klass was the arch nemesis of Stanton Friedman (died 2019). Klass was famous for conclusively showing the MJ-12 documents were a hoax. He was able to demonstrate Truman's signature was a photocopy of a genuine signature. In 1983 Klass formulated what he called The UFO Curse: "No matter how long you live, you will never know any more about UFOs than you know today. You will never know any more about what UFOs really are, or where they come from. You will never know any more about what the U.S. Government really knows about UFOs than you know today. As you lie on your own death-bed you will be as mystified about UFOs as you are today. And you will remember this curse."

JAMES OBERG

James Oberg is one of the founding members of CSICOP. He's wound down a lot of his skeptical activism in the last two decades. During the height of the Internet Usenet era, Oberg was a prolific critic of Ufology. Oberg served with the US Air Force in the early 1970s and then went on to work for the Defense Department. He helped design ARPANET, the forerunner to the Internet. He's fluent in Russian and long worked for NASA as an expert on the Soviet space program. Oberg retired from NASA in the late 1990s. Post-retirement NASA tapped Oberg to write an official agency rebuttal to the Fox network's ridiculous 2001 special alleging NASA hoaxed the moon landings. The project was terminated when NASA decided addressing the conspiracy mongering gave it attention and authority it did not deserve.

ROBERT SHEAFFER

Robert Sheaffer has been a long-time UFO skeptic. Sheaffer, Philip Klass, and James Oberg founded the Committee for Skeptical Inquiry's UFO committee back when it was originally known as CSICOP. Since 2010, Sheaffer has been writing the *Bad UFOs* blog. It's probably the UFO skeptic's blog of record. It's replete with skeptical material about classic UFO cases.

Eight Alien Species Visiting Earth

The number of space alien races supposedly visiting Earth is staggering. Although there seems to be no agreed upon number in the world of Ufology, the number seems improbably high. Val Valerian in his *Matrix* series asserted 70 separate alien species are visiting Earth. The website Disclose.tv ups it to 82.

ANDROMEDANS

We are told Andromedans are aliens that come from one or more stars in the Andromeda constellation. Although others claim they come as far as the Andromeda galaxy. If so, they've got really fast warp drives. They are said to resemble your typical grey alien but are more human sized. They seem to be the invention of "Contactee" Alex Collier. Since the 1990s Collier has been writing something called the "Letters From Andromeda". Their letters assure us they are here to defend us from the evil Reptoids and they really, really want us to drive electric cars.

The grey alien image can be found everywhere.
Photo Credit: Karl Mamer

GREYS

The Grey alien has become the standard alien form. Everyone has seen the image a billion times in movies, on TV, on bumper stickers, and on cereal boxes: big head, lithe body, oversized eyes, slits for a nose. H.G. Wells in the late 19th century wrote about aliens that matched the classic grey in many ways. For example, in his book *The First Men in the Moon* he described the indigenous moon folk as having big heads, large black eyes, and grey skin. Prototypical grey aliens were starting to be featured on 1950s and early 1960s era TV shows like *The Outer Limits* and *The Twilight Zone*. The Grey was brought into vogue among UFO believers following the alleged 1961 abduction of Betty and Barney Hill. Another boost was given to the form by *Close Encounters of the Third Kind.* Greys seemed initially imbued with the qualities of the Contactee aliens of the 1950s. They were kind aliens here to warn us or help us through various crises. The *X-Files* and the off-the-wall writings of Milton William Cooper (died 2001) started to cast the greys in a more nefarious light. They are variously working for the New World Order, the Illuminati, or another controlling alien species.

INSECTOIDS

This one seems to have been largely cooked up in the 1990s. Some people have alleged to have encountered mantis-like aliens. Many of the encounter tales are from dubious abduction reports. In the world of Ufology, these mantis aliens are ostensibly working with the Reptoids to dominate humans.

LITTLE BLUE MEN

Aliens described as "Little Blue Men" are featured in Whitley Strieber's *Communion*. Strieber claims he was abducted from his cottage by such beings. Little blue men sightings can trace their origin to a hoax pulled by three Michigan youths in 1958. Inspired by UFO reports and a popular novelty song of the time called "Little Blue Man" by Betty Johnson, the young men created a costume out of long underwear, boots, and a football helmet with blinking lights. They painted their costume a dark blue and hung out roadside trying to startle drivers. Although people were clearly seeing a human-sized alien, surprisingly a number of witnesses reported the alien was only two feet tall.

PLEIADIANS (AKA PLEJARENS)

The Pleiadians, as the name suggests, are from worlds around stars in the Pleiades star cluster. They seem to trace their origin back to the Contactee days. They're an updated version of the Nordics. Modern UFO mythology posits they are a race of "star seeders" that helped beget the human race. Billy Meier has tried to turn the Pleiadians into his own cottage industry. He says they prefer to be called "Plejarens" and pretty much only want to talk to him. He claims the Plejarens warn him of all sorts of impending disasters, gave him a ray gun, and even took him back in time to see dinosaurs. Very few people in Ufology take Meier seriously. Astronomers give dim prospects to the chances of super advanced alien life emerging on a planet around a star in the cluster. The Pleiadians stars are only about 150 million years old.

MOTHMEN

The Mothman is reputed to be a flying humanoid space alien that appears to warn people of looming disasters. The alien was popularized by Fortean writer John Keel in his 1975 book *The Mothman Prophecies*. The first reports emerged in 1966. A year later the Silver Bridge collapsed, fixing in the minds of Mothman proponents the alien's nature as a harbinger of doom. In 2011, Mothman sightings started to pop up in the Chicago area.

NORDICS

Early aliens supposedly seen by the Contactees of the 1950s weren't that alien. They are typically described as being tall blonde people and referred to as "Nordics". Alleged sightings of Nordics dwindled over the decades. However, Travis Walton claimed to have seen Nordics during his abduction experience in 1975. Swiss UFO "experiencer" Billy Meier has claimed for decades his so-called Plejarens are Nordic type aliens.

REPTILOIDS

In the late 1990s, David Icke began popularizing the notion that the world is run by a shadowy conspiracy of shape-shifting reptiles from the Alpha Draconis star system. Most of a certain age will recognize Icke's lizard conspiracy as being lifted from the 1980s *V* mini-series about the world being run by a shadowy conspiracy of space reptiles who take human form. The notion of lizard men "among us" originates in the writings of Helena Blavatsky from the 19th century. Blavatsky founded the Theosophy religion. Although she existed a long time before the advent of the UFO myth, many early Contactees were followers of Theosophy and repurposed her ideas. Much of Ufology's core dogma is directly lifted from Blavatsky. In one of her books, she posited the notion of lost "root races" like Atlanteans and Aryans. One of her supposed progenitor races was an egg-laying Lemurian race.

Eleven UFO Types

When we went to the moon, we didn't send wholly different Lunar Excursion Modules. We used one basic model. For some reason, our alien visitors enjoy variety in terms of their landing craft. In the mind of UFO believers, there is no end to the UFO types visiting our planet. Naturally, if UFOs were simply the result of people misidentifying things they see in the sky, it's not at all surprising people would see all kinds of oddly shaped things.

ACORN-SHAPED

The UFO that supposedly crashed in Kecksburg, Pennsylvania and spirited away by the government was described by witnesses as being acorn shaped. The problem is the only witness of this acorn-shaped UFO was a man who recounted his sighting 25 years after the fact when he was a ten-year-old boy.

BELL-SHAPED

It has been claimed the Nazis had developed or were developing a bell-shaped UFO before the war ended in 1945. Cranks, many of the hollow or flat Earth variety, believe Nazis absconded with a fleet of bell-shaped UFOs and are hiding out in an underground base in Antarctica.

BOOMERANG-SHAPED

The boomerang-shaped UFO entered public consciousness in 1997 when people in Phoenix, Arizona misinterpreted military flares as a single, structured object.

CIGAR-SHAPED

It would seem flying saucer mania had fallen out of favor by the 1960s. That didn't mean an end to UFO sightings. It just meant people started interpreting a wider variety of anomalous sightings as alien spaceships. Sightings of cigar-shaped UFOs started to come into vogue. It's generally believed sightings of the U2 and SR-71 spy planes are the genesis of the cigar-shaped UFO.

CUBE-SHAPED

In 2020, a couple of UFOs described as "cube-shaped" showed up in the popular media. In one case a US military pilot took a photo of a silvery "something". In another case a commercial pilot took a short video from the cockpit of his A320 of a silvery "something". In both instances the UFOs were described as being cube-shaped. *Star Trek* fans would instantly recognize this as suggestive of a Borg invasion. However, the object photographed by the US military pilot doesn't actually appear cube shaped. Both instances were better explained on the *Metabunk* site by commercial helium balloons that likely got loose.

EGG-SHAPED

A 1964 close encounter in Socorro, New Mexico by police officer Lonnie Zamora was said to involve an egg-shaped UFO and two aliens dressed in white coveralls. Due to the main witness being, by all accounts, an upstanding community member and the existence of trace evidence like burned brush and fused sand, UFO groups have called this one of the best UFO cases they have on their books. In correspondence by the New Mexicans for Science and Reason group, it was revealed a pair of mining grads from New Mexico Tech at the time liked to go out to the desert and use dynamite to launch empty oil drums into the sky. They would wear white coveralls when they did it. They had just lit the fuse on the dynamite when Zamora came along. They quickly fled as they didn't want to get caught and expelled. Naturally, they kept as silent as possible about their hijinks until they graduated.

RODS

An excellent argument against UFOs is with everyone having a high-definition video camera in their pocket, there should be a lot more videos of UFOs. The other side of that is people inexperienced with cameras are now capturing a wider variety of out-of-focus objects and calling them UFOs. In the 2000s, people started capturing fast moving insects on their cameras which would resolve as fuzzy rods. Many pronounced these a new kind of UFO.

SAUCER

The term flying saucer got its origin in a news article covering what is considered the first modern day UFO sighting. Pilot Kenneth Arnold reported seeing circular objects during a solo flight in 1947. The press quickly dubbed what he saw "flying saucers." The expectation that UFOs should be saucer shaped has remained fixed in the public consciousness for over sixty years.

Photo Credit: Karl Mamer

SPORT MODEL

Bob Lazar, a self-employed film processor, claims he was hired by the US government to work at Area 51. He has offered many fantastic stories, including having obtained advanced science degrees no one has ever been able to verify. While at Area 51, he claims his job was to help reverse engineer alien saucer technology. One of the otherworldly vehicles he saw at Area 51 is a saucer he described as a "sport model", which was presumably a small two-person (or alien) UFO.

TRIANGLE

Triangle-shaped UFOs came into vogue in the late 1980s with the so-called Belgian UFO Wave. Hundreds, if not thousands, of people in Belgium, spurred on by media stories, started to report seeing mysterious lights in the sky that they interpreted to be running lights on triangle-shaped UFOs. Despite thousands of people reporting they had seen triangular UFOs, not one person managed to capture a good quality photograph or video. Naturally any three anomalous lights one sees in the sky would form a triangle. When primed by the media to interpret such lights as part of a single craft, the most likely explanation for triangular UFOs doesn't need one to reach for the ET hypothesis.

TIC-TAC SHAPED

Navy pilots who captured a blobby infrared video during a 2004 test flight described the object they saw in their infrared camera as being shaped like the Tic-Tac candy.

Ten Claims of Physical Evidence of UFOs and Aliens

Despite multiple UFO crashes and thousands of reports of sightings and abductions, no one has ever produced the one thing that would prove the reality of alien visitors beyond a shadow of a doubt: indisputable physical evidence. This could take the form of a piece of alien technology, some bit of metal the CIA missed during a crash clean up, an alien implant, or an alien body. There are, of course, claims people have such evidence but it's just never really forthcoming in an open and independently testable way.

ANGEL HAIR

Since at least 1561 people have been reporting silky strands falling from the sky. Those of a religious bent have associated it with angels or visitations by the Virgin Mary. Ufologists believe it's produced by UFOs. The association between UFOs and this substance started in 1959. After a UFO sighting in Évora, Portugal, locals reported finding this so-called angel hair. Analyses of angel hair indicate people are just finding spider silk, usually produced by spiders that use it to migrate through the air.

ROGER LEIR'S ALIEN IMPLANTS

Dr. Roger Leir (died 2014) was a podiatric surgeon who was convinced he had removed alien implants from up to a dozen abductees. The few implants he ever had independently tested came back decidedly of this Earth. Leir hung a lot on test reports from the New Mexico Institute of Mining and Technology that indicated some of the implants contained metals one can find in meteors. That, of course, doesn't mean the metals came from meteors. You might find, say, iron in meteors. Finding iron in your backyard is not indicative the iron is from space. One might wonder how a small bit of metal got under an alleged alien abductee's skin anyway. For many doctors it's not surprising. People fall down, injure themselves, and bits of just about everything can get into a wound and then heal over.

CROP CIRCLES

Crop circles are a perfect example of the argument from ignorance. A cereologist can't imagine how humans might be able to stamp out a complex pattern in a farmer's field overnight, therefore it has to be UFOs trying to send Earth a message. Crop circles got their start in the "UFO nest" phenomenon of the 1960s. Beginning in 1966, Australians started reporting odd circles in reed swamps. This inspired a couple UK pranksters to create crop circles in British fields using rope and a wood plank. This then inspired others to create ever more intricate and beautiful designs. This also inspired a whole generation of anomaly hunters to declare themselves cereologists. Many cereologists proudly assert they cannot be fooled. They can tell the difference between crop circles made by pranksters and those made by aliens. Unfortunately, cereologists time and time again pronounce hoaxed circles as genuine. Some cereologists have claimed analysis of flattened plants within the circle show anomalies not present in the plants outside the circle. However, their studies have been criticized for lacking double-blind, independent testing.

THE FALCON LAKE INCIDENT

In 1967, Stefan Michalak was prospecting for silver near the town of Falcon Lake, Manitoba. He claimed to have seen a saucer land in front of him. Defying all better judgment, he walked up to the craft and touched it. The craft took off, burning Michalak's gloves, hat, shirt, and chest in the process. The burns on his shirt and chest manifested as a grid of dots which Michalak claimed matched a grid-like exhaust port on the saucer. Medical examiners had no explanation for the burn marks and Michalak tested positive for radiation. An investigation of the site discovered high levels of radiation although it was discovered it was from Precambrian rock with a rich vein of radium. Michalak was also noted to be wearing an early glow-in-the-dark watch that used radium to illuminate its watch face.

Michalak's radiation readings may well have been a product of natural contamination and his watch. An initial RCMP police report was skeptical of Michalak's story. For example, he showed the responding officer his burned hat but had no good answer why his head wasn't also burned. How does one get burned from head to chest by hot exhaust gas but the hot gas mercifully doesn't burn your face and neck? He refused to let the officer examine him further for possible injuries. The officer reported he couldn't smell alcohol on his breath but Michalak did appear to be similar to someone who had "over indulged".

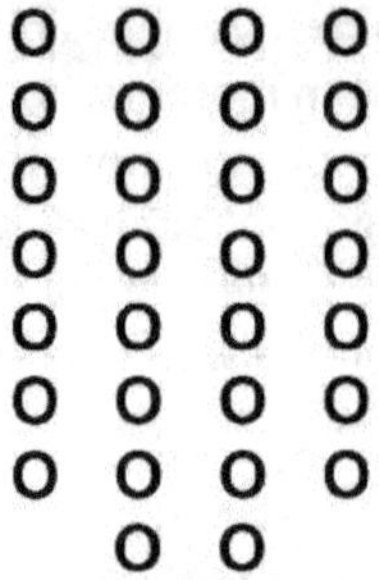

Illustration of the burn mark pattern.
Illustration: Karl Mamer

BETTY HILL'S DRESS

Almost everything we know about the Betty and Barney Hill abduction comes from hypnotic sessions conducted by Boston therapist Benjamin Simon. There were, however, two claims of physical evidence of their abduction. Betty Hill (died 2004) claimed their car's trunk had been marked by several shiny circles. When the Hills brought a compass near the circles the needle would spin rapidly. Oddly no one who interviewed the Hills ever verified this. Hill also claimed the dress she wore that night was covered with a mysterious pink powder. Conveniently the powder all blew away when she hung the dress up outside to examine it. She did save the dress and noted the pink powder had left stains. The dress and the stains have been analyzed at least twice. Nothing unusual was found.

RENDLESHAM FOREST

In the early morning hours of December 26, 1980, US Air Force security personnel from a UK air base entered the nearby Rendlesham forest to investigate the possibility of an aircraft crash. The security personnel reported seeing anomalous lights which they attributed to a UFO landing. It's worth noting many of those who participated in the investigation have seemingly embellished their stories over the years. Later in the day Air Force personnel returned to the area and reported three impressions in the ground and burn marks on trees. A UK Forestry Commission employee called in to take a look recognized the burn marks as axe cuts Forestry

Commission workers use to mark trees. The three impressions were easily recognizable as the work of rabbits. A day or two after the sighting, the deputy base commander Lieutenant Colonel Charles Holt returned to the site with a Geiger counter. His readings were within background levels, although Ufologists make a big deal out of a few readings that popped up closer to the high end of background levels.

STARCHILD SKULL

For years Ufologist/cryptozoologist Lloyd Pye (died 2013) lugged around the malformed skull of a long deceased five-year-old boy. He would tell anyone drunk enough to listen that it was the skull of a human-alien hybrid! If you squint really hard, you can maybe see the skull looks a bit like a grey alien. However, a number of medical doctors recognized the skull's shape as a result of hydrocephalus. DNA testing was performed on the skull twice, once in 1999 and once in 2003. In 1999 the testing revealed the skull had both X and Y chromosomes, which indicates the child had both a human father and mother. This completely eliminated the alien hybrid hypothesis. The 2003 test was able to determine the child belonged to haplogroup C, characteristic of Native American parentage.

STAR JELLY

People have been finding odd gelatinous globs on the ground for centuries. Quite early on people associated this jelly with comets and dubbed it "star jelly". Similar to angel hair findings, sometimes people stumble on amorphous bits of goop on the ground following a UFO sighting and attribute it to the work of aliens. There's no one single explanation for star jelly finds. Star jellies that have been analyzed have been discovered to be *Myxarium nucleatum* (a fungus that lives on decaying wood) and even frog remains.

TRANS-EN-PROVENCE

Near Trans-en-Provence, France a farmer witnessed a saucer landing in or near his field. The French version of MUFON known as GEPAN (*Groupe d'Etude des Phenomenes Aerospatiaux Non-Identifies*) investigated the landing field. According to their investigation they discovered what looked like landing marks and believed something heavy had made them. As well they found burnt soil in the area and concluded the soil had undergone heating around 1,000 Fahrenheit. Various anomalies were found with vegetation near the landing site. For example, nearby plants had less chlorophyll than expected. The supposed compressed landing marks seem less mysterious when you discover they were found on a road near a military base. A police investigation reported the landing marks resembled tire tracks. One would expect some heavy vehicles around a military base. No? The plant anomalies seem like just that. Anomaly hunting. There's no established reason why a saucer landing should rob nearby plants of chlorophyll and the GEPAN report does not rule out natural explanations. Perhaps, plants near roads used by smog-bleaching military vehicles aren't as healthy.

THE UBATUBA MAGNESIUM

In 1957, near the Brazilian town of Ubatuba, witnesses reported seeing a UFO crash into the Atlantic Ocean. A local physician collected what he thought was metal expelled by the craft before the crash. One of the metal fragments was studied by a Brazilian government lab. It was determined to be magnesium. Ufologists claim that analysis determined the magnesium to be 100% pure, which doesn't occur in nature and can't be manufactured by any process on Earth. So aliens. The 100% purity claim, however, appears to be a misreading of the original analysis that found "no contaminates". This can simply mean no contaminants beyond what is normally expected in naturally occurring magnesium. That 100% purity finding is also disputed by later studies. The Condon Committee in the 1960s found the magnesium to be "less pure than the regular commercial metal produced in 1957". A 2004 study likewise found "no evidence that the specimens are of extraterrestrial origin".

Eight Cases of UFOs Sighted by Pilots

If there was ever a gold standard for a UFO witness, it's an aircraft pilot. A pilot supposedly has superior observational skills and knows exactly what is in the friendly skies. Pilots who are also skeptics dispute the notion pilots are better UFO witnesses. According to James McGaha, a UFO skeptic and a retired USAF pilot, pilots can be unreliable observers of aerial phenomenon because they are trained to over interpret objects in the sky. In other words, everything unidentified in the sky should be treated as a possible aircraft and a potential collision danger. It's better to think Venus is an airplane and look foolish than think an airplane is Venus and die.

KENNETH ARNOLD

Kenneth Arnold (died 1984) kicked off the world of Ufology with his 1947 sighting of nine objects over Mt. Rainier in Washington State. On June 24, he was overflying the area helping in a search for a downed military plane. Around 3 PM he saw nine objects flying in a V-formation at what he thought was an astonishing speed. When Arnold landed, he told airport workers about his sighting. Some suggested he had seen some kind of new military aircraft. The airport workers quickly spread his story. The next day Arnold was interviewed by the *East Oregonian* newspaper. He suggested he had seen some kind of new remote-controlled missile. His description of the objects seems to have varied a bit within his initial interview. He described what he saw as saucer-like. He also described the objects as "flat like a pie-pan and somewhat bat-shaped" and "like a pie plate that was cut in half with a sort of a convex triangle in the rear". Ultimately, the press shaved off these details and reported Arnold's sightings as "flying saucers". The legend was born. It has been suggested Arnold saw a flock of pelicans. From a distance they would resemble the bat-wing saucer shape Arnold reported.

MANTELL UFO INCIDENT

On January 7, 1948, control tower operators at the US Army Godman Airfield noticed a very bright white object hovering in the sky. The Army dispatched four F-51 Mustang fighters to investigate. Two of the pilots described the object as small and impossible to identify. Three of the four pilots called off their pursuit but Captain Thomas Mantell pressed on. Mantell lacked oxygen but attempted to climb beyond 25,000 feet. He quickly passed out. He died when his plane spiraled into a farmer's field. A number of legends emerged, including the pilot got into a dogfight with the UFO and was shot down. It's believed the control tower and pilots saw a Skyhook balloon. Other observers in the area had reported sighting a weather balloon. The Skyhook program was also secret at the time and no one at the airbase would have been made aware such a balloon was in the vicinity.

CHILES-WHITTED UFO SIGHTING

The Chiles-Whitted UFO sighting in 1948 is considered the first sighting by commercial pilots. It prompted the Air Force's initial investigation into UFOs called Project Sign. Two Eastern Airlines pilots flying a DC-3 near Montgomery, Alabama on July 24 reported seeing a glowing red UFO zip past their plane. It appeared to them as a wingless cigar-shaped aircraft, glowing as bright as a magnesium flare, with two rows of windows. The pilots filed their sighting with the Air Force. J. Allen Hynek believed the pilots had seen a meteor.

THE COYNE INCIDENT

In 1973, around 11 pm, US army helicopter pilot Maj. Larry Coyne and his crew witnessed a bright red light approaching their craft. Coyne and his crew believed the light was on a collision course and braced for impact. Luckily, it didn't happen. Coyne reported the UFO was following them and zapped them with some kind of green tractor beam. Philip Klass noted the date and time of the Coyne Incident coincided with the Orionids meteor shower, which typically begins at 11 pm.

THE 1976 TEHRAN UFO INCIDENT

In September 1976, Iran's Mehrabad Air Force Base started receiving some panicky calls by locals about a brilliant light in the sky. The Iranian Air Force dispatched two F4 Phantoms to investigate. According to a report prepared by a US Air Force officer in Iran (this was during the Shah era...) the pilots were unable to get near the UFO as it receded as quickly as they approached. Space history expert James Oberg notes this is exactly what one would think if you were trying to chase a bright astronomical object in the sky. One of the F4 pilots tried to fire an AIM-9 missile at the UFO but his fire control system locked up. Both pilots returned to base after experiencing mechanical problems. Iranian news made a big deal out of the incident the day after and it quickly entered American news and UFO lore. The *National Enquirer* ran with the story, declaring it one of the best UFO cases ever. Mechanical problems common to Iranian planes were spun by the Ufologists as evidence of astounding alien technology able to jam military radar.

FLIGHT JAL 1628

On the night of November 17, 1986, over Alaska, a Japan Airlines crew witnessed two smaller UFOs rendezvousing with an alien mothership. The mothership was the size of two aircraft carriers and shaped like a walnut. Ufologists consider this one of the best and most credible sightings. The primary witness, Captain Kenju Terauchi, was a former military pilot and had logged more than 10,000 hours of flying time. What is usually not mentioned about Terauchi was he was a fervent UFO believer before the 1986 sighting. He had reported four UFOs previously. Terauchi was the only one who reported the giant mothership. His crew only reported some anomalous lights. Other aircraft were also in the vicinity and were asked by air traffic control to investigate the sighting. No other pilot reported seeing anything unusual.

AMERICA WEST AIRLINES FLIGHT 564

Pilots flying at night from Nevada to Texas in 1995 reported seeing anomalous lights at about 30,000 feet. The pilots reported to air traffic controllers the object was cigar shaped with a strobing light. The object did not appear on their radar or the radar of air traffic control. It's not unknown, given the right conditions, for particularly bright stars to appear as if they are strobing.

SWISSAIR FLIGHT 127 NEARLY HIT BY A UFO

In 1997 pilots of a Swissair 747 flying over New York reported to air traffic control an encounter with a fast, white object that they thought nearly hit their plane. Interestingly the encounter took place about a year after TWA Flight 800 went down in the same area. The TWA 800 flight itself was the subject of a number of conspiracies, notably one involving an errant missile. The pilots themselves suggested they were buzzed by either a "rocket" or a UFO. The flight occurred within days of the peak Perseids meteor season. Phil Klass suspected the pilots witnessed a meteor. What appeared to the pilots as a close call is a feature of fireball-like meteors. According to Klass such meteors "always appear to be very much closer than they really are".

Nine Questions to Ask a Nazi UFO Crew

Canadian-German Holocaust revisionist Ernst Zündel (died 2017) was convinced the Nazi elite escaped capture at the hands of Allied occupational forces. They flew UFOs to Antarctica where they established a secret underground base. Zündel once tried to raise money for an expedition to the South Pole in search of this Nazi base and their UFOs. His plan was to overfly the South Pole in a plane with highly visible swastika markings so Nazis manning AA weapons would immediately know they were friends. If you wanted a seat on this plane, you just had to send Zündel $10,000. It's unknown if a single ticket was ever sold. As part of his fundraising efforts, Zündel sold books, lecture tapes, and, most prized of all, an official UFO Investigator pass! Not only did the pass officially identify you as an official investigator of UFOs, specifically an official investigator of Nazi UFOs, but the pass included a number of questions (in English and German) you could ask a Nazi UFO pilot. Handy. I mean, who isn't going to be left speechless after meeting a Nazi UFO pilot?

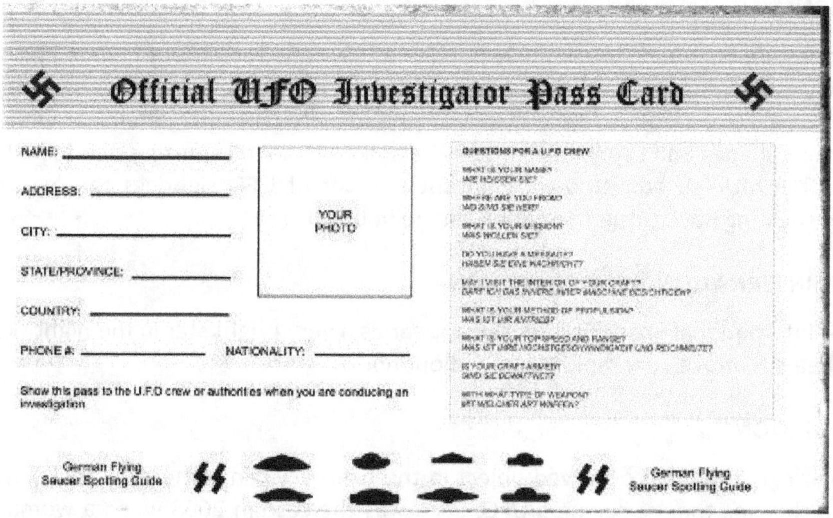

A reproduction of Zündel's UFO Investigator Pass
Illustration: Karl Mamer

1. WHAT IS YOUR NAME?/WIE HEISSEN SIE?
2. WHERE ARE YOU FROM?/WO SIND SIE HER?
3. WHAT IS YOUR MISSION? /WAS WOLLEN SIE?
4. DO YOU HAVE A MESSAGE?/HABEN SIE EINE NACHRICHT?
5. MAY I VISIT THE INTERIOR OF YOUR CRAFT?/DARF ICH DAS INNERE IHRER MASCHINE BESICHTIGEN?
6. WHAT IS YOUR METHOD OF PROPULSION?/WAS IST IHR ANTRIEB?

7. WHAT IS YOUR TOP SPEED AND RANGE?/WAS IST IHRE HÖCHSTGESCHWINDIGKEIT UND REICHWEITE?
8. IS YOUR CRAFT ARMED?/SIND SIE BEWAFFNET?
9. WITH WHAT TYPE OF WEAPON?/MIT WELCHER ART WAFFEN?

Eight and a Half Things Frequently Mistaken for UFOs

CHINESE LANTERNS

The "sky lantern" variant of traditional Chinese lanterns are in essence a small paper balloon with a candle inside to provide lift. They are set aloft during various Chinese festivals. In places where people are not accustomed to small candle lit balloons set aloft, UFO reports can result. Tricksters have also used sky lanterns to fool people into thinking they were witnessing a UFO.

DRONES

Drones are becoming increasingly popular as toys and for various private business applications. Drone swarms used to create novel nighttime light shows can appear as a single, solid object to the eye. Many people are currently unfamiliar with the various shapes and capabilities of drones and have been reporting them as UFOs. In 2019 MUFON admitted a significant number of UFO sightings they've been investigating have turned out to be simple hobby drones.

THE INTERNATIONAL SPACE STATION

The International Space Station can appear as a very bright star in the night sky. It appears to move very rapidly and without noise.

THE MOON

Amazingly the most observed object in the night sky can sometimes trick people into thinking they're seeing a UFO. Such was the case in 2008 when a woman in Wales called police to report a bright stationary UFO hovering in the sky. She claimed it had been hovering for at least 30 minutes before making the call. When the police came to investigate, they determined she had simply seen the moon.

THE U2 AND THE SR-71 BLACKBIRD

A 1997 CIA study suggests in the 1950s and 1960s roughly half of UFO reports could be attributed to overflights by U2 and SR-71 spy planes.

SR-71 Blackbird on display at Seattle's Museum of Flight.
Photo Credit: Karl Mamer

SATELLITE FLARE

As low-Earth satellites orbit, they can reflect sunlight back towards the ground. The Iridium satellite constellation is particularly known for this. Their flaring can be so bright they can sometimes be seen during daylight hours. To astronomers these flares are a form of annoying light pollution. To those inclined to interpret mysterious lights in the sky as alien visitors, these flares can result in UFO reports.

VENUS

Venus shines quite brightly just above the horizon. It can outshine background stars. Many times people have reported Venus as some sort of bright, hovering UFO. It has been suggested President Jimmy Carter's 1969 UFO sighting was Venus. Carter denies he mistook Venus for a UFO. He felt as an amateur astronomer he could tell the difference.

THE HALF: SWAMP GAS

Claiming skeptics handwave away all UFO sightings as misidentified swamp gas has become something of a cliché in the world of Ufology. The only time swamp gas has ever been used by a serious skeptic to explain a UFO sighting was by J. Allen Hynek. Hynek offered glowing swamp gas as *one* of many possible explanations for a mass UFO sighting in Michigan in 1966. The press focused on the swamp gas hypothesis and the meme was born.

Six Proposed Non-Extraterrestrial Origins of UFO

The majority of people who think UFOs are best explained by some unknown entity usually come down on the "extraterrestrial hypothesis". That is, UFOs are structured craft piloted by alien beings from another world. Most skeptics, however, accept the most likely explanation for the UFO phenomenon is a combination of misidentifications, hoaxes, and tricks of human perception. There are, however, people that have explanations for the UFO phenomenon that are as fantastic and unconventional as the ET hypothesis.

UFOS ARE SECRET WEAPONS

It's certainly plausible that stealth aircraft and now drones are being confused for alien spaceships flying around our skies. However, there are those who believe UFOs are evidence that the world's militaries have weapon systems so incredibly advanced, they're indistinguishable from alien flying saucers. Some assert the Nazis were working on antigravity UFOs near the end of the war and that technology was seized by the Allies. Others claim UFOs are offshoots of Tesla inventions long suppressed by oil interests. A problem with this idea is UFOs are frequently reported near large urban centers. A government that wanted to keep these weapons secret wouldn't so regularly expose them to people armed these days with high quality camera gear.

UFOS ARE CRAFT FROM UNDERWATER CIVILIZATIONS

Atlantis, we're told, sunk to the bottom of the ocean. Somehow, Atlanteans were able to not drown, adapt to live at crushing depths, and develop a flourishing civilization. UFOs come not from outer space but inner space. The biggest proponent of this idea was Ivan T. Sanderson (died 1973). He popularized the idea in his 1967 book *Invisible Residents: The Reality of Underwater UFOs*. Sanderson was fond of claiming, without any kind of citation, that half of UFO reports involve UFOs being seen over water or coming from/going to a body of water.

UFOS ARE SPACE AMOEBAS

This idea was put forward by Trevor James Constable (died 2016). Constable was a UFO writer and con man who ripped off the Malaysian government in the early 1990s by selling them bogus rain making technology. In the 1950s, Constable suggested that flying saucers acted "more like animals than anything else". He reasoned then UFOs were actually creatures that lived in our upper atmosphere or on the fringes of space. They were giant amoebas. The problem here is anything living that high up in the atmosphere is exposed to a constant bath of cosmic rays from the sun.

UFOS ARE TIME TRAVEL DEVICES

Many have considered maybe UFOs are really just humans visiting from the future. This would naturally explain why so many space aliens look human, understand our language, can breathe our air without difficulty, and are quite immune to billions of novel pathogens. It would also explain why they don't actually interfere with humans today. They need to stay passive observers and not disrupt the timeline.

EXTRATERRESTRIALS ARE ULTRATERRESTRIALS

This idea gained currency in the 1960s. Maybe UFOs aren't space aliens with the technology to cross vast distances but maybe they are beings who can travel between different universes or exist on different "vibrational planes". John Keel (author of *The Mothman Prophecies*) was an early proponent of the ultraterrestrial hypothesis. Dr. Jacques Vallée has suggested UFOs, angels, fairies, ghosts, and the like are really all just manifestations of ultraterrestrials.

UFOS ARE PSYCHIC PROJECTIONS

Carl Jung at one point suggested UFOs were an example of humanity projecting their internal states outward. Many have taken this to mean human consciousness is creating the UFO phenomenon via some kind of psychic means. UFOs are merely one of a long line of heavenly objects humanity psychically conjures into the skies. Angels, the Virgin Mary, fairies, and the like are also a product of this collective psychic channeling. Of course, most of Jung's ideas of human cognition have always been highly speculative and backed by no research. It's not much of a theory on which to base UFOs. Likewise, ESP has never been demonstrated to exist. Trying to explain one unknown phenomenon with two other unknown phenomena doesn't get you very far.

Six Proposed Explanations for Cattle Mutilations Other than Space Aliens

Cows sometimes die out in the field. A whole ecosystem starts to turn that cow corpse into food. In the 1970s, a cow mutilation panic began to sweep America, seeking to explain these fully explainable deaths with fantastical claims. Today, so-called cow mutilations stories are mostly provided as evidence of alien visitation. As Ufologists see it, eyes and genitals are removed with surgical precision and the cow is mysteriously drained of blood. They will intone there's no possible way the cow's condition could have come about naturally. Therefore, aliens. It's a classic argument from ignorance. Ufologists aren't the only group to have a fantastical explanation for so-called cattle mutilations. Here are six other fringe explanations.

CHUPACABRAS

The chupacabra ("goat sucker" in Spanish) is a cryptid said to attack and eat livestock in Puerto Rico and increasingly mainland America. The chupacabra story is a relatively modern invention. Today dead cows are just as likely to be blamed on a chupacabra attack as ghastly experiments by space aliens.

GOVERNMENT AGENTS IN MYSTERIOUS BLACK HELICOPTERS

Some cattle mutilations have been associated with sightings of black helicopters. Supposedly these cows are involved in some kind of clandestine government experiment. It is unknown why governments don't simply purchase their own herd of cows to harvest cow lips and genitals.

JACK THE RIPPER

A few people have noted there are similarities to the *modus operandi* of Jack the Ripper and the cattle mutilations. Jack the Ripper's victims and incidences of cow mutilations share elements like a lack of blood, organs removed with surgical precision, and random wanton mutilation.

THE NATIONAL INSTITUTE OF HEALTH

This theory was put forward in the 1990s when Mad Cow Disease hit the headlines. One author suggested the National Institute of Health was sneaking into farmer fields at night and testing cows for Bovine Spongiform Encephalopathy. Evidence for this claim is supposedly that mutilated cows are tested and found to have been injected with anticoagulants and drugs.

RANDOM HUMAN DEVIANTS

Apparently after cow-tipping loses its fun, some deviant individuals, particularly those with surgical skills, have taken to sneaking into farmer's fields at night and slicing and dicing cows.

SATANISTS

The evangelical set view cattle mutilations as evidence real, live Satanists are operating in their communities. This was a strongly held "theory" during the Satanic panic of the 1980s. The claim was Satanists drain the cows of blood and remove their sex organs for use in, as a Kansas politician put it, "their fertility rituals". Some communities began to fear cattle mutilations were a prelude to some kind of planned attack on elected officials by these Satanic cults.

Seven Supposed UFOs in Works of Art

Caravaggio's "Martha and Mary Magdalene" (1598)
Photo Credit: Karl Mamer

There's a popular notion among some Ufologists that artists from the medieval and renaissance periods hid images of UFOs and space aliens in their paintings. Ostensibly they were trying to communicate their knowledge of space aliens or they were simply representing these alien artifacts in their work under the assumption they were of divine origin. Somehow art historians through the years have simply missed these inclusions. Of course, art historians understand what the images actually mean. Although it would seem like a UFO or a space alien to someone who gets all his or her information from the History channel, art historians are well aware they are actually religious iconography contemporary viewers of the work would have well understood.

ANNUNCIAZIONE BY CARLO CRIVELLI

This painting from 1504 depicts a bearded god appearing in the sky and zapping Mary with some kind of beam. Riding the beam down towards Mary is what appears, to UFO believers, a circular spaceship. This is understandable if one is only looking at low resolution versions of the painting via Google Images. If one were to find a higher resolution version and zoom in a bit on the object riding the divine beam, it's pretty clear the object is a white dove. Doves and beams of light are common elements of Christian iconography.

THE CRUCIFIXION FRESCO FOUND IN THE VISOKI DECANI MONASTERY

A monastery in Serbia features a 14th century fresco depicting the crucifixion. In the sky, at each corner of the fresco, are representations of the sun and Moon that are stylized as chariots bearing heavenly riders. The chariots bear a passing resemblance to primitive Sputnik-like satellites or some kind of space capsule in a fiery reentry phase. UFO believers claim the artist was depicting alien spaceships. One would hope space aliens have advanced beyond 1950s Sputnik technology. Many Byzantine-era crucifixion scenes have stylized depictions of the sun and Moon in the corners with heavenly beings inside. Art historians recognize these kinds of depictions as carry overs from Greek and Persian traditions.

EXALTATION OF THE EUCHARIST BY BONAVENTURA SALIMBENI

To UFO believers, Bonaventura Salimbeni's *Exaltation of the Eucharist* looks a lot like the Christian god and Jesus screwing a pair of antenna wires into a metallic globe. There appears to be a camera lens emerging from the bottom of the globe. There's also a Sun-like disk on the globe that UFO types ignore. Ufologists believe the painting depicts a satellite. Salimbeni painted this work in the 17th century. Longitude and latitude had been worked out by then. Art historians understand the painting to represent the Christian god and Jesus using wands to draw longitude and latitude lines. The so-called camera lens is simply the moon, complimenting the depiction of the sun on the other side. The metallic nature of the globe is simply the artist trying to render shiny blue oceans.

MADONNA AND CHILD WITH THE INFANT ST. JOHN BY UNKNOWN

The Madonna and Child with the Infant St. John was painted around the end of the 15th century. The painter is unknown although art historians have narrowed it down to either Sebastiano Mainardi or Jacopo del Sellaio. It's a standard adoration painting from the time: Mary is looking down at a baby Jesus cuddling with toddler John the Baptist. UFO believers see a circular UFO hovering in the sky over Mary's left shoulder. Art historians identify it as a radiant cloud. This is common iconography for works of art dealing with adoration and annunciation themes.

THE MIRACLE OF THE SNOW BY MASOLINO DAPANICALE

The Miracle of the Snow was painted in the early 15th century. It depicts a legend from the 4th century. According to the legend Mary told the Pope to build a church on a hill in Rome. He would know the right hill because it would have snow on it. Rome doesn't get a lot of snow even in winter and the divine directive was for summer. This painting depicts a number of very dark clouds hovering over a patch of snow. Jesus and Mary are depicted above one of the clouds. To UFO believers, the clouds resemble a fleet of flying saucers with Jesus and Mary surfing atop the largest. The artist Masolino wasn't great at painting clouds but his style was common to other paintings he had done.

THE THEBAID BY PAOLO UCCELLO

The Thebaid was painted by Palol Uccello in the mid-15th century. It's a busy, busy painting. It rolls up several depictions of the lives of various saints. One part of the painting depicts St. Jerome. St. Jerome was reputed to have spent the balance of his life living in a cave near Bethlehem. In the painting, St. Jerome is beholding a vision of Jesus on the cross. If you look really closely, there's a red disk-shaped object. If you're a UFO believer, you might think it's a classic Adamski-style saucer parked in St. Jerome's cave. If you're anyone else, you'd probably instantly recognize it as a hat worn by cardinals of the church.

THE TOMB OF PTAHHOTEP

Many UFO sites reproduce a rather blurry photo of a relief painting found on the wall of an Egyptian tomb dating back to the 5th Dynasty. The tomb was the final resting place of Ptahhotep, an adviser to Pharaoh Djedkare Isesi. The relief depicts servants offering various goods Ptahhotep would need for the afterlife. Among the servants is what UFO believers claim is a grey alien! Proof space aliens helped build the pyramids! Sure, in the blurry photo distributed on many sites, it does look like a rather sinister grey with the stereotypical big black lidless eyes. However, if one accessed versions of the relief in focus and at a higher resolution it's simply a depiction of a lotus blossom in a jar.

Ten Locations of Underground Alien Bases

ANTARCTICA

The South Pole is the location of a number of improbable things, including a base for Nazi holdouts who are developing super advanced weapons and an entrance to the hollow earth. Some in Ufology roll it all into one. Nazis are guarding an entrance to the hollow earth where aliens have created a parallel civilization.

CALGARY, ALBERTA, CANADA

A UFO sighting around Calgary in 2020 caused Scott Waring, creator of the popular *UFO Sightings Daily* website, to suggest to local media organizations that this was "conclusive proof" there's a UFO base under Calgary. No media organization bothered to ask how anomalous lights in the night sky offered such proof.

DULCE, NEW MEXICO

Near the town of Dulce, New Mexico, there's an entirely mythological underground base where humans and aliens are working together to experiment on human-alien-animal hybrids. The myth's origins come from UFO hunter Paul Bennewitz (died 2003). Bennewitz appears to have been the target of a US Air Force intelligence campaign to hide stealth fighter development. The Air Force's aim was to get the UFO community to associate stealth plane sightings with UFOs. The crazier the stories the UFO community could generate, the better. Bennewitz stepped up to the task. Since his death, other UFO writers have expanded on the base myth, telling tales of an underground battle between space aliens and Delta Force troops sent in to rescue kidnapped humans.

LAKE ONTARIO

In the book *Underground Alien Bases*, author Commander X asserts there's a UFO base somewhere under Lake Ontario. Commander X suggests this has something to do with the so-called Great Lakes Triangle (a freshwater version of the Bermuda Triangle). However, that's supposedly located in Lake Michigan.

LOS ALAMOS YY-II

According to a 1987 letter written by John Lear, there's a secret base known as YY-II that's part of the Los Alamos National Laboratory. YY-II, which is sometimes referred to as the Ice Cave, is where survivors of crashed saucers are taken.

MALIBU, CALIFORNIA

In 2014, Jimmy Church, host of the *Fade to Black* podcast, published images of an oval-shaped underwater feature known as the Sycamore Knoll. The knoll is over two miles wide. It's located about six miles off the coast of Malibu (34° 1'23.31"N 118° 59'45.64"W). From one angle it looks like a building with a round roof and Greek columns out in front. Church and the follow-on UFO community have excitedly declared it an underwater UFO base.

MOUNT RAINIER

Mount Rainier is where Kenneth Arnold claimed to have spotted an armada of flying saucers in 1947. It gave birth to the term and the whole field of Ufology. Naturally, many suggest Mount Rainier is home to a UFO base. There's one problem. Mount Rainier is an active volcano. It seems like a bad idea to build a UFO base inside a volcano.

MOUNT SHASTA

An 1899 book written by Frederick Spencer Oliver called *A Dweller on Two Planets* spins a tale of how survivors of Lemuria set up an outpost named Telos under Mount Shasta in California. Since its publication, Mount Shasta has become an epicenter of all sorts of paranormal beliefs. In the mythology of UFO believers, Mount Shasta is a base for Reptoids.

SUPERSTITION MOUNTAINS

The Superstition Mountains is a mountain range in Arizona. It is best known to some as the supposed location of the Lost Dutchman's Mine. Others claim if you search hard enough you can find a portal to hell in the mountain range. Some UFO types believe it's home to a UFO base. This seems based on a 1971 abduction account by a man named Brian Scott. Scott claims while on vacation near the Superstition Mountains he was beamed aboard a UFO and taken to a base under the mountain range.

TORONTO, ONTARIO, CANADA

UFO author Commander X claims there's an ancient and possibly abandoned alien UFO base under downtown Toronto. Although it's suggested the aliens long ago packed up and moved, they apparently left the lights on. It's claimed the base radiates a strange magnetic anomaly and this explains why the area has the highest rate of traffic accidents in the city. Actual traffic stats indicate the area doesn't have an abnormally high rate of accidents, however.

Seven Buzzwords in Ufology

ABOVE TOP SECRET

Ufologists assert there's a US classification level that's even more secret than Top Secret. It is called Above Top Secret. Not true! By federal law, there is no level of secrecy above Top Secret.

DARK SIDE HYPOTHESIS

This term was introduced by John Lear to suggest the government and aliens are conspiring for some malevolent reason.

DISINFORMATION

UFO types seem to believe nearly everything they don't personally believe is true is some kind of disinformation plot.

EXTRATERRESTRIAL HYPOTHESIS

This is usually shortened to ETH. It's the suggestion the UFO phenomenon is best explained by space aliens piloting craft.

EBE

Short for Extraterrestrial Biological Entity. So, another way to say space alien. Believers in the Roswell crash claim EBE was the name given to an alien that survived the crash.

MISSING TIME

An apparent feature of an abduction experience or some other kind of close encounter. Abductees report large spans of time have passed without noticing after an alien encounter.

STRUCTURED CRAFT

A buzzword usually thrown around by UFO witnesses to imply what they saw was not swamp gas or some kind of visual illusion but some kind of physical flying object.

SPACE

Seven Problems with Astrology

NOT EVEN SPOOKY ACTION AT A DISTANCE

There are no plausible mechanisms by which planetary alignments could affect a person's personality or future. Some proponents suggest maybe the gravity of the planets has some effect on the development of the fetus. This claim is best understood in the context of the inverse square law. Gravity's effect decreases by a square of the distance. No matter where Mars is in the sky, it exerts the same gravitational strength as an 18-wheeler truck that might be parked on the street outside of the hospital delivery room.

WHY THE MOMENT OF BIRTH AND NOT THE MOMENT OF CONCEPTION?

If a person still seems to think the alignment of the planets offers some kind of spooky action at a distance, why is the moment of birth at all significant? Why not the moment of conception or when the brain starts to form in the womb?

ASTROLOGY HASN'T KEPT UP WITH PROCESSION

When you take procession into account, the sun is no longer found in the constellations that delineate the zodiac. For example, a person born in the 21st century on September 30 would traditionally fall under the sun sign of Libra. Due to procession, that person should be considered a Virgo according to the position of the sun.

SPURIOUS CORRELATIONS

The date you were born not only determines what sun sign you're under but when you start school. Data suggests children who enter school several months older than their younger classmates tend to excel at sports. Such early social factors can have profound effects on things like personality and confidence.

ASTROLOGY OFFERS NO PREDICTIVE ABILITY

A theory is only useful if it makes useful predictions. Failed astrological predictions are always conveniently explained away.

LACKS INTER-RATER RELIABILITY

Any good test or classification system should have good inter-rater reliability. That is, no matter who does the assessment, the assessments should agree to a high degree. Blinded tests of astrologers demonstrate a complete inability to reliably match people to their sun signs. As tests of the Barnum effect demonstrate, if you give people born under different sun signs the same description of one sun sign's supposed associated personality traits, people are highly likely to agree it matches them.

THERE ARE ONLY 12 KINDS OF PEOPLE

Each horoscope represents the future of some 650 million people. Seriously?

Five Bad Arguments for a Moon Hoax

There are many reasons why a hoaxed moon landing is nigh impossible. Hundreds of thousands of people worked on the project and not one has ever come forward going "yeah, I was building something that just couldn't get to the moon". Recent missions to the moon have taken excellent images of the landing sites that match up with "the official story". As people have noted, it was easier to go to the moon than to orchestrate a massive coverup that has lasted five decades. Here are five really bad arguments that the moon landings were hoaxed.

THE ASTRONAUTS TOOK MORE PHOTOS THAN THEY HAD TIME TO TAKE!

Moon hoax cranks claim astronauts did not spend enough time on the moon to account for all the photos they brought home. Six missions spent a total of 4,834 minutes on the moon and brought home 5,771 photos. To moon hoax cranks they seemed to be doing nothing but taking a photo every 50 seconds on the moon (5,771 photos divided by 4,834 minutes). Yet, the "official story" is they were doing other stuff like collecting moon rocks, setting up laser reflectors, and hitting golf balls. The first error cranks make is missing that there were two astronauts on the moon and each had a camera. That doubles the average amount of time they had to take photos. But, like anyone who visits someplace exotic, you bring back a lot of pictures and also manage to do a lot of other stuff. You can manage this because you don't take one photo, wait a minute, take another photo, wait another minute, take another photo. You take a bunch of photos in one spot, you go do something else, and then you take a bunch more photos. The cameras on the moon could take 2 pictures a second. If they were truly doing nothing BUT taking photos, they should have brought back over half a million photos in those 4,834 minutes on the moon.

THE MOON SET DESIGNERS FORGOT TO REMOVE A LABEL FROM ONE OF THE FAKE MOON ROCKS!

In one photo of a moon boulder, it looks like there's a rather sharp C carved into the rock. Moon hoax cranks claim everyone knows Hollywood set designers label props in such a fashion. So, it looks like the set designers accidentally set the rock label side up. There's no evidence any set designer labels props in such a fashion. The reality is the photo is a scanned version of an original. The C is a bit of hair curled up on the plate.

THE PHOTOS WERE A LITTLE TOO GOOD!

This claim goes 1) no one has ever seen a bad shot taken from the moon 2) you should expect to see some bad photos 3) it's too improbable all were so good therefore hoax. This argument suggests conspiracy cranks have looked at all 5,771 photos taken on the moon and judged them all perfect. It's trivial today to find the complete archive of NASA moon photos and find some that no one would judge perfect. Obviously, the handful of photos we're familiar with are the best of the best. NASA didn't want to release photos with lens flare or reflections as part of its press kit.

An Apollo command module on display at Expo 67 in Montreal.
Photo Credit: Orval Mamer

FOOTAGE OF CHATTER BETWEEN THE MOON AND EARTH HAS NO TIME DELAY

The argument here is footage you see of the astronauts talking to NASA should have a time delay of about two second before NASA responds. So, when Armstrong says something, it takes about a second for that to reach Earth. When NASA responds it takes another second for that response to get back to Armstrong. Cranks claim in all the footage we see there's no such time delay. Why it's like NASA is in the room next door! Like many things packaged up by NASA for the press, stuff got edited. If you're trying to send a punchy and engaging clip to air on the 11 PM news, you might want to edit out those time delays.

WHEN THE ASTRONAUTS RETURNED FROM THE MOON, YOU NEVER SAW THEM SMILE

Moon hoax cranks suggest in post-mission interviews, the astronauts seemed sullen, chose their words carefully, and were nervous. The implication is they knew they were part of a massive lie and weren't super happy about it. Obviously, NASA picked astronauts for their ability to get the job done, not because they made for great interview subjects. The men were tired, disoriented, and probably coming off massive adrenaline highs. It's not surprising they weren't willing to engage in a lot of witty repartee.

Seven Simple Proofs of a Round Earth

WATCH A SHIP SAIL OFF INTO THE HORIZON

On a flat Earth, a ship will simply fade away. On a round Earth the mast will slowly sink because of the curvature.

FIND A MUSEUM WITH A FOUCAULT PENDULUM

The Foucault pendulum was created by Léon Foucault (died 1868) as a way to conclusively demonstrate that the planet rotates. If you set up a Foucault pendulum at different latitudes, the pendulum will demonstrate different precession rates. This can only be explained if the earth was a globe.

Foucault pendulum at the Detroit Science Center
Photo Credit: Karl Mamer

GOOGLE FOR PHOTOS

A wide variety of space-faring nations have photos and videos showing a round Earth from space.

LUNAR ECLIPSES

If you observe a lunar eclipse, a distinctly round shadow crosses over the face of the moon. If Earth were a flat disk, there's no way the earth could cast such a shadow. It would be a shadowy line.

GO TO THE TOP OF A BUILDING

Ever notice you can see more distant objects as you go from the bottom to the top of a building? That's because as you go up, you see more of what is hidden by the curve.

REPEAT ERATOSTHENES' SHADOW EXPERIMENT

Greek mathematician Eratosthenes figured out the earth was round back in 240 BC. His experiment was pretty simple and anyone can reproduce it. Place sticks of equal length in different locations that are at least a few hundred miles away from each other. At the same time of day, measure the shadow. On a flat Earth the shadows will be equal length. On a round Earth the shadows will be different because the sun is hitting the sticks at different angles.

LOOK FOR THE SOUTHERN CROSS IN NORTH AMERICA

Ever notice those Australians talk about a constellation they call the Southern Cross? Have you ever noticed you're incapable of finding it by looking at a night sky in most parts of North America? That's because as you head south from the equator on a round Earth, you get to see different stars hidden by the curvature. If the earth were a disk, you'd see the same stars as Australians.

Six Stupid "Proofs" of a Flat Earth

THAT GUY WHO TOOK A SPIRIT LEVEL ON A PLANE

A YouTube flat Earth star who goes by the name of D. Marble thought he had the ultimate mic drop proof the earth is not a sphere. In 2017, he took a spirit level on a plane. A spirit level is one of those "rulers" with a little air bubble that lets you know how badly your homemade desk is out of whack. Based on the rather confused way he thinks planes fly on a curved Earth, he believes as the earth curves underneath, a pilot has to keep dipping the nose down to stay flat and level with the ground. After the plane reached cruising altitude, he shot a 20-minute time lapse video of his spirit level sitting on his tray table. For the duration of the video, gasp, the bubble stayed dead center. In his view he fully demonstrated the plane was, in fact, traveling straight and level, and not at all dipping its nose down to compensate for the earth rapidly curving away. Hence, the world is flat and every pilot on the planet is part of a conspiracy of silence. Of course, planes don't have to keep flying down to stay level with the receding Earth. If they did, putting payloads into space would be incredibly cheap. Just keep flying straight and you'll sail right off the planet. As it turns out, there's a thing called gravity that keeps the plane perpendicular to the center of Earth's gravity.

THE UN FLAG

Flat Earthers, like any group of conspiracy theorists, always seem to think the conspirators believe the best place to hide is in the open. Or maybe they think it's hardly worth keeping a secret if you can't tease the truth with obscure hints. Keeping the myth alive that the earth is round requires a massive level of international cooperation. So the United Nations must be involved in some way. To do their part of teasing the truth or hiding in the open, the UN flag doesn't, shockingly, feature a round Earth but a flat Earth that curiously matches one of dozens of flat Earth models. The UN flag is not a flat Earth map. It's actually an azimuthal equidistant projection. For doubters, the flag is also divided into 33 sections. So, like, I guess the Freemasons are behind this too.

NASA ADMITS THE EARTH IS FLAT

Flat Earthers thought they had the smoking gun evidence that even NASA admits the earth is a flat, non-rotating sphere. In 1988 NASA published a paper about a FORTRAN program for an aerodynamic model that assumes a "linear aircraft model for a rigid aircraft of constant mass flying over a flat, nonrotating Earth". See. Flat. Nonrotating. How could one deny it? Of course, the reality is if you're flying short distances, like across a dry lakebed test range, you can strip out all the complicated math and assume you're flying across a flat surface.

DE NILE

If the world were round, the Nile, which flows north, would have to continually flow uphill. Since it demonstrably flows north, the world can't possibly be round. I know. What?

TRAVEL TO ANTARCTICA IS BANNED

In the views of many flat Earth believers, Antarctica is not a frozen continent at the South Pole. It's a giant ice wall that rings the outside of the flat Earth disk. Travel there is ostensibly banned because people would discover the truth. Of course, you can book cruises to Antarctica. And nothing would prevent you from chartering a private flight over the South Pole.

MARINERS USE FLAT MAPS

America's first radio evangelist Wilbur Glenn Voliva offered many bizarre proofs of a flat Earth back in the 1920s and 1930s. He claimed it was obvious sailors accepted the world was flat because their navigational maps were flat, not round.

Twelve Hollow Earth Proponents

Since the 17th century, various people have proposed that the earth is hollow. It was not an unreasonable idea until mathematician Charles Hutton established the earth's density based on data from 1774's Schiehallion experiment. His calculations clearly established the earth was not hollow. A hypothesis as delightful as a hollow Earth was too good for some to ever let go of. People imagine there are whole civilizations living inside the earth. What Earth's interior looks like varies with the telling. Some view the interior as a dark, cave-like realm. Others posit there's a sun at the center of the hollow Earth and the interior is a lush, jungle paradise. Hollow Earth proponents today are a small and dying group.

EDMOND HALLEY

Edmond Halley (died 1742) is famous for discovering Halley's comet's periodicity. In 1692, he proposed the earth was composed of a series of concentric shells. Each inner shell rotated at a different rate which, for him, explained variations in compass readings. Halley speculated each shell had its own atmosphere and was lit by some kind of luminous gas affected by Earth's magnetic field.

JOHN CLEVES SYMMES JR.

John Cleves Symmes Jr. (died 1829) was a US army officer during the War of 1812. In civilian life he went into business but failed. In 1818, he began promoting a variation of Halley's idea of a hollow Earth with concentric interior spheres. Symmes' big addition was the notion the North and South Poles featured entrances to this hidden realm. However, Symmes proposed these openings were massive, each over 4,000 miles wide. The holes were actually so large that as you traveled towards them, you'd not even notice you were going down a hole. The transition would be very gradual. Symmes spent most of the rest of his life trying to convince people the earth was hollow. His wife and ten kids lived in poverty as he traveled America giving lectures and self-publishing pamphlets. Symmes tried hard to get established scientists like Sir Humphry Davy to sign on but no one took him seriously. His proposed entrances to the underground were derisively referred to as "Symmes Holes" by real scientists.

PRESIDENT JOHN QUINCY ADAMS

President John Quincy Adams (died 1829) was the driving force behind the creation of the Smithsonian Institute and a national observatory. He also fell for John Cleves Symmes' ideas of a hollow Earth and urged Congress to fund an expedition to find the opening at the North Pole. By the 19th century it was well established the planet was not hollow. Congress denied Adams funds, understanding the trip was foolish. Why Adams fell for this is something of a mystery. Historians suggest

Adams' intense curiosity about the natural world may have motivated him to fund any kind of expedition to the North Pole.

JEREMIAH N. REYNOLDS

Jeremiah N. Reynolds (died 1858) was a Symmes devotee. He was also a newspaper editor and amateur explorer. He was tapped by President Adams to lead the expedition to the North Pole to find Symmes' Hole. Despite Congress refusing to fund the quest, Reynolds managed to raise private funds from a wealthy investor for a journey to the South Pole. The trip set off for the South Pole from New York in 1829. Reynolds and his team made it to Antarctica but realized they were not actually provisioned for a land expedition. They didn't bring enough food, having assumed Antarctica had enough wildlife they could live off of. On their return journey, the crew mutinied and dumped Reynolds off in Chile. Edgar Allan Poe parodied Reynolds' voyage in his novel *The Narrative of Arthur Gordon Pym of Nantucket*.

CYRUS TEED

Cyrus Teed (died 1908) was a self-professed alchemist and cult leader in the late 19th century. He founded a commune called Koreshan Unity which was based out of Florida. The group practiced celibacy, which predictably led to dwindling numbers as members died off. Hollow Earth was one of his cult's tenets. Teed was one of the first to suggest we're living inside a sphere. Centrifugal forces keep us anchored to *terra firma* as the planet spins. Looking up means we're looking towards the center of Earth, with the sun and the entire universe in the center. Teed got Ulysses Morrow (a newspaper editor) to test his theory. Morrow constructed a 12-foot long steel-and-wood frame encasing a spirit level. He called this measuring device a rectilineator. Over four months, Morrow and a group of cult members meticulously moved the rectilineator along a four-mile path and took measurements every eighth of a mile. The goal was to determine if the earth was flat, convex, or concave. At the end of the experiment, Morrow believed his data firmly established the earth was concave, that is rising upwards towards the horizon, as one would expect if you were living inside a sphere. Morrow's results have survived. And indeed, they show exactly what Morrow expected. Which is the problem. Morrow assumed his rectilineator was rigid but a reconstruction showed anything but. The device sagged as it was dragged along and Morrow likely tossed out any measured sag that didn't agree with expected results and only accepted results that agreed with expected results.

WILLIAM REED

William Reed (died 1920) published a book in 1906 called *Phantom of the Poles*. He made the bold claim no one has ever made it to the North or South Pole because they simply don't exist. He subscribed to Symmes' view that as you approach either

Pole, the world curves inwards and unless you were paying strict attention, you'd not even realize you were now following the curve into the earth. His book intended to put Hollow Earth on a more scientific foundation but, similar to creationists, it consisted of questions he didn't think science had answers for and then tried to show how Hollow Earth neatly accounts for them. He suggested, for example, science couldn't explain what causes the Aurora Borealis or tidal waves or why meteors seem to fall more frequently near the Poles. Therefore, hollow Earth.

RICHARD SHARPE SHAVER

Richard Sharpe Shaver (died 1975) was the author of the incredibly influential *Shaver Mysteries* stories that appeared in *Amazing Stories* magazine during the 1940s and 1950s. Shaver, who was likely suffering from schizophrenia, believed he was a victim of an ancient race of Lumerians and minions known as the Deros or "detrimental robots". The Lumerians made their home in a vast realm under the earth's surface and occasionally left the planet in rockets to parlay with evil space aliens.

WALTER ISIDOR SIEGMEISTER (AKA RAYMOND W. BERNARD)

Walter Isidor Siegmeister (died 1965) was an early American raw food advocate in the 1950s. He spent some time in South America in the 1950s seeking to create a race of superhumans in the jungles of Brazil. When he returned to the USA he changed his name to Raymond W. Bernard and began promoting his ideas of a hollow Earth. Siegmeister appears to be a fan of *The Shaver Mysteries*. Siegmeister was one of the first to link the emerging UFO phenomenon to civilizations living in the hollow Earth. He published *Flying Saucers from the Earth's Interior* in 1960 and *The Hollow Earth* in 1964. Siegmeister is probably the most widely read hollow Earth proponent. His books were heavily promoted for decades in UFO zines of the 1960s, 1970s, 1980s, and 1990s.

MOSTAFA ABDELKADER

Mostafa Abdelkader is a mathematician who picked up on Teed's idea that when we look up, we're not looking out from the surface of a sphere (the Copernican model) but we're inside a sphere and looking inwards towards the center. Abdelkader wondered how that would all work mathematically and published his results in a 1982 paper titled "Mapping Outer Space Into a Hollow Earth". Abdelkader realized the proposition that light travels in a straight line at a fixed rate is actually an untestable hypothesis. What if light actually moves in curves and speeds up as it moves out from the center of a universe inside a sphere? As you move towards the center of the sphere you get smaller and smaller and you move slower and slower. In Abdelkader's model, everything at the top of a building is a bit smaller and moves a bit slower than what's on the ground floor. It's similar to Relativity's idea that as you move faster, your clock slows and you get smaller relative to a stationary frame of reference. You don't notice this change in space and time in your local frame of reference. Likewise, just as you can never reach the speed of light, because you need more and more energy, in Abdelkader's cosmology you can never reach the center because you move slower and slower. Abdelkader's paper was an interesting exercise in teasing out assumptions and turning them, literally, inside out to show you can perfectly map reality. Reality is ultimately an assumption either space is infinitely big as you move out into the universe from the surface of a sphere or space is finite and contained in a sphere but you can never ultimately reach the center. The only real point of objection that can be made to Abdelkader's model is while you can map the Copernican model to a hollow Earth universe, the math is enormously more complicated to keep light moving in curves. If you subscribe to the assumption of parsimony, the Copernican model is mathematically simpler and, hence, the better model of reality.

ERNST ZÜNDEL

Ernst Zündel (died 2017) was a neo-Nazi and active Holocaust denialist. He was born in Germany but dodged the German draft and moved to Canada at age 19. Zündel was eventually deported back to his home nation of Germany where he served time in prison for inciting racial hatred. While living in Canada, he formed his own publishing house and cranked out "scholarly" works like *The Hitler We Loved and Why*. A large part of his output throughout the late 1970s was works about Nazis having an ongoing UFO base at the South Pole and a thriving civilization inside a hollow Earth. Zündel tried to raise funds for an expedition to find this Nazi base. His plan was to lease a 747, adorn it with Nazi themed livery, and then overfly the base. If you wanted to be part of this historic venture, you just had to pony up $9,999. It's unknown if anyone bought a ticket on Air Hitler but the flight never took place.

RODNEY CLUFF

Rodney Cluff is probably the best-known amplifier of Walter Siegmeister's hollow Earth ideas today. Like Mitt Romney's father George Romney, Cluff was born in a Mormon colony in Mexico. After completing his mission, he settled in Arizona where he encountered Siegmeister's book *Hollow Earth*. In Cluff's view, the hollow Earth offers a ready answer to "what happened to the lost tribes of Israel?" Apparently, their god stashed them inside Earth. Cluff has advocated his idea on *Ancient Aliens* and *Coast to Coast* as well as his book *World Top Secret: Our Earth IS Hollow!* In 2006, Cluff and two others were actually planning something called the "North Pole Inner Earth Expedition". Cluff seemed keen to "have the privilege of visiting his lost cousins of the Ten Tribes". Alas the expedition was put on hold when one of the organizers died of cancer. The plug was pulled a few years later when the financial backer pulled out after his electric car company tanked.

DAVID ICKE

What doesn't David Icke believe in, besides reality? Icke is famous for his wild-eyed beliefs that the world is run by shape-shifting lizards. The origin of these lizards seems to vary. At times they are space aliens. At other times they rule below in a hollow Earth. The latter idea comes from his book *The Biggest Secret*. Icke is also not content to believe the earth is hollow. Icke claims the moon is hollow and is actually an alien space station.

Eight Planets That Don't Exist or are Not Likely to Exist

NIBIRU

In various works, ancient astronaut crank Zecharia Sitchin (died 2010) wrote about a planet he called Nibiru. Every few thousand years, Nibiru passes close to Earth. As it draws near, beings called the Anunnaki (a name lifted from Babylonian mythology) pay a visit to Earth and carry off our gold. According to Sitchin, knowledge of Nibiru and these ancient astronauts became codified in most of our world religions, including works in the Bible.

THE SATURN-LIKE OBJECT

Amateur astronomer Chuck Shramek (died 2000) took a photo of comet Hale-Bopp. His photo included a bright object he didn't recognize. He checked his computer's star catalog to see if it was a star. Because he didn't actually know how to use his software, it didn't turn up an 8.5 magnitude star called SAO141894. So, he rang up Art Bell and claimed he had photographic proof that something which he described as a Saturn-like object (SLO) was "following" the comet. Shramek's reputed SLO dovetailed with the mythology of the Heaven's Gate cult that space Jesus was coming to get them. Cult members drank poison so they could join the SLO.

VULCAN

To explain peculiarities in Mercury's orbit, 19th century astronomers suggested there was a small planet wedged in between Mercury and the sun. They called it Vulcan. Occasionally people reported to have found Vulcan but their claims always proved to be in error. Vulcan was abandoned as a possible planet when Mercury's orbit was fully explained by Einstein's Theory of Relativity. Should you find another planet in the outer solar system and be tempted to call that one Vulcan, that name will be rejected by the International Astronomical Union. The IAU has that name reserved for any other planet that we might happen to discover between Mercury and the sun.

NEMESIS

Though not technically a planet, Nemesis was hypothesized to be a brown dwarf that orbits the sun and swings by Earth every 26 million years. Nemesis was conjectured to explain why mass extinctions seemed to happen every 26 million years. More recent analysis of extinction data indicates the 26-million-year cycle is spurious, hence most scientists have abandoned the Nemesis hypothesis as a tenable explanation for extinction events.

MALDEK AND HEDRON

Contactee George King, founder of the Aetherius Society, claimed aliens told him the asteroid belt is a remnant of an Earth-sized planet called Maldek. The Maldekians blew up their planet through the careless use of atomic technology. We know the asteroid belt wasn't once a planet that blew up because there simply isn't enough matter in the asteroid belt to have once constituted a planet. The asteroid belt has about 4% of the moon's matter. Interestingly, King invented this tragic tale of Maldek a few months after the release of a movie called *The Mysterians*. What was the plot of *The Mysterians*? Oh well, it's a movie about aliens who came from a planet that was in the asteroid belt but was destroyed by nuclear weapons. A number of years later Elizabeth Clare Prophet's Church Universal and Triumphant cult seems to have lifted Maldek from King and incorporated it into their own theology. And because one former planet is not enough to explain the small bit of matter that comprises the asteroid belt, the cult seems to have added a precursor planet to Maldek called Hedron. Hedron appears in a 2005 book written by Prophet (died 2009) and her husband. As the name suggests, Hedron was a planet of Hedonists and was destroyed by "the Lords of Karma" because the Heddites just enjoyed life a little too much. Now, if Hedron sounds familiar to *Doctor Who* fans it's because Hedron was a planet the Sixth Doctor (the Colin Baker incarnation) visited in a comic book adaptation published in 1986.

TYCHE

Back in 1999 some astronomers suggested there was a gas giant out beyond the Oort Cloud. They called it Tyche. Tyche would explain the existence of certain long period comets. Based on the planet's size and distance, NASA realized it should have been detected by its Wide-field Infrared Survey Explorer (WISE) telescope. After an exhaustive search of the data, NASA conclusively ruled out the planet's existence.

NEITH

Several astronomers of the 17th and 18th centuries thought they had spotted a moon of Saturn. It was dubbed Neith, after an Egyptian goddess. By the end of the

19th century astronomers concluded Neith was likely one of several misidentified stars.

Eight Steady State Proponents

Although one is inclined to think of the Steady State model of cosmology as predating Big Bang, Steady State was proposed in response to Big Bang. Steady State recognizes the universe is expanding, yet ultimately remains unchanged due to new matter being continuously created to fill in the gaps left by galaxies flying apart. Despite the overwhelming support Big Bang enjoys in the scientific community, there are, or were, some scientists that stood by Steady State long after it was time to abandon the model.

JAMES JEANS

James Jeans (died 1946) was a British physicist. He and Arthur Eddington were some of the first physicists in the UK to tackle the topic of cosmology. Jeans' major discovery was the Jeans' Length. This is the critical radius of an interstellar cloud. A cloud smaller than the Jeans Length will condense to form a star. Jeans was also the first to propose a steady state cosmology based on the continuous creation of matter. He suggested this model back in 1928.

FRED HOYLE

Astronomer Fred Hoyle (died 2001) introduced the idea of stellar nucleosynthesis. It was a groundbreaking theory that helped explain the origin of chemical elements beyond hydrogen, helium, and lithium. Hoyle also did not like Big Bang and gave astronomy the Steady State alternative. In fact, Hoyle coined the term Big Bang as a way to disparage the idea.

HERMANN BONDI

Hermann Bondi (died 2005) did a lot of important work on the General Theory of Relativity and black holes. He was the first to accurately describe the nature of gravitational waves. He is also a co-author of Hoyle's Steady State model. Bondi abandoned Steady State as observational data started to come down on the side of Big Bang.

THOMAS GOLD

Thomas Gold (died 2004) was another co-author of Hoyle's Steady State. Gold was also a big proponent of a fringe theory that oil doesn't come from organic sources but comes from an abiotic process that continues today. Gold believed this means the earth has an unlimited supply of oil. Gold remained a champion of Steady State well into the late 1990s.

ALBERT EINSTEIN

Albert Einstein was also skeptical of Big Bang initially, even though his own General Theory of Relativity predicted it. When Big Bang's originator Father Georges Lemaître discussed Big Bang with Einstein, Einstein was reputed to have told Lemaître "Your calculations are correct but your physics is abominable." In 1931, Einstein worked on his own Steady State model but abandoned it.

GEOFFERY BURBIDGE/JAYANT NARLIKAR

Geoffrey Burbidge and Jayant Narlikar worked with Hoyle in the early 1990s to propose a quasi-steady state model. The original Steady State model could not explain cosmic background radiation. Quasi-steady state tries to explain that the cosmic background radiation is actually normal starlight being interfered with by cosmic dust. Quasi-steady state also proposes the Big Bang is right but we're only observing a local Big Bang. The universe is eternal but expanding from the creation of new matter from smaller local Big Bangs.

NEAL ADAMS

Neal Adams is a comic book artist largely known for his work on Batman. Adams has long proposed a local version of Steady State. In his view planets themselves are expanding as matter within planets gets created spontaneously. Adams believes this explains why the continents seem like they used to fit together. Not because of plate tectonics but because they were a single landmass when Earth was smaller and our planet expanded like a balloon in response to matter continually being created in the center. Adams rejects that his expanding planet hypothesis violates well-established theories of gravity and electromagnetism.

Nine Problems with Electric Universe Cosmology

Proponents of the pseudoscientific Electric Universe Cosmology (EUC) hold the view that gravity's role in cosmology has been grossly overstated and the universe is better explained by the effects of electricity and electromagnetism. Whereas Steady State has a few real scientists working within the halls of academia still poking at it, you'll find EUC is the domain of amateur scientists, crank engineers (drink!), flat Earthers, fans of Immanuel Velikovsky, and climate change denialists. High kooks only need apply!

SCIENCE REALLY HASN'T CONFUSED THE EFFECTS OF ELECTROMAGNETISM WITH GRAVITY

Our theories of gravity and electromagnetism have been astoundingly well established by observation and experimentation. Our modern world simply wouldn't work if for decades scientists have been attributing certain effects to gravity instead of electromagnetism as EUC proponents nakedly assert.

THE ELECTRIC UNIVERSE DOESN'T ADD ANYTHING

For a theory to supplant another well-established theory, it needs to explain everything the old theory explains and explains what the old theory can't. EUC doesn't do this.

OBJECTS WITH DIFFERENT ELECTRICAL CHARGES DON'T FALL AT DIFFERENT RATES

If electromagnetism was behind what we think of as gravity, it would stand to reason positively charged objects should fall at a different rate as negatively charged objects. This has never been observed and EUC proponents can point to no experiment that demonstrates it.

YOU CAN'T LEVITATE IN A FARADAY CAGE

If electromagnetism has a primary role in what attracts two bodies, one should be able to put a person in a Faraday cage and watch them float. You never see this.

MOST MATTER IS ELECTRICALLY NEUTRAL

EUC proponents offer no good explanation how electromagnetism exerts a force on matter that's electrically neutral.

CLAIMS GRAVITY IS A LOCAL PHENOMENON DEFIES OCCAM'S RAZOR

Our theory of gravity has a principle advantage: you only need one formula. It doesn't matter if you're in space or standing on Earth. EUC proponents try to claim gravity actually takes over when you need to describe how matter behaves on Earth but you need to invoke EUC as you zoom out.

EUC OFFERS NO EXPLANATION AS TO WHERE ELEMENTS CAME FROM

Our theory of gravity offers a demonstrable explanation for the creation of elements. Elements above lithium up to iron are a result of nuclear fusion in the center of a star. That requires gravity. A lot of it. If stars are not powered by fusion but are big balls of electricity as EUC proponents believe, where did elements come from? EUC offers no explanation.

GAPS ARE NOT PROOF FOR EUC

A big part of EUC is cataloging aspects of the universe the standard model can't explain and then suggesting physics is wrong and therefore EUC wins by default. This is a very common feature of almost every pseudoscience. Instead of offering positive evidence for a claim, the crank spends a great deal of time conflating not being able to explain everything with not being able to explain anything.

THE ANCIENTS KNEW! WHAT?

EUC cranks love to dip into pseudoarchaeology and claim modern science is so focused on this gravity thing they've missed that EUC was obvious to the ancients.

Four Modern Proponents of Geocentrism

Geocentrism is the long-disproven idea that the earth is at the center of the solar system (or universe) and everything else revolves around us. From a religious point of view, it's comforting to know we're actually at the center of everything. Otherwise, you have to confront the reality that there's a lot of universe out there that a god needlessly created simply to raise up humanity in an unremarkable corner of that universe.

WALTER VAN DER KAMP

Walter van der Kamp (died 1998) was a Dutch-Canadian teacher. He worked as a schoolmaster in Canada in the 1960s and self-published a paper arguing for geocentrism titled "The Heart of the Matter". He distributed this to various Christian institutions. He founded the Tychonian Society, named after Tycho Brahe who worked out a geocentric model in response to Copernicus' heliocentric model. Van der Kamp's main issue with heliocentrism seemed to be that Copernicus never experimentally demonstrated his model was true. Likewise, scientists since Copernicus have never, to his satisfaction, experimentally demonstrated the truth of the Copernican model. It's not clear how he reconciled the reality that we can fling probes to Jupiter and Saturn with a high degree of accuracy based on this kooky heliocentric assumption. That would seem a valid test.

Statue of Copernicus in Warsaw.
Photo Credit: Karl Mamer

GERARDUS BOUW

In 1984, Walter van der Kamp gave up running his Tychonian Society and leadership passed to Gerardus Bouw. Bouw is an astronomer and computer scientist. He moved the society to his hometown of Cleveland and renamed it the Association for Biblical Astronomy. Bouw expanded the group's reach by hosting geocentrism conferences. In 1992 Bouw self-published a 400-page defense of geocentrism titled *Geocentricity*. In it Bouw argues the heliocentric model has led to moral degeneracy. Famous heliocentrists like Copernicus and Kepler were men of low moral regard. He conveniently ignores that Tycho Brahe had a reputation for being a glutton. By way of proof of geocentrism, he notes the Bible has more than two dozen passages that talk of the sun going down, setting, or rising. In his view, the true literalist interpretation is the earth is fixed and the sun is doing all the moving. This seems to be a literalism bridge too far for even groups like Answers in Genesis.

MARSHALL HALL

Marshall Hall self-published a book in 1991 called *The Earth is not Moving*. In his view, and as the book title suggests, the earth is not moving. He claims there's simply no proof of that. He seems to be unaware of Foucault's pendulum. Or stellar parallax. Or space travel. He calls the Copernican Model the "Copernican Counterfeit" and he has a special loathing for Kepler. He passionately believes Kepler murdered Tyco Brahe. His evidence? Kepler's mother was once accused of being a witch. Anyone whose mother is a witch is probably also a murder. QED. This whole rotating Earth thing is "all part of a Satan-empowered, anti-Christ religious Cabal".

ROBERT SUNGENIS

Robert Sungenis is a traditionalist Catholic. He's so traditionalist he wants the Vatican to reverse its modern stance on Galileo and re-condemn the man for proposing the heliocentric model. Oh yeah, he also wants the church to stop trying to reconcile with Judaism and go back to blaming Jews for everything. (The Southern Poverty Law Center describes him as "one of the most rabid anti-Semites of the radical traditionalist movement.") He came to greater prominence in 2014 when his film *The Principle* hit the theaters (or more accurately "theater", as only one cinema screened it). His film argues for the geocentric model. Somehow Sungenis was able to hoodwink mainstream scientists into giving interviews. Participants reported they were led to believe they were helping an indie filmmaker make a general documentary about cosmology. Sungenis also managed to get *Star Trek*'s Kate Mulgrew to provide narration. Mulgrew distanced herself from the film upon learning its true intent.

Nine Alien Structures on Other Planets Alleged to Exist

Pareidolia is a phenomenon of perception. Humans tend to interpret random stimuli (usually visual or audio) as matching some known pattern. Typically, this might be a human face. In the world of Ufology, spending hours poring over highly pixelated or third-generation NASA photos to find something that looks like a manufactured artifact has proven profitable for a number of Ufologists.

FACE ON MARS

The face on Mars is a mesa in a region of Mars known as Cydonia. Back in the 1970s, NASA released images from the Viking 1 orbiter. Its caption for the area noted the mesa "resembles a human head". NASA stressed it was a trick of light and shadow. Richard Hoagland, a fan of *Star Trek* and conspiracy theories, has been beating the "face is real" drum since the 1980s. Later Mars missions have captured multiple, high resolution photos of the face and found it's just weathered rock. That has not stopped the face on Mars industry. The face no longer looks like a face because apparently NASA bombed the site before taking higher resolution images.

ZIGGURAT ON THE FARSIDE OF THE MOON

Back in 1999, someone going by the online name of "kksamurai" released a doctored scan of an Apollo 11 image of the far side of the moon. The already blurry image was made to look like there was a primitive ziggurat on the far side. The image was part of a book kksamurai was going to publish called *Let's get back To The Moon I think we missed something*. That book was never published. The images bounced around for a number of years until *Ancient Aliens* talking head Mike Bara discovered it on a gaming discussion board and ran with it. The region has been imaged in later years by both NASA and Japan. These later, higher resolution images reveal no ziggurat.

PHOBOS MONOLITH

In 2009, Buzz Aldrin gave an impassioned speech for continual space exploration in a C-SPAN interview. He quipped there is a monolith on Mars' moon Phobos. Charitably, Aldrin was trying to suggest there are many astounding natural wonders we can discover. However, many cranks took this to mean Aldrin was announcing to the world (on C-SPAN of all places) there was an artificially constructed monolith on Phobos (something right out of *2001: A Space Odyssey*). Cranks don't get that terms in science sometimes have specific definitions that are different from the more broadly known definition. For example, in geology (Apollo astronauts all got extensive geology training) monolith doesn't refer to a tall, flat rectangular object of intelligent creation. It simply means a single large stone. Ayers Rock (Uluru) is probably the best-known example of a natural, terrestrial monolith.

A ROBOT HEAD ON THE MOON

In 2006, Richard Hoagland began circulating a picture of what looked a bit like a severed robot head lying at the bottom of a lunar crater. The photo was taken by the Apollo 17 crew. The version Hoagland was circulating was highly cropped and zoomed in on. Claims have been made that the moonwalkers went into the crater and retrieved the head. That's impossible as no astronaut ever entered a moon crater. NASA viewed the danger as too great.

A TOWER, CASTLE, AND SHARD ON THE MOON

In March 1996 Richard Hoagland invited major news organizations to the National Press Club in Washington D.C. He promised actual NASA scientists were ready to reveal suppressed evidence of alien structures on the moon! The press might have ignored him but he used the early Internet's Usenet system to promote his conference. News orgs got calls from readers demanding they show up and cover Hoagland's press conference. *The Washington Post* dispatched a reporter who came away feeling Hoagland was "basically a kook". Hoagland's so-called NASA scientist was a contract engineer (drink!). The suppressed evidence amounted to photos that had been in the public domain for decades. Try as they might, no reporter at the conference could see the towers, castles, and shards Hoagland insisted were easy to see in moonscape photos.

A SPACESHIP AND FACE ON PLUTO

The New Horizons mission delivered some stunning photos of Pluto. Of course, it delivered reams of images into the hands of cranks who could blow up and "enhance" photos to find all manner of objects that look vaguely like whatever the crank wants to see. Scott Waring, long time webmaster of the popular *UFO Sightings Daily* website, made a bit of a splash by being first to market with a face. Sort of. This one you really have to look at it to see the face. He also claimed to find a heart-shaped alien spaceship parked on Pluto. When planetary geologists pointed out the feature identified as a spaceship is over 4 miles long, the response was "and?"

Four Astronauts with Crank Ideas

GORDON COOPER

Gordon Cooper (died 2004) was one of the original Mercury 7 astronauts. Before volunteering to be put in a small pillbox and shot into space he was a test pilot and flew for the Air Force. He claimed to have seen a number of UFOs during his life and went to his grave believing the US government was covering up what it knew about alien visitors. Cooper's most famous UFO claim was a 1957 sighting of a silent saucer-shaped object. The story goes that the saucer made a landing on a dry lakebed being set up as an aircraft test site. Cooper's retelling of the incident has varied over the years. In one version he reported to not have been on site to witness the landing but was told of the landing by two crewmen setting up film equipment. In another version he said he witnessed the landing himself. In some versions the crew saw the saucer land. In other versions the crew simply saw the saucer float on by. The incident was reported to the Air Force. Based on photos taken by the crew, the Air Force determined they had witnessed a weather balloon that was known to be operating in the area. UFOs weren't Cooper's only crank pursuit. In 1978 Cooper lent his celebrity to a company called American Consumer Incorporated that was hawking a device you could install on your car to turn air into energy. The FTC told Cooper the product was a scam and advised him to stop helping the company promote the device. In Cooper's final days, he claimed while orbiting Earth in his Mercury capsule, he used secret sensors to detect and map sunken treasure ships in the Caribbean. He recreated the map and shared it with a friend who was a professional treasure hunter. Cooper's supposed treasure map was turned into a Discovery channel reality show called *Cooper's Treasure*. It ran for two seasons, starting in 2017. No treasure-laden ships were ever found.

Gemini 7 capsule on display at Expo 67
Photo Credit: Orval Mamer

JAMES IRWIN

James Irwin (died 1991) was Apollo 15's lunar module pilot and the eighth person to walk on the moon. Irwin and his fellow crewmates had hatched a plan to take a collection of stamps to the moon and then sell them to a stamp dealer on their return. The idea was they were going to use the profits to set up trust funds for their children. NASA put the kibosh on that plan, however. After retiring from NASA, Irwin became a born-again Christian and toured churches proclaiming he had seen the light while in space. Irwin, who held a Master of Science degree and should know better, came to believe the *Genesis* creation myth was real. He was convinced he could find evidence of Noah's ark on Mount Ararat. He led several expeditions and overflights but found nothing.

BRIAN O'LEARY

Brian O'Leary (died 2011) never made it into space but he was training as a scientist-astronaut with NASA in the late 1960s. O'Leary was a planetary scientist and was considered for a possible role for a Mars landing in the 1980s. O'Leary resigned before completing his training and accepted an invitation by Carl Sagan to lecture at Cornell. In the late 1970s and 1980s, O'Leary took a bit of a departure from science and started investigating remote viewing and near-death experiences. He dabbled in free energy crankery. He co-produced a DVD called *Miracle in the Void: Free Energy, New Science, Consciousness & The Earth*. In his 60s he was diagnosed with skin cancer and tried to treat it with pseudoscientific methods. He almost died of a heart attack after participating in a ceremony that involved imbibing ayahuasca.

EDGAR MITCHELL

Edgar Mitchell (died 2016) was an aeronautical engineer (drink!) and the sixth person to walk on the moon. Mitchell was into woo during his NASA days. He is famous for conducting an unapproved remote viewing experiment during his Moon voyage. Mitchell visualized a series of numbers and friends back on Earth tried to psychically determine the numbers. Mitchell seems to think the experiment was a success because his subjects not only failed to guess his numbers but did worse than chance. In psi-world, if you "fail up" like that, the experiment is still considered evidence for a made-up phenomenon called psi-missing. Post-NASA Mitchell established the Noetic Institute which sponsors research into a wide range of super woo topics. In his final days, Mitchell became a voice for UFO cranks. You can't get much more legit than an actual NASA moonwalker. The problem was, other than Mitchell's own personal conviction in the reality of UFOs, he did not have one shred of evidence to back his claims of government cabals hiding the UFO reality.

Thirteen Urban Legends about NASA and Space Exploration

You hear so many wild-eyed conspiracies about NASA that it's easy to miss some more mundane and amusing urban legends that have sprung up around NASA and space exploration in general.

TEACHER-ASTRONAUT DODGES TWO BULLETS

Teacher Barbara Morgan was Christa McAuliffe's backup on the ill-fated Challenger mission. There is a legend that Morgan was actually the primary candidate for the flight but was bumped at the last minute. A number of years later, as the legend goes, Morgan was picked again to be on the Columbia which broke up on reentry in 2003. In reality, Morgan was selected to be on a mission after the Columbia launch but that flight was scrubbed following the disaster.

THE CURIOSITY MARS ROVER DISCOVERS A MESSAGE FROM GOD

Online news source the *Daily Currant* reported in 2013 the Curiosity rover imaged a rock on Mars with a message from the Christian god indicating that it was real. The *Daily Currant* was an *Onion*-like parody site. Several Christian groups missed the site was a work of parody and excitedly repeated this ultimate proof of their god's existence. The site ceased publication in 2016.

NASA CAN LOCK YOU UP IF YOU TALK TO ALIENS

There's a legend that there's a law on the books that lets NASA lock you up if you talk to aliens. If one actually asks for the law, the legend teller might cite the so-called "Extra-Terrestrial Law". The actual law is the "Extra-Terrestrial Exposure Law". This is a real law from the Apollo days. It gives NASA the right to quarantine anyone who touches or comes close to a person or thing that has been in space. So, astronauts returning from the moon, moon rocks, and the like.

MARS WILL APPEAR AS BIG AS THE MOON

This one crops up periodically. It has been circulating since at least 2003. The legend is Mars and Earth will be in some unusually close conjunction and sky watchers will be treated by a once-in-a-lifetime opportunity: the night sky will appear to have two moons, one being Mars.

THE BUDGET BUSTING SPACE PEN

The legend goes that during the 1960s, NASA discovered pens don't work in zero-G. They embarked on a vastly expensive program to develop a pen that could write in space. The amount of money NASA spent seems to grow with each telling, but usually the cost of the space pen program is cited in the hundreds of millions of dollars. The legend then ends with this boner: And what did the Soviets do? They used pencils. NASA rejected pencil because bits of pencil lead floating around a space cabin is a bad thing. They did take a specially designed pen into space. However, the Fisher Pen Company spent its own money (about $1 million in 1960s dollars) to develop the pen. It easily recouped the outlay in the free advertising and bragging rights.

THE SECRET CHALLENGER TAPE

The space shuttle Challenger exploded on January 28, 1986. There is some evidence not every crew member was killed by the explosion. In 1991 the *Weekly World News* (a grocery store tabloid) ran text of what was purported to be deleted by NASA from the official transcript of the cockpit voice recorder. At some point someone retyped the *Weekly World News* transcript and put it online. There it circulates today. The alleged transcript's source (a grocery store tabloid) had long been stripped out. People have been left to imagine they've stumbled upon a major leak.

NEIL ARMSTRONG UTTERED THE MYSTERIOUS WORDS "GOOD LUCK, MR. GORSKY"

In the mid-1990s an urban legend began circulating on the Internet that Neil Armstrong at one point during his moonwalk uttered "Good luck, Mr. Gorsky". As the story goes, Armstrong had always been tight lipped about the nature of this utterance. However, after decades of fielding questions about it, in 1995 Armstrong finally opened up. As a child he overheard his neighbor's wife (presumably Mrs. Gorsky) berating her husband and claiming she'd only have sex with him if "the kid next door walks on the moon!" Armstrong has gone on the record that this quip is a myth and no recordings of Armstrong on the moon have ever turned up with him uttering those words. Many have suggested the story has as its origins a Buddy Hackett routine.

THE LOST COSMONAUTS

There are a number of legends about the Soviets losing more cosmonauts in space than they've let on. Collectively these legends are referred to by some as The Lost Cosmonauts myth. For example, in 1959 newspapers began to carry reports that the Soviets had unsuccessfully tried to launch four astronauts into space. Each launch ended in disaster. In the early 1960s, two brothers in Italy claimed they had picked up transmissions from more than half a dozen failed Soviet space missions. They released what they asserted were actual recordings of these cosmonauts succumbing to space disasters. Skeptics point out the radio equipment the brothers were using was incapable of picking up and recording the alleged transmissions. As well, the terminology used by the "lost cosmonauts" doesn't match known Soviet terminology.

THE NAVAJO GAVE NASA A WARNING TO TAKE TO THE MOON

The story goes NASA trained astronauts on a part of a Navajo reservation that resembled the moon. One Navajo elder asked NASA if the astronauts could take a message to the moon on behalf of native Americans. NASA felt this would be a great gesture and good PR so they agreed. A NASA official recorded the elder's message, which he spoke in his native Navajo. When the official asked if the elder could translate, the elder refused. The official got a few other locals to translate. The message was "Watch out for these guys, they come to take your land." This legend was lifted from a Johnny Carson monologue that aired on *The Tonight Show* two days after the Apollo 11 landing.

NASA HAS PLANS TO ELIMINATE THE YELLOWSTONE SUPERVOLCANO VIA GEOENGINEERING

Yellowstone park sits on a massive caldera (basically a supervolcano). Should it blow, it would wipe out most of our civilization in North America. That's a pretty scary idea. Geologists don't believe the caldera will blow for many, many more lifetimes. However, that doesn't stop legends from popping up from time to time that NASA is worried the caldera is about to blow and will soon embark on a massive geoengineering project to save the day. The legend has its genesis in a NASA analysis that supervolcanos are a more likely threat to life on Earth than a dinosaur killer asteroid. People have interpreted "more likely" to mean "imminent". A NASA scientist did speculate on a BBC documentary that the risk from the caldera could be minimized by drilling into the caldera and pumping in water. This would also double as a geothermal electrical plant and produce a lot of cheap energy. However, this plan is certainly not in the works and any such project would seek to reduce the risk slowly, over thousands of years.

TRUMP OFFERED ANCIENT ALIENS HOST TOP POST

In 2018, President Trump announced his intent to create a new branch of the US military called the Space Force. A legend emerged that *Ancient Aliens* host Giorgio A. Tsoukalos was offered the role of Secretary of the Space Force by Trump. Like many modern urban legends, this one got its genesis in a satirical article posted on a science-oriented news site called *A Science Enthusiast*.

NASA CONFIRMED EARTH HAS TWO MOONS

In 2016, astronomers reported the discovery of an asteroid (2016 HO3) in a highly irregular orbit around the sun. It also happens to travel around the sun in an orbit close to Earth's orbit. Sometimes the asteroid is ahead of Earth and sometimes it falls behind. This gives it the appearance of orbiting Earth. The mechanics of this are extremely hard to picture. Fortunately, NASA's JPL has a video simulation available on YouTube that illustrates this odd dance. Some news sites have oversimplified the discovery and boiled it down to "Earth has a second moon". 2016 HO3 most certainly does not meet the definition of moon, however.

NASA HAD TO RENAME ITS URINATION CONDOMS SIZES TO PROTECT EGOS

If you need to go in a space suit, no problem. Your John Thomas is strapped into, in essence, a condom you can wizz into. It's rather critical that astronauts are fitted for the right sized condom. As the legend goes, the suit condoms came in three sizes: small, medium, and large. Very few astronauts were willing to admit they required anything but a large. So NASA renamed the sizes large, gigantic, and humongous. It would appear this legend has its genesis in the astronauts' own humor about having to admit they're a small or medium. According to Apollo 11's Michael Collins, astronauts themselves would informally refer to a small as "extra large" and the medium as "immense".

Eight Times Scientists Thought They Might Have Evidence of ETs

Although there could be up to two trillion galaxies in the universe and it seems sensible to believe life has gotten a foothold someplace else in the universe, so far we've found no evidence of any form of life outside of Earth. There have, however, been some times when scientists thought they had evidence of life beyond our atmosphere.

PULSARS

Jocelyn Bell and Antony Hewish discovered the first pulsar in 1967. The regularity of the radio emissions (about once every 1.3 seconds) made them consider it could be a signal from another civilization. That was not their first or strongest hypothesis but it was one they entertained. They jokingly cataloged the first pulsar LGM-1, which stood for Little Green Men 1.

VIKING'S LABELED RELEASE EXPERIMENT

Both Viking 1 and 2 had four experiments intended to detect life in the Martian soil. Three of the four came up negative. However, Viking's labeled release (LR) experiment produced interesting results. In essence, the experiment injected nutrients into a soil sample and looked for the release of gas that would be a sign the nutrient had been metabolized. The LR experiment indeed saw a release of gas. As a control, the soil was sterilized and the nutrients were reinjected. If the gas release was from a non-life chemical process, there should be a similar release of gas. There was none. The suggestion is some form of life was killed by the sterilization and no longer metabolizing the nutrient. Given the three other experiments found no evidence for life or even presence of organic compounds, the general consensus is the LR experiment was a result of some unknown non-biological reaction.

WOW! SIGNAL

In 1977 an Ohio State University radio telescope was being used to conduct early SETI work. The name comes from the word "Wow!" written on a printout of the anomalous signal. The signal came from the direction of the Sagittarius constellation and had some features that could be interpreted as originating from an extraterrestrial civilization. No one has yet come up with a natural explanation for the signal's strength. However, the signal was never repeated and lacked evidence of modulation.

ALH84001

ALH84001 is a piece of Mars rock found in Antarctica. In 1996, analysis of a sample suggested it contained microscopic bacteria fossils. President Bill Clinton, who had previously expressed more than a passing interest in UFOs, trumped the possible discovery of alien life. A widely distributed picture taken by a scanning electron microscope sure did seem to feature a fossil of a creepy crawly worm. However, these fossils were smaller than the smallest cellular life known on Earth. Later analysis determined the features interpreted as fossils had explanations that didn't require a life process.

WE'VE GOT A SIGNAL. AND IT'S LOOKING GOOD.

In the summer of 1997, SETI thought they had an excellent candidate signal. SETI's chief scientist called the group's poobah Seth Shostak at home telling him "We've got a signal. And it's looking good." The signal was strong and passed the first test. If you rotate the radio telescope away and then rotate it back, is the signal still there? Yup. The SETI team started to review a document they call "the protocols". What should a team do if they've confirmed an extraterrestrial signal? Alas. Further checks determined the signal was not coming from a distant solar system. The signal was telemetry from a NASA solar research satellite.

SHGB02+14A SETI@HOME

Did you ever have one of those SETI@Home screen savers running on your computer, helping crunch SETI data? Did you ever wonder if they found something? Well. Sort of. In 2004 project scientists made up a list of the 200 best candidate signals and scanned for them again. A signal known as SHGb02+14a was the only one they found a second time. It looked good in some ways. The frequency was in what was known as the "waterhole" which is a quieter band of spectrum. If you want to send a clear signal across space, this is a logical frequency to use. Arguing against it, the signal is weak and there are no stars within 1,000 light years of the signal's direction. There's no reason to suggest it's anything other than random radio fluctuations.

OUMUAMUA

Oumuamua is the first interstellar object known to have passed through our solar system. At first it was thought to be a comet but its speed, trajectory, and lack of a coma indicated the object came from outside our solar system. Analysis suggested it came from the direction of the star Vega. At its given speed that would be a 600,000 year journey. However, Vega was not in the same part of the sky at that time. There are no good candidates for nearby star systems. It's estimated Oumuamua has been in transit for over a billion years. Observations indicated Oumuamua increased its acceleration in a way that couldn't be explained by gravitation. A 2018 paper suggested the unexplained acceleration was because Oumuamua was an alien solar sail. That was rejected by scientists at large who noted Oumuamua was tumbling, which would make it a pretty ineffective solar sail.

UNUSUAL STELLAR FLUCTUATIONS

In 2017 a star known as Tabby's Star was observed to experience periodic dimming. There was no known natural explanation for the dimming. Several natural hypotheses were offered however some suggested we could be observing ETs building some kind of Dyson structure around their sun. Later observations have cast doubt on that hypothesis.

Seven Alternative Forms of Life

Life as we know it comes down to carbon and water. Carbon has the ability to form stable bonds with multiple elements, including itself. Carbon is the basis for millions of large and complex molecules used as the building blocks of life. Water is an excellent solvent. It helps provide cells structure, and it allows the transport of important substances to other parts of the cell or body. Carbon and water all come together in a solar system's "Goldilocks Zone". However, outside the Goldilocks zone, elements that are normally in states unusable by life can be in states that allow them to approximate carbon and water's favorable qualities. Here are several proposed alternative chemistries for life.

AMMONIA-BASED LIFE

In 1954, the great evolutionary biologist J. B. S. Haldane suggested liquid ammonia might be an alternative to water. Liquid ammonia, like water, can dissolve many organic compounds. On the downside, liquid ammonia doesn't have good surface tension. Exobiologists have their doubts liquid ammonia could hold organic compounds together long enough for the emergence of a self-reproducing system.

ARSENIC-BASED LIFE

Phosphorus is a key element for life on Earth. It's a big component in fertilizer. It's also a key structural element for DNA. Unlike most other elements that make up life, phosphorus is incredibly rare and not readily available for primordial life to make use of. Those who argue life is rare in the universe point to phosphorus as a reason. Not all hope is lost. Arsenic has many similar properties to phosphorus. Many scientists have suggested life could use arsenic in places where phosphorus is not available. In 2010, a team of NASA-funded scientists released a paper suggesting they had discovered a terrestrial bacterium that used arsenic in place of phosphorus. Unfortunately, reviews of the 2010 findings determined the NASA team had probably not discovered an arsenic-based form of life. The results were more suggestive that the team had found a highly arsenic-resistant bacterium.

BORON-BASED LIFE

Boron is similar to carbon in that it can form complex molecules. However, compounds made from boron tend to be explosive in an Earth-like oxygen atmosphere and they're unstable in water. A boron-based organic molecule would be more stable in atmospheres lacking oxygen. The biggest downside to life evolving naturally to use boron is the element is rare in the universe.

METAL-BASED LIFE

Some metals like germanium are able to form complex molecules. A life-form based on metal would have cell-like structures made from a variety of metals linked together by oxygen atoms. The problem faced by any metal-based life form would be a lack of a mechanism for self-replication. Metal may be in the DNA of some totally righteous musicians but it's unlikely nature could create metal-infused DNA.

METHANE-BASED LIFE

In 1981, Isaac Asimov suggested life could use liquid methane as a water-like solvent and use lipids in place of proteins. There is a lot of liquid methane in the universe. Titan appears to have lakes full of it. There is a hint of evidence of a methane-based form of life on Titan. Astrobiologist Chris McKay has theorized higher levels of hydrogen than theory predicted in Titan's upper atmosphere is a signature of methanogenic life. However, he has cautioned there are other explanations.

SILICON-BASED LIFE

Star Trek's season 1 episode "The Devil in the Dark" featured a silicon-based lifeform. It introduced to many the idea that life didn't necessarily have to be based on carbon. Silicon is right below carbon on the periodic table and can form long molecular chains. Like carbon, silicon is reasonably abundant in the universe. Silicon is not water soluble and silicon life would have to exist on a cold planet with liquid ammonia.

VITAMIN-A BASED LIFE

Plant life on Earth uses chlorophyll. Chlorophyll absorbs red and blue light but reflects green. Hence, plants appear green. There's a hypothesis that early life on Earth (and potentially life elsewhere) used retinal (Vitamin A) to harvest light energy. Retinal absorbs yellow light and reflects red and green light, which appears purple. Haloarchaea are a class of archaea that thrive in water saturated with salt. They use retinal to transform light energy into chemical energy.

6A
SCIENCE &
PSEUDOSCIENCE

Seven Things Nikola Tesla Didn't Actually Invent

Statue of Nikola Tesla found in Niagara Falls, Ontario.
Photo Credit: Karl Mamer

Inventor Nikola Tesla has become something of a cult figure to garage inventors, perpetual motion types, and people who feel their unique brand of genius will never be recognized in their own time. In their view Tesla was this lone inventive genius ignored by his peers and not given his proper place in the history of everything. His potential was never fully realized owing to his reclusive ways, the jealousy of rivals, and establishment fears Tesla's inventions would empower (literally) the masses. Tesla was most certainly not a recluse unsure of himself in the public spotlight. Tesla was a showman, hung out with the New York elite, and liked to throw large public birthday celebrations in honor of himself. For a time, Tesla was able to attract big money investors like J. P. Morgan and Westinghouse. His ideas weren't ignored because of establishment fears. Tesla tended to oversell his ideas and he got a rep for not producing the results he promised. Investor money dried up, particularly during the Great Depression. One thing that has fueled the Cult of Tesla is an inability, or unwillingness, of cult members to separate Tesla's actual accomplishments (some 300 legit patents) from his pure braggadocio. If he said it, he did it. It's well documented, however, Tesla had a habit of insisting he was capable of inventing something but saying it in such a way as the listener (usually a reporter looking for a good story) could interpret it as Tesla had already built it. The Cult of Tesla types have repeated Tesla's self-aggrandizing tall tales as if they were fact and have attributed to him some quite astounding inventions that he never actually invented.

BALL LIGHTING GENERATOR

Ball lighting is a hypothetical form of lightning. No one has ever convincingly captured it on video "in the wild" although people claim to have seen it. There is, however, a legend Tesla could create ball lighting on demand. This legend gained some recent legs when the popular web comic *The Oatmeal* promoted it in a cartoon about Tesla's overlooked contributions to modern society.

DEATH RAY

Starting in 1931, Tesla would throw himself public birthday parties. Each year he would invite the press. The press was happy to come because Tesla would wow them with tales of astounding inventions. At his 1934 birthday bash, he told gathered reporters he had invented a death ray. Technically he didn't call it a death ray. He called it a "teleforce". At other times he described it as a "death beam". The press, however, went with "death ray" and so it goes. Technically his death ray or death beam wasn't any kind of ray gun. According to Tesla, it was a particle accelerator that would fire a steady stream of microscopic matter at tremendous speeds and energies. He tried to get funding from J. P. Morgan Jr. but failed. Tesla thought his teleforce was a purely defensive weapon, a perfect defense against bombers that terrified many Londoners during World War I. There is no record of him having drawn up plans for his death beam, let alone built any kind of prototype. However, by his 1937 birthday bash, he was claiming he had actually built and tested the device. If he did, no assistant working for him ever saw his death beam or even plans for the device.

ELECTRIC CAR THAT RUNS ON COSMIC ENERGY

This one seems to be pure fantasy. It got started in the early 1980s over early computer bulletin board systems of the day. The story is told from the point of view of Tesla's nephew Peter Savo. Savo was taken on an extended car journey by Uncle Nikola in what appeared to be a standard Pierce Arrow automobile. Tesla let him drive the car for a bit. He noted despite using an ignition key to start the engine, the car ran silently, even after getting it up 90 mph. Peter drove it for at least 50 miles at that breakneck speed. When he asked what was powering the car, Tesla responded it extracted a mysterious radiation from "the ether". There are a few problems with this story. There's no record of Tesla having a nephew named Peter Savo. Since Einstein laid down special relativity, it's become clear there is no such thing as the ether. Finally, there's zero documentation in the form of plans or workshop notes about such a spectacular device. In Tesla's later years, when he was given to telling tall tales to eager reporters, he did offer reporters a story that he was able to harness cosmic rays and "caused them to operate a motive device." But that's about as close as you can get to Tesla actually claiming he built a cosmic-ray powered car.

EARTHQUAKE MACHINE

Tesla invented a steam powered generator in 1893 (known as an "electro-mechanical oscillator"). Through incredibly violent steam powered up-and-down vibrations, his oscillator would generate an electrical current. It was a more efficient way to produce electricity than the reciprocating steam engines which were, at the time, used to turn generators. However, Tesla's oscillator was quickly made obsolete by even more efficient steam turbines. How a power generator got turned into an earthquake machine is complicated. By the 1930s, Tesla was financially ruined by the Great Depression. His best days were behind him and he liked to brag a bit about his past accomplishments. As the legend goes, Tesla admitted a test of his oscillator caused a small earthquake in New York. It was only his quick thinking, that is taking an ax to his machine, that saved New York from terrible destruction. He did go on to assert if he had set it up properly, he could have used it to create frequencies that resonated with nearby buildings and cause them to collapse. There's no good evidence, other than a humble brag about being the guy who both nearly destroyed and saved New York, that he ever tried to turn his oscillator into an earthquake generator. Years later, the cast of *MythBusters* tried to see if they could create an oscillator that could generate frequencies that could collapse a bridge. They failed.

FLYING SAUCER

You will find many steampunk-looking blueprints online purporting to be genuine patent illustrations Tesla drew up for a flying saucer powered by some kind of exotic energy source. Oddly, you won't find a copy of the actual patent or even a patent number anywhere. The claim is that Tesla patented the world's first flying saucer in 1928. An interesting choice in words as that term didn't come into use until the late 1940s. Tesla did patent a flying machine in 1928 ("Apparatus for Aerial Transportation" US patent 1,655,114). However, it looks nothing like a UFO. Its exotic method of lift is a propeller.

THOUGHT CAMERA

Tesla might have been a good engineer (drink!) but he didn't know much about human biology. He assumed your thoughts caused reflections on your retinas. From there he reasoned he could create a lens that could resolve these images and take pictures.

WIRELESS POWER

In 1902 Tesla designed and built an experimental radio tower, known as the Wardenclyffe Tower. His goal was to create a radio tower capable of broadcasting wireless transmissions across the Atlantic to England and even around the world. The earth being a sphere makes it hard to get radio waves to bend around the planet. That and the inverse square law. The power of your broadcast diminishes with the square of the distance. Tesla knew this but didn't consider it a barrier. Much like his ideas about knocking down buildings by creating resonance frequencies, Tesla thought he could use his tower to leverage the atmosphere and the planet's own resonance frequencies to broadcast messages and pictures all around the globe. Get the planet hummin' to your tune and it would do the work for you carrying your message. The biggest problem is Tesla had no actual experimental evidence Earth has such a resonance frequency. His investors noticed this. Investors ignored his broadcast tower plan in favor of Marconi's wireless telegraph system. With investors unwilling to back his unfinished radio tower, Tesla doubled down. Not only would this tower allow the broadcast of messages and images, he could use it to broadcast electricity. Tesla had previously demonstrated he could light up neon tubes and even some incandescent lights a short distance with a Tesla coil. But he was never able to demonstrate the effect could scale beyond his backyard. Tesla never finished the tower and the project essentially bankrupted him for the rest of his life.

Ten Energies and Fields that Don't Exist

AETHER

Medieval scientists posited the space above Earth was filled with a substance called aether. It filled space between the stars and planets and was possibly the material out of which celestial bodies were made. Scientists of the 17th and 18th centuries understood light as a wave and resurrected the aether concept to explain how light seemed to propagate through vacuums. There was this invisible and hard to detect "luminiferous aether". The Michelson-Morley experiment conducted in 1887 was an attempt to detect the aether. When the experiment failed, it was the first piece of strong evidence that aether simply didn't exist. Einstein's 1905 Special Theory of Relativity explained light without needing a medium like an aether to propagate.

AURA

An aura is an imagined human energy field. Although woo proponents will assert it is an ancient idea, the concept was really started in the 19th century by Theosophist Charles Leadbeater (died 1934). Author Christopher Hill (died 1997) introduced the idea auras radiate solid colors in his 1977 book *Nuclear Evolution: The Rainbow Body*. Today it is taken as received wisdom everyone has an aura of a specific color and the color is indicative of one's health or spiritual advancement. New Agers will offer Kirlian photography as proof positive auras exist. Kirlian photography involves taking a picture of an object while running high voltage current through it. Remarkably, a camera can capture the discharge of this energy. Blinded tests of psychics with a supposed ability to see these auras clearly demonstrate they're unable to detect them.

CHI

Chi is a Chinese life energy and the absolute favorite buzz word of alt med practitioners and New Age proponents everywhere. We are to believe chi can get out of balance or blocked and illness is caused by this. There are various ways to fix your chi, like practicing certain martial arts exercises or acupuncture. Chi is supposed to flow along meridians in the body and if you stick acupuncture needles you can unblock this energy. Rooms themselves can have bad chi which can be improved by a form of geomancy called feng shui. Moving some furniture around can improve a room's chi flow. The problem is no two feng shui experts will ever agree on the best place to put your couch. The Michelson-Morley of chi detection was a paper by 9-year old Emily Rosa called "A Close Look at Therapeutic Touch". It was published in a 1998 issue of the *Journal of the American Medical Association*. Rosa got 21 practitioners of therapeutic touch (a form of chi therapy that has taken hold in the field of nursing) to engage in blinded trials. All practitioners were confident they could detect this energy, even if blinded. Her paper demonstrated they did no better than chance.

ECTENIC FORCE

Spiritualists of the 19th century learned to put on a great show during séances. Tables and horns were seen to levitate. Mysterious rapping sounds by an invisible hand were heard. Several psychical researchers of the day proposed something called an ectenic force to explain how objects in the room moved about. The notion was first proposed by French psychical researcher Count Agenor de Gasparin. If one were lucky, one could photograph this force, which revealed itself as ectoplasm. The photos of the day weren't particularly convincing. Most featured easily identified objects like cheesecloth, muslin towels, and in one case a rubber glove offered as examples of ectoplasm.

ORGONE ENERGY

Freudian psychoanalysis Wilhelm Reich (died 1957) believed he had discovered a primordial cosmic energy which he called "orgone energy". It's a kind of life energy but it's also radiated by non-organic matter. Almost any phenomenon Reich didn't think had an explanation was actually explainable by appealing to orgone energy. For example, why is the sky blue? Orgone energy is blue and the atmosphere is positively dripping with it. Being a doctor type, Reich was bound and determined to figure out how his orgone energy could be used in the pursuit of crank medical cures. He invented a machine he called the Orgone Accumulator. It was a box a person could sit in. The outside of the box supposedly absorbed ambient orgone energy and then focused it on the person sitting in the box. Reich gave Einstein a miniature version. Reich claimed the device would grow warmer through no means other than accumulation of Orgone energy. Einstein subjected it to a few tests. He quickly determined anomalous heat readings were caused by the temperature gradient in the room. Einstein suggested to Reich that was the source of his measurements and then asked Reich to not engage him on the matter in future or use his name in any kind of promotional materials. Reich obeyed Einstein's wishes but still made various claims about physical and psychiatric health benefits derived from sitting in his magical box. Eventually the FDA came calling and forbade him from selling his devices. He ignored the order and wound up in jail for contempt of court. He died in jail after almost completing his sentence. The prog rock band Hawkwind released a song in 1972 called "Orgone Accumulator" as a nod to Reich.

MORPHIC RESONANCE

Parapsychologist Rupert Sheldrake believes organisms transmit and receive information via something called a morphic resonance. This neatly explains why plants and animals seem to have instincts and memories. A species' collective memory is stored in these resonances. Brains, then, are not so much organs that store memories but are like TV receivers that can tune into these morphic resonances as they propagate through a morphic field. He developed this idea in his 1981 book *A New Science of Life*. Naturally, there's zero evidence such an energy exists. While we don't completely understand how brains store memories, we've got some good leads. There's zero reason to believe memories are stored outside of the brain and invent new entities like a morphic resonance.

N-RAYS

French physicist Prosper-René Blondlot (died 1930) thought he discovered a new form of radiation while trying to polarize X-rays. He dubbed it N-rays after his research institution the University of Nancy. Blondlot and many other scientists believed they saw N-rays emanating from an ever-increasing number of objects, both inanimate and living. Many other scientists, try as they might, couldn't replicate the findings. *Nature* sent American physicist Robert W. Wood (died 1955) to Blondlot's lab to get to the bottom of the N-ray issue. His hunch was Blondlot and others were imagining things. While Blondlot was running a demonstration on his apparatus, claiming to unmistakably see N-rays, Woods removed a prism from the machine. This was an essential part of Blondlot's apparatus. Despite the lack of the prism Blondlot kept reporting seeing N-rays. The concept of N-rays has become a cautionary tale about how experimenter bias can cloud your judgment.

PIONEER ANOMALY

Pioneer 10 and 11 are space probes launched in the 1970s. They flew by Jupiter and Saturn. Their speed is high enough for them to escape the sun's gravitational pull. Like the Voyager probes, they're destined to wander interstellar space until used as target practice by Klingons. The probes were expected to be slowed down by the sun's gravity well but measurements in the 1980s indicated they were slowing down more than theory predicted. Many people jumped on this as evidence of previously unknown effects or the need to re-write Newtonian dynamics. However, by 2012 it was pretty well established their power source, which uses radiometric isotopes, was radiating more on the side of the probe facing away from the sun. When taken into account, it fully explained the probes' unexpected deceleration.

RED MERCURY

Red mercury is a mythical substance capable of releasing huge amounts of energy. It can be used to compress a nuclear bomb's plutonium core and create a massive thermonuclear explosion. Conspiracy cranks believe it's what allowed governments like the USA to miniaturize their nuclear warheads. The red mercury legend got started in the 1970s, likely by con artists trying to rip off terrorists who were seeking fissionable materials on the black market. The legend reached critical mass in the 1990s when Manhattan project scientist and father of the neutron bomb Samuel T. Cohen (died 2010) asserted the substance really existed. No one is sure why he suddenly started spreading this myth, although he was rather elderly at that point. Cohen was opposed to governments signing nuclear non-proliferation treaties. In his view, red mercury allowed nations to make thermonuclear weapons as small as baseballs. Therefore, no non-proliferation treaty could ever have a workable verification mechanism if a nation's secret arsenal could be hidden on the shelves of Dick's Sporting Goods. Red mercury's mythical properties aren't all bad, however. Some believe red mercury radiates astounding healing energies or allows you to summon a jinn. In 2009, a rumor took hold in Saudi Arabia that Singer sewing machines were powered by red mercury. People started buying them up, paying thousands of dollars for the machines.

TORSION FIELD

A torsion field is a field pseudoscientists claim can use quantum spin to instantly transmit information across vast distances. The notion was cooked up in a lab in the Soviet Union in the 1980s. The lab was shut down when it was discovered project directors were embezzling their government funding. Scientists call the idea nonsense but like most science-y nonsense, it has been warmly embraced by free energy cranks, Ufologists, and psychics.

Nine Outrageous "Free Energy" Claims

"In this house we obey the laws of thermodynamics." — Homer Simpson

Perpetual motion is a physical impossibility. But that hasn't stopped people from fooling themselves or fooling others into thinking they've unlocked a simple secret that has long eluded the best minds in science. Free energy does not always imply perpetual motion, however. It can also refer to some new but dubious power source (like cold fusion) that mainstream science has supposedly missed or ignored (usually at the behest of the oil companies).

Photo Credit: Karl Mamer

BHĀSKARA'S WHEEL

Bhāskara's wheel is an early, if not the first, example of someone claiming to have invented a perpetual motion machine. It was invented by Indian mathematician Bhāskara II in 1150. It is what is known as an overbalanced wheel. Spokes radiate from an axis. At the end of each spoke is a capsule filled with mercury. As the wheel spins the mercury capsules slosh downward with the motion of the wheel and give it torque. But like anything spinning, some of the energy will be converted to heat and noise as it interacts with the atmosphere and the wheel will eventually stop rotating.

REDHEFFER'S GENERATOR

In 1812, Charles Redheffer began touring his perpetual motion machine across the American North East. He claimed his generator could power a range of devices hooked up to its gears. He first displayed his machine in Philadelphia and charged admission to view his wonder. He also sought investors. When experts turned out to verify that the machine could do what he claimed, he wouldn't let them near it. He said he was afraid they would damage it. Despite having to observe the machine at a distance, it became obvious the machine's cogs were turning the wrong way. The machine it was hooked up to was clearly powering Redheffer's device. Redheffer tried to run his scam later in New York. Robert Fulton came out to investigate. Wandering around the building the machine was housed in, Fulton discovered an old man on the floor above the machine manually turning a hand crank to power it.

ETHERIC WAVE ACCUMULATOR

In the 1910s, an MIT graduate by the name of Harry Perrigo claimed he had invented a machine that could tap energy from the air. He called it the Etheric Wave Accumulator. He created various versions of his wave accumulator but for the most part his device consisted of an array of copper wires. These wires broke up some kind of atmospheric wave action and created power. He applied for a patent but was denied. The patent office rightly classified it as a perpetual motion machine. He managed to demonstrate his device before Congress in 1917 but still couldn't convince anyone to give him a patent. He did attract some investors but his major financier pulled his funding after Perrigo was unable to give him a machine he could have independently tested. Undaunted, Perrigo tried to prove the reality of his device in 1927 before a panel consisting of a reporter from the *Kansas City Star*, a member of the Better Business Bureau, and a Missouri securities commissioner. Perrigo claimed his device was wired to a car and was powering its engine. Perrigo's device was exposed as a fraud when the reporter poked around the car and discovered an array of batteries hidden under the car's rear seat.

THE MINTO WHEEL

Did you ever have one of those drinking birds? It's in essence a glass tube with a bulb at either end. There's a liquid inside the tube. One of the bulbs is done up to look like a bird's head complete with a beak made from felt. The differences in temperature on either end of the tube causes the liquid to flow between the bulbs causing the bird to teeter back and forth. The Minto Wheel, invented by Wally Minto in 1976, extends this relationship into the form of a wheel. Minto suggested the bulbs around his wheel could be filled with either propane or freon. As the sun heats the bulbs at the bottom of the wheel, it forces the gas into the bulbs at the top. The new weight at the top causes the wheel to spin. The cycle continues. Minto himself never really sought to profit from his Minto Wheel. He made the plans public domain. And, technically it is a true "free energy device" insofar as anything solar powered is a free energy device. However, there are far more efficient ways to do work with solar energy.

THE PATTERSON POWER CELL

Chemist James A. Patterson announced in 1989 that his Patterson Power Cell could produce 200 times more energy than put into the system using a form of cold fusion. Most scientists doubted Patterson's invention as, like most reports of cold fusion, the expected and absolutely necessary byproducts could never be detected. Although some labs were able to replicate Patterson's results, no one has been able to reliably reproduce them.

BLACKLIGHT POWER

In 1991 Randell Mills, a Harvard-educated medical doctor, announced a cornerstone of quantum mechanics was wrong. He claimed he discovered a hydrogen atom's electron can dip below its ground state and that's a source of never-ending clean energy. He formed a company called Blacklight Power and set up attracting investors to build an energy cell based upon his "findings". Most physicists, including some Nobel prize winners, suggested Mills was mistaken about his discovery. A state lower than the ground state would contradict well-founded theories about quantum mechanics. If it were even possible, someone besides Mills would have noticed the phenomenon long before him. Despite Mills claiming to have created something assuredly impossible, in the early 2000s he was able to attract a lot of high-profile interest from companies and investors who should have known better. Mills reported in 2008 that his company had developed a prototype powercell and it would have a commercially available system within a couple years. More than a decade later, you still can't buy one.

THE ORBO

Steorn was an Irish company that claimed they had invented a perpetual motion machine based on the principle of the self-winding watch and magnetic fields or something. Steorn never revealed what principle they were exploiting for free energy. They did have a nifty name for it, however. They called this miracle product the Orbo. They announced their breakthrough to the world in a full-page ad in *The Economist* magazine (that's a $80K ad buy). The ad asserted that despite the machine appearing to violate the conservation of energy, no less than eight independent scientists had validated the machine's reality. Who were they? Steorn wouldn't say. Naturally, people were skeptical. Steorn challenged skeptics to form their own panel and test the Orbo. Over 400 scientists applied to be on the jury. Remarkably Steorn let 22 scientists onto the panel to investigate. The panel came to the unanimous conclusion that Steorn was unable to demonstrate its claims. In 2007 Steorn was set to demonstrate the Orbo to the public at the Kinetica Museum in London but cancelled when the company reported the device wouldn't work due to heat from the museum's lighting. Uh huh. The company went out of business in 2016, despite having raised well north of $25 million from really stupid investors. The closest the company came to offering a free energy machine was a $1,200 phone charger that could supposedly charge itself. It's not clear how many people were able to order the product and if anyone ever actually got one. There are certainly no Steorn phone chargers being offered on eBay today.

GENEPAX WATER ENERGY SYSTEM

Hydrogen makes for a pretty clean burning fuel, as when it burns in air it leaves behind just water as an exhaust. However, there are no natural sources of hydrogen gas on Earth. You need to separate it from good old water. It takes more energy to break up a water molecule than you get from burning the freed hydrogen. If you have a lot of excess solar or wind energy generated during low demand periods, you can use it to separate hydrogen from water, store the hydrogen, and then later burn that. Hence, hydrogen is seen as a way to store energy, not a source of net new energy. That distinction gets lost on many people. So it's no surprise several media organizations, including Reuters and *Engadget*, bit down hard when in 2008 Japan's Genepax Water Energy System reported it had invented a car that used nothing but water for fuel. Reuters, which is normally a pretty sober media organization, wrote "The car has an energy generator that extracts hydrogen from water that is poured into the car's tank. The generator then releases electrons that produce electric power to run the car". People who understood there is no free lunch trying to turn water into a source of hydrogen did a bit of digging. The company seemed to have only registered its website domain a few weeks before making the media rounds with its water-fueled car. You don't go from groundbreaking primary research to working prototype in less than a month. About half a year after their press tour, the company closed up shop, claiming the cost of development had grown too large.

RCA AIRNERGY

RCA used to be a leading American corporation involved in radio, television, and music. Today it's three letters owned by a consortium that licenses the name out to other manufacturers seeking a bit of instant brand recognition. Someone at the 2010 Winter Consumer Electronics Show started showing off a device called the RCA AirNergy. This little device could ostensibly harvest all the Wi-Fi signals that vibrate invisibly around us and use them to charge a battery. You could not only recapture all that unused energy your own home router was putting out but you could also charge your phone on Starbucks' dime! So, free energy! *Popular Science* (which should know better) and the website *Engadget* gave it rave reviews. However, people who understood things like the inverse square law, how much energy is actually put out by Wi-Fi routers, and how much energy actually falls on a tiny antenna did some quick back-of-the-envelope calculations. Assuming the AirNergy worked at 100% efficiency (which is impossible), it would take over 30 years for the device to charge a typical phone battery. The device was supposed to hit the market by summer of 2010. However, nothing ever ended up on the shelves. To this day, no one knows if the RCA AirNergy was a hoax (an expensive one given the cost of a booth at the CES) or the product of an ambitious marketing type who licensed the RCA name and gambled that an engineer would figure it out later.

Fourteen Inventors of Impossible Devices

OTIS CARR

In the mid-1950s Otis Carr (died 1982) announced he was building a flying saucer that would transport people to the moon in less than a day. His flying saucer was powered by something he called the Utron Electric Accumulator, which was supposedly based on the work of Tesla. Carr's claims were a hit with the Contactee movement of the time and he was able to raise considerable funds from followers. One of his investors was the owner of an amusement park. Carr got him to finance a prototype he promised he would live demo on the grounds of the park during the spring of 1959. After that he would have ready a full-sized flying saucer which he would fly to the moon in December of that year. The day Carr was supposed to demo his amusement park prototype, he claimed he got mercury poisoning and cancelled. Naturally, his Moon flight never took place. Carr was eventually arrested and convicted for selling unregistered securities. He served six months in jail. He apparently ended his days penniless and living in a Pittsburgh slum.

NORMAN LORIMER DEAN

John W. Campbell (died 1971), editor of *Astounding Science Fiction*, fell hard for claims by Norman Lorimer Dean (died 1972) that he had invented a reactionless drive. A reactionless drive is, in essence, a rocket engine that creates forward motion without a propellant. In other words, you can go forward without shooting anything out the back. If the drive is a closed system, this is a violation of Newton's Third Law. Campbell seemed not to understand the Dean Drive was actually impossible. He trumpeted it as a breakthrough in the September 1960 issue of *Astounding Science Fiction*. Dean demonstrated his drive in front of Campbell by putting it on a bathroom scale and turning it on. It began to vibrate. A lot. The scale itself indicated the device decreased in weight. Campbell took this apparent loss in weight as evidence the drive was exhibiting some kind of antigravity effect. As John G. Cramer suggests in a 1996 article in *Analog* magazine "Campbell served principally to demonstrate that bathroom scales are not reliable instruments for investigating antigravity..." If you vibrate the hell out of your bathroom scale, the scale's needle is going to move around a lot too. Sci-fi author and long-time *Byte* magazine columnist Jerry Pournelle (died 2017) contacted Dean to arrange his own test. Dean refused unless Pournelle pre-paid him for the device and promised him he'd win a Nobel prize. Pournelle politely refused.

BILL HOLLOWAY

Bill Holloway thought he found a surefire treatment for his Fibromyalgia called Penta water. It was a new form of "restructured micro-water" that no scientists had ever noticed in hundreds of years of experimenting with water or mathematically predicted based on known chemistry and physics. Holloway soon started selling it. The health benefits of Holloway's Penta water multiplied: the water's microclusters were absorbed by cells more quickly, it could protect your DNA, and, in Holloway's words, "it may even be the fountain of youth." Those kinds of commercial claims begged legal action. A 2003 US class action suit alleged false advertising and forced Holloway to stop making health claims about Penta water.

ELIZABETH HOLMES

In 2003, Elizabeth Holmes founded a company called Theranos to develop and market a "lab on a chip". It would be a small automated machine that could test for a range of diseases using a tiny amount of blood (which ranged from 2-10 drops of blood). Holmes was able to raise hundreds of millions of dollars and formed partnerships with Safeway and Walgreens to offer blood testing services using her device. She called her invention the Edison. The problem was she was never able to create a working version of the Edison. Her explanations for how her device would work have been described as "comically vague". The company was able to perform some testing services, but, as it turns out, testing was not done on her vaunted "lab on a chip" but done using standard equipment found in a typical blood testing lab. The company eventually went belly up in 2018. Holmes was indicted in that year on multiple counts of wire fraud. As of this writing, her trial is ongoing.

JOHN HUTCHINSON

John Hutchinson has claimed for decades he's the discoverer of a method for creating antigravity. He calls this the "Hutchinson Effect". He alleges he discovered the secrets of antigravity in 1979 while playing with high voltage equipment. On YouTube, you will find thoroughly unconvincing videos demonstrating his antigravity effect. It's plainly obvious in many of the videos he's simply dropping things in front of a camera turned upside down. Curiously, no one has ever been able to reproduce his results, including Hutchinson himself. He's never been able to produce his effect on demand. Hutchinson has been able to, however, transmute his video stunts and babble about Tesla, zero-point energy, the Casimir effect, and government conspiracies into a large number of appearances on reality shows over the years.

TIMOTHY JOHNSTON

Wouldn't it be great if you could put a small pill in your gas tank and significantly increase your gas mileage while lowering your emissions? Timothy Johnston of Australia started marketing such a pill in 2004. He called it the Firepower Pill. One small pill in your gas tank would magically treat the gasoline in such a way as your engine would burn "more of the heavier elements... increasing power and fuel economy". Although the product idea is dubious from the get-go, he was able to raise some $100 million from investors for his Firepower company. His company was also able to secure various grants from the Australian government, which gave Johnson's company credibility it didn't deserve. Firepower used funds it raised to support various Australian sports teams and consequently attracted big names in Australian sports as follow on investors. His attempts to associate his product with Australian sporting prompted questions. Despite being a huge promoter of Australian sports and the darling of the Australian government, the company had no miracle pills to sell in Australia. Why advertise so much in Australia for a product you can't buy in Australia? People began to look into Johnson's claims. For example, he bragged that various professional racing leagues used his pill. No one could find a racing league that allowed any such fuel additive. Johnson promoted his company's joint ventures with Shell and General Motors. Both companies had no idea what Johnson was talking about. Anyone even requesting a pill to test was threatened with a big money lawsuit.

WOLFGANG VON KEMPELEN

In the 18th century, Wolfgang von Kempelen (died 1804) toured Europe with his chess playing machine dubbed The Turk or The Mechanical Turk. Von Kempelen's chess playing machine consisted of a large box out of which emerged the torso of an automaton dressed like an Ottoman gentleman, complete with robe and turban. A chess board was placed on the box. The automaton's hand could move chess pieces around the board. The Mechanical Turk was able to defeat rather good chess players wherever von Kempelen took it. It was only after von Kempelen died and the Mechanical Turk was sold to Johann Nepomuk Mälzel that the secret was revealed. The interior of the machine had space for a human chess player who was able to control the movements of the automaton.

EDWARD LEEDSKALNIN

Edward Leedskalnin is the builder of Florida's fabled Coral Castle. Despite the name, the Coral Castle was not actually built out of coral. It was built from slabs of a sedimentary rock known as oolite. No coral reefs were damaged in the making. As legend has it, Leedskalnin single-handedly built the Coral Castle south of Miami, Florida in hopes of winning the love of a woman. Although called a castle, it's probably better described as a rock garden featuring megalithic structures. Leedskalnin opened up his park in 1921 and continued to add to it until his death in 1951. Leedskalnin was reputedly only about 5 feet tall and weighed no more than 100 pounds. So many wondered how a man of such diminutive proportions could have quarried and moved around so many weighty megaliths (the heaviest of which weighs about 30 tons). Kooks have proposed everything from magic to anti-gravity. The problem is Leedskalnin himself, though vague in detail, was quite clear how he did it: leverage. Leedskalnin's friend, associate, and building contractor Orval Irwin self-published a book in 1996 called *Mr. Can't Is Dead! The Story of the Coral Castle* in which he explained Leedskalnin's various techniques. As well, we have photos of

Leedskalnin using a homemade tripod-and-pulley system to move a large stone block.

JAMES MCCORMICK

In an age of suicide bombers, an inexpensive, easy-to-use bomb detector would be a godsend to police, military, and border security. The UK's James McCormick answered the call by selling governments around the world, which included the Iraqi government, dowsing rods for $60K a pop. His company Advanced Tactical Security & Communications, in essence, took a novelty golf ball finder with a swiveling antenna, put its own label on it, changed the handle to accept "substance detection cards", and called it the ADE 651. McCormick's company assured users that the device, depending on what card was inserted, could detect bombs, drugs, weapons, and money. It could penetrate lead and concrete. Oh, it didn't even need a battery. It was said to be powered by the static electricity generated by the operator. When tested by Israel and the US Army, it became obvious the device not only did not work but could not possibly work. It has been estimated hundreds of Iraqis died because checkpoint guards were issued the devices and completely missed bomb-laden cars they would have normally found via a simple search. McCormick was eventually arrested and sentenced to a decade in jail for fraud in 2013.

JOSEF PAPP

Josef Papp (died 1989) first came to the larger attention of the Montreal press when he announced he had invented a submarine capable of traveling 300 mph underwater. A few days after making that bold pronouncement he disappeared and mysteriously turned up in France. He told French authorities he had taken his submarine there to prove it could travel as fast as he claimed. Oddly no one could find his submarine. He claimed he sunk it so it wouldn't fall into the hands of the Soviets. French authorities were doubtful of his story, particularly when they found airline ticket stubs in his pocket, indicating he actually just flew from Montreal to Paris. A couple years after his submarine hoax, he went on to try and convince people, including investors, that he had invented a car engine that used fusion power. Richard Feynman turned up to one of his demonstrations. Feynman noticed the engine had something plugged into it that looked a lot like an electrical cord. Feynman suspected it was charging an internal battery. Papp got incredibly upset when Feynman unplugged it. Papp vigorously denied it was a charging cable. It was there merely to supply power to various instruments. No one ever got to find out as Papp's machine exploded during the demonstration and killed a nearby observer.

FRANK ROBERT WAKE

In the 1950s, men who were suspected to be gay were considered a security risk to many governments. Canada's RCMP (the fabled "Mounties") was tasked with the job of eliminating gay people from civil service roles. The Mounties normal procedure was to follow men that were suspected of being gay. The approach was time consuming and expensive. The Mounties asked Frank Robert Wake, a psychology professor at Carleton University, for a more efficient way of detecting gay people. Wake assured the Mounties he could build a device that could identify gay men based on physiological responses to gay pornography. The device was dubbed "the fruit machine". Suspect men were sat in the device and shown pictures of naked men and women. The subject's pupils were measured for responses. If pupils dilated while viewing a picture of a naked man, it was an indicator the subject was gay. There were all kinds of problems with Wake's approach, notably there was no control for pictures of varying brightness. The Canadian government spent tens of thousands of dollars on Wake's machine but, fortunately, it never got beyond the prototype stage. Test subjects were initially told the machine was meant to measure stress. When the machine's real purpose got out, Wake's list of test subjects quickly dried up.

MAURICE WARD

Maurice Ward (died 2011) was the UK's favorite rogue inventor for a time. Ward was a former hairdresser who invented something he called "Starlite". Starlite is an intumescent material that expands as heat is applied and insulates whatever it coats. Ward loved to demonstrate its effectiveness by coating an egg and then trying to cook it with a blowtorch. After several minutes under direct flame, he'd crack open the egg to demonstrate the yolk was still raw. Although Ward tried to get several major industries to take an interest in his Starlite, he never landed any contracts as he was unwilling to leave samples so companies could rigorously test it. Ward never shared his recipe before he died. Starlite was pretty impressive for an amateur chemist but it's not the miracle insulator that has eluded science for generations. There are already many products on the market, like aerogels, that do exactly what Starlite was claimed to do. Having passed in 2011, the media today invariably mentions Ward "took the secret of Starlite with him to the grave." His secret recipe doesn't seem all that secret. Mick West, creator of the *Metabunk* site, cooked up his own version which was likewise able to insulate a raw egg from a blowtorch.

NEIL YOUNG

MP3s disturb some audiophiles as it's a lossy compression scheme. To crunch a raw music file (like a WAV file) down to something one-tenth the size, MP3 compression throws out frequencies. That "sounds" bad until you realize lossy compression casts aside frequencies that are utterly beyond the range of human hearing. Still, audiophiles like Neil Young are convinced MP3s are missing something the "tuned" ear can hear. To address this, Young crowd funded his PonoPlayer project. His PonoPlayer would support the lossless FLAC format. Audiophiles rejoiced until it was revealed blinded A/B tests demonstrated people couldn't tell the difference between a FLAC song being played on the Pono and a modern iTunes 16-bit/256Kbps music file being played on an iPhone.

PEAR CABLE

For years stereo aficionados used plain old copper zip cord to connect their speakers. No one noticed any kind of degradation of sound until companies realized they could sell super expensive stereo cables to audiophiles as long as their packaging had a lot of science-y terms and they could get a few stereo magazines to write glowing reviews. The nadir of this speaker wire mania came in the early 2000s when Pear Cable released a brand of high-end home stereo speaker cables called the Pear Anjou. And when I say high end, I'm talking about a $7,000 cable. It has been shown through A/B testing almost no one can tell the difference between sound coming out of speakers wired with moderately high-end speaker cables and a coat hanger jammed in the holes. In 2007, James Randi offered Pear Cable the opportunity to apply for his $1 million challenge. Although not technically a psychic claim, the notion that anyone could tell the difference in a blinded A/B test between the Pear Anjou cables and a standard set of decent quality stereo cables would be something approaching a paranormal ability. The company declined, alleging Randi was the one perpetrating the hoax.

Nine and a Half Claims of Time Travel

CELL PHONES THROUGH TIME

In 2016 Apple head Tim Cook joked he had spotted an iPhone in Pieter de Hooch's 1670 painting *Young woman with a letter and a messenger in an interior* while touring Amsterdam's Rijksmuseum. Although he was only joking (the messenger was holding some kind of letter), there have been several photos, paintings, and early films that people have seriously claimed are examples of time travelers using cell phones long before the technology was invented. This is problematic on many fronts, the least of which is that cell phones need cell phone towers and a whole infrastructure to make them work. In 2010, the net was abuzz about a clip from newsreel footage taken in 1928 outside of Grauman's Chinese Theatre. It appears to show an elderly woman walking through the crowd talking on a cell phone. Film maker George Clarke discovered the clip while reviewing bonus material that accompanied a DVD of Charlie Chaplin's 1928 silent film *The Circus*. Clarke, unable to determine what the device was, suggested the woman was a time traveler. Experts familiar with technology of that era have identified the device as an early hearing aid. In 2011, someone posted a clip to YouTube of the 1948 movie *Fort*

Apache claiming it offered proof of time travel. The clip shows a brief scene where Henry Fonda appears to be looking at an iPhone. One doesn't have to be a film historian to understand he's looking at a small leather-bound notebook. In 2017, *Vice* did a piece on an 1860 painting by Ferdinand Georg Waldmüller titled *The Expected One*. It shows a woman walking along a path. Her head is down and looking at what some believe is an iPhone. Art historians disagree and have identified it as simply a depiction of a prayer book.

Young woman with a letter and a messenger in an interior (1670)
by Pieter de Hooch
Photo Credit: Vincent Steenberg, released to public domain by via Wikimedia

BILLY MEIER TAKEN BACK IN TIME TO SEE DINOSAURS

Swiss UFO Contactee Billy Meier has told a large number of tall and improbable tales to his followers over the years of encounters with space aliens. He calls these space aliens Plejaren. Apparently, they have time travel technology. The Contactee's followers like to point to a historical archive of cryptic utterances by Meier's Plejaren friends that turned out to be predictions of events that have actually come to pass. If we had only listened! Few outside of Meier's circle, including super credulous Ufologists, have been impressed. For example, followers pointed to a cryptic prediction Meier made in 1958 about "monstrous flooding" foretold the 2004 Indonesian tsunami. This is *Bible Code* level stuff. But, wait, is there proof of time travel to the past? Why, yes! Meier says in the 1970s the Plejarens took him back in time to see dinosaurs. As proof to his followers, he produced a blobsquatchy photo of a *Pteranodon*. Skeptics figured out the picture is an exact match of an illustration in a 1972 dinosaur book called *Life Before Man*. It's a safer bet he simply took a fuzzy photo of the illustration than his improbable tale of time traveling space aliens who helped him take a photo that aligns perfectly with a painting.

IRAN HAS TO DENY IT HAS INVENTED A TIME MACHINE

In 2013, Iran's Fars News Agency, which describes itself as "Iran's leading independent news agency", passed a story about an Iranian scientist who had invented a time machine. The inventor was cited as Ali Razequi who was apparently running something called Iran's Center for Strategic Inventions. The story gets a bit hazy after that. In the Fars report, Razequi spoke of a machine that was the size of a PC and could "predict details of the next 5-8 years". Maybe something was lost in the translation of "tech speak" by an idiot reporter and then further degraded by translation into English. It's also possible Razequi, like a long line of perpetual motion inventors, simply fooled himself into believing he unlocked some Hari Seldon-like way to compute the future. A few days later, an Iranian government minister denied Iran had come into possession of a time machine.

THE SAFETY NOT GUARANTEED GUY

In 1997, John Silveira was working at *Backwoods Home* magazine and had to come up with a fake classified ad to help balance a column. He penned an ad that read "Wanted: Somebody to go back in time with me. This is not a joke... You'll get paid after we get back. Must bring your own weapons. Safety not guaranteed." The ad got quite a bit of play in traditional media and later the Internet after it appeared in a gag on the *Jay Leno* show. Silverira included his own PO box in the ad. Much to his surprise he got people applying for the job as assistant time traveler. Silveira says he's gotten at least 1,000 responses since the ad first appeared. Many of the respondents recognized the ad was a joke. Others chose to threaten him. Some just liked to list a scary number of weapons they could bring for the journey. However, more than a few responses were genuinely heartfelt. Sadly, many came from people who had lost a loved one to some tragedy and wanted to go back in time to help undo the person's death.

SCI FI STORY BECOMES AN URBAN LEGEND

Science fiction author Jack Finney wrote a short story in 1951 called "I'm Scared". It's a story of a 19th century man named Rudolph Fentz who is suddenly transported through time to New York city in the 1950s and killed by a motorist. For some reason, this short story started to spread in the 1970s as an urban legend. Passers of the legend tell of a man found dead in Time Square, a victim of a hit and run. When his body was taken to the morgue, he was dressed in clothing that had gone out of style almost a hundred years ago. A search of his pockets revealed items that suggested he had come from the 19th century. Items included old banknotes, a bill from a liberty stable that had long ago gone out of business, and a letter dated June 1876. The man also had a business card which identified him as Rudolph Fentz. More mysteriously, all the items on Fentz's body appeared to be new, showing no sign of decades of wear. When police searched for possible relatives of Fenz they found his daughter-in-law. She told a strange tale of her father-in-law vanishing without a trace one night in 1876. The first appearance of Finney's short story told as true was in a 1972 issue of a paranormal magazine called *Journal of Borderland Research*.

TIME TRAVEL EMAIL SPAM

For a few years, starting in 2002, people started getting a bizarre spam email by a person claiming to be a time traveler stuck in 2002. The emailer, who went by the name Bob White, wanted to buy replacement parts for his broken-down time machine. He was offering various forms of payments like "2002 currency" or precious metals. White was apologetic that he was tapped out of "Galactic Credits". Regardless, he was desperate for a dimensional warp generator and atomic elements unknown to our time named barosian and kryoken. Many thought it was some kind of early phishing ploy. The email was crafted in such a way as it begged to be replied to. If you responded, that indicated you were an active email account. As it turns out, upwards of 100 million such emails were pumped out over the course of a couple years. The trans-time spammer left a rather obvious clue to his real identity: an email address with a Whois-able domain name. People looked it up, found it was registered by a person in the Boston area named Robert Todino Junior. The Whois record also provided a phone number. People called it. They mostly got through to the spammer's father, Robert Todino Senior. Todino Senior regrettably informed callers the spam was indeed coming from his son and his son has "psychological problems". Those who managed to get through to the son reported Todino Junior did nothing to lead them to believe his father's opinion was wrong. Todino Junior himself owned his own email marketing firm called "RT Marketing". Just before he went on his time machine parts quest, he was fined by the state of Massachusetts for spamming people with ads for free government grants and something he called "detective software". At least one semi-enterprising individual named Dave Hill turned the tables on the time traveling email spammer and set up his own web store offering the parts Todino was looking for.

JOHN TITOR'S QUEST FOR AN IBM 5100 COMPUTER

In November of 2000, an anonymous user of an Internet bulletin board called the Time Travel Institute claimed he was an actual time travelling soldier from 2036. He identified himself as John Titor, born in 1998 in Florida. He was taking a brief side trip to the year 2000 to pop in on some relatives but his quest was to travel back to 1975 to acquire an IBM 5100 computer. Ostensibly this was part of a plan to save the world from the Year 2038 problem, which in Titor's timeline was only a couple more years away. A lot of his story was faintly reminiscent of the 1995 movie *12 Monkeys*. Titor attracted quite a fanbase with his tales of a looming civil war that was to start in 2004 and would culminate in World War III in 2015. Obviously neither of these events happened. His descriptions of his time machine and posting of intriguing but grainy pictures of his device (along with photocopies of its user manual) became as close to time traveler porn as you're going to find on the net. Ultimately, not much of his story was believable. His story was inconsistent at times, suggesting there were multiple writers behind the Titor character. Titor did seem to have one intriguing bit of knowledge. He knew about a secret, undocumented feature in the IBM 5100. The computer, which was in the early "luggable" class, had a built-in emulator that could run software intended for IBM's larger, more expensive machines. IBM never publicized this out of a fear that customers would buy smaller, cheaper IBM 5100s and cannibalize sales of its more expensive big iron. Some early IP tracking and detective work by an actual private detective eventually suggested the Titor character was the creation of an entertainment lawyer in Florida who later set up a company to sell Titor-themed merch (as well as trademark the name "John Titor"). It was found his brother was a computer scientist, which explained why the Titor character might have known about the undocumented emulator in the IBM 1500 as well as why the character could spew enough realistic sounding techno babble to thrill the denizens of the time travel forum.

THE VATICAN CHRONOVISOR

François Brune (died 2019) was a French Catholic priest. In 2002 he published a book titled *The Vatican's New Mystery*. In it he wrote about a Benedictine monk/exorcist in Venice named Pellegrino Ernetti (died 1994) who said he had, with the aid of Enrico Fermi and Wernher von Braun, built a device that let the user see into the past. Ernetti dubbed the device a chronovisor. Brune, who was no stranger to writing uncritically about the paranormal, credulously accepted the reality of this device. Ernetti told Brune he used it to document a wide range of historical events found in the Bible including the crucifixion of Jesus. Ernetti even produced a photograph he took of the scene displayed to him on the chronovisor. It later turned out to be a photograph of a crucifix by a sculptor named Lorenzo Coullaut Valera (died 1932). According to Brune, the Vatican forced Ernetti to make a false confession that he lied about the device and then seized the device upon the monk's death. It remains tightly locked away in the Vatican.

WOULDN'T YOU GO BACK IN TIME AND BUY APPLE STOCK?

One would think if there were time travelers in the modern day, one or more would get super rich playing the stock market. A time traveler from the future would probably know the next Apple or when to buy bitcoin. Since 2003, mystery mongering publications and websites have told a story of a man named Andrew

Carlssin who in 2002 was able to parlay $800 into a stock portfolio valued at over $350 million in only two weeks. The man was immediately arrested by the SEC on suspicion of insider trading. His defense? He was a time traveler. He sits in a jail cell today because he refuses to give up his sources for insider stock tips. Some sites give this story's source as Yahoo News. Yahoo was a pretty respectable dot com back then so why would such a preeminent company just make up a story? The truth is Yahoo News did pass on this story in 2003. Yahoo, however, didn't write the story. Yahoo News simply reposted a story made up by the *Weekly World News*. People didn't look too closely at the actual byline and assumed this was legit.

THE HALF: TWEETING FOR TIME TRAVELERS

One of the best ways to either prove you're a time traveler or determine a person is a time traveler is demonstrating detailed and specific knowledge of the future. People are great at making vague predictions. The law of large numbers suggests if you get enough people making specific predictions, a handful will eerily come true. However, any one individual shouldn't be able to demonstrate a series of highly accurate predictions. In 2014, astrophysicist Robert Nemiroff and his students at Michigan Technological University embarked on a project to search social media for signs certain individuals demonstrated specific knowledge about the future. They looked for Twitter and Facebook users that seemed to have foreknowledge of two specific events: the election of Pope Francis as pope and the discovery of Comet ISON in 2012. They came up empty. In hopes of getting one or two shy time travelers to come forward, they came up with a pair of hashtags they verified had not been used prior to 2012: #ICanChangeThePast2 and #ICannotChangeThePast2. If there were any time travelers in this timeline that could jump back before 2012, they asked them to inject one of those hashtags into their Twitter feed pre-2012. None, however, chose to jump back in time and bat signal their existence.

Four Organizations Pseudoscientists Love and Hate

It has been observed many times that pseudoscientists are vocal about their distrust and loathing of government agencies and scientific bodies and yet seem to crave the approval of the same such organizations. Here are four that are typically in the woo crosshairs yet are still cited if they seem to offer any approval.

THE CIA/FBI

Conspiracy types will claim three letter agencies like the CIA and the FBI are behind a wide range of nefarious acts. They're the stormtroopers of the one world government and simply can't be trusted. Yet many conspiracy writers will allege they were former members of these organizations, therefore much of their baseless information must come from secret documents they saw when they were working on the inside. Trust them. Others will spin tales about the super important work they did for these groups. For example, psychics love to claim they used their talents to spy on behalf of the CIA. So, their powers must be real.

FDA

The FDA is enemy number one to the alt med industry. It's in the service of the medical establishment to hide simple, natural cures and keep people sick. Yet many purveyors of alt med nostrums and cranks medical devices will proudly label their products as FDA Registered or FDA Cleared.

THE IVORY TOWER

Cranks accuse academia of ignoring their astounding breakthroughs, having knowledge filters, and outright suppressing proof of whatever cranks think is real. Yet, you'll find credentialism rife within their fold. It's not hard to find some paranormal proponent, Ufologist, or creationist going by the title "Doctor" and implying they have a PhD. They are either very coy about where they earned their degrees or it's obvious their degrees are from a well-known diploma mill.

NASA

NASA is frequently in the crosshairs of Ufologists, flat earthers, and free energy cranks. Yet, many within those same groups will tout UFO photos being supposedly declared genuine by NASA photo analysts, their free energy machines passing NASA testing, or research published by NASA proves the earth is really flat.

Eight Pseudo Journals

ANSWERS RESEARCH JOURNAL

In the early 1980s, the state of Arkansas passed a law requiring "creation science" to be taught in public schools. The law was challenged and struck down in the landmark case *McLean v. Arkansas Board of Education*. A judge ruled creation science wasn't an actual science. Of the many reasons the judge ruled creation science failed to meet the characteristics of actual science is creationists weren't publishing their so-called science in recognized peer-reviewed scientific journals. The *Answers Research Journal* is one of a handful of supposed peer-reviewed journals that emerged in the wake of the court's rather definitive ruling on what constitutes a science. Of course, "peer review" in this case means review by so-called scientists who, according to *Answers Research Journal*, "support the positions taken by the journal". You know what real scientific journals don't do? They don't tell prospective scientists the answer they need to find before doing the actual science. If the name of the journal sounds vaguely familiar that's because it's a Ken "Answers in Genesis" Ham project.

CHAOS, SOLITONS & FRACTALS

Chaos, Solitons & Fractals bills itself as a journal of "nonlinear science". It is published by Elsevier. Elsevier publishes roughly 2,500 generally legit scientific journals. *Chaos, Solitons & Fractals* ... not so much. It came to the attention of many when it was noticed this obscure journal had an impact factor that dwarfed serious and better-known mathematical journals. Impact factor is one measure of a journal's importance. In essence, impact factor indicates how many citations a journal's articles receive in other papers. More is more better. It is science's version of Google's venerable Page Rank. A little digging discovered *Chaos, Solitons & Fractals* had published over 300 papers by its founder and editor-in-chief Mohamed El Naschie. While not illegal, it's a bit like editing your own Wikipedia page. El Naschie didn't stop there. He tended to cite his own papers which were previously published in *Chaos, Solitons & Fractals*. Again, not illegal but it's a red flag if you're largely the only person in a field publishing on a topic. The papers El Naschie was approving and publishing in his journal have been variously described as "undisciplined numerology larded with impressive buzzwords", "I get crud like this in my email every day. I delete it without comment", and "of no scientific or mathematical merit". Hmm. In 2009, El Naschie stepped down as editor, but not before suing the journal *Nature* for libel. *Nature* had suggested *Chaos, Solitons & Fractals* was publishing his papers without adequate peer review. El Naschie lost, possibly due to the fact he acted as his own lawyer, and possibly due to the fact he couldn't himself prove his papers had undergone any proper peer review beyond informal discussions with colleagues.

DENOVO

In 2012, Bigfoot researcher Melba Ketchum announced she had sequenced Bigfoot DNA. Alright. People responded, "publish your findings in a scientific journal". So, she did. She published her findings in 2013 in a journal called *DeNovo Journal of Science*. People responded, "we've never heard of that journal." And they were right. The journal seems to have come into being only about nine days before Ketchum published her paper. And it had published only one paper, which was Ketchum's. Her paper was also behind a $30 paywall. Since *DeNova*'s "de novo" creation, the journal has only published one subsequent edition in 2014. Similar to the 2013 edition, this one had only one paper written by Ketchum. This one is not behind a paywall, however.

JOURNAL OF 9/11 STUDIES

The *Journal of 9/11 Studies* kicked off in 2006 with BYU's Steven Jones as editor. Those who remember the Pons and Fleischmann days of cold fusion may recognize Jones as a player in that clap trap. Jones was researching an actually legit form of cold fusion known as muon-catalyzed fusion. Pons and Fleischmann famously reneged on a deal with Jones to jointly submit their papers on cold fusion to *Nature* and instead took their findings directly to the press. During the fallout, Jones emerged as the sober, moral center of the narrative. Oh boy, what happened? After 9/11, he seems to have thrown his cred into the toilet and became a talking head for 9/11 Truthers and launched the *Journal of 9/11 Studies*. BYU was embarrassed by this nuttery and placed him on paid leave. Jones retired from BYU not long after. *Journal of 9/11 Studies* seems to have sputtered out in 2015. In that year it published just a single article which is a law review article. This paper argues engineers and scientists who have established the actual cause of the towers' collapse would not pass a Daubert test but Truther cranks would.

MANKIND QUARTERLY

Mankind Quarterly (*MQ*) describes itself as an anthropology journal devoted to "the science of man". That sounds reasonable, until you notice the journal's editor-in-chief is Richard Lynn. Lynn was for a time an emeritus professor of Psychology at Ulster University. He had that title stripped from him for his involvement with the White Nationalist movement. *MQ* publishes articles that argue for "hereditarianism", a belief that race is a key factor affecting IQ. You'll find many of the papers published in *MQ* are funded by the Pioneer Fund. The Pioneer Fund was founded during the heyday of America's flirtation with eugenics. The fund uses its money to promote studies in race and eugenics.

MEDICAL HYPOTHESES

Medical Hypotheses is another questionable journal published by Elsevier. Elsevier publishes several high impact medical journals like *The Lancet* and *Cell*. *Medical Hypotheses* is an odd one out. It doesn't offer "traditional" peer review. It seems to be a dumping ground for cranks looking to gain some respectability by having their paper associated with a publisher that's known for producing high quality peer-reviewed journals. *Medical Hypotheses* has published papers on topics ranging from rank HIV denial to a finding that nasal congestion can be effectively treated by masturbating. It also published a paper by Jacques Benveniste claiming he could basically transmit homeopathic "water memory" over the phone.

JOURNAL OF COSMOLOGY

Top journals seem to have pretty parsimonious names: *Nature, Science, Cell, Astrophysical Review*. So it's forgivable if the media gets excited about a paper about space aliens or the reality of quantum consciousness when told it's published in what sounds like a totally legit journal: *Journal of Cosmology*. The journal's website looks like it was designed by web devs who have never upped their HTML game since 1998. This might suggest to some this is not an established and storied science journal. The first issue is from 2009 and it generally lacks what you'd expect from an actual scientific journal: scientific papers. There are about a dozen commentaries suggesting life on Earth could not have possibly evolved and therefore panspermia. The journal went downhill from there: dipping into quantum consciousness, "astro-theology", and the idea that pandemics are from space. The rationale for that seems to be if viruses weren't from space, why does it take so long to develop vaccines? Wut? Issue 27 begins with this serious, sober observation about the nature of the cosmos: "GREEN LIVES MATTER! So do BLACK LIVES, WHITE LIVES, and YELLOW LIVES! HUMAN LIVES MATTER!" Wut wut?

JOURNAL OF SCIENTIFIC EXPLORATION

The *Journal of Scientific Exploration* (*JSE*) is published by the Society for Scientific Exploration. Both are devoted to paranormal and fringe topics, ranging from ESP to cryptids to UFOs. The *JSE* has been publishing since 1987 and claims to be peer-reviewed. Although the journal is a cover-to-cover woo fest, articles are at least well written. The journal is edited by Stephen E. Braude. Braude is a retired philosophy professor who has a strong interest in mediumship and séances.

Ten Hints It's Not Science but Pseudoscience

Photo Credit: Karl Mamer

A LOT OF EVIDENCE IS ANECDOTAL

As some skeptics like to say, the plural of anecdotes is not "data". Pseudoscientists offer anecdotes and personal testimony but don't collect data under rigorously controlled conditions.

A LOT OF EVIDENCE IS "GOD OF THE GAPS"

A pseudoscientist will try to dazzle people with how impossible something is to explain via conventional science and then offers his/her pseudoscience as the only possible explanation.

A LOT OF THE EVIDENCE IS SUBJECTIVE

Sometimes science needs to rely on subjective evidence. Does a certain pill reduce pain? Pain is a highly subjective experience. Real scientists work hard to find clever ways to remove subjectivity from their experimental design. Pseudoscientists make no such efforts and will usually suggest it's simply impossible.

CANNOT SET A CLEAR GOAL POST

Pseudoscientists have two problems here. They can rarely express what constitutes a definitive experiment for their claim. Likewise, they can't establish any kind of experiment that might falsify their claim.

DOESN'T ACTUALLY FOLLOW THE SCIENTIFIC METHOD

A scientist observes a phenomenon, develops a hypothesis, develops a way to rigorously test the hypothesis. The scientist then accepts, rejects, or modifies the hypothesis based on the outcome of experiments. Pseudoscientists tend to believe a story that seems to explain a given phenomenon is sufficient.

EXTRAORDINARY CLAIMS ARE NOT BACKED UP BY EXTRAORDINARY EVIDENCE

A lot of pseudoscience seems to defy well-tested scientific principles. If the hypothesis is correct, it would invalidate whole branches of well-established science. Scientists are unwilling to throw relativity, Newtonian mechanics, or evolution into the dumpster because one difficult-to-repeat experiment shows a very small effect size.

LACKS PREDICTIVE ABILITY

A well-tested scientific theory makes predictions. A theory that doesn't say exactly what you should see given a set of conditions is useless.

TAKES CLAIMS DIRECTLY TO THE PUBLIC AND MEDIA

Pseudoscientists rarely do what real scientists do: they submit their papers to peer-reviewed journals. If their work is rejected, they address the shortcomings. Pseudoscientists will issue press releases, tweet, put up YouTube videos, speak at paranormal conferences, or write popular books. Rejection by peers is attributed to a vast conspiracy to hide the truth.

USES A LOT OF UNNECESSARY JARGON

Real science uses jargon for accuracy and specificity. Pseudoscience uses it to mislead or simply sound science-y.

ZERO PROGRESS

Real breakthroughs in science tend to follow a common trajectory. It goes from "huh, that's funny" to an experiment that independent labs can duplicate. If that shows promise, labs will develop research programs. After ten or twenty years, universities have whole departments teaching and researching it. Pseudoscience never gets beyond the quest for a single replicable experiment. Pseudoscientists across generations keep going over the same well-trodden ground.

Six Unusual or Notable Independent Scientists

Most scientists work for institutions, like universities, and get funded from various outside sources. Independent scientists are those that fund their own research and work outside of the institutional system. Being independent of outside funding means they can pursue atypical avenues of research.

AUBREY DE GREY

Aubrey de Grey is a leading proponent of life extension (which is sometimes referred to as transhumanism). De Grey holds a bachelor's degree in computer science and at one point worked for Sinclair Research, the company that made the low-cost ZX80 computer in the early 1980s. Since then de Grey has been actively involved in researching and promoting life-extension strategies. De Grey became self-taught in biology and eventually wrote and published *The Mitochondrial Free Radical Theory of Aging* in 1999. Cambridge awarded him a "PhD by publication" for his work on this book. De Grey popularized the idea that we might be able to end or ameliorate the aging process by reducing damage to our mitochondria from free radicals. Since the publication of his book in 1999, mainstream science has found very little support for any of de Grey's life-extension strategies. A 2005 review concluded none of his strategies have been shown to "extend the lifespan of any organism, let alone humans."

RAY KURZWEIL

Ray Kurzweil made his fortune a couple times over. He invented the first practical OCR system that was capable of recognizing text written in almost any standard font. He sold that to Xerox in 1980. Two years later he partnered with Stevie Wonder to found Kurzweil Music Systems. His company produced a line of highly regarded and innovative synthesizers. He sold that to Young Chang in 1990. Through much of the 1990s, Kurzweil turned his mind to creating technology that could help people who have difficulty reading (the blind and people with dyslexia and ADHD). In 1999, Kurzweil published *The Age of Spiritual Machines*. In this book he developed his ideas of "the singularity". Similar to Moore's Law, Kurzweil hypothesized a Law of Accelerating Returns where AI starts to increase exponentially to the point where machines surpass human intelligence. What happens then is anyone's guess. Since 2012 he's been working with Google to help create computers that understand natural language.

JAMES LOVELOCK

James Lovelock is best known for developing the Gaia hypothesis in the 1970s. The Gaia hypothesis views the earth as a single, self-regulating system that uses a combination of living organisms and inorganic matter to maintain conditions favorable for life. The idea is wildly popular with many in the environmental movement but has not gained much traction in mainstream science over the decades. At best most scientists have accorded it the status of "romantic metaphor" until Lovelock is able to demonstrate an actual mechanism by which the planet maintains this supposed homeostasis.

ALEX TSAKIRIS

Alex Tsakiris founded a company called Mind Path Technologies that made a remote-control device that let presenters control aspects of their multimedia presentations. He sold his company in the late 1990s and seems to have devoted his life to proving scientific materialism is completely wrong. He started a project called Open Source Science and ran two experiments trying to prove dogs are psychic. Open Source Science doesn't seem to have done much since. Tsakiris has also been the long-time host of the *Skeptiko* podcast, which he uses to promote his idea of mind/body dualism, near death experiences, mediums, hyperdiffusionism, UFOs, Covid denial, and various conspiracy theories.

CRAIG VENTER

Craig Venter is a biotechnologist and might be best known for creating a private company that mapped the human genome in 2000. He managed to do it years ahead of the publicly funded Human Genome Project and at a fraction of the cost. In 2006, Venter founded the J. Craig Venter Institute with the aim of creating the world's first synthetic life form.

STEPHEN WOLFRAM

Stephen Wolfram founded Wolfram Research which developed the widely used Mathematica software package. He's been using income from his software to fund his own research into what he calls a "new kind of science". Computation is not just a tool to be used by other sciences but should be studied as a natural phenomenon itself via the scientific method. He believes from that study we'll realize the universe's building blocks can be described by a set of simple computational rules. He's recently started the "Wolfram Physics Project" which is trying to represent the laws of physics using a simple set of symbols. His ideas have failed to impress mainstream scientists who find his ideas more descriptive than quantitative.

Thirteen Scientific Scandals

BOGDANOV AFFAIR

The Bogdanov brothers (Igor and Grichka) are twins. They rose to fame in the 1970s and 1980s as heart-throb science presenters on French television. The brothers published a popular science book titled *God and Science* in 1991. Shortly after its publication they were accused of plagiarizing the work of an American astronomer and inflating their academic credentials. The dust jacket on the Bogdanov book indicated the brothers held PhDs in physics. This was not the case. It would be years before either was awarded a PhD. Grichka was awarded a doctorate in Math in 1999 and his brother received his in Physics in 2002. Papers written by the brothers in 2002 came under scrutiny by several physicists who couldn't make any sense out of their work. It was first assumed their papers were hoaxes, meant to test the peer review process of mainline physics journals. The brothers denied their papers were hoaxes but seemed unable to answer key questions about their math. As one physicist put it about one of their papers, it was "a mishmash of superficially plausible sentences containing the right buzzwords in approximately the right order." They made bizarre claims like a Foucault pendulum swings in line with the point in space of the universe's singularity and they were able to math their way past the Big Bang's "Planck era" (10^{-43} seconds after the Big Bang). Much debate about the purpose and merit of the brothers' papers flew across Internet message groups used by working physicists. After one physicist published a draft of Grichka's PhD thesis on his own website along with his criticism, the Bogdanov brothers lawyered up and had the scientist arrested for copyright violations. Many scientists were outraged by the brothers' tactic. An open letter was signed by some 170 scientists protesting the use of a criminal court to ostensibly silence scientific criticism by a peer.

THE BONE WARS

The Bone Wars refers to a rivalry between Edward Drinker Cope and Othniel Charles Marsh. They were two 19th century American paleontologists who spent years using underhanded methods, like bribery and the destruction of fossils, to outdo each other. Cope came from a family of means. He lacked a formal degree but fashioned himself as your classic gentleman scientist. Marsh came from more humble beginnings but was a Yale scholar. Each seemed to find something to hate envy in the other. Their enmity was lit when Marsh pointed out an obvious error Cope had made while reconstructing a dinosaur skeleton. Cope had affixed the head to what was actually the tail. It was a bone war from that moment on. Both ended up ruining themselves financially competing to see who could discover more new and wondrous dinosaur fossils. Each used their political and institutional influence to try and destroy the reputation of the other. They accused each other of plagiarism and mismanagement of research funds. The upside of this rivalry was the discovery of over 100 new species of dinosaur including favorites like the Triceratops and the Stegosaurus.

Photo Credit: Karl Mamer

CRAP GMO SCIENCE

The anti-GMO crowd in 2012 were starting to oil up the wheels on the "I told you so train" after French molecular biologist Gilles-Éric Séralini published a study that suggested rats fed a diet of GMO corn had an increase in cancerous tumors. The article was published in *Food and Chemical Toxicology*, which is an established mid-level journal on food safety. Séralini's publication was accompanied by an unusually slick media campaign, which included press packets with pictures of rats with oversized tumors, and hints that a popular book and movie were in the works. None of this typically accompanies the publication of a scientific paper. It starts to make sense when you understand Séralini is the founder of an anti-GMO lobby group and part of his research was funded by a supermarket chain that had "bet the farm" on selling organic food. This also wasn't Séralini's first kick at the GMO can, either. He had already published at least two other papers suggesting GMO corn presented health risks. Both earlier papers were judged to have statistical problems and conclusions not supported by their evidence. When scientists tucked into his 2012 paper, they discovered numerous and obvious flaws. Again, his statistical analysis was questionable. Standard statistical analysis didn't actually show any difference between the control group and the experimental group. As well, many researchers familiar with the rats Séralini used pointed out they were prone to developing tumors, notably if they're fed an unrestricted diet. Séralini's paper did not indicate the amount of corn the rats were actually being fed. Séralini's effect size was also a little too hard to believe. GMO corn has been in the human food chain for a long time. If it was as deadly as Séralini was alleging, humans would be, as one critic put it, "dropping like flies". *Food and Chemical Toxicology* retracted the paper a year after its publication date.

HASTINGS RARITIES AFFAIR

The Hastings Rarities Affair was the product of a hoax perpetrated by British ornithologist George Bristow (died 1947). Bristow claimed to have discovered some 29 bird species native to the UK that were previously unknown to fellow ornithologists and bird watchers. They were all discovered within a 20-mile radius of Hastings. By way of proof, Bristow collected and preserved examples of his astounding finds. He pocketed a considerable sum subsequently selling those finds to collectors. After Bristow's death, sightings of these "rarities" dropped off. A statistical analysis published in 1962 suggested Bristow's ability to find so many rare birds in such a limited geographical area was impossible. It was later determined Bristow had simply imported birds and then later claimed to have found them around Hastings.

JAPANESE PALEOLITHIC HOAX

Shinichi Fujimura became something of a sensation in Japanese archeological circles in the 1980s and 1990s. Fujimura was an amateur archaeologist but earned the respect of professional peers who were researching Japan's paleolithic era. Fujimura had an uncanny ability to find stone tools and other artifacts at dig sites. He garnered the nickname "Divine Hands". Fujimura's many finds required the rewriting of Japanese textbooks about the period. The best evidence before Fujimura entered the picture was that the first human habitation of Japan was no earlier than 35,000 BC. Fujimura's findings, by his divine hands, pushed that back to 700,000 BC! By the late 1990s, his unbroken run of remarkable finds started to draw skeptical attention. Although he worked with a team on various digs, only he seemed to be able to find stone artifacts. Some of his findings didn't make sense. He produced artifacts supposedly buried in a pyroclastic flow. That is, he claimed to have found stone tools intact despite having been enveloped by lava. In 2000, a Japanese newspaper investigating accusations of fraud set up a hidden camera at one of Fujimura's dig sites. It captured Fujimura shamelessly planting items he would then later "discover". Fujimura copped to the hoax. An investigation by the Japanese Archaeological Association concluded in 2004 that all 168 sites excavated by Fujimura were completely faked.

LANCET PUBLISHES ANDREW WAKEFIELD'S PAPER LINKING AUTISM TO VACCINES

In 1998, the UK's *The Lancet*, one of the top medical journals in the world, published a paper by Andrew Wakefield that suggested there was a link between the MMR vaccine and autism. Wakefield's paper crystalized a growing but nebulous fear among parents that childhood vaccines posed a risk to their newborns. What many didn't realize was Wakefield was being funded by a lawyer (to the tune of about £435,643) seeking to start a class action lawsuit in the UK against vaccine manufacturers. Many problems with Wakefield's paper, undisclosed conflicts of interest, and breaches of medical ethics were uncovered by journalist Brian Deer. Wakefield's sample size was only twelve children. Wakefield failed to disclose he had also filed a patent on his own single dose measles vaccine, meant to be a supposedly safer alternative to the MMR jab. *The Lancet* eventually retracted the paper. An investigation of allegations of ethical breaches raised by Deer resulted in Wakefield losing his license to practice medicine in the UK.

RESEARCHER INVENTS 908 PATIENTS

Jon Sudbø was a researcher in Oslo. Sudbø's study of 908 cancer patients indicated ibuprofen offered smokers an effective way to reduce the risks of oral cancer. He submitted his finding to *The Lancet* and the editors fast tracked it for publication. One Norwegian reader of Sudbø's paper noticed in the fine print it claimed patient data had come from Norway's cancer database. The problem was the database had yet to go online. It was impossible for Sudbø to have used the data. Further digging showed 250 of the patients in the paper had the same birthday. That just never happens. As it turned out Sudbø faked all 908 patients including their age, weight, and diagnosis.

SCIENCE BY PRESS CONFERENCE

A hallmark of pseudoscience is "science by press release" or "science by press conference". That is the researcher bypasses traditional peer review and takes his/her ideas or results directly to the public. One immediately thinks of Richard "Face on Mars" Hoagland. Hoagland booked the National Press Club in Washington D.C. back in 1996 on the promise "scientists" would reveal undeniable proof there were alien structures on the moon (spoiler: he revealed nothing of the kind). Real scientists, who should understand peer review, are sometimes guilty of science by press conference. The first real example was in 1989 when scientists Stanley Pons and Martin Fleischmann held a press conference to announce they had successfully achieved cold fusion, despite their paper still being in the review process and not formally published. Brigham Young University researcher Steven Jones was, at the time, also working on a known form of low-grade cold fusion called muon-catalyzed fusion. Pons and Fleischmann had entered into an agreement with Jones to simultaneously publish their papers. Pons and Fleischmann broke the spirit of that agreement by giving a press conference on their own research before either paper was formally published. What followed was what could charitably be described as a circus. Different labs scrambled to try and reproduce their results. Some claimed they were able to. Others could not. In no case was any lab able to produce the telltale signature products of fusion. This caused a few pro-cold fusion scientists (who should have known better) to suggest maybe they had stumbled on some kind of new physics.

SCHÖN SCANDAL

German physicist Jan Hendrik Schön was working at Bell Labs in the late 1990s. He was studying ways to use organic compounds instead of silicon in semiconductors. In the early 2000s, Schön started to report amazing breakthroughs. He reported he was able to use organic materials to create single-molecule transistors that demonstrated aspects of superconductivity. His results were published in both *Science* and *Nature*. Other labs tried to replicate Schön's methods but failed. That resulted in people taking a much closer look at his papers. Bell Labs launched an investigation. It was noticed he had repeated the same data for different tests. The "noise" in his data was identical. He reported his data at a level of precision that was actually beyond his ability to measure. Schön was unable to produce his original data as he claimed he needed hard drive space and deleted it. Schön was found to have engaged in misconduct and was let go. His own university in Germany stripped him of his PhD. The investigation raised questions about the quality of peer review at topflight journals like *Science* and *Nature*. The co-authors on Schön's paper were absolved of misconduct but many wondered why? This prompted the American Physical Society to issue new rules that held co-authors accountable on papers they put their names on.

SOKAL AFFAIR

Many scientists in the 1990s were starting to notice it was becoming fashionable among academics outside of the hard sciences to frame science as simply just another human narrative. All facts were subject to a certain human point of view, and your opinion of the truth of, say, relativity was as valid as a physicist's. It wasn't so much that scientists were suddenly becoming the least cool people at faculty mixers. It was starting to have a profound effect on the public understanding of science. For example, founder of the Discovery Institute Philip Johnson (died 2019) started to use this postmodernist paradigm as part of his wedge strategy against the scientific consensus on evolution. Alan Sokal sought to demonstrate the lack of any kind of rigor within postmodern academic circles. Sokal wrote up a nonsense paper called "Transgressing the Boundaries: Towards a Transformative Hermeneutics of Quantum Gravity" and in 1996 he submitted it to *Social Text*, a high-profile postmodernist journal. Sokal's paper made the bold assertion that reality did not exist. It was written in such a way as to flatter the journal's editors while being patently gibberish. It was also salted with a number of inside jokes instantly recognizable to anyone with a physics degree. Sokal's paper was accepted for publication. It was obvious the paper was not sent out for peer review by anyone with a background in physics. A few weeks later Sokal revealed his paper was a hoax. *Social Text*'s editors initially alleged Sokal had just regretted submitting his paper (his actions were "a folding of his intellectual resolve") and this was his odd way of trying to retract. Sokal's hoax was generally applauded by those in the hard sciences and condemned by academics working in the humanities. Debate remains today whether Sokal's actions were unethical or a necessary wakeup call to a branch of academia that didn't quite "get" science.

SOUTH KOREA'S STEM CELL DEBACLE

Hwang Woo-suk was a professor of biotechnology at South Korea's Seoul National University (SNU). SNU is considered South Korea's Harvard. Between 2004 and 2005, Hwang published two breakthrough articles in the journal *Science* claiming he had created human embryonic stem cells through a cloning process. This breakthrough meant, for the first time, patient-specific stem cells could be created. There would be no fears of rejection as patient-specific stem cells were cloned from the very cells of the target patient. Hwang rapidly became a celebrity in South Korea. The South Korean government promised to invest untold sums of money to turn the nation into the "stem cell hub" of the world. Cracks started to emerge in Hwang's research. It was revealed the source of his biological material was from his own graduate students. This was considered an ethical violation (you don't get your own low-level employees to "volunteer" for medical experiments). A few months later it turned out Hwang's reports of creating stem cell lines from a cloning process were fabricated. Not one cell line could be genetically linked back to the donor. Hwang was bounced from his lofty position at South Korea's "Harvard". For several years large swaths of South Korea's biomedical research were subject to deep suspicion by the rest of the world's scientific community.

STUDY 329

Study 329 was a clinical trial conducted by British pharmaceutical company SmithKline Beecham between 1994 and 1998. SmithKline Beecham was trying to test the safety and efficacy of prescribing an antidepressant marketed as Paxil for adolescents. On paper, the study looked good. It was a double-blinded, multi-center study conducted over a number of years. Its intent was to conduct the largest ever investigation of the efficacy of prescribing SSRI drugs to teenagers and children. The study's results showed Paxil was no better than placebo. More disturbing, the study found high incidences of suicidal behavior among users. These findings, however, were ignored by the pharma company's marketing wing. SmithKline Beecham hired a marketing firm to ghostwrite an article and submitted it to the *Journal of the American Academy of Child and Adolescent Psychiatry*

(*JAACAP*). The article, published in 2001, focused on the drug's safety and downplayed the high placebo response rate and the suicide link. After the *JAACAP* publication, the pharma company launched a campaign to suggest doctors should "off label" prescribe Paxil for teens with depression. SmithKline Beecham was smoked out when the families of adult patients who had been prescribed Paxil and died of suicide sued the drug company. A BBC *Panorama* investigation uncovered Study 329. UK and later American investigations into SmithKline Beecham's activities resulted in the company paying $3 billion in fines.

THE TASADAY: STONE-AGE TRIBE?

As the story goes, in 1971 a Filipino hunter penetrating deep into the Philippines' Mindanao rainforest stumbled on an isolated tribe of "stone-age" people called the Tasaday. It was reported this unknown tribe lived a primitive existence, separated from the rest of humanity for at least 2,000 years. They lived in caves, used stone tools, and wore leaves for clothes. Even more remarkable, they had no clue there existed other human beings. The Tasaday were depicted as a peaceful people with no words in their language for war or weapons. They were living in a veritable garden of Eden. It was exactly what the 1971 peace movement zeitgeist required. *National Geographic*, which was still a respectable magazine at the time, ran with the story. Access to this tribe was first severely limited and then eventually cut off by a Ferdinand Marcos crony named Manuel Elizalde (died 1997). The few scientists who had been allowed contact with this supposedly isolated tribe noticed things at odds with Elizalde's claims. For example, the tribe's alleged diet was not adequate to sustain them. Their language was a dialect of a language spoken by other tribes in the area. Although distinct, it was, at best, a product of no more than 200 years of isolation. Not much more was learned before Elizalde completely banned further contact by researchers. Elizalde was put in charge of a multi-million-dollar fund set up to protect the Tasaday. After Marcos was deposed, Elizalde fled. It is likely he feared it would be discovered he had personally squandered at least $1 million of the fund. Researchers were allowed to visit the Tasaday again in the mid-1980s. Things start to get weird after this point. A Swiss journalist named Oswald Iten was able to interview tribe members. He reported being told by members of the tribe they were forced to appear more primitive and isolated than they were by Elizalde. Allegedly Elizalde told them if they appeared poor, they'd get money. Iten himself was accused in court, by members of the tribe, of perpetrating his own journalistic hoax. They were what they claimed to be: a "stone-age" tribe. A Manila court agreed with them. Many anthropologists who have visited with the Tasaday since the mid-1980s paint a picture that's someplace between outright hoax and Elizalde's claim of a lost tribe of cave people. The stone axes that were trotted out for journalists in the early 1970s were fakes. Bamboo implements were also found being used by the tribe. Bamboo was not native to their area. The closest source of bamboo was from cultivated fields. This was solid evidence the group had been occasionally trading with local farmers. A picture emerged of a tribe that likely branched off from other tribes at the start of the 20th century and went deeper into the jungle. They had likely lived reasonably isolated for a couple generations, although they had infrequent trading contact with farmers. Tales of their isolation for thousands of years and their Eden-like existence were made up by Elizalde for reasons that can only be speculated upon.

Seven Ways to Determine the Age of Stuff Beyond Radiocarbon Dating

Creationists seem to think the only way to date something is via radiocarbon dating. They like to point out how if you totally botch your sample collection (like you try to date a living organism), you get a widely inaccurate reading. And, of course, they'll point out radiocarbon dating can't give you dates beyond 40,000 years so it's simply impossible for scientists to claim the world is billions of years old. Creationists are likely not aware there are many different dating techniques employed by geologists and archaeologists that do not rely on radiocarbon dating and many easily give accurate dates beyond 40,000 years.

DENDROCHRONOLOGY (ACCURATE TO ABOUT 11,000 YEARS)

Trees produce a growth ring every year. In dryer seasons the ring is thinner. In wetter seasons, the ring is thicker. Scientists use a pattern of tree rings from different trees to form reference chronologies for a given area. Tree ring dating is accurate down to a year. It's useful for double checking the accuracy of radiocarbon dating.

POLLEN DATING (CAN GIVE RELATIVE DATES BEYOND 40,000 YEARS)

Similar to trees, plants release pollen on a yearly basis. This forms layers of pollen in rock or other strata. Radiocarbon dating is usually used to determine an initial absolute date and then one can "count down". By using radiocarbon dating to establish the initial absolute date, one can date earlier layers that cannot normally be dated due to the limitations of radiocarbon dating.

THERMOLUMINESCENCE DATING (ACCURATE TO 100,000 YEARS)

Thermoluminescence dating is useful for dating clay objects and even stone tools. The principle here is background radiation penetrates the clay and knocks electrons out of the orbit of the various atoms in clay. These electrons are quickly captured by crystalline structures. When an object is made from clay and fired in a kiln, these crystalline structures release all their captured electrons. The kiln resets this clock to zero. The process then starts again. Archaeologists can reheat pottery sherds, measure the electrons released as light, and get a good measure of how long ago the object was made.

AMINO ACID RACEMIZATION (ACCURATE TO ABOUT 1 MILLION YEARS)

Amino acid racemization can be used to date things that were once alive. Living organisms are made up almost exclusively of left-handed amino acids. However, as soon as an organism dies right-handed amino acids start to replace the left-handed variety in the nonliving organic remains. Since we know this rate of replacement, we can measure the ratio of left-handed to right-handed amino acids. Higher the ratio, the longer the item has been dead. This process can get archaeologists accurate dates beyond carbon dating's 40,000-year limit.

PALEOMAGNETISM (ACCURATE TO ABOUT 600 MILLION YEARS)

Paleomagnetism is useful for dating ancient lava flows and other rocks where radiometric dating methods can't be used. Geologists have been able to create a very precise timeline of reversals of the earth's magnetic field. Magnetic minerals found in lava flows act as frozen compass needles and preserve the orientation of the poles. Depending on the "angle of dip" found in the mineral's frozen compass needle, geologists can then cross reference the mineral's polar orientation to a known timeline of the North Pole's migration.

FISSION TRACK DATING (ACCURATE TO 2 BILLION YEARS)

When uranium in rocks decays, the expelled particles leave tracks in the rocks. Geologists measure the uranium present in a rock. Since we know the decay rate of uranium, geologists can easily count up the fission tracks to understand how much uranium was present when the rock was formed, and then calculate the age of the rock.

URANIUM-LEAD DATING (ACCURATE TO 4.5 BILLION YEARS)

Uranium-lead dating is one of about half a dozen methods for dating rocks using a process called radiometric dating. The principle is that most rocks form with radioactive elements. These elements have known decay rates. For example, given enough time, a pile of uranium will decay into a pile of lead. The rate of decay is called a half-life. Given a starting quantity of uranium, we have very good measures of how long it will take for half of the uranium to decay into lead. So given a rock that a geologist knows has not been exposed to new sources of uranium contamination, one can measure the ratio of uranium to lead. More lead, older the rock. Given the known half-life of uranium, it's simple math to work back to when the rock was formed.

Seven Independent Lines of Evidence for Common Descent

Evolution has been described as the unifying principle of biology. Very little in biology would make sense if evolution were not true. A key reality of evolution is the concept of common descent. Humans and mice or whales and petunias, if you go back far enough, share a common ancestor. The beauty of evolution is how well different fields of science since the time of Darwin have converged on the same inescapable conclusion: life evolved over millions of years and everything alive today shares a common ancestor.

ALL LIFE SHARES THE SAME BASIC GENETIC CODE

Years before DNA was unraveled, evolutionary biologists predicted, given common descent, life must share some kind of universal genetic code. This prediction from evolution was critical in helping Watson and Crick unlock the secret of DNA in 1962.

ATAVISM

Atavism is the reappearance of a structure found in a species' distant evolutionary ancestor. For example, whales and dolphins evolved from land-dwelling mammals with limbs. Whales and dolphins today have no need for limbs. From time to time, these aquatic mammals are found with rare genetic mutations that produce vestigial hind limbs. Chickens, like all birds, are modern day dinosaurs. Modern chickens do not have teeth. However, chickens still have genes that allow them to grow teeth if these genes are turned on in the lab. The only way to make sense of these hind limbs in whales and the ability to induce chickens to grow teeth is common descent.

BIOGEOGRAPHY

Evolution takes place not only over time but also over a limited geography. You will find more examples of closely related species in a similar geographic area. For example, while marsupials dominate Australia you will find very few examples in the Americas. Fossils of common ancestors again should be found in a similar geographical area. New fossils of transitional species should be found closer to fossils of their ancestors without large spatial jumps.

CYTOCHROME C

Cytochrome c is an essential protein used by all life on Earth. It has a necessary role in electron transport in mitochondria. *Cytochrome c* is made up of 100 amino acids. Of those 100 amino acids about 34 are key. These 34 amino acids must be present and arranged correctly. Evolution has highly conserved that arraignment. One can take *Cytochrome c* from a human cell and put it in a wheat cell and as long as those critical 34 amino acids are in place, wheat will happily use human *Cytochrome c*. What is of interest to understanding evolution is the remaining 66 amino acids. Those have been free to mutate. Evolution would predict if you compare two species, the more recent the common ancestor the greater the similarity between these mutated amino acids. Sure enough, not only do you find this in nature but when phylogenetic trees based on *Cytochrome c* are compared to phylogenetic trees worked out based on homology and the fossil record, the match is brilliant. There is no good reason for this match other than evolution.

ENDOGENOUS RETROVIRUSES

Genomes are littered with bits of DNA from viruses that have infected the germline over millions of years. These are known as endogenous retroviruses. About 1% of the human genome's genetic material comes from ancient insertions. Evolution would predict if you compare two species, the more recent their common ancestor the more similar their genomes should be in terms of endogenous retroviruses. There's no good reason for this pattern other than common descent.

HOMOLOGY

Evolution doesn't just invent new organs and structures out of thin air. Evolution works with what it has. Homology predicts as species separate the differentiating species adapt existing structures to meet new survival challenges. For example, bats have wings. The bones of bat wings are homologous to the front flippers of whales. A more remarkable example are the bones of the human ear. Three small bones in the ear are responsible for transmitting sound from the eardrum to the inner ear. Two of these bones are homologous bones that form the jaw of lizards. There is no good reason bat wings and whale flippers or ear bones and jaw bones should share the same arrangement unless all evolved these structures from a common ancestor.

VESTIGIAL ORGANS

Throughout biology, we find animals with vestigial organs and structures. These are organs or structures that have either no function or reduced function. Some vestigial organs and structures with reduced function, from a design perspective, are over engineered. For example, ostriches are flightless birds yet have massive wings. Their wings certainly help them maintain balance while running yet there is no good reason wings should be used to serve this purpose. Vestigial organs and structures only make sense in light of common descent.

Six Examples of Speciation Taking Place Before Our Eyes

APPLE MAGGOT FLY

Before Europeans migrated to North America, the apple maggot fly lived solely on a diet of berries from hawthorn shrubs. When European colonists started planting apple orchards some of the flies started to live in apple trees and developed a taste for apples. Today we have two variations, one that still dines on the hawthorn berry and one that dines on apples.

CENTRAL EUROPEAN BLACKCAP

The Central European blackcap bird makes it home in Germany and Austria during summer. Until the mid-20th century, the species migrated to Spain during winter. Today about 10% of the population flies to the UK, taking advantage of plentiful food provided by backyard bird feeders. In 2009, it was discovered the Central European blackcap is becoming two different species. The birds that take the shorter flight to the UK have evolved different wing shapes and beaks sizes.

GOATSBEARD

Goatsbeard is a flowering plant. Three different species were introduced to America from Europe in the early part of the 20th century. When these different species swapped pollen, they produced sterile hybrids. However, around the 1950s, it was discovered two non-sterile hybrids species emerged.

ITALIAN WALL LIZARDS

In 1971, researchers introduced several breeding pairs of Italian wall lizards to a small island off the coast of Croatia. The lizard's traditional source of food was insects. However, their new home, while rich in plants, did not offer a bountiful supply of insects to feed on. When researchers checked back some 30 years later, they found the lizards had adapted to the new plant diet and had evolved novel gut structures to aid with the digestion of leaves. Other changes include a more powerful bite (aiding in stripping plants of their leaves), a larger head, and a decrease in territoriality.

RICHARD LENSKI'S E. COLI

Richard Lenski started with a population of E. coli bacteria in 1988. By 2010 he had bred 50,000 generations in his lab. He was able to demonstrate step wise genetic adaptations (the kind creationists say are impossible) that allowed the bacteria to live on a novel food source (citrate).

THREE-SPINE STICKLEBACKS

A single species of three-spine stickleback fish was introduced into a Swiss lake about 150 years ago. In that time, the species has split into two. One species still lives in the lake, while another group of fish split off and radiated into the lake's feeder rivers.

Ten Dumb "Disproof" of Evolution

BILLIONS OF GENERATIONS OF FRUIT FLIES HAVE BEEN CREATED IN LABS WITHOUT A NEW SPECIES EMERGING

Fruit flies are incredibly useful for studying genetics. A female fruit fly lays dozens of eggs every day. Scientists can quickly generate thousands of generations in a short period of time to see the effects of changes in certain genes. Given how useful the fruit fly is, scientists want to keep it that way. No scientist working with fruit flies needs to waste limited research dollars to prove what every evolutionary scientist knows is fact: mutations and selection can produce new species. There's no goal post an experiment could cross that would make a creationist accept the reality evolution produces new species and beneficial mutations.

THE BANANA: THE ATHEIST'S WORST NIGHTMARE

This one was put forward by Ray Comfort, possibly one of the dumber celebrity creationists working the circuit. In one of his "Way of the Master" videos widely circulated on the Internet in the 2000s, he knowingly informed viewers the banana is too well "designed" for human benefit to have evolved naturally. God did it! Consider: It fits perfectly in the human hand, it's easy to eat, it's easy to see when it's ripe, and its wrapping is biodegradable! Comfort was clearly unaware grocery store bananas have these features because they were selected for those features by banana farmers. Real, wild bananas have none of those features.

CELLS AND PROTEINS COULD NEVER HAVE EMERGED BY CHANCE

The argument is things like cells and even proteins necessary for life are so amazingly complex, with so many parts that need to be in just the right place, chemicals sloshing around in a warm pond could have never just randomly produced a living cell. Creationists like to cite Fred Hoyle's assertion that "the chance that higher life forms might have emerged in this way is comparable to the chance that a tornado sweeping through a junkyard might assemble a Boeing 747 from the materials therein." Creationist rarely add Hoyle offered the argument in favor of panspermia not creationism. Hoyle was no ally to Christians. For example, he thought Big Bang was accepted uncritically by scientists because it supported the religious idea of creation. While it is correct cells and other complex building blocks of life cannot come together by pure chance alone, that's not a feature of evolution. Complex structures evolve from a random process guided by natural selection. It's very unlikely you'll get a royal flush from the first hand you're dealt. But if you can select cards that are more like a royal flush, throw away cards less like a royal flush, and select new cards to replace them, within a few dozen iterations you'll hold a royal flush.

EVOLUTION IS A THEORY IN CRISIS

This was a popular tactic back in the 1990s, pushed mostly by the Discovery Institute. The tactic seems to have been borrowed from efforts by Big Tobacco to sew doubt about the link between smoking and cancer. The idea is there are serious doubts and disagreements between scientists over how evolution works and they are keeping this from the public. Any theory on such shaky grounds should never be taught in public schools! Within any broad scientific theory, there is certainly debate over things like the degree to which known mechanisms contribute to the phenomenon, but such debate cannot be conflated with divisions fatal to the theory. Another tactic is to trot out recent science that establishes what Darwin got wrong. If biologists keep proving Darwin wrong, why do scientists keep claiming evolution is a fact? The obvious problem with this is Darwin's *Evolution of Species* is not treated as an infallible holy book. No one expected Darwin to have gotten it 100% right from the outset. Those who argue the "theory in crisis" line have been described as "incompetent, dishonest, or both".

LIVING "FOSSILS" DISPROVE EVOLUTION

The claim here is there are animals alive today that resemble ancestors known only via the fossil record (e.g., jellyfish and coelacanths). Superficially they appear similar. If these ancestor species have been on Earth for millions of years, they should show some amount of evolution. And yet zero evolution has taken place! The short answer is certain forms are well adapted to their environment. If that environment has remained stable for millions of years, there likely is little selection pressure to change the form in broad strokes. Turkish creationist Adnan Oktar (who goes by the pen name Harun Yahya) is behind the most amusing (and expensive) attempt to argue this point. In 2006 he published an 800-page book titled *The Atlas of Creation*. It is loaded with side-by-side pictures of fossils and what Oktar identifies as their living equivalents. In each case, Oktar asserts no evolution had taken place. Oktar took it upon himself to not only self-publish this book but mail thousands of physical copies to universities throughout the United States and Europe. People who received the book started to look into the source of Oktar's photos. It's not cheap to publish an 800-page book full of photos you have to license. As it turns out, investigators determined many of the photos were used without permission. More amusingly, a few photos trying to show there has been no evolution in the insect kingdom were actually photos of fly-fishing lures. This isn't really how you science. It might come as no surprise that outside of the USA, the highest concentration of creationists is in Oktar's home nation of Turkey.

THE SECOND LAW OF THERMODYNAMICS PROVES ORDER CAN'T EMERGE

Creationists claim nature, without the constant tweaking of a god, can only go from order to disorder. The Second Law of Thermodynamics proves that! By this logic very well-ordered snowflakes could not form. The Second Law applies to closed systems without energy being inputted into the system. The Sun-Earth system is not a closed system. Order can emerge locally at the cost of disorder someplace else. One wonders how a preacher with a PhD from a diploma mill figured this barrier out when no physicist ever once looked at the Second Law of Thermodynamics and published a landmark paper describing the same problem. People who offer this claim do so out of an arrogance that assumes they understand complex topics in physics better than actual working physicists.

THE VAST MAJORITY OF MUTATIONS ARE FATAL

A key facet of evolution is descent with modification. Modification comes, partially, from genetic mutations. Creationists assert the vast majority of mutations are fatal. Therefore, it should be almost impossible for any species to develop enough beneficial mutations to evolve into a new species. The reality is most mutations are neutral and beneficial mutations are commonly observed in the lab.

IF WE ARE DESCENDED FROM MONKEYS, WHY ARE THERE STILL MONKEYS?

It's amazing how often this one comes up despite good answers for all reading levels being a Google away. It is a straw man idea of evolution. It's predicated on the ideas that parent species immediately die out after a new species emerges and evolution is a straight line. One species begets another in a linear fashion. This is akin to saying, "if that man is your father and the guy next to him is your uncle as you claim, why are they still alive?" Also, evolution doesn't suggest humans evolved from monkeys. Humans and monkeys share a common ancestor.

WERE YOU THERE?

Ken Ham of the Answers in Genesis group seems to believe this is the ultimate mic drop. The implication is science is limited to only that which can be witnessed in real time. Since no one has seen evolution in action (we have), you cannot say with certainty it actually happened. Therefore, creationist perspectives are as valid! This is like saying unless a crime has eyewitnesses, you can't convict someone based on overwhelming evidence of his/her guilt.

WHAT USE IS HALF AN EYE?

Similar to the problem creationists have with the evolution of the cell, creationists believe complex structures like eyes or the bacterial flagellum can't evolve step wise. Darwin, anticipating such objections, worked out eye evolution in *The Origin of Species*. He showed that in nature there are examples of eyes at every stage of evolution, from simple cells that are sensitive to light to animals with increasingly complex adaptations that lead to the mammalian eye.

Ten Long Debunked "Proofs" of a Young Earth

Most creationists are young Earth creationists. Young Earth creationists posit life, Earth, and the universe were winked into existence by a Harry Potter like god about 6,000 years ago. The 6,000-year figure comes from a 17th-century calculation by James Ussher (died 1656). Naturally, that assertion defies dozens of lines of evidence that the earth, the solar system, and the universe are far, far older. But they have their "proofs".

THERE'S NOT ENOUGH DUST ON THE MOON

Creationists claim an old Moon would have accumulated so much dust over billions of years, the Apollo astronauts would have never been able to land and walk around on the moon's surface. The LEM would have quickly sunk into the dust, entombing the astronauts. And yet, as the astronauts discovered, the moon only has a few inches of dust. Obviously, the moon, the earth, and the solar system are only thousands of years old. The problem with this is the amount of dust expected on the moon is based on a single calculation by geologist Hans Pettersson from the 1950s. He went to the top of a mountain, used an air filter to collect dust, and based his calculation on this single sample. His estimate was orders of magnitude too high. Geologists contemporary to Pettersson's time came up with a much lower accretion rate. Earth itself has actually only accumulated less than 2 inches of space dust over its lifetime. This claim took root in creationism in 1974 with the publication of Henry Morris' *Scientific Creationism*. Morris relied on Pettersson's outdated calculations, ignored contemporary ones, and failed to reveal Pettersson offered a range of accretion rates. Morris simply picked Pettersson's top estimate.

THE MOON'S RECESSION RATE PROVES A YOUNG EARTH

This one comes to us from Kent "Doctor Dino" Hovind. His idea here is the moon is receding from the earth at a rate of 6 inches per year. If you go back only 1 million years, the moon would be a mere 95 miles away from Earth. The tidal forces would have wiped out all life on Earth. The problem with this argument is Hovind doesn't grasp that the moon and Earth were one body during the early years of the solar system's formation. The earth was hit by another body causing a chunk of Earth to coalesce into the moon. The moon likely coalesced out of this cleaved off matter about 150,000 miles from Earth. It was never that close and there's no logical or scientific reason for Hovind's extrapolation.

THE RATE THE EARTH IS SLOWING DOWN PROVES A YOUNG EARTH

Creationists assert the earth is slowing at a rate of one second per year per year. If the earth were really over 4 billion years old, the length of Earth's day would be about ten minutes back then. The planet would be spinning so fast Earth would have flattened out like a pizza. Since we're living on an oblate spheroid and not a flat Earth, the earth must be young! This "disproof" is courtesy of Walter T. Brown, a young earth creationist who runs an outfit called Center for Scientific Creation. A major problem with Brown's claim is the earth is not slowing at a rate anywhere near one second per year per year. The actual measured value is 0.005 seconds per year per year. No one is quite sure where Brown got the one second figure. The two sources he cites for that figure are an issue of *Popular Science* magazine and an issue of *Readers' Digest*. Putting aside these are not reliable sources for specific scientific data, neither of the works he cited supports his value. Both sources give the correct figure of 0.005 seconds per year per year.

THE DECAY RATE OF EARTH'S MAGNETIC FIELD PROVES EARTH CANNOT BE BILLIONS OF YEARS OLD

Thomas Barnes (died 2001) was a relativity denialist and a young Earth creationist. His greatest contribution to creationism was his enduring claim that if you take the decay rate of the earth's magnetic field, known since the early 19th century, and extrapolate back, the earth's magnetic field would be so strong millions of years ago life on Earth could not possibly exist. He published his findings in a 1973 book titled *Origin and Destiny of the Earth's Magnetic Field*. Barnes' claim has been identified as problematic on several fronts. He made the unwarranted assumption the decay rate has been uniform. He assumed measurements from the early part of the 19th century were as accurate as today's. He didn't take into account error bars. Finally, he based his idea on a 1934 theory of Earth's magnetic core that was discarded by science 16 years before Barnes published his own book.

IT'S NO COINCIDENCE THE EARLIEST HISTORICAL RECORDS GO BACK 6,000 YEARS

This one is frequently trotted out by Kent Hovind. If one defines historical records as ancients recording their own history via a writing system, then that necessarily goes back to the invention of writing in the 4th millennium BC. However, humans have been recording their activities for much longer. For example, the Jiahu symbols have been dated to the 6th millennium BC. Hunters recorded their activities via cave paintings that are over 20,000 years old.

THE OLDEST TREE IN THE WORLD DATES TO THE TIME OF NOAH'S FLOOD 4,000 YEARS AGO

Creationists believe Noah's flood ended about 4,000 years ago. So, it's not remarkable to them that the oldest trees date back to the time the flood waters were supposed to have receded. The problem is that's simply not true. There are a handful of living trees that are far older. For example, there's a 9,550-year-old spruce in Sweden. There are also clonal colony trees far more ancient. In Utah there is a clonal colony of trees that shares the same root system. It has been a going concern for about 80,000 years.

POLYSTRATE FOSSILS PROVE THE GEOLOGICAL COLUMN WAS FORMED RAPIDLY

Creationists will point to the existence of what they claim are fossilized trees that are found running through multiple layers of the geological column. How could a fossilized tree be present through layers that were supposedly laid down over millions of years? To creationists, this is evidence that the geological column was formed rapidly by Noah's flood. Polystrate fossils are no mystery to scientists, however. These fossils were well understood by 19th century geologists. These trees are an example of a rapid mineralization process that is similar to the petrified well phenomenon. Polystrate fossils are indeed found in layers that were rapidly laid down (rapidly from a geological time frame). However, the layers are not from a global deluge but by a nearby river that slowly overflowed its banks and started to bury the tree in sediment. These kinds of fossils are universally found in formations featuring coarse-grained rocks, indicative of the type of flood.

Polystrate fossil from the Joggins Formation in Nova Scotia.
Photo Credit: Michael C. Rygel via Wikimedia Commons

POPULATION GROWTH RATES BETTER SUPPORT A YOUNG EARTH

An outfit called Creation Ministries International asserts if modern humans emerged 1 million years ago, given an average growth rate of the human population at 0.5% per year, there should be trillions of humans. However, if the Bible is correct and there were only six surviving humans 4,500 years after the flood, a 0.5% growth rate would net out to about the six billion people we have on Earth today. It's math, atheists! Where Creation Ministries International got this average growth rate is a mystery. It appears they simply began with a starting point of six people, a 4,500-year time frame, the end point, and then calculated the required yearly average growth rate. That creationists simply made up a number to fit a hypothetical curve is the least of this young Earth proof's problems. It's pretty clear from historical and archeological records there was a lot of building going on 4,500 years ago. It's unlikely six people were able to build the Great Pyramid of Giza, the Pyramid of Khafre, and the subterranean Hypogeum temple on Malta some 4,500 years ago. It's unclear how six people did that and then created enough cultural products to make historians think there was a flourishing civilization in the Indus Valley at the time. Oh, China records the Xia imperial dynasty from that time. You can't have an Emperor without subjects. Lots and lots of subjects.

RECENTLY FORMED POLONIUM HALOS IN ANCIENT GRANITE PROVE A YOUNG EARTH

Granite is rock that comes from magma that has cooled down over millions of years. It's pretty old. Polonium is a radioactive element with a half-life of 3 minutes. Young Earth creationist Robert Gentry (died 2020) thought he found traces of decayed polonium (in the form of ring-shaped "halos") in Precambrian granite. In Gentry's view, it is impossible for this short-lived element to become trapped in ancient granite during its molten phase, decay, and then mark the inside of the rock with a tell-tale halo. Gentry published his findings in his 1992 book *Creation's Tiny Mystery*. Creationists really loved this for three reasons. First, it was a genuinely new claim. Creationism has been recycling the same old debunked "disproofs" for decades. Second, Gentry had a legit degree in nuclear physics. Third, this problem with young polonium halos in old rocks was super science-y! Checkmate, atheists! Gentry failed to recognize polonium is a byproduct of radon gas which is ever present where you find granite. He didn't take into account granite has numerous cracks and fissures radon gas can enter, decay, form polonium, which then decays and leaves halos in the rock.

"TIRED LIGHT" PROVES THE UNIVERSE IS YOUNG

It's pretty inescapable that galaxies that are millions of light years away require a universe older than 6,000 years. There are many ways creationists attempt to deal with this. Some simply suggest their god created the light in transit. Other creationists are uncomfortable with that dodge as it suggests an inherently dishonest god. Another approach is to posit the universe is much, much smaller. Using Cepheid stars to measure the distance to galaxies is flawed in some handwavy way. Those unwilling to devote time to disprove the basic physics behind the Cepheid star method have only one thing left to them: maybe light has slowed down over the years. This is variously known as "c-decay" or "tired light". It was first proposed in a 1981 book by Barry Setterfield titled *The Velocity of Light and the Age of the Universe*. Setterfield argues 6,000 years ago light was millions of times faster. For some reason, in only six thousand years, light decayed at a rate never seen by any other natural phenomenon and, curiously, tapered off right around when we could accurately measure the speed of light. Setterfield failed to grasp that the speed of light isn't some kind of arbitrary speed limit set by a god. It's wrapped up in a little equation that reads $E = mc^2$. If you change the value of "c", you change the amount of energy released by stellar nuclear fusion. Six thousand years ago our sun would have output 800 million times more energy and cooked Earth, Adam, Eve, and that talking snake to a cinder.

Seven Scientific Careers Hampered by Proxmire's Golden Fleece Award

William Proxmire (died 2005) was a Democratic senator from Wisconsin. He was an influential senator on the Appropriations Committee. Between 1975 and 1988 he would bestow a monthly Golden Fleece Award to people and agencies he believed were wasting government funds. Occasionally he'd award it to reputable scientists receiving public research dollars. It was clear Proxmire took no time to understand the research. He and his staff looked for studies that just seemed silly or inconsequential and then would allege the researchers were lining their pockets with government money and turning in nonsense work. A number of scientists given the award found their careers and ability to apply for future funding impeded by the sudden notoriety. Here are seven scientists who fell victim to Proxmire's "award".

ELLEN BERSCHEID AND ELAINE WALSTER (1975)

Ellen Berscheid and Elaine Walster were the inaugural "winners" of Proxmire's Golden Fleece award. Berscheid and Walster had spent over a decade studying interpersonal relationships and the nature of love. Their research has been useful in developing family counseling techniques. Proxmire felt they were "squandering $84,000 to try and find out why people fall in love". Like that's a bad thing. After Proxmire's award, the researchers were inundated with threatening letters and phone calls. Funding was cut off due to the stigma associated with the reward. The stress and embarrassment for being singled out as one of the supposed worst abusers of public dollars led to the end of Berscheid's marriage.

RONALD HUTCHINSON (1975)

Ronald Hutchinson was studying environmental and biological causes of aggression and the effects drugs, alcohol, and coffee can have on it. Part of his research was determining if jaw clenching in rats and monkeys was a way to measure anger and aggression. It's hard to study something if you can't come up with an objective way to measure it. Hutchinson's work was at times funded by NASA. Being able to understand the causes of aggression and measure warning signs were deemed critical if NASA was going to send people on long space voyages. Proxmire, not understanding the scope of the research, quipped in the text of his award "In view of the transparent worthlessness of Hutchinson's study of jaw-grinding and biting by angry or hard-drinking monkeys, it's time we put a stop to the bite Hutchinson and the bureaucrats who fund him have been taking out of the taxpayer." Proxmire went on various radio and TV shows and alleged Hutchinson was getting rich from these funds. He did understand the vast majority of any scientific grant gets paid directly to the university, not the researcher. Hutchinson's funding dried up. He didn't take it lying down. He sued the senator. The case actually made it all the way to the Supreme Court. The court ruled that "parliamentary privilege" does not extend to the press releases and newsletters that politicians issue. Proxmire paid Hutchinson $10,000 and offered a somewhat non-apology from the Senate floor.

HARRY RUBIN (1975)

Harry Rubin was a professor of medicine at Southern Illinois University. At the behest of the Department of Health, Education, and Welfare he had completed a study on how alcohol affects sexual response. Given that in the early 1970s it was estimated some 24 million Americans had tried pot, a similar study of marijuana was in order. In 1975, the department earmarked Rubin additional funding. The funding was in keeping with the White House's own directives on studying the effects of drugs and pornography on sex crimes. Rubin's research was intended to establish ways to objectively measure baselines. To not put too fine a point on it, it did involve giving male subjects pot, showing them pornographic movies, and then measuring certain physiological changes in their penis. Proxmire was not impressed. He gave Rubin a Golden Fleece award. The press and politicians had a field day with the so-called "sex-pot study" and allegations public dollars were being spent to build a "federal love machine". Rubin's funding was cut off by order of Congress and Rubin's family was threatened.

ROBERT JOHNSTON AND ROBERT KRAUT (1980)

In 1980, Proxmire awarded social psychologists Robert Johnston and Robert Kraut a Golden Fleece award for government funds supposedly being used to figure out why bowlers smile. Many primates bare their teeth as a threat display. Humans are the odd ones out. We show our teeth, via a smile, as a sign of happiness. It has been postulated smiling in humans has developed as a social cue, versus a physiological response to something joyous. Johnston and Kraut felt bowlers offered an excellent test. Bowlers spend some amount of their time with their backs turned to teammates, friends, and associates. A strike or spare should be an outcome that produces a joyous physiological response. If smiling is a physiological response, they should break out into a smile upon getting the strike or spare. What Johnston and Kraut found was that bowlers were more likely to break out into a smile when they turned to face their teammates and friends. This supported the notion that smiling is a social phenomenon. Many had noted that Proxmire tended to overwhelmingly pick on social scientists like Johnston and Kraut because their research was somewhat easier to read. Papers in hard sciences were likely incomprehensible to Proxmire and his scientifically illiterate staff. When researchers like Johnson and Kraut figured this out, a response was to start making their papers *less* readable by the general public and people like Proxmire.

MICHAEL DOMJAN (1988)

In the 1980s, psychologist Michael Domjan was studying mate selection. His subject was the Japanese quail. This kind of research can have a number of practical uses, from understanding how to increase the chances of endangered species mating to helping farmers increase mating behavior among livestock. Proxmire couldn't grasp any of that and dropped one of the last Golden Fleece awards on Domjan before the senator retired. Domjan didn't quite relish his research suddenly being made fun of in the press. He believes it hampered his ability to present his research effectively at conferences.

Eleven Truly Remarkable Works of Science that Won an Ig Nobel Prize Between 2010 and 2020

The Ig Nobel Prize today is awarded for "science that first makes you laugh and then makes you think". The Ig Nobel Prize has become a booster and ally for the kind of science William Proxmire lambasted with his Golden Fleece award. Good science doesn't start off trying to build a better consumer product or a new wonder drug. Research can go off in seemingly odd directions but can have future real-world applications if you spend a bit of time thinking about it. Interestingly the Ig Nobel Prize wasn't always a booster for "curious" science. It started off in the early 1990s mocking bad science. For example, the 1991 Chemistry award went to homeopath Jacques Benveniste. The economics prize went to Michael Milken. By the early 2000s, the prize does seem to have changed focus. Here we'll look at some interesting papers from the Ig Nobel's "make you laugh and make you think" era.

ENGINEERING: USING A REMOTE HELICOPTER TO COLLECT WHALE SNOT (2010)

Whales get diseases. It's not easy to get them to hold still to get a syringe in them to take a blood sample. Researchers Karina Acevedo-Whitehouse, Agnes Rocha-

Gosselin, and Diane Gendron pioneered a method for using a drone to collect "exhaled breath condensate" as a way to diagnose if a whale pod is infected with a disease that could lead to population decline.

MEDICINE: PEOPLE SOMETIMES MAKE BAD DECISIONS WHEN THEY NEED TO PEE (2011)

The Medicine prize went to two groups of researchers. One group was able to demonstrate people required to control their bladders performed better on tasks associated with impulse control. For example, when trying to hold it in, people made better monetary decisions. This has implications beyond asking people to pick stocks while doing the pee pee dance as people tend to make bad choices under pressure from visceral drives. For example, it's always a bad idea to go grocery shopping when you're hungry. Discovering ways to control impulsive behavior in the face of powerful drives could have profound positive effects on human health. Another group identified negative effects on attention and working memory. The degradation in performance was on par with someone being awake for 24 hours. Studies like this can contribute to ways to decrease accidents among people who may not get convenient bathroom breaks like truckers, pilots, and factory workers.

NEUROSCIENCE: DEMONSTRATED BRAIN RESEARCH TECHNIQUES USING AN MRI COULD SHOW BRAIN ACTIVITY IN A DEAD SALMON (2012)

Neuroscience has been using functional MRI (fMRI) studies to correlate blood flow in the brain to brain activity. The idea is when neurons become active, more blood flows to those neurons. One could hypothetically predict what a person is about to do before they do it based on the neurons that start to activate. A major issue with fMRI is there is a lot of noise in the data, some of it caused by the machine itself. Researchers Craig Bennett, Abigail Baird, Michael Miller, and George Wolford put a dead salmon from the grocery store into their machine to establish a baseline. They went through all the normal steps to test a human subject: asked it questions and showed it pictures of social situations. Again, this was a dead salmon. The source of the data naturally was purely random noise from the machine. However, they took an extra step. They ran the data through statistical correction methods used by about 40% of researchers. These corrections are usually done to eliminate false positives. When run through these corrections, they were able to demonstrate statistically significant results. In a dead salmon. Their paper had a real impact on how fMRI data was corrected. Within two years of publishing only 10% of studies were using problematic statistical correction identified in the dead salmon study.

ARCHAEOLOGY: A SHREW'D APPROACH TO SELF EXPERIMENTATION (2013)

Sometimes scientists are required to eat their own dog food. An experiment might be so dangerous or onerous, it would be hard to find a motivated test subject. So the scientist tests his/her hypothesis on him/herself. Such was the case for researchers Peter W. Stahl and Brian D. Crandall who were trying to answer a question: if ancient humans were consuming small animals whole, what bones should we expect to be fully digested and what bones should be left over in middens? This would naturally require a human to swallow, whole, a dead animal. So, they stepped up, swallowed a parboiled shrew, and then later examined their own feces. They published their findings in their paper "Human Digestive Effects on a Micromammalian Skeleton".

PSYCHOLOGY PRIZE: THE MENTAL HEALTH EFFECTS OF OWNING A CAT (2014)

Jaroslav Flegr, Jan Havlíček, Jitka Hanušova-Lindova, David Hanauer, Naren Rama Krishnan, and Lisa Seyfried shared in this prize for their work on an actual potential public health concern. Is it mentally hazardous for a human being to get infected with a common cat parasite called *Toxoplasma gondii*? When the parasite infects mice, mice have a reduced fear of cats, making it far easier for cats to consume the mice and then get infected with *Toxoplasma gondii*.

Photo Credit: Karl Mamer

Cats are the parasite's ultimate medium for reproduction. There have been some hypotheses that *Toxoplasma gondii* infections in humans, with cats as the vector, can affect behavior in dangerous ways. The scientists won an Ig Nobel for research that suggests a link between *Toxoplasma gondii* and depression as well as lower IQs in men.

CHEMISTRY PRIZE: HOW TO UNBOIL AN EGG (2015)

Researchers Callum Ormondel, Colin Raston, Tom Yuan, Stephan Kudlacek, Sameeran Kunche, Joshua N. Smith, William A. Brown, Kaitlin Pugliese, Tivoli Olsen,

Mariam Iftikhar, and Gregory Weiss won a 2015 Ig Nobel for figuring out a way to, in essence, unboil an egg. When you boil an egg, the proteins in the egg unfold and then refold into a tangled mess. Scientists unfolding proteins to make pharmaceuticals sometimes find their proteins fold in ways they did not intend. To counter that, they find processes that are difficult and expensive but severely reduce the chances of misfolds. If there was a fast and simple method for reversing unintended folds — a biological control+z key — drug companies could employ cheaper methods and then quickly fix up the misfolded proteins.

ECONOMICS: DETERMINING BRAND PERSONALITIES OF ROCKS (2016)

Marketers had long talked about a brand's personality, but no one had really tried to define it and figure out a way to measure it. In 1997, social psychologist Jennifer Aaker conceived of the idea that the so-called Big Five Human Personality Scale could be applied to commercial brands (like Apple or McDonald's) and be measured. Her measurement methods became very popular with advertising agencies. However, over the years, people began to suspect some part of a brand's personality as measured by Aaker's framework was an artifact of the testing process, not people's actual perception of a consumer brand. To test this, researchers Mark Avis, Sarah Forbes, and Shelagh Ferguson applied Aaker's methods to rocks. Rocks are not brands. No one is exposed to advertising for rocks. They have no commercial value to most people. They found Aaker's methods generated brand personalities for rocks, suggesting some or all of a brand's actual personality is an artifact of the testing method.

ECONOMICS: EFFECTS OF CROCODILE CONTACT ON CASIO GAMBLING (2017)

Matthew Rockloff and Nancy Greer won an Ig Nobel for their paper "Never smile at a crocodile: Betting on electronic gaming machines is intensified by reptile-induced arousal". Problem gambling can lead to a wide variety of negative social ills. To help understand how arousal can affect a person's gambling behavior the researchers needed something that would produce a highly negative arousal affect in those identified as problem gamblers. Making a person hold a crocodile seemed as good a way as any. They found problem gamblers who were in a state of negative arousal (having been forced to hold a crocodile) were more likely to bet less.

NUTRITION PRIZE: CALORIC INTAKE FROM EATING HUMANS (2018)

James Cole in his paper "Assessing the Calorific Significance of Episodes of Human Cannibalism in the Paleolithic" tried to measure the calories offered by a human body. There's debate within the scientific community about the role of cannibalism in early *Homo sapiens* and Neanderthals. Were humans and Neanderthals eating each other for calories or was it ceremonial? Cole's paper tries to establish humans are, relative to other large fauna available to Paleolithic hunters, not the most efficient animals to hunt and consume. This lends support to the notion ancient cannibalism was ceremonial and our early human ancestors, including the Neanderthals, had much more complex beliefs and societies than is commonly held.

MEDICAL EDUCATION PRIZE: HELPING SURGEONS CLICK (2019)

Doctors Karen Pryor and Theresa McKeon won an Ig Nobel for demonstrating you could train up orthopedic surgeons using a clicker, a device usually used by pet owners to train their dogs and cats. Apparently, it works nearly as well on surgeons. The clicker provides a clear reinforcing signal that is independent of language, judgment, or vocal tone.

MATERIALS SCIENCE PRIZE: YOU CAN'T MAKE A KNIFE OUT OF FROZEN POOP (2020)

Metin Eren, Michelle Bebber, James Norris, Alyssa Perrone, Ashley Rutkoski, Michael Wilson, and Mary Ann Raghanti published a paper titled "Experimental replication shows knives manufactured from frozen human feces do not work". The paper sought to test an often-reported ethnographic account from the 1950s of an Inuit man who made a knife out of his own feces and used it to kill a dog and turn it into, literally, a dog sled. They were unable to create a feces knife capable of slicing through a hunk of meat (not dog meat, however). The study suggested that given government policy directed at Inuit populations are partially based on anthropological findings, there are risks tainting the scientific literature with what amounts to folklore.

Six Examples of the Nobel Disease: Scientists Who Won Nobel Prizes and Turned to Pseudoscience

BRIAN JOSEPHSON

Brian Josephson won a Nobel prize in Physics in 1973 for his work in the field of superconductivity. He seems to have used his Nobel money to pursue his interest in Transcendental Meditation and the paranormal. At one point Josephson vouched for the reality of "psychic" Matthew Manning's ability to use telekinesis to move a compass needle (an easy stage magic trick). He is also an early proponent of tying quantum mechanics to psychic phenomena.

LUC MONTAGNIER

Luc Montagnier won a Nobel in Physiology or Medicine in 2008 for discovering HIV. Within a few years of receiving his Nobel prize, he became convinced DNA emits a low frequency electromagnetic wave that imprints itself on water. He believes you can record these supposed EM waves and then blast them at another vial of water. The targeted water will then replicate the DNA's structure. Montagnier has also signed on with elements of the anti-vaccine crowd. In 2012, he was a speaker at Autism One, which is a kind of the Lollapalooza for cranks who still think vaccines cause autism.

KARY MULLIS

Kary Mullis (died 2019) won a Chemistry Nobel in 1993 for his invention of the truly remarkable technique of amplifying DNA via the polymerase chain reaction (PCR). He was also an HIV denialist. He made many comments in the 1990s indicating he believed the well-established reality HIV causes AIDS was a hoax. He believed doctors and scientists who participated in the 1970s "war on cancer" initiative made up HIV/AIDS to fuel another national paranoia to pay for their BMWs.

LINUS PAULING

Linus Pauling (died 1994) was the only person in history to win two unshared Nobel prizes. He won a Nobel prize in Chemistry 1954 and the Peace prize in 1962. In the late 1960s Pauling started advocating for the health benefits of megadosing vitamins. He coined the term "orthomolecular medicine" as a nice science-y label for the act of shoving vitamin-laden horse pills in your mouth. He single-handedly created a multibillion-dollar industry offering negligible benefit to its consumers. Pauling also advocated for eugenics. He felt people with genetic diseases like sickle cell should get a brand on their foreheads as a deterrent to people who might otherwise choose to mate with them.

WILLIAM SHOCKLEY

William Shockley (died 1989) is probably the poster child for the Nobel Disease. Shockley won a Nobel Prize in Physics in 1956 for his co-discovery of the transistor. Life would be radically different today if not for his work. After winning, he packed up his bags and moved to Palo Alto to start his own company. Why Palo Alto? It was close to his mom. It is said after winning the Nobel his personality changed. He became an autocratic and paranoid boss and drove away his best engineers (who went on to found Intel and transformed sleepy Palo Alto into Silicon Valley). He also became a raving proponent of eugenics and racism. He believed Black people should be paid money to sterilize themselves. In 1981, Shockley won a defamation suit he brought against the *Atlanta Constitution* newspaper. He won but he was paid only $1 in damages as his reputation was so far in the toilet the jury reasoned the newspaper's libel didn't do much to make his reputation any worse.

JAMES WATSON

No one has thrown, arguably, one of the greatest scientific careers in modern history in the toilet so effectively as James Watson. Watson won a Nobel Prize in 1962 for the discovery of DNA. Since then, he's made a lot of wingnut comments about race. In 2000, he argued there's a link between sex drive and the amount of melanin in your skin. In 2007, he started shooting his mouth off that Africa had dim prospects as Black people were less intelligent than white people. By way of evidence, he suggested simply talking to people who work with Black people. They'll tell you. Desk meet face. After his 2007 comments, Watson was "retired" from his position at the Cold Spring Harbor Laboratory.

Ten Odd Inclusions on the Discovery Institute's "A Scientific Dissent from Darwinism"

"A Scientific Dissent from Darwinism" is a statement put together by the Discovery Institute in 2001. It circulated it among scientists and the science adjacent (doctors, engineers) and asked them to sign on if they agreed with the following statement: *We are skeptical of claims for the ability of random mutation and natural selection to account for the complexity of life. Careful examination of the evidence for Darwinian theory should be encouraged.* Technically, the complexities of life are not the result of only selection acting on random mutation. There are other drivers of evolution like mate selection and genetic drift. Any evolutionary biologist could sign such a statement and still support the modern theory of evolution. Not many, however, fell for it. Some did. The most recent list has over 1,000 signatories. One might at least go "wow, 1,000 evolutionary biologists signed?" Well, not so fast. You'll find about 39 engineers, 55 chemists, 28 physicists, 21 mathematicians, and 7 computer scientists on that list. What they understand about evolutionary biology is questionable. It is exactly like 1,000 biologists offering their doubts about the efficiency of a new type of wing for an aircraft. They are certainly smart people but they are too out of their field to offer any relevant commentary on the relative advantages of different airfoil designs. Here are ten even more questionable signatories who seem not particularly qualified as experts on evolutionary biology.

BERNARD D'ABRERA

Bernard d'Abrera (died 2017) is described on the list as a "visiting scholar" with the British Museum. D'Abrera had a BA in History and Philosophy of Science. While he was a visiting scholar with the UK's Natural History museum, that isn't actually the British Museum.

GARY ACHTEMEIER

Gary Achtemeier is listed as a PhD in Meteorology. According to his LinkedIn he works as a Meteorological scientist for the US Forest Service. Meteorology is, of course, the study of the earth's atmosphere and various weather phenomena. It has very little interface with evolutionary biology.

RICHARD ANDERSON

Richard Anderson is listed as an Assistant Professor of Environmental Science and Policy at Duke. Impressive! Anderson, however, no longer works at Duke. And he's not a biologist. His PhD is in Geography and Environmental Engineering. He also has a M.S.E and a B.S.E. in Mechanical Engineering.

PHILLIP BISHOP

Phillip Bishop is listed as a Professor of Kinesiology at the University of Alabama. He appears to be a Professor Emeritus currently. His personal LinkedIn also lists him as a "Porfessor [sic] of Exercise Science" at Liberty University. Oddly he doesn't list his educational achievements on his LinkedIn page and even his U of A profile doesn't list it either. I'm confident he knows a lot about exercise. Who knows what this guy knows about evolutionary biology?

MARCOS N. EBERLIN

The list simply describes Marcos N. Eberlin as a "Professor" and a member of the Brazilian Academy of Science. Eberlin is actually a chemist. He also believes in the reality of Noah's ark and that it held 18,000 animals including dinosaurs.

BRADLEY R. JOHNSON

Bradley R. Johnson is listed as having a PhD in Materials Science. His institution is listed as the University of Illinois. It's true he is a materials scientist. He received his degree from the U of I, certainly. But he doesn't hold a position there as the list might imply. His career, while illustrious, has had nothing to do with biology.

MICHAEL J. KAVAYA

Michael J. Kavaya is listed as a "Senior Scientist" for NASA. Cool. So, what does he know about evolutionary biology? Maybe not a lot. He seems to be helping NASA study wind.

FORREST MIMS

Forrest Mims has no formal scientific training, though he's listed as an Atmospheric Researcher. He holds an undergraduate degree in the liberal arts. He became interested in electronics and wrote several best-selling books for electronic hobbyists. His books were sold by Radio Shack for a time. In terms of his atmospheric research, he seems active in figuring out ways to measure ozone in the atmosphere and has invented several devices to do just that. He's certainly done a lot in his life but he's not an authority on the evolution of life.

DENNIS D. RATHMAN

Dennis D. Rathman is listed as a staff scientist working at MIT's Lincoln Laboratory. MIT! Those guys are smart right? But is he an evolutionary biologist? No. He has a doctorate in Physics and seems to be working on creating computer chips.

HANSIK YOON

Hansik Yoon has a PhD in Fiber Science. What is that? That's the world of textiles. His publication record indicates he's doing some serious next level shit in the field of metasurface nanostructures. But nothing in his education or field of research qualifies him to assess the reality of evolution.

Fifteen Taboo Topics and Their Tabooness Rank

Scientists seeking grants for a particular line of research sometimes refer to having to overcome something they call "the giggle factor". Topics like sex (notably sex in space), the search for extraterrestrial intelligence, and even asteroid defense are generally given as examples of topics high on the giggle factor scale. Woo topics like researching Bigfoot and ghosts are so high on the giggle factor scale that woo proponents claim science has deemed these topics taboo. In 1982, psychologist Robert Rosenthal (the man who coined the term "File Drawer Effect" in 1979) investigated scientists' attitudes toward woo topics to determine what topics were high on "tabooness". His findings were as follows (in order of "most taboo" to "least taboo"):

1. Ghosts
2. Alien Abductions
3. The Loch Ness monster
4. Astrology
5. Bigfoot
6. Religion
7. Reincarnation
8. UFOs
9. Psychokinesis
10. Graphology
11. Precognition
12. Hypnosis
13. Language in apes
14. Mental telepathy
15. Self-fulfilling prophecies

OILS
POWER
HERBS
ALTERNATIVE
MEDICINE

Seven Deadly Alt Med Practices

When people dip into alt med as a "compliment" to conventional medical treatment, many wonder "oh, what's the harm?" The harm is twofold. First, the complementary treatment is simply unsafe. Second, the patient forgoes conventional medical treatment in favor of a useless alt med treatment. Studies indicate people who forgo or delay conventional medical treatment for cancer have a death rate six times greater than patients who follow the science. These are some of the most notorious, and deadly, alt med practices.

CHELATION THERAPY

Chelation therapy is a legitimate medical intervention for people poisoned by heavy metals. Chelation therapy involves the injection of chemicals into the bloodstream. The chemicals bond with heavy metals like lead and mercury. They then get excreted in the urine. The anti-vaccine crowd believes autism is a result of heavy metal exposure, supposedly from both childhood vaccines and our super toxic environment. Crank doctors began offering chelation therapy to desperate parents of autistic children a number of years ago. There have been at least two deaths linked to the alt med use of chelation therapy. In one case, crank doctors killed a 5-year old autistic boy. In another case a 53-year woman who believed she was suffering from cardiac problems had chelation therapy administered by her naturopath. She died within minutes of the drugs being administered.

COFFEE ENEMAS FOR CANCER

There are many suggested therapeutic effects for enemas in the world of alt med. The worst is Gerson therapy's promise it will cure your cancer. Like most alt med claims, this one is based on the idea that illness is caused by toxins and the cure is getting those toxins out of the body. Max Gerson was convinced administering coffee enemas was the best way to remove cancer-causing toxins. Although Gerson died in the late 1950s, his therapy was continued by others after his death. Several deaths from infections have resulted directly from the treatment. And, of course, many people with cancers that can be effectively cured via legitimate medical treatments have died telling themselves Gerson therapy is totally working. Jessica Ainscough, Australia's so-called "Wellness Warrior" blogger, is probably the most high-profile victim of substituting Gerson Therapy for actual science-based medicine. In 2008 Ainscough was diagnosed with a soft tissue cancer in her arm. Her oncologist recommended amputating her arm. Instead she turned to Gerson therapy. In her blog she boasted the therapy was not only working but she could see her cancer tumors emerging through her skin and disappearing. Contrary to what she was telling people on her blog, her cancer was spreading. When it became obvious to her Gerson therapy was not working, she did seek conventional medical therapy but by then it was too late. She died of cancer in 2015.

JILLY JUICE

Jilly juice is a drink made from salt and fermented cabbage. It's the creation of Jillian Mai Thi Epperly who promoted its ostensible healing powers on her Facebook group in 2017. Her group attracted some 50,000 followers who seemed to fervently believe in her questionable medical theory that most disease is caused by a fungus called *candida*. The way to fight candida was to drink gallons and gallons of Jilly juice. Not only will Jilly juice cure what ails you, but it will apparently retard the aging process and turn gay people straight. Bruce Wilmot was a follower of Epperly's Facebook group and decided Jilly juice was the way to treat his pancreatic cancer. He downed mason jar after mason jar of the stuff believing it would cure him. It didn't. He died.

OZONE INJECTIONS

Using ozone to disinfect drinking water was used in Europe in the 19th century. For a brief period, it was used to disinfect wounds. Ozone's use as a medical disinfectant was quickly discontinued when it was observed ozone killed human tissue as readily as it killed pathogens. Despite medical science knowing for more than a century that ozone does more damage than good, people happily pay naturopaths to be poisoned. The treatment was banned by the FDA in 2016. Naturopaths continue to offer it in Germany. Naturopaths claim ozone, if injected subcutaneously or through the vagina, can cure cancer. Since 2012, five people in the USA seeking a miracle cancer cure have died from ozone treatments.

REBIRTHING

There's this idea that if you can simply be born again, and not in the religious sense, well maybe it offers a chance for a do over. If conducted in a safe manner, it's goofy looking but harmless enough. However, rebirthing turned deadly when rebirthing therapists tried to cure a 10-year-old girl of an attachment disorder by wrapping the girl in sheets and covering her with pillows. For 70 minutes, despite the girl's cries she couldn't breathe, they tried to push her from this simulated womb. They ended up suffocating and crushing her to death.

SWEAT LODGES

Who doesn't love spending a bit of time in the sauna? Hot, steamy air can be comforting, especially in winter when you have a head cold. Of course, the alt med crowd will take anything that's simply relaxing and turn it into a money-making operation offering all kinds of bogus health claims. If they can then tie it together with "ancient wisdom", all the better. Such is the case with the rise of sweat lodges in the late 1990s and early 2000s. New Agers took the practices of indigenous Americans and re-packaged them as ways to detox and attain some kind of heightened spiritual awareness. So, basically, they've found ways to get rich white people to pay thousands of dollars to sweat in a hut in the middle of a desert. In one high profile case from 2009, three people died after they participated in James Arthur Ray's sweat lodge retreat in Arizona. In another case an Australian man died in 2004 after he entered a sweat lodge. Although he started to exhibit signs of distress, his friends assumed he was astral traveling. In total, between 1996 and 2009, there have been at least eight deaths attributed to these New Age sweat lodge practices.

WHOLE BODY CRYOTHERAPY

Whole body cryotherapy involves entering a small chamber cooled to -148°F for two to four minutes. Alt med proponents assert it is effective for sore muscles, weight loss, and depression. Science has found no evidence that exposing your naked body to extremely low temperatures has any beneficial effect other than transferring $75 from your wallet to the cash register of the spa owner. Whole body cryotherapy started in Japan in the late 1970s. It did not become popular in the USA until the 2010s. The use of a cryotherapy chamber has resulted in at least one death in 2015 when a spa worker, using a cryo chamber alone and after hours, got trapped. Interestingly, she actually suffocated to death before freezing to death. Her coworker found her frozen body the next morning.

Eleven Bizarre or Questionable Medical Practices

ALKALINE DIET

Your body is pretty good at keeping your pH level pegged right around where it needs to be for good health. There's a branch of alt med that believes many diseases (from cancer to heart disease) are really caused by your body's pH level getting out of whack and becoming too acidic. The treatment is to eat a diet of "alkaline foods" like fruit and vegetables and avoid "acidic" foods like meat and cheese. Although a diet with a lot of fruits and vegetables is generally a good thing, proponents of the alkaline diet don't stop there. They suggest one can cure cancer by eating an alkaline diet. Although one might be able to change the pH level of your urine based on what you're eating, the reality is you can't change your body's pH level through food. Real dieticians call the idea bonkers.

APPLIED KINESIOLOGY

Applied kinesiology should not be confused with kinesiology. The latter is a legitimate science while the former is quackery. People who practice applied kinesiology promote their ability to diagnose illness by tugging and pushing on people's limbs and poking their muscles. Based on the resistance the quack thinks he or she detects while pushing on the patient's arm or leg, he/she can diagnose diseases and what allergies a patient might be suffering from. If this sounds like limb chiropractic, you wouldn't be wrong. Applied kinesiology was invented in 1967 by chiropractor George Goodheart (died 2008). Studies of inter-rater reliability (i.e., do two or more practitioners come up with roughly the same diagnosis?) have shown no correlations better than chance.

COLORPUNCTURE

Colorpuncture is a pseudoscientific belief that you don't have to stick needles in acupuncture points, you can just stimulate them with colored lights. Colorpuncture was invented by German naturopath Peter Mandel in the 1980s. Mandel thought cells in the body communicate with each other through photons. Those who practice colorpuncture believe they can then inject light into cells with basically a multicolored flashlight that will tell the cells to communicate with different areas of the body and promote healing. Even weirder, naturopaths who believe in colorpuncture will use Kirlian photography like a doctor might use an X-ray.

EAR CANDLING

There are many bizarre ways people think they can remove toxins from your body. Not many involve lighting a fire over your ear. The claim is if you light a special hollow candle placed over your ear, it creates a vacuum and will remove built-up wax in your ear, clear your sinuses, and, of course, detox you. The problem is the candles don't create any kind of vacuum. Even if they did, the vacuum needed to suck sticky wax out of your ear would be tremendous and probably damage your ear in the process. While no one has ever died burning wax right next to their face, numerous ear candlers have experienced the joy of burns, having hot wax drip down their ear canal, and requiring multiple visits to the doctor to have the candle wax removed.

FUNCTIONAL MEDICINE

Functional medicine seems to be a recent catch all term used by those who practice various forms of alternative medicine. The term originated with chemist Jeffrey Bland in 1991 when he started his Institute for Functional Medicine. People practicing functional medicine suggest their approach is to understand the cause of disease and treat the individual. This is in contrast to the approach of science-based medicine which seeks to understand the cause of disease and treat the individual.

HAIR ANALYSIS

There's real hair analysis which can be used as a noninvasive way to determine things like exposure to heavy metals. Naturopaths, however, use it as a Ouija board for diagnosing all manner of illness and nonexistent allergies. It has been used by the alt med industry since at least the 1980s. In 1985, Dr. Stephen Barrett, author of the *Quack Watch* website, published an article in the *Journal of American Medicin*e detailing his investigation of labs offering hair analysis. He sent hair from two healthy teenagers to thirteen labs offering the service. Each came back with wildly different findings and recommendations for food supplementation to correct imagined problems like kidney disease.

LYMPHOTHERAPY

Lymphotherapy is the idea you can treat disease by massaging the lymph nodes. The technique was created by a French doctor in 1918 and was for a time popular in Italy in the 1960s. You will still find alt med types that promote it today as a cure for insomnia, fatigue, headaches, cellulite, and pain caused by pregnancy. Alt med types, never tired of finding ways to kill us all, promote it as an effective way to prevent being infected with the flu.

MIRACLE MINERAL SOLUTION (MMS)

It would seem people are willing to drink bleach if they're told drinking bleach is a cure big pharma doesn't want you to know about. Such is the case with chlorine dioxide, which has been repackaged by alt med types as "Miracle Mineral Solution". The rebranding of bleach as a miracle cure all is primarily the efforts of former Scientologist Jim Humble. In 2006 he published a book called *The Miracle Mineral Solution of the 21st Century* and started selling bottles of the crap. MMS, if quaffed in large quantities, supposedly makes one proof against HIV, autism, Lyme disease, Parkinson's, cancer, and acne. If you're not outright killed by drinking MMS, the only effect of drinking it is diarrhea and vomiting. Of course, proponents assure the person poisoning themselves those are signs MMS is working. Humble converted his MMS business into a church and now takes the position that MMS cures nothing but merely "supports" the body's own super amazing ability to cure itself of HIV and Parkinson's.

NEURO-LINGUISTIC PROGRAMMING (NLP)

NLP was developed in the 1970s by Richard Bandler and John Grinder as a form of psychotherapy that can treat phobias, psychosomatic illnesses, and learning disorders. Others have wildly expanded the scope of NLP by suggesting it can improve eyesight and prevent the common cold. NLP involves several different techniques in an effort to reprogram the human brain. One technique involves something called "swish". A person about to embark on some unwanted course of action is taught to learn a visual cue that lets him/her disrupt their behavior. Another technique involves recalling past traumatic experiences but in a dissociative state. Unlike a lot of alt med treatments, NLP has a veneer of plausibility. Many of its techniques seem reasonable. However, proponents' claims of being able to treat a wide range of mental and physical disorders with little to no evidence of efficacy has NLP squarely in the realm of pseudoscience.

ROLFING

In the 1940s, biochemist Ida Rolf (died 1979) came up with an idea the human body needs to properly align itself with Earth's gravitational field for optimum health. A body misaligned with the earth's gravity exhibits symptoms of disease due to an improperly aligned human energy field and shortened fascia (a fibrous bodily tissue that attaches muscles and organs). Technically, standing aligns your body with gravity, which is straight down. But, whatever. Rolfing, as it's called, involves massage and bending body parts in an attempt to lengthen your fascia. There's no evidence you can do this. And it's not clear how any Rolfing technique realigns the body with gravity.

THERMOGRAPHY

It's probably fair to say not all women enjoy getting mammograms. It's a pretty invasive procedure of a fairly personal part of a woman's body. So one can imagine if a less invasive method of breast cancer screening were available, many women would choose it. Some naturopaths answered that call. They offer thermography as a method for screening for breast cancer. This basically entails using a heat sensitive camera, the same kind of camera used in airport arrival lounges to check people's temperature as they step off a plane. The theory is since tumors require more blood and there is greater metabolic activity, a thermographic camera should show heightened skin temperature near a tumor. The problem is there is zero evidence thermography can detect tumors this way. In the words of the FDA: "There is no valid scientific data to demonstrate that thermography devices, when used on their own or with another diagnostic test, are an effective screening tool for any medical condition including the early detection of breast cancer or other diseases and health conditions." The FDA even issued a warning letter in 2011 to everyone's favorite alt med health pusher Joseph Mercola and his "health center" for marketing a thermographic camera as "the newest safe cancer screening tool".

Five Modern Alt Med Devices of Questionable Value

Given people seem to be fully content purchasing homeopathic sugar pills to treat their vague pains, it seems more than unnecessary some alt med practitioners try to engineer and sell medical devices to treat those same constellations of nonspecific symptoms. But no. History is replete with funny looking quack devices that promise cures by supposedly bathing people in radio waves or, worse, X-rays. Although we should know better, or at least government health authorities should know better, quack medical devices persist. Instead of being made out of wood and brass, they're made out of brushed aluminum and beveled glass. Dials and knobs are replaced by touch-sensitive LCD screens. Here are some of the more bizarre alt med devices that have been marketed to people in recent years.

DETOXING FOOT BATHS

Most of these footbaths use a pair of metal electrodes that run a current through the water. Sounds perfectly safe. The theory goes that the current creates positive ions that attract negatively charged toxins in your body. Just put your feet in a bath of positively charged water and pretty soon, like a magnet, the water will attract and draw out the toxins. Many are impressed by what seems like undeniable results. After a few minutes of putting your feet in the bath, the water turns a sickly brown. It's "clear" proof the machine is working and your body has been purged of "toxins"! Oddly enough if you run the foot baths without your feet in them, the water turns the same sickly brown. Where did the toxins come from? Electrical engineers have demonstrated that the electrodes are made from cheap metal that rusts quickly and this results in the discolored water. Tests of the so-called toxins reveal nothing more than iron oxides (i.e., rust) and whatever salts the foot bath maker recommends adding to the water. The salt, of course, helps speed the rusting process.

MAGNETIC BRACELETS

The idea sounds plausible enough. Put a magnet over an area of your body that is injured or causing you pain, like arthritic joints, and the magnet will attract iron-rich red blood cells to the area and speed healing. Because more blood. The problem is the iron in human blood is so microscopic the weak magnets contained within these bracelets manifest zero attractive force on blood.

BALANCING BRACELETS

Wouldn't it be great if you could slap on a bracelet and have the balance and poise of an Olympic gymnast? Sure. A lack of balance typically isn't what plagues most of us but any advantage in the gym couldn't hurt. Back in the first decade of the 2000s, many people started selling such bracelets. Elite athletes were seen (sometimes paid to be seen) wearing them. One could hardly go to a mall without seeing a floor kiosk offering these $50 pieces of plastic. These bracelets usually featured some kind of cheap hologram sticker that the manufacturers claimed had some kind of magical quantum properties that resonated with your body's natural energy field. Double-blinded tests of these devices, some with the hologram sticker unknowingly removed, revealed they gave no measurable benefit for any tasks whatsoever.

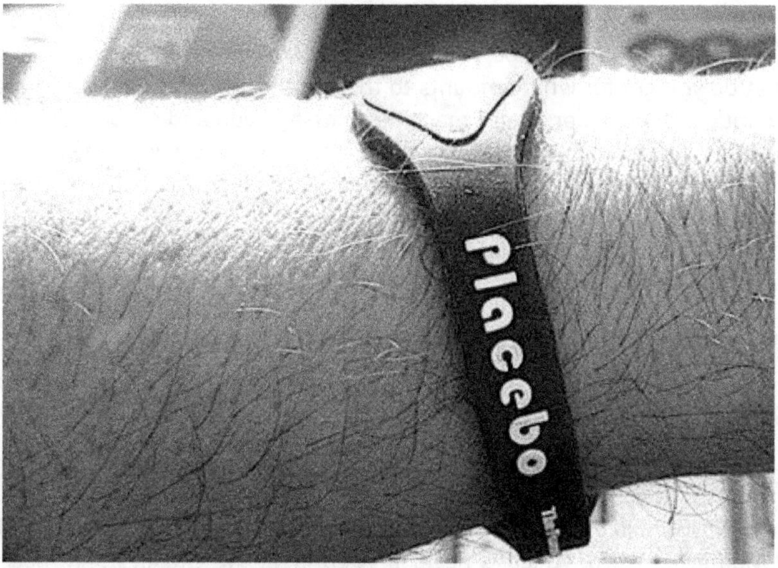

Placebo band courtesy of Australia's Skeptic Bros.
Photo Credit: Karl Mamer

JADE VAGINA EGG

Until Gwyneth Paltrow's "wellness" company Goop was sued in 2018, one could buy a $65 jade egg that was claimed to impart a wide range of health effects. If you shoved it up your hooha after letting it charge in moonlight, a jade egg could do everything from regulate menstrual cycles to prevent depression. Don't get me wrong, Goop didn't STOP selling them after being successfully sued. Goop just stripped away all the health claims and now sells it as a kind of free weight for your pussy. And they upped the price by about $30. Is it a good idea to put a jade egg up your lady hole? Well, given the porous nature of jade, and the difficulty of sterilizing the vag rocks after each use, almost all right-thinking gynecologists strongly advise not putting a rock "up there". The risks (infection) seem to outweigh the benefits (nothing).

SPINAL DECOMPRESSION MACHINES

Chiros have done a pretty good job of cornering the mindshare market on back pain in the same way Kleenex cornered the mindshare market for tissue and Xerox cornered the mindshare market for photocopies. Have pain in your lower back? Many people immediately prescribe themselves the treatment: visit the chiro! Nothing helps a placebo treatment along like elaborate theater and expensive machines. Enter the spinal decompression machine. These space-age looking machines aren't cheap. They run north of $200K. But chiros have reported billings of $600K or more in just one year for the machines. A course of treatments with the machines can involve up to 28 one-hour visits. If your chiro bills about $100 an hour for services, that's $2,800 you're out of pocket unless your insurance covers it. What do you get for what amounts to the price of an excellent cruise vacation? Not so much. Using rollers these machines massage you and supposedly pull your spinal discs apart a bit. Given these machines command such a high price and they're amazingly popular with chiros and patience alike, you'd think there would be loads of good data on their efficacy. Well, like almost everything in the chiro alt-universe, the data is piss poor. If you look at the best surveys, and "best" is used in a relative sense, you're guaranteed some temporary lower back pain relief.

Eleven Bizarre Homeopathic Treatments

A PIECE OF THE BERLIN WALL

Hey, feeling forsaken or bullied? A dilution of an actual piece of the Berlin Wall might be just the thing to cure you. Some homeopaths also claim it has cured asthma. Who would even prescribe something this ridiculous? If you're to take the *Daily Mail* at its word, which is never a good idea, the Queen's pharmacist was selling it at one point.

A BLACK HOLE

This homeopathic remedy was created when a homeopath aimed his telescope at Cygnus X-1 and then dipped the viewing end in a jar of alcohol. The homeopath reported people who were given it felt a sensation of being drawn inward and felt like there was a lump or stone in their stomach. It's not clear how this made anyone feel better or why you may want to take this.

CHECK YOUR EMAIL

Luc Montagnier won a Nobel prize in 2008 for co-discovering HIV. A year later he started making claims that if he placed bacterial DNA in water and then highly diluted it, he could detect residue electromagnetic signals from the DNA molecules no longer present in the water. But there's just one catch, the DNA has to be from harmful bacteria. Non-pathogenic bacteria apparently don't transmit EM radiation. In an interview with Canada's CBC television network, Montagnier denied what he's studying is anything like homeopathy. Despite his denial, it is completely like homeopathy. If that weren't silly enough, Montagnier went on to claim he could record these EM signals, save them as a WAV file, email them, and the person receiving the email could play the WAV file on a loop for about an hour to a tube of water. If the water was then run through PCR amplification, the water would test positive for the original DNA.

DOG POOP

Homeopaths won't, of course, offer you a dog poop pill. They've got a fancy Latin name for it: *Excrementum caninum*. That makes it all better. Homeopaths suggest it can treat mental problems like low self-esteem and self-disgust. I have no doubt people with low self-esteem can sometimes be, sadly, cajoled into consuming things made from dog poop.

A LATEX CONDOM

It really does seem homeopaths will try to dilute almost anything. Case in point, one homeopath gave volunteers a dilution of a latex condom. Volunteers reported effects ranging from increased clumsiness to barking like a dog.

NEW WATER

This one isn't just water. It's water that has diluted "new water". You might wonder what that actually is. Apparently, it's water created by mixing hydrogen and oxygen and then igniting it to create water vapor. So. New water. It's unclear what new water can treat, although one patient given it reported experiencing a very itchy vagina.

PLAGUE LOCUST

Feeling like you're in need of a bunch of new friends or you have digestive issues? Why not try a homeopathic preparation of plague locust?

PLUTONIUM

A site called Homeopathy for Health in Africa offers what is claimed to be a homeopathic dilution of plutonium (one of the deadliest substances on Earth). One wonders how anyone not working in a government lab could get a hold of any amount of plutonium and how one might go about diluting it safely. The site asserts it can be used to treat radioactive pollution and cancer. Pass.

SALIVA OF A RABID DOG

In 2018, a Canadian homeopath in British Columbia boasted on her website she treated a preschooler with behavioral problems by giving him two homeopathic tablets made from the diluted saliva of a rabid dog. This was so nutty sounding that even the British Columbia Naturopathic Association filed a complaint against the quack. The homeopath's defense was an admission homeopathy is just water. She was quoted as saying "I have news for you people, homeopathy either works or it doesn't. If it doesn't, what does it matter what it's made of, dog saliva, elephant's dung or badger's pubic hair, it's so diluted that it's only water after all, or a placebo, a sugar pill." Yeah. What does it matter?

THE SOUTH POLE OF A MAGNET

If you have a patient coming in complaining of levitation (you read that correctly), giving them "attenuations of media saturated with emanations of the [South] pole" is just the thing to cure them. It apparently also works for ingrown toenails.

AN X-RAY

Take a vial of alcohol, expose it to an X-ray, and then dilute. Homeopaths believe it increases sexual desire in men and is effective for dry throat and as a treatment for bone cancer.

Cancer "Cures" from A to Z You Might Want to Stay Away From

ANTINEOPLASTONS

Since the 1980s, Dr. Stanislaw R. Burzynski has been using something called antineoplastons to treat cancer patients in his Texas clinic. Antineoplastons are what Burzynski calls peptides isolated from your urine. Unlike many alt med types, Burzynski publishes extensively as part of a phase II clinical trial that never seems to end. Be warned, participating in his clinical trial is not free. You can help him science for $7,000 a month. What do you get for that? In the words of *The Lancet* "To date, there have been no randomised, controlled trials showing efficacy of antineoplastons. However, side-effects of this therapy can include serious neurological toxicity."

BIORESONANCE TUMOR THERAPY

People who offer this therapy think hooking a patient up to what looks like a current tester can detect cancer and cure the disease in all its forms. The claimed method of action is using electromagnetic waves. The evidence-vacant premise is cancer cells (along with other diseased cells) emit unusual EM radiation and this can be detected by the bioresonance device. Then, by beaming corrective magnetic waves, you can make the diseased cells go "poof!" A 1997 double-blinded study ("Efficacy trial of bioresonance in children with atopic dermatitis") cautioned "[due to] the high costs and false promises caused by the promoters of this kind of therapy, it is concluded that [bioresonance therapy] has no place in the treatment of children with atopic dermatitis."

CAT'S CLAW

Cat's claw is bark from a vine that grows in parts of South America. It's promoted by quacks as a cancer- and AIDS-fighting botanical from the rainforests of Peru that was used by the ancients! In tests on immune cells in labs, cat's claw stimulated certain immune cells and DNA repair processes. Alt med types like to run wild with very preliminary results out of labs. There's a long way to go before you can say a substance also works in the human body. What they won't tell you about these lab results is cat's claw was also shown to aid in the survival of leukemia cells. And that's not good.

DIMETHYL SULFOXIDE (DMSO)

Dimethyl Sulfoxide (DMSO) is an industrial solvent. It's not particularly toxic. Medically it is interesting because it can easily and rapidly penetrate the skin and cell membranes. Anything dissolved in DMSO will be carried with it. It can be a way to target drugs by topically applying DMSO. By itself, it has some use as a topical painkiller. Cancer quacks have long touted it as a cancer fighter. There is zero evidence DMSO can be used to treat cancer. Worse, DMSO can interfere with some forms of chemotherapy. DMSO generally entered the quack pharmacopoeia after it was touted as a miracle cure on an episode of *60 Minutes* in 1980. In the early 1980s, *60 Minutes* was one of the highest rated shows and each episode reached about 70 million viewers. The segment painted the FDA as the bad guy, denying approval to a simple cure for cancer and various intractable pains. It prompted the US government to hold hearings as to why the FDA wasn't approving DMSO for a range of conditions advocates claimed efficacy for.

ESCHAROTICS

Escharotics are corrosive salves people apply to their skin. At the turn of the 20th century, they were a popular way to remove warts and skin lesions but have long been replaced by safer and more effective methods. Quacks claim by applying an escharotic salve to an area of the skin above a tumor, the salve will draw out the tumor. The only thing the salve does is burn away useful skin, leaving a scar.

FRESH CELL THERAPY

Fresh cell therapy involves injecting cells from a sheep embryo into a cancer sufferer. The idea is if, say, you have kidney cancer, you inject kidney cells from a sheep embryo into a person's butt. Fresh cell therapy has been kicking around since the 1930s. It was popularized by Swiss doctor Paul Niehans (died 1971). He attempted to treat Pope Pius XII, King Ibn Saud, and Charlie Chaplin with fresh cell therapy. Pius XII was apparently so impressed by Niehans' quackery that he appointed him to the Papal Academy of Sciences. There is zero evidence fresh cell therapy can treat anything.

THE GREEK CANCER CURE

The Greek Cancer Cure was developed by Athenian microbiologist Dr. Hariton-Tzannis Alivizatos (died 1991). He claimed to have developed a blood test that could diagnose the presence and location of cancer. Once diagnosed, Alivizatos prescribed a magical serum that would cure the patient. Alivizatos boasted he had a nearly 100% cure rate but studies showed most of his patients died within a few months of his treatment. The American Cancer Society sent a doctor to Greece to investigate. Posing as a cancer sufferer, the doctor obtained a sample of Alivizatos' serum. It was found to contain not much more than Vitamin B and water.

THE HALLELUJAH DIET

You might be surprised to discover there are a number of "Bible-based" diets (going by names like the Daniel Fast and the Eden Diet). Most promise miraculous weight loss. However, the Bible-based Hallelujah Diet layers on an additional miracle. It cures cancer. It was developed by Reverend George M. Malkmus based on his reading of *Genesis 1:29*. Malkmus read it as a directive that people should eat a diet of raw food and avoid meat. Okay. He added his Hallelujah Diet helped him fight a colon cancer diagnosis, making a baseball-sized tumor disappear. To support his diet, Malkmus started marketing something called Barley Green. He promoted several dubious benefits of his diet that resulted in the FDA ordering him to stop suggesting it was effective against cancer (and a range of other health problems).

INSULIN POTENTIATION THERAPY

Insulin potentiation therapy is based on the idea that injections of insulin can boost the effectiveness of chemotherapy. The idea was first introduced in a 1986 study published in everyone's favorite crank magnet medical journal *Medical Hypothesis*. The article suggested insulin is so effective in boosting chemotherapy that doctors can reduce chemo dosages by as much as 90%. Chemotherapy is not pleasant and that would certainly be a win. However, there have been no longer term studies demonstrating that. Evidence consists of a handful of case studies and one small short-term controlled study.

JUICE THERAPY

It is claimed going on a diet of only fruit and vegetable juice for several days can detox your body and rid you of cancer. The problem with taking nice fiber-laden vegetables and rendering them into juice is you actually lose the fiber, which in combination with a whole lot of other things can help prevent cancer. There is no evidence juicing can remove cancerous cells and toxins.

KREBIOZEN

Krebiozen was a bogus cancer elixir popular in the 1950s and 1960s. Quacks offering it sold it for $1,000 a dose. Those who were able to analyze it found it was little more than creatine dissolved in mineral oil. Krebiozen was developed by Yugoslavian doctor Stevan Durovic. Durovic claimed amazing results with cats and dogs. No one might have heard of it unless Andrew Conway Ivy (died 1978) had not started stumping for Durovic. Who was Ivy? Ivy was in some ways a cross between Carl Sagan and a real-life Dr. Marcus Welby M.D. Ivy was part of the Nuremberg trials that prosecuted Nazi doctors. The media hailed him as the guardian of ethics in science. Unfortunately, his ethics went out the door when he met Durovic. Ivy started testing Krebiozen on a number of cancer patients. He told the press he had achieved a remarkable breakthrough in treating cancer. He failed to tell anyone nearly half his patients had actually died of cancer and he had falsified his reports. In other words, Ivy, the great ethicist, was claiming patients who had actually died of cancer under his care were alive and well. Owing to Ivy's standing, many institutions took claims about Krebiozen seriously and tried to replicate his results. Predictably, no other doctor or hospital was able to duplicate Ivy's results. Ivy and Durovic were eventually put on trial for mislabeling drugs but were acquitted by a hung jury.

LORRAINE DAY'S 10-STEP PROGRAM

Lorraine Day is a doctor who was diagnosed with breast cancer. As she tells it, after a revelation from her god, she came up with a 10-step program (involving diet, getting enough water, and stress reduction) to beat her cancer. And it can beat yours too! Just buy her books, lecture tapes, and supplements. Now you might think "great, another self-deluded medical professional doing the maverick doctor thing". What makes Day particularly odious is not just her claim her questionable cancer treatment is rubber-stamped by the Christian god, but she's a pretty outspoken Holocaust revisionist. She claims Jews control the world. She was also a vocal apologist for neo-Nazi of Ernst Zündel (died 2017). Yeah, a real piece of work this one.

MOXIBUSTION

Moxibustion involves burning dried mugwort over a human body. It's a popular but ineffective treatment used in East Asia for a wide variety of ailments. Quacks offer it as a complement to cancer treatments. They promote it as a way to alleviate the nastier symptoms of chemotherapy. There are a lot of really crappy, unblinded trials that suggest moxibustion is effective in reducing some side effects. However, placebo can't be ruled out.

NONI JUICE

Noni juice is a super expensive juice made from a Polynesian fruit. A single cup of the stuff will run you about $7. People selling it promote it as having astounding levels of antioxidants and it can prevent or treat cancer. Studies have shown supplementing with antioxidants has no effect on cancer prevention. As well, you can get as much or more antioxidants from eating an apple.

OXYGEN THERAPY

Oxygen therapy involves dubious means to increase the oxygen content of the blood in the belief this will aid the body in fighting cancer. It's based on something called the Warburg effect. As cancer cells multiply, they outstrip the ability of blood vessels to supply them with oxygen. So cancer cells turn to glycolysis for energy. At this point, quacks make a rather frightening leap in logic. The best way to fight the disease is to overwhelm cancer cells with oxygen. Okay. How do you drown cancer cells in oxygen? Simple. Inject hydrogen peroxide directly into tumors or blow compressed oxygen into a person's vagina or rectum. The reality is cancer cells grow just fine if supplied with oxygen. You can't kill off tumors this way.

POLY-MVA

Poly-MVA is offered by Merrill Garnett, a former dentist. It is a mixture of vitamins, amino acids, and palladium. Garnett's hypothesis is that by quaffing this reddish-brown substance, it can alter the electric charge of cancer cells and invoke apoptosis. There is, of course, no peer-reviewed evidence drinking palladium and a magical mix of vitamins can do any such thing. This stuff isn't cheap. A bottle goes for about $280 on Amazon.

QUARTZ

New age proponents make endless claims about the healing properties of crystals like quartz. In the view of New Agers, disease is a result of blocked or misaligned chakras. Crystals and their magical energies can unblock or realign these chakras. Of course, there is no evidence chakras even exist. You will find various people who promote rose quartz as proof against breast cancer. For example, for only $19 (plus $15 in shipping), you can buy a "Breast Cancer Fighter" bracelet on Amazon featuring rose quartz. Rose quartz, we're told, "has a strong effect on the breast area". Oh. Well then.

RIGVIR

Rigvir is a pseudoscientific cancer treatment that was approved in 2004 by the Latvian health ministry. This came as a bit of surprise to a number of medical associations in Latvia that found no scientific basis for Rigvir's approval. Approval was withdrawn in 2019 after protest by the medical community. Rigvir's makers claim the substance contains an echovirus that targets cancer cells. Using viruses to fight cancer is a current avenue of research but there are zero approved treatments in Europe or North America. Tests of Rigvir have revealed the amount of echovirus contained in bottles was far, far less than advertised by the manufacturer.

SHARK CARTILAGE

Somehow it got into the public mind that sharks don't get cancer. They do. Many who subscribed to the belief sharks are cancer proof believed it was because shark skeletons are all cartilage, not bones. So, the math is simple. Eat shark cartilage if you want to avoid cancer. Studies have found people with cancer who take shark cartilage have no measurable effect on their life expectancy or quality of life. The stuff hasn't even been able to work in mouse models. There is zero point to killing dwindling shark populations to make pills that do nothing.

T-UP

T-Up was a concoction of aloe vera and cesium chloride. In the 1990s a man by the name of Allen J. Hoffman was peddling it to people dying of cancer and AIDS. You could buy a $75 bottle of the stuff or pay for an intravenous treatment that would cost you $12,000. T-Up was marketed to desperate people with the slogan "You do not have to die". (Probably a better slogan would be "you do not have to die with assets you can leave your family".) Hoffman, who was not a doctor, treated over 3,000 people with what was effectively hand lotion you can purchase at a dollar store mixed with an exotic salt. By 2000, he was in the sights of the FDA. He was charged with, and later found guilty of, introducing an unapproved new drug into interstate commerce with the intent to defraud the U.S. Food and Drug Administration. He was sentenced to 46 months in prison.

URINE THERAPY

There are people who believe drinking your own urine is about the healthiest thing you can do. It is claimed drinking your own pee can ward off cancer. If you need a little help, there are clinics in Mexico that will treat your cancer with urine. Proponents reason since some tumor proteins can be found in urine, by drinking urine you can cause your bodies to form antibodies that will snap into action the next time you come down with a bout of lung cancer.

VENUS FLY TRAP

Some alt med types claim an extract derived from the Venus fly trap plant is effective in treating skin cancer. You can buy a 1-ounce bottle of the stuff for $28 on Amazon. Although some quacks who tout Venus fly trap extract as a cancer cure are vague about its method of actions, others point out it contains a substance called plumbagin which has been investigated as a treatment for cancer. However, trying to get plumbagin out of a Venus fly trap plant has to be the least cost-effective way of doing it. Walnuts, for example, are a cheap and ready source.

WALNUTS

Walnuts, particularly the black walnut variety, have a high concentration of phenolic compounds. Studies have found diets rich in such compounds are associated with lower risks of cancer. A 2011 study found cancer cells in test tubes when exposed to walnut extracts were prevented from multiplying. However, that was one study. There are thousands of papers that suggest some chemical in some food has some benefit in a test tube. Very few translate into effective cancer treatments in the body. There have been no studies to suggest mowing down on walnuts will result in any additional anti-cancer benefits.

714-X

714-X is the invention of Canadian hematologist Gaston Naessens. According to the American Cancer Society, Naessens moved to Canada after he was convicted in France of practicing medicine without a license. The drug's name comes from his birthdate. He was born July 14 (714) and X is the 24th letter in the alphabet which then represents his birth year of 1924. Naessens claims 714-X's method of action is attacking a cancer-causing biological lifeform called a somatid. The problem is somatids are invented out of whole cloth. There's no evidence the lifeform even exists. 714-X is a mixture of camphor, salt, and ethanol. Although the substance has been banned in the USA, a judge in Canada ruled in 2006 that it is legal for Naessens to manufacture it and sell it to cancer sufferers under Canada's compassionate drug use laws.

YAM

Many tout wild yam as effective in fighting breast cancer. The reasoning is wild yam has a chemical called diosgenin which mimics estrogen. Alt med types claim this can protect one against breast cancer. There's no evidence of this, however.

ZOETRON THERAPY

Zeotron therapy involves running people through a machine called a Zoetron. The machine features a magnetic ring. For $15,000, all in advance, operators will not only use the machine to detect cancer in your body but fix it with the super and mysterious power of magnetism! Studies of the machine have found the magnet far too weak to have any effect on anything in the human body.

Seven Poor Arguments Against Water Fluoridation

Fluoridating a town's water supply is an incredibly cost-effective way to improve the dental health of citizens. It's estimated for every dollar invested in fluoridating the water supply, the average citizen saves about $38 in dental care. Many of the benefits go directly to the poorest people in society who cannot afford regular visits to the dentist or even toothpaste. It's a huge win. But there have long been cranks that oppose fluoridation based on junk science. In recent years, many cranks touting scary YouTube videos have been successful in convincing their city councils to ignore the advice of every health expert under their employment and cease water fluoridation programs.

THE EPA OPPOSES IT!

The Environmental Protection Agency actually doesn't oppose water fluoridation. Cranks claim this because in 1999 the head of a local chapter of the National Treasury Employees Union representing some employees of the EPA wrote a grievance letter claiming his union brothers and sister opposed fluoridation.

EUROPEANS DON'T FLUORIDATE THEIR WATER!

The argument seems to go if socially conscious nations like Sweden, Denmark, and Finland don't fluoridate their water, why should we? What do they know we don't? It's true many European nations don't add fluoride to their drinking water. They add it to their salt instead. As well, these populations have equal access to dental care and better surveillance in schools.

FLUORIDE IN WATER LOWERS IQ!

Several studies (a 2006 study by the NRC, a 2014 New Zealand study, and a 2017 study in Mexico) have found no link between safe levels of fluoride in drinking water and claims of lower IQ in children. "Developmental Fluoride Neurotoxicity: A Systematic Review and Meta-Analysis" from 2012 is often cited by anti-fluoridation cranks as the smoking gun. In reality the study compared the IQs of children drinking water with fluoride at the recommended safety levels and the IQs of children exposed to drinking water with levels well above the recommended dose (due to natural causes or industrial spillage leaking into the groundwater supply).

THE NATIONAL RESEARCH COUNCIL OPPOSES IT!

In 2006 the National Research Council published a report on fluoride in drinking water. The study does not oppose fluoridating water supplies. The anti-fluoridation crowd simply cherry picks scary quotes from the document to make it seem like the NRC is strongly opposed to water fluoridation. One of the report's authors is quoted as saying in response to the misuse of his report "I do not believe there is any valid, scientific reason for fearing adverse health conditions from the consumption of water fluoridated at the optimal level."

EVERYONE KNOWS THE NAZIS ADDED FLUORIDE TO THE WATER IN DEATH CAMPS TO MAKE PRISONERS DOCILE!

Apparently, starvation and being surrounded by barbed wire wasn't enough to take the fight out of Jewish prisoners in death camps. They needed fluoride added to their drinking water. There's zero documentation the Nazis ever added fluoride to the water being piped into death camps. The source of this legend seems to stem from the writings of an early anti-fluoridation crank named Charles E. Perkins. He based it solely on an undocumented claim made to him by an anonymous worker at an unnamed German chemical company that it had plans to fluoridate the water in occupied countries.

YOU NEED INFORMED CONSENT TO PUSH A DRUG ON PEOPLE!

If you define a drug as a chemical to prevent a disease (like tooth decay), fluoride is a drug. Is it moral to force everyone to consume such a drug? This may be the only argument in the anti-fluoridation arsenal that makes you stop and think. For a moment. A lot is added to our drinking water for our health benefit. Chlorine, for example, is added to water to prevent a range of water borne diseases. The government also mandates iodine be added to table salt to prevent disease like goiter, dwarfism, and mental retardation. Vitamin D is added to milk to help prevent things like rickets in children. Folic acid is added to flour. When consumed by a pregnant woman it can help prevent spina bifida in a child. Indeed, we could remove all of these additives and let rich people supplement themselves. But the reality is these things are added because they are a low-cost public health win for all, notably people lower on the economic ladder who can't afford pricey supplements.

THE WORLD HEALTH ORGANIZATION OPPOSES IT!

Anti-fluoridation cranks will point to various World Health Organization studies that seem to call on governments to remove fluoride from water. What is missed, or quote mined, is the WHO recommends removal of unsafe levels of naturally occurring fluoride from water, not removing safe levels in drinking water.

Six Superfoods That Don't Live up to Their Hype

The notion of "superfoods" seems to have emerged in the late 1990s and early 2000s. The idea is a certain food has an extremely high level of some substance that has been shown to have health benefits. Other superfoods are promoted as having almost everything you need for good health. In most cases, however, "superfood"" is merely a marketing term and a clickbait headline. Very few so-called "superfoods" get described that way in the scientific literature. Only about 68 articles in PubMed use the term. And many of those are simply papers investigating popular claims about a certain fruit or vegetable being a "superfood".

AVOCADOS

Promoters of this as a superfood focus on the fruit being an excellent source of good fats that help prevent heart disease. They're also an excellent source of calories. One avocado has as many calories as a typical chocolate bar. People who think this is a cancer fighter ignore the fact that the possible cancer fighter is an extract of the avocado seed. No one actually eats the seeds. A major downside of avocados is they have a terrible environmental footprint. Similar beneficial fats can be had by consuming oily fish, nuts, and sunflower seeds.

BLUEBERRIES

In 1991 the USDA published a list of the antioxidant content of foods. Blueberries topped the list and quickly became, arguably, the first "superfood". Although the evidence was weak that antioxidants had a role in cancer prevention, the alt med crowd ran with it. Suddenly blueberries were more than a nice summer treat or an ingredient in an excellent pie. They were a weapon in fighting cancer, memory loss, aging, heart disease, and Alzheimer's. Like anything in the world of alt med, if a bit of something provides some benefit then shoving megadoses in your mouth must have even more benefit. Three decades later, the data on antioxidant supplementation is mixed. Supplementing has little effect on reducing cancer rates and may put one at greater risk if a person has cancer.

CHIA SEEDS

These taste-absent seeds from Central and South America hit the big time when it was discovered they were particularly high in Omega-3 fatty acids. There's evidence to suggest well balanced diets that regularly incorporate foods high in Omega-3 fatty acids are particularly healthy. So, we're talking about the so-called Mediterranean diet. Studies indicate that although the seeds are high in Omega-3 fatty acids your body isn't very efficient at breaking the seeds down. You need to eat a lot of the seeds (and consume more calories in seeds) to get the same benefits found in a smaller portion of an oily fish. While certain traditional diets are thought to be healthier than others, there's very little evidence simply parachuting a ton of Omega-3 fatty acids into your own particular diet is of any benefit.

GOJI BERRIES

The evidence for the supposed benefits of goji berries is slim to none. People will, however, swear they're a cornerstone of Chinese medicine. This is evidence of exactly nothing. But, hey, there's an anecdote about a Chinese dude who lived to be over 200 years old by making the berry a part of his diet. Of course, there is no evidence whatsoever any human has lived that long. Evidence of the berry's benefits are, at best, based on bad studies that look at consuming concentrations so high you can't possibly ingest that amount by sprinkling the berries on your granola or drinking goji berry juice.

KALE

Kale is great, whether in salad or crispy chip form. It's high in iron, vitamins, and antioxidants. But, like blueberries, consuming more antioxidants than your body can actually use at any given moment will not necessarily give you benefits. Kale is great for adding variety to your diet but there are no studies that show kale provides any additional benefits if added to a normal healthy diet. Adding more antioxidants to your diet than your body actually needs is, as one health expert puts it, "like putting more gas in your fuel tank and expecting your car to go faster." It doesn't work that way.

WHEATGRASS JUICE

The only hype this "superfood" lives up to is its name. It tastes just like it sounds. There's zero scientific evidence that wheatgrass is beneficial in any way other than it has nutritional content similar to green leafy vegetables that tend to taste better. Alt med people assert drinking wheatgrass juice is a cure-all based on the observation cats eat grass when they're sick. As well, it contains chlorophyll which, umm, is beneficial to plants but there's no evidence humans need to consume chlorophyll to maintain their health.

Thirteen Examples of Food Woo

AUTISM AND ADHD DIETS

You will find numerous gurus who promise distraught parents these disorders can be treated by diet. This typically means eliminating things like dairy, gluten, and processed foods. It also means megadosing on various vitamins and fish oil. Jenny McCarthy swears her son recovered from autism via one of these restrictive diets plus helpings of "anti-fungals for yeast". Parents who believe their children show improvement under these diets tend to not mention in the same breadth their child is undergoing conventional therapy as well.

DIETARY SUPPLEMENT HEALTH AND EDUCATION ACT OF 1994

In 1991, Congress was planning legislation that would give the FDA increased powers to regulate the supplement industry. The supplement industry lobbied vigorously against the law. It also blitzed the voting public with erroneous warnings the new law would give the FDA the power to ban the sale of vitamins. Mel Gibson too came out against the bill. He appeared in a nutball commercial depicting his arrest by heavily armed SWAT police because he was taking Vitamin C. The campaign proved highly effective. The bill died on the vine. In its place, senators Orrin Hatch and Tom Harkin sponsored and helped pass a pro-supplement bill called the Dietary Supplement Health and Education Act (DSHEA). DSHEA allows supplement makers to market a wide variety of substances as dietary supplements. As long as the maker makes no claim about any specific medical benefits, DSHEA allows the product to be sold without needing FDA approval. Makers can claim their supplements support certain structures (like a bodily organ) or certain functions (like blood circulation or an immune response) if they have evidence. So, ad copy like "Vitamin D supports a healthy immune system" would be legal under DSHEA but "Vitamin D prevents the flu" would not. Unlike drug makers, under DSHEA, supplement makers don't have to prove the safety or efficacy of their supplement to the FDA. They merely have to assure the FDA they established the safety and efficacy of their structure/function claim. The kicker is, they do not have to reveal any of that to the FDA or the public before they can market their product. It's a bit like getting a course credit simply because you assured the professor you did the work and passed an exam the prof never saw. What constitutes "evidence" for a supplement's safety and efficacy is left up to the manufacturer. Until something goes horribly wrong. The FDA's sole power is its ability to ban a supplement only after it kills people. And the onus is on the FDA to prove the supplement is dangerous.

EAT A CHEMICAL FREE DIET

One sees the boast "our food is chemical free" quite frequently in health-food restaurants. Since food and everything that is matter is made of chemicals (e.g., chemical elements), one can only imagine a "chemical free" restaurant offering patrons things like bowls of heat or glasses of the strong nuclear force. Food Babe blogger Vani Hari is probably the best-known proponent of this kind of crankery. "There is just no acceptable level of any chemical to ingest, ever" has been her rallying cry.

Photo Credit: Karl Mamer

She demonstrated the growing popularity of chemical phobia in 2014 when she was able to shame the Subway restaurant chain into stopping the use of a chemical called Azodicarbonamide in the production of its bread. What's so bad about Azodicarbonamide? Well, she pointed out this chemical was also used in the production of yoga mats. She let her readers connect the dots ("so, uhhh, you're saying Subway bread is made from yoga mats?"). Azodicarbonamide in very large quantities is toxic but as a food additive (about 45 parts per million) it is harmless. Hari's next target was Starbucks. She accused the coffee chain of not having real pumpkin in its seasonal pumpkin spice lattes. Hari seemed unable to grasp that pumpkin spice has never contained pumpkin but it is simply a spice *for* pumpkin-based foods like pie. It's a bit like wondering why does baby powder not contain any ground up baby?

GLUTEN-FREE DIETS

There's a small group of people that suffer from an autoimmune disorder called Celiac disease. It causes some pretty horrible pain and other unpleasantries if the sufferer eats glutenous food products like bread. Over the last twenty years, a class of gluten avoiders who call themselves "gluten sensitive" has emerged. These self-avowed gluten sensitives are in such large numbers (about 30% of Americans will purposefully consume gluten-free foods) that manufacturers have been eagerly supplying the market with all manner of gluten-free products at a premium price. This has actually not been a bad thing for those with Celiac disease as it has expanded their food choices and they can enjoy things like pizza, cupcakes, and beer again. Those who don't necessarily call themselves gluten sensitive follow diet gurus who assert wheat is the cause of the obesity problem. In an effort to lose weight, they've switched to gluten-free bread products. Studies have shown one cannot achieve weight loss by removing gluten from your diet. Boy, if the gluten woo only stopped there. Quacks have emerged who prey upon parents of kids with autism, claiming autism can be cured by introducing a gluten-free diet.

GMO FEAR MONGERING

Penn Jillette was given to quip that once the food woo crew coined the phrase "Frankenfood" it was pretty much game over for the GMO industry. GMOs have been a tough sell at times. "GMO", of course, refers to foods that are classified as genetically modified organisms. Unlike traditional hybridization techniques, GMOs are produced by directly injecting beneficial genes from one organism into another. The food woo crowd finds it particularly disturbing that could mean, say, a fish gene being part of a GMO tomato. The woo crowd blames GMOs on the rise of childhood allergies, cancer, and autism. They spread fears that GMO makers are destroying traditional farming methods and are making family farms dependent on their genetically engineered seeds, locking them into expensive, one-sided contracts. This is rather at odds with the reality farmers are one of the most powerful political lobbies in America. If farmers were so hard done by Big Agri, they would be dumping crops in protest on the steps of Congress. More scientifically literate anti-GMO proponents like Canadian geneticist David Suzuki seem to take a page out of the anti-vaccine playbook. Suzuki and his ilk repeat claims GMOs need to be banned until we have solid research on their long-term safety. They rarely, if ever, articulate a goal post of how much research is good enough to prove GMOs are safe. The reality is, there have been at least two dozen long-term, multi-generational studies of GMOs and none have ever turned up any hint GMOs are unsafe over the long term.

IRRATIONAL ABOUT FOOD IRRADIATION

The irradiation of food is one of the greatest technologies ever developed to keep our food supply safe. However, because it involves beaming X-rays and gamma radiation at our food, large swaths of the public fear irradiated food. In reality, the process is a win-win. It kills pathogens and improves shelf life, leading to less food waste. The major public objection is the fear the process leaves food radioactive. It does not. Food is only exposed to radioactive waves, it's not salted with radioactive materials. It's a bit like suggesting sun-dried tomatoes should be radioactive because they've been exposed to the ionizing radiation of the sun for so long.

MICROWAVING YOUR FOOD MAKES IT TOXIC

Microwave ovens heat your food by beaming microwaves through your food and exciting the molecules. That produces heat that cooks your food. Microwave radiation is not the same as ionization radiation. It can't alter the genetic codes inside the cells that make up the foods you put into your ovens. But people seem to think that's the case. Microwaving your food genetically and chemically alters food, producing toxins you then consume. Yeah, that doesn't happen. Microwaving sometimes leaves some food soggy but that's about as bad as it gets. Microwaving actually has some inherent advantages over most stove top cooking methods. Multiple studies have shown microwaving tends to be the best way to heat up food and preserve vitamins, folic acid, and antioxidants.

PROBIOTICS

We're at the very beginning of learning how bacteria in our gut affects our health. What we do know is taking antibiotics can kill off bacteria that seems useful. This then allows harmful bacteria to colonize our gut. A simple check against that is to take one of any number of probiotic foods made with cultures of good, live bacteria. Beyond helping prevent antibiotic-induced diarrhea, there is very little evidence probiotics give any additional benefit in otherwise healthy people. But, hey, the lure of selling expensive yogurt to people using vague structure/function claims ("supports a healthy immune system!") is very attractive to marketers. Naturopaths likewise run crazy with probiotics. They promote it as a way to boost your immune system and fight diseases like cancer, autism, Crohn's disease, diabetes, and so on. There's zero evidence for any of this.

RAW FOOD DIETS

Raw food advocates believe uncooked food is healthier than cooked food. A diet high in fruits and veggies traditionally eaten raw (like celery or baby carrots) is certainly healthy. And some cooking methods (like boiling vegetables) can rob some foods of vitamins. If raw food advocates only stayed within the realm of fact, their claims might be best countered by pointing out cooking actually increases available vitamins in some foods and in those foods that lose vitamins a diner can make up for the loss by simply consuming a *slightly* larger portion (like one extra baby carrot). As well, cooking has additional health benefits, like preventing food poisoning. Raw food advocates, however, go beyond the factual effects of cooking and make wildly unsupported claims. For example, they will knowingly tell you cooking destroys enzymes present in the food your body needs to help digestion. The reality is your body does not rely on any consumed enzymes to digest food. Raw food advocates will also use the naturalistic fallacy: humans in their natural state ate raw food and they did just fine. But today we're beset by all these health problems so cooked food is to blame. That is simply an unfounded assertion. There's no evidence humans were healthier before we learned to control fire and cook food. In fact, cooking increased the energy availability in foods and let us actually do stuff other than barely subsisting on anything we could cram into our mouths.

RAW MILK

Milk is delicious but it also is an excellent medium for the proliferation of the nastiest bacteria that infects our food supply (*Salmonella*, *Listeria*, and *E. coli* to name just a few). If you live right next door to your cow, drinking raw milk that goes from teat to table isn't so bad. But in a world where most people live nowhere near a farm, most people don't get that chance. The longer raw milk sits, the greater the chances it can become tainted with a deadly pathogen. Before pasteurization, improperly handled raw milk killed tens of thousands of people a year. Most sane, rational countries today require milk to be pasteurized. Some nations, like France, have exceptions for cheese production although such cheeses are made safe via irradiation. There is a branch of raw food advocates that believes raw milk provides astounding health benefits not found in milk made safe by pasteurization. Raw milk advocates like to point to an article written by a Dr. J. R. Crewe in 1929. We are told Crewe worked for the Mayo Foundation, which was the forerunner of the Mayo Clinic. Crewe's article "Real Milk Cures Many Disease" promotes raw milk as an effective treatment for a range of diseases including stomach cancer, diabetes, and anemia. Crewe's evidence? Just a lot of personal anecdotes. Naturally, there is no scientific evidence that raw milk is a cure for any of these diseases or offers significant health benefits when compared to drinking pasteurized milk.

VITAMIN C

Vitamin C megadosing was proposed back in the 1970s by Nobel disease sufferer Linus Pauling. He suggested taking hundreds of times more Vitamin C than the RDA could prevent the common cold and heart disease. Pauling's idea was based on the amount of Vitamin C that gets produced naturally by gorillas. Research indicates, at best, mega doses of Vitamin C can shorten the duration of a cold in people undergoing high levels of stress. Vitamin C is water soluble and for the most part you simply excrete that which your body doesn't need. It has been often pointed out people taking expensive megadoses of Vitamin C are simply creating very expensive pee.

VITAMIN D

Vitamin D is one of those vitamins people feel like maybe they're not getting enough of. If we expose our skin to the sun our body makes its own Vitamin D. Many parts of the world also fortify milk with Vitamin D. Many people have the mistaken impression they're not getting enough Vitamin D because they don't go outside much or don't drink milk. North American health authorities have, however, detected no such deficiencies in populations other than breast-fed babies. This is no reason to supplement, particularly by the mega dose. Since this is a fat-soluble vitamin, consuming more Vitamin D means any excess gets stored in your fatty tissue. There are woo proponents that don't understand that mega dosing can lead to a buildup of Vitamin D your body can't easily get rid of, which then results in Vitamin D toxicity. Health crank Gary Null discovered the reality of this back in 2009 when he quaffed two of his own company's "power" supplement drinks. Little did he realize the company he hired to make the slop had dumped in 1,000 times more Vitamin D than the recommended dose and it nearly killed him. Reportedly Null felt unwell after consuming his first power drink and then drunk another thinking it would make him feel better. He just poisoned himself further. On the other side of the woo coin we have something called the Marshall Protocol. This is a diet developed by an electrical engineer (drink!) to fight chronic inflammation. According to the Marshall Protocol, our poor health isn't a result of not getting enough Vitamin D but getting too much. The claim is that bad bacteria can co-opt Vitamin D and cause infections. So, you need far less Vitamin D than that junky old science-based table of RDAs. That idea, however, flies in the face of actual research that shows a relationship between Vitamin D deficiencies and infections.

SUGAR

Sugar in most of its forms is blamed for a wide variety of health problems, from obesity to ADHD. Sugar certainly adds additional calories to food that we don't need. Much is made, however, about high-fructose corn syrup (HFCS) these days. Curiously, table sugar (sucrose) for a while was the great health concern. Many health gurus dubbed it "white death" and fructose was considered a healthy sugar, based on its association with fruit. HFCS is enormously cheap to make and food processors, like soda makers, have been using it in their products since the FDA approved HFCS in the early 1980s. HFCS makes it more economical to sweeten foods that typically were too expensive to sweeten with sucrose. Over the last few decades, there's been a 25% increase of sugar in our processed foods. Your body does process HFCS differently from sucrose. Diet gurus have seized on this as being the one true cause of the obesity epidemic (versus we're getting 25% more unneeded calories). Multiple studies, however, have found HFCS has no measurable difference in weight gain beyond the calories it contributes.

British Chiropractic Association's Six Worst Lines of Evidence for Childhood Chiropractic Treatments

In a 2009 article for the UK's *Guardian* newspaper, science journalist Simon Singh took chiropractors to task for not having "a jot of evidence" to support chiropractic treatments for children. The British Chiropractic Association (BCA) responded by suing Singh for libel. Much criticism was directed at the BCA. Real scientists respond to claims a field has a lack of evidence by providing the evidence, not launching a lawsuit. The BCA eventually responded to that criticism with a statement that there's loads of evidence for the safety and efficacy of treating children with their bone tugging techniques. The BCA published a list of 29 publications that supposedly provided evidence. Real scientists quickly weighed in on the BCA's 29 "jots". What the BCA offered was quite laughable and basically ended up making Singh's case for him. The BCA withdrew their libel case in 2010 when British judges made it clear they viewed Singh's comments as legally permissible fair comment that would be weighed against the evidence.

THE BCA'S OWN CODE OF PRACTICE

Number one on their list of "research about the effectiveness and safety of chiropractic treatment for children" (their exact words) is a link to their own code of practice. That is not research. It's unclear how this supports their positions, beyond the fact that the code asserts British chiropractors follow "evidence-based care".

AN ARTICLE ABOUT MEDICAL AND ACADEMIC MISCONDUCT

This citation seemed to be little more than a "but her emails" type deflection. It's an attempt to suggest real medical science is full of fraud and therefore... what?

THREE PAPERS ON THE RISK OF PAINKILLERS

These all seemed to be another "but her emails" inclusion. Painkillers like ibuprofen and aspirin are a class of drugs called nonsteroidal anti-inflammatory drugs (NSAIDs). They're great for reducing pain, decreasing inflammation, and fighting a fever. They are not without their risks. This is no mystery and lends no support to the idea that chiropractic for children is safe and effective.

THREE ARTICLES ABOUT THE EFFICACY OF OSTEOPATHY

Osteopathy is not a branch of chiropractic "medicine". It's a bit like suggesting hypnosis is effective because studies on meditation indicate they help with stress relief. Chiropractors have, themselves, rejected negative osteopathic studies as evidence of chiropractic treatments being bogus. You can't have it both ways, boys.

A NEGATIVE STUDY

One study on the BCA list that "strongly suggested" spinal manipulation was effective in reducing bed wetting did not actually achieve statistical significance. In other words, there was no statistical difference between the control and treatment groups. Most scientists call this a negative study, not a "win".

SMALL STUDIES WITH POSITIVE RESULTS NEGATED BY LARGER STUDIES

Regarding the proposition that spinal manipulation is effective in treating childhood asthma, the BCA listed a weak pilot study but failed to cite another study that employed a larger test group and blinding. That study showed spinal manipulation had no effect on childhood asthma. As evidence spinal manipulation is effective for colic, the BCA listed two case reports and four studies lacking blinding and placebo controls. Again, they conveniently failed to cite the best designed study on the subject came up negative. To some, the BCA appeared less than forthright in their own assessment of the efficacy of chiropractic treatment for children.

Nine Fake Diseases

ADRENAL FATIGUE

In 1998 naturopath James Wilson identified a condition he termed adrenal fatigue. Wilson suggested people experiencing a constellation of nonspecific symptoms like a general malaise, poor sleep, and craving salty foods are actually suffering from worn out adrenal glands. These glands fail to produce sufficient quantities of adrenaline and cortisol. The cure? Loads and loads of vitamins. And stop drinking all that coffee! No established medical organization recognizes adrenal fatigue as a real thing, however. A 2016 systematic review of studies published on adrenal fatigue ("Adrenal fatigue does not exist: a systematic review" by Flavio A. Cadegiani and Claudio E. Kater) flat out concluded "there is no substantiation that 'adrenal fatigue' is an actual medical condition. Therefore, adrenal fatigue is still a myth." It's just a garbage diagnosis by naturopaths to sell you expensive vitamins.

BRA CANCER

In 1995, Sydney Ross Singer published a book called *Dressed to Kill.* In it, he argues wearing bras affect the lymphatic system and increases the likelihood of a woman getting breast cancer due to a buildup of "toxins". Singer's data was based on unblinded surveys and didn't adjust for known risk factors like age. A few studies have been conducted to test this link but have found no evidence for it. It's not surprising that despite the complete lack of evidence, the claim lives on via the Internet and causes many women who have breast cancer to grieve they may be responsible for their condition. Multiple cancer organizations have published statements letting women know there is no link between bras and cancer.

IRLEN SYNDROME

Irlen syndrome was identified and named in 1983 by psychologist Helen Irlen. The belief is people with reading difficulty are suffering from a problem involving the brain's ability to process certain wavelengths of light. The claim is if a person with Irlen syndrome is trying to read in a room under certain lighting conditions (for example, fluorescent lights), the person's brain gets over excited, visual perception is distorted, and words or numbers on the page look different. The solution, provided by Irlen, is to issue people expensive glasses that filter out light wavelengths that supposedly cause these "brain quakes". Blinded tests of the efficacy of these lenses, however, have shown no effect whatsoever on reading ability. A rare joint statement by several ophthalmology groups, which included the American Academy of Ophthalmology and the American Association for Pediatric Ophthalmology, warned the "expense of such treatment is unwarranted and may divert resources from evidence-based treatment."

ELECTROMAGNETIC HYPERSENSITIVITY (EHS)

Some people think they can detect various radio waves being produced around them by electronic equipment and it's making them sick. Cell phones, cell towers, Wi-Fi routers, computer monitors, and "dirty electricity" in the home cause people to experience health problems that include headaches, poor sleep, fatigue, rashes, and achy muscles. All manner of bogus EMF absorbing products are marketed to these people. You can actually find some EMF-safe communities around the world. For example, the town of Green Bank, West Virginia (population 147) is a popular destination for people who believe they suffer from EHS. The town is in the middle of something called the U.S. National Radio Quiet Zone. The area is set up to reduce interference with telescopes at the National Radio Astronomy Observatory. Blinded studies have demonstrated people are unable to detect EMF fields and readily report suffering from the effects of EHS even when no source of electromagnetic radiation is present. One amusing demonstration of the placebo effect behind EHS was reported in South Africa. In 2009, a cell phone company erected a tower in a Johannesburg neighborhood. Residents almost immediately started reporting nonspecific symptoms typical of electromagnetic hypersensitivity. A meeting was called by the residents, demanding the company remove the tower. The company revealed to the residents that it had not been turned on during the time the residents were claiming to experience symptoms of "tower radiation".

FAN DEATH

There is a widely held belief in South Korea that leaving an electric fan on while you sleep can result in your death. The mechanics of this are a bit hazy, although many Koreans subscribe to the hypothesis that a running fan leads to death through asphyxiation. The fan blows back exhaled CO_2 molecules into the face of the sleeper, depriving him or her of oxygen. It's an idea but that's not what fans do. To help prevent fan death, electric fans in Korea are typically sold with a timer dial that will turn the fan off after a set number of minutes, ensuring the sleeper is not killed by the running fan.

MORGELLONS

Morgellons is a mythological disease where people believe they suffer from painful sores caused by unusual and unnatural fibers growing out of their skin. Real medical doctors recognize Morgellons as a form of delusional parasitosis. The fibers found come from people simply scratching sores and rashes and accidentally transferring pocket lint into their open sores. The "disease" was named in 2001 by a mother who thought her child was suffering from sores caused by the body's production of these mysterious fibers. Doctors could find nothing wrong with her son and suspected the mother was suffering from Munchausen's by proxy. The mother refused to believe her son was perfectly fine and just needed to stop scratching himself so much. She did her "research" and decided this was a disease and called it Morgellons after an old French term for black hairs. She founded the Morgellons Research Foundation. Morgellons has become the favorite diagnosis of chemtrail and Wi-Fi cranks.

MULTIPLE CHEMICAL SENSITIVITY (MCS)

If there is one fake disease that actually has had a pro-social outcome it's probably the belief that some people suffer from hypersensitivity to chemicals. In the 1990s a growing number of reports of MCS had many companies and entertainment venues asking people to tone down the use of perfume and cologne. MCS is not a recent phenomenon. Claims of MCS have existed since the 1950s. In the 1980s Sick Building Syndrome was a popular concern. MCS came into wider public consciousness following reports of Gulf War Syndrome. That issue seems to have gotten conflated into a wider belief some people are hypersensitive to certain chemicals, smells, or the imaginary off gassing produced by polyester ties. Like many fake diseases, MCS is characterized by people complaining of everyday gripes like fatigue, headaches, and dizziness. Blinded trials of people suffering from MCS show they react as strongly to placebos as they do to chemicals supposedly causing their symptoms. The list of substances that are claimed to trigger MCS is as long as a list of anything that humans make in a factory. That list includes perfumes, plastics, synthetic fabrics, and paint.

SPONTANEOUS HUMAN COMBUSTION

Spontaneous human combustion is the notion that people at times just burst into flame. Evidence is a smoking corpse that doesn't seem to have a source of ignition. The notion was suggested as far back as the 18th century. A study of cases that have been popularly ascribed to death through spontaneous human combustion generally share a constellation of similarities. Supposed victims tend to be non-mobile due to age or obesity. Many are smokers or were near a burning candle. Medical forensics usually determines the person died in their sleep, which led them to drop a burning cigarette or knock over a candle. These low intensity sources of fire over time start what is known as the wick effect. In essence a corpse's own body fat helps maintain a low smoldering fire that slowly consumes the body and clothing. Spontaneous human combustion is almost always suggested by lay people when confronted by a deadly fire that has an unknown cause. However, much controversy erupted in 2010 when an Irish coroner actually recorded an elderly man's death as resulting from spontaneous human combustion.

WINDMILL CANCER

In 2019, Donald Trump, while speaking at a National Republican Congressional Committee conference, promoted the bogus idea that windmills cause cancer. Trump has not been a fan of windmills since losing a 2015 lawsuit that tried to stop a wind farm development near his golf course in Scotland. In 2019, Trump was ordered to pay almost $300,000 in legal fees to the government. Trump also sort of missed what cranks are actually claiming about the health effects of living near a windmill. Cranks generally don't use Trump's term but prefer Wind Turbine Syndrome. Wind Turbine Syndrome was coined in 2009 by pediatrician Nina Pierpont. She believes people living near wind turbines are more likely to experience a cluster of nonspecific symptoms like headaches, poor sleep, visual blurring, and vertigo. She initially blamed the low frequency sounds wind turbines emit. Studies pointed out the vast majority of wind farms are located near sources of noise people have been living with for decades, like highways, aircraft, and farm machinery. In response, Pierpoint seems to have backed off infra-noise as the cause of people's symptoms. She now suggests wind turbines create a kind of alternating air pressure that the body interprets as seasickness.

Ten Breatharians

Breatharians are people who claim the only thing they need to sustain themselves is the life energy they extract from fresh air and sunlight. They can go months or years without food. Given the inherent danger of going without food for extended periods of time, Randi refused to allow such claimants for his million-dollar challenge.

PARACELSUS

Paracelsus (died 1541) was a Swiss physician and alchemist. He is known as the "father of toxicology". He gave us the quote "the dose makes the poison". It's a handy line to throw at anti-vaxxers who proclaim the dangers of adjuvants found in some vaccines. The Rosicrucians seem to hold Paracelsus in high regard, describing him as a "Monarch of the Empire of Wisdom". A Rosicrucian book from the 17th century titled *Comte De Gabalis* suggested Paracelsus took no food but lived on a "half scruple of Solar Quintessence".

MARTHA TAYLOR

Martha Taylor (died late 17th century) became something of a local sensation in the UK's Derbyshire county around 1662. Following an accident, Taylor was bed ridden and refused all food and drink. News got out about Taylor's supposed ability to survive without food or water.

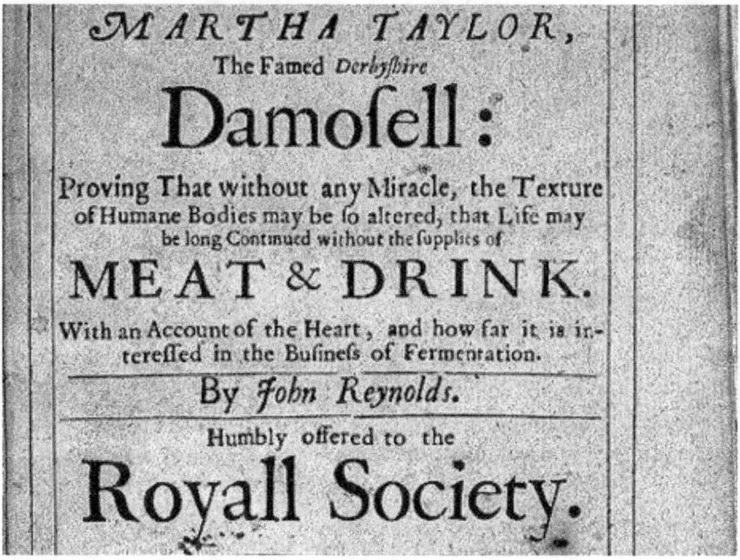
MARTHA TAYLOR,
The Famed Derbyshire
Damosell:
Proving That without any Miracle, the Texture of Humane Bodies may be so altered, that Life may be long Continued without the supplies of
MEAT & DRINK.
With an Account of the Heart, and how far it is interessed in the Business of Fermentation.
By John Reynolds.
Humbly offered to the
Royall Society.

Photo Credit: Wellcome Library, London. Wellcome Images images@wellcome.ac.uk

Many considered this a religious miracle and she began to attract pilgrims. One believer, John Reynolds, discounted religious explanations and thought her ability to go without food or water was physiological. According to Reynolds, doctors watched her for a period of 40 days and detected no fraud. He submitted a report on Taylor to the Royal Society in 1669 titled *A Discourse upon Prodigious ABSTINENCE: OCCASIONED By the Twelve Months FASTING OF MARTHA TAYLOR, The Famed Derbyshire Damosel: Proving That without any Miracle, the Texture of Humane Bodies may be so altered, that Life may be long Continued without the supplies of MEAT & DRINK.* In it he suggested several bodily mechanisms by which a person could sustain themselves without food. Reynolds astutely noted Taylor was a virgin and virgins had "seminal humors" which may supply the blood with all the nutrients the body needs.

THERESE NEUMANN

Sister Therese Neumann (died 1962) was a German nun who claimed to have miraculously cured her own blindness and appendicitis. She was also stigmatic and a breatharian. She claimed for the last forty years of her life she subsisted on nothing but the eucharist. This was put to the test in 1927. Medical personnel studied her for fifteen days. She lost about ten pounds during the examination period, suggesting she couldn't actually sustain herself eating nothing but eucharistic bread. Joe Nickell has suggested she was likely engaged in pious fraud.

WILEY BROOKS

Wiley Brooks (died 2016) was probably the first to popularize Breatharian pseudoscience in North America. He appeared on an early 1980s reality TV show called *That's Incredible*. The show featured a lot of paranormal topics. The show billed him as the man who took no food or water but could lift 1,000 pounds! Brooks founded the Breatharian Institute of America to teach others how to live off of sunlight and air. In the 1990s, the *Seattle Times* reported he was offering two-day seminars for $150. Around 2015, he upped his fees to $100,000. It's not clear how many signed on at that premium rate. In 2000 he was selling a product called "Fresh Air in a Bottle: Liquid Air (Not Water)." What might normally look like water to the uninitiated was actually air condensed into a liquid form. Normally that would require keeping it at a temperature of −317.83 °F. However, Brooks claimed he invented a new process to liquify water at room temperature. Evidence suggested Brooks himself wasn't able to live a Breatharian lifestyle. He was famously discovered exiting a 7-Eleven with a bag full of hot dogs and Twinkies.

MICHELLE PFEIFFER

Actress Michelle Pfeiffer revealed in 2013 she fell in with a breatharian cult in the 1980s. Pfeiffer was just at the start of her acting career and was convinced by a pair of cultists she could keep her trim figure by living off of air. Pfeiffer admitted in retrospect it wasn't the best idea and the couple were extremely controlling. She got out when her then husband was researching the Moonies. After reading a definition of a cult, she realized she was in one.

ELLEN GREVE

Australian Ellen Greve has been a leading advocate of "pranic nourishment" (that is, living without food or water) since 1998. Today she goes by the name Jasmuheen. She made a big splash in the world of total idiocy when she released a six-part video *The Legend of Atlantis: Return of the Lightmasters*. The Australian investigative news show *60 Minutes* got her to agree to test her claim she can live on just light and air. They sealed her in a hotel room and had her monitored by a doctor. After 48 hours her condition had so deteriorated the doctor asked her if she wanted to continue. She suggested obvious symptoms of acute dehydration and high blood pressure were being caused by the bad air in the hotel room. All agreed to continue the test at a mountainside retreat where apparently the air was better. Her condition continued to deteriorate forcing the doctor monitoring her to call the test off. Despite failing on national TV, Greve managed to attract thousands of followers.

LANI MARCIA ROSLYN MORRIS

Lani Marcia Roslyn Morris died in 1998 following the advice of an Australian breatharian cult. She visited the cult run by a husband and wife team to learn their techniques. After a week of taking no food or water, she became paralyzed and then eventually died. The husband and wife team, Jim and Eugenia Pesnak were convicted in 1999 of manslaughter.

PRAHLAD JANI

India's Prahlad Jani (died 2020) claimed he went without food or water for seven decades. He attributed this ability to the Hindu goddess Amba pouring an invisible elixir through a hole in his palate. Jani was tested twice under controlled conditions, once in 2003 and once in 2010. In 2003 he was observed in an Indian hospital for a period of ten days in what was described as a sealed room. He ostensibly had neither food nor water for that period and was not observed to have urinated or defecated. He was observed to have lost some weight. Neither test involved a trained magician creating the controls or in monitoring Jani. Interestingly he was allowed access to water while bathing. Although researchers assured skeptics he was monitored 24/7 via CCTV camera, he was actually allowed to move out of the camera's field of view.

RAM BAHADUR BOMJON

In 2006 the Discovery Channel aired the documentary *The Boy with Divine Powers*. It featured Ram Bahadur Bomjon, then a Nepalese teenager who claimed he could go months without food, water, or sleep. He was dubbed "the Boy Buddha". The documentary filmed him over a period of 96 hours. Although they were able to film him during daylight hours, and he certainly did not eat or even move when the camera was filming, the documentary makers were forced to turn their cameras off at night. At times during the day they were required by Bomjon's family members to cease filming and move their cameras. Bomjon attracted thousands of followers who left donations in the form of money and jewelry. He became a bit of a cash cow for his village. A bit more than a decade later, Bomjon started to attract less favorable attention when followers accused him of physically and sexually abusing them.

HIRA RATAN MANEK

Hira Ratan Manek was the subject of a 2011 documentary titled *Eat the Sun*. The documentary explored the phenomenon of "Sun gazers". These are people who stare into the sun and believe they can draw all their sustenance from sunlight. Manek boldly proclaimed even if you paid him a billion dollars he would not eat. Amusingly, Manek was caught by the filmmakers eating in a San Francisco Indian restaurant. It's not clear if he was offered a billion and *one* dollars to enjoy such a meal.

CRYPTIDS

Four Paranormal Explanations for Bigfoot

Bigfoot seems to be a wide-ranging primate. People have sighted Bigfoot in every state in the USA, with the exception of Hawaii. The question arises, for a species that seems to exist almost everywhere, why isn't there any definitive physical evidence? Bigfoot proponents, troubled by the lack of expected evidence, often resort to special pleading. The expected evidence is unavailable because of some special feature or ability that lets it avoid human detection.

BIGFOOT IS AN INTER-DIMENSIONAL SHAPE SHIFTER

Bigfoot hunter Jon-Erik Beckjord (died 2008) spent years trying to find Bigfoot after a chance encounter earlier in life. Unable to ever find hard physical evidence, he came to the conclusion the creature had a range of amazing powers to avoid detection. Bigfoot can warp in and out of our physical reality as well as bend light around it to turn invisible.

BIGFOOT IS AN EXTRATERRESTRIAL

Why is there no fossil record of large primates in North America? We can find, for example, Gigantopithecus teeth in Asia but we can't find a single tooth of any large primate matching the description of Bigfoot in North America. The obvious explanation is Bigfoot is a recent arrival to this planet, having stepped off a UFO. The idea goes as far back as 1973 when UFO researcher Stan Gordon started to note reports of people seeing Bigfoot alighting from flying saucers.

BIGFOOT TOO BELIEVES IN THE AFTERLIFE

Why have we never, in a century of looking, found a single piece of a Bigfoot corpse lying in the middle of a forest? Some Bigfoot proponents believe other Bigfoots take time to bury their fallen. The problem with this is burial tends to increase the likelihood of preservation.

BIGFOOT IS A QUANTUM BEING

How come Bigfoot can be seen by some people and not others? Why does Bigfoot seem to disappear when you actively go looking for it? Bigfoot is quantumly entangled with humans. Bigfoot can sense our energy and decohere when humans approach. The idea is championed by Ron Morehead, author of the so-called "Sierra sounds" (recordings of what he claims to be Bigfoot babble).

Five States with the Lowest Number of Bigfoot Sightings

The Bigfoot Field Research Organization (BFRO) maintains a database of Bigfoot sightings going back to at least 1920. According to their site, every state in the USA has at least one so-called credible sighting report except for Hawaii. It is possible, of course, there are Bigfoot sightings in Hawaii but BFRO *a priori* rules them out as credible. The following states have ten or fewer sightings:

Delaware: 5 sightings

Rhode Island: 5 sightings

North Dakota : 6 sightings

Nevada: 9 sightings

Vermont: 10 sightings

Nine Alleged Bigfoot Hoaxes

Photo Credit: Karl Mamer

JACKO HOAX (1884)

In 1884, the *Daily Colonist*, a newspaper in British Columbia, ran a story about rail workers who captured a creature described as four feet tall and resembling a gorilla. They named it Jacko. The rail workers turned it over to the authorities in Yale, British Columbia who then placed it in a jail cell. Within days of the article running, some 200 people turn up at the jail demanding to see the creature. They were met by a rather confused jailer who had to explain to them they were hoaxed. Competing newspapers of the day called the *Daily Colonist* out for falling for the story and not checking its sources. The story was largely forgotten until it was rediscovered in the 1950s by a news reporter who seems to have accepted the story as true. He shared it with some fellow cryptid enthusiasts as a possible early Bigfoot report. The story was then incorporated into Ivan T. Sanderson's 1961 book *Abominable Snowman: Legend Come to Life*. Bigfoot writers have ever since been citing the hoaxed article as authentic.

RAY WALLACE AND THE BIRTH OF BIGFOOT (1958)

Ray Wallace (died 2002) loved to play jokes. He found his stride in 1958 when he carved a pair of big feet, strapped them on, and strode around a logging camp he owned in Humboldt County, California. He was trying to prank his workers. He was more than successful. The loggers began to describe the person or thing leaving the tracks as "Big Foot". A newspaper reporter writing of the tracks' discovery changed the term to "Bigfoot".

WOMAN KIDNAPPED BY BIGFOOT (1976)

Cherie Darvell and two other friends were out in Humboldt Country, California apparently trying to film a documentary about Bigfoot. Darvell was allegedly kidnapped during the making of the film. The other two filmmakers went back to town and reported her abduction to the police. An expensive search ensued but to no avail. Within a couple days of the search commencing, Darvell walked into a nearby resort asking for help. She claimed Bigfoot had taken her to be its bride. Witnesses described her condition as being remarkably clean for a person who had been mauled by a Bigfoot and had spent at least two days seeking rescue. All concerned suspected a silly hoax meant to stoke publicity for their documentary, which was released in 1977 as *Wildman of the Woods: An Encounter with Bigfoot*.

BIGFOOT PANIC IN PA (1985)

After reports of strange noises and smells in his neighborhood made the local news, Craig A. Brashear of Enola, Pennsylvania decided he wanted people to think it was because Bigfoot had decamped near his town. He made a homemade Bigfoot costume and stood on the side of a road early in the morning. Multiple drivers caught brief but terrifying glimpses of Brashear in costume. News spread like wildfire and the town went into panic mode. People were locking their doors and windows at night for fear Bigfoot would come crashing through. And because this is America, people started arming themselves. Brashear's hoax didn't last long. An anonymous tipster called police and notified them they knew Brashear had recently constructed a Bigfoot costume and, well, maybe they should give him a talking to.

THE SONOMA HOAX (2005)

In 2005, the Showtimes series *Bullshit!* (starring Penn and Teller) produced a hoaxed video of a Bigfoot stomping around Sonoma, California. The video was claimed to be created by Mark Nelson who billed himself as an "amateur naturist" (this would technically mean he was a non-professional nudist). The producers salted it online to see how the Bigfoot community would react (the material would later be used for a *Bullshit!* episode on Bigfoot). The Bigfoot Field Research Organization (BFRO) fell for it, proclaiming the video authentic. In their words "We've seen plenty of hoaxed footage over the years as well. With that said, we are confident the Sonoma footage is not fake...This figure is most likely a real sasquatch..." Penn and Teller stepped forward to reveal they were behind the hoax. BFRO tried to save some face by alleging the magical duo were lying. BFRO suggested the video was genuine and Penn and Teller were just trying to sow doubt.

ICEBOX BIGFOOT (2008)

Bigfoot hunters Rick Dyer and Matthew Whitton claimed they found a Bigfoot corpse in a Georgia forest. Somehow, they were able to get 500 lbs. of dead weight Bigfoot into a large rectangular freezer. No one was able to examine the frozen cryptid except for fellow Bigfoot enthusiast Tom Biscardi. Biscardi pronounced it authentic. Published photos convinced almost no one. Many instantly recognized it for what it was: A popular Bigfoot costume available for online order. Maybe hoping to make the costume look more "ex-alive", Dyer and Whitter filled the suit with roadkill guts. Dyer and Whitter later admitted their hoax was intended as a joke.

BLINKY AND EWOK (2010)

Bigfoot made an odd appearance on Les Stroud's *Survivorman* cable show, courtesy of self-avowed Bigfoot expert Todd Standing. On a sixth season show, Standing aired what he alleged was high quality video of two radically different looking and thoroughly unconvincing Bigfoot creatures. One was dubbed Blinky because about the only action in the video is a Bigfoot blinking as it gazes through some foliage. One was dubbed Ewok because it resembled a static Ewok doll. Daniel Falconer, a special effects expert who helped create creature effects for *The Lord of the Rings* and *King Kong*, wasn't super impressed. He and a colleague analyzed the videos. They concluded the Ewok Bigfoot was just a prop. Blinky was a bit more interesting. Falconer used some morphology software to overlay Standing's face onto Blinky's. Falconer's overlay makes it clear Blinky's eyes, lips, and nose were actually Standing's. The blink was likely created by a known After Effects trick.

HANK THE LATEX BIGFOOT ATTRACTION (2012)

A few years after the icebox Bigfoot hoax, Rick Dyer rolled out another dead Bigfoot named "Hank". This one he claimed to have personally shot dead in Texas. Yes.

Texas. A place where eight-foot tall creatures covered in thick fur are sure to thrive in August. Apparently perpetrating one lame Bigfoot hoax is not enough to convince true believers to be skeptical. Dyer took this hoax on the road and charged admission to view the body. It's reported he made nearly $500,000 in ticket sales. Eventually a prop maker in Washington came forward, admitting he had made the Bigfoot out of latex and camel hair per Dyer's instructions. He didn't realize he was participating in a hoax for money. He believed he was making a movie prop. Similar to Dyer's 2008 hoax, as events unraveled, Dyer explained he was pulling another joke. However, he maintains he really does have a Bigfoot corpse on ice. He's just not allowed to show anyone.

RANDY LEE TENLEY DIES FAKING A BIGFOOT SIGHTING (2012)

Randy Lee Tenley was apparently trying to convince passing motorists on a stretch of Montana highway that they had seen a Bigfoot. One night he put on a ghillie suit and started jumping out in front of passing cars. He didn't live long. He was struck by a car and then run over a second time by another car. A Montana Highway Patrol officer described the attempted hoax as "one of the dumbest things I've ever seen."

Eight Things Mistaken for Bigfoot

BEARS STANDING ON THEIR HIND LEGS

Bears are known to walk upright at times. Standing on their hind legs gives them a better view and sniff of their surroundings. It's not surprising there's significant overlap between areas with large numbers of Bigfoot sightings and the North American bear's habitats.

BEARS WITH MANGE

In 2007, BFRO published a trail cam photo of what the group thought was a juvenile Bigfoot. The Pennsylvania Game Commission identified the creature as a bear with mange.

AN AFRICAN SLAVE

Probably one of the most horrific stories Bigfoot types keep passing along is the sad story from the 19th century of Zana. Zana was an African woman in the Russia Caucasus region who was held captive by local nobleman Edgi Ganaba. Ganaba exhibited her as an oddity and allowed men to rape her. Later Russian monster hunters merged her story with tales of almas (the Russian Bigfoot). Others over the years have speculated Zana was a yeti. Zana gave birth to at least three children as a result of the rapes Ganaba permitted. Bryan Sykes (died 2020), a Professor of Human Genetics at the University of Oxford, was able to obtain DNA from some of Zana's descendants. He determined her offspring were fully human and from their mitochondrial DNA he determined that Zana was of African origin. There's no mystery here that needs to keep being told. The only story that needs to be told of Zana is one of the unspeakable cruelties she had to endure at the hands of a so-called nobleman.

AN AMATEUR SHAMAN

Gawain MacGregor describes himself as a shaman who wanders around the forest wearing animal pelts as part of a sacrament. He does it to reenact a ritual from *The Epic of Gilgamesh*. This brings him closer to nature. Hey, whatever floats your boat. His sacramental wandering was captured on video by a North Carolina resident on the night of August 4, 2017. The video hit the Facebooks and the local news. Many reports came in it was an eight-foot tall Bigfoot. Out of an abundance of caution, police forces in the area reminded Bigfoot hunters not to try and shoot any monsters they encounter in the woods as "you'll most likely be wounding a fun-loving and well-intentioned person, sweating in a gorilla costume".

A HIKER

In 2014, Canada's CBC network reported a story of Myles Lamont (a wildlife biologist) who, while hiking on a mountain peak near White Rock, British Columbia, spotted and then videoed what he thought was a Bigfoot moving in the valley below. The video reveals a moving black dot on a field of snow. That's it. The biologist reasoned it had to be a Bigfoot because he was familiar with bears and the black dot didn't look like a bear. Lamont also reasoned it couldn't be another hiker as he couldn't see if the moving black dot had a backpack and snowshoes. A couple weeks later a man who refers to himself as Ridgewalker Pete told the CBC he was in that exact area hiking. It was kind of his thing and it's likely Lamont mistook him as a Bigfoot from about 7,000 feet away.

OTHER BIGFOOT HUNTERS

In 2017, a trio of amateur Bigfoot hunters headed out into the woods of Oklahoma. Oklahoma isn't what you'd call a Bigfoot hotspot. But, whatever. One of the hunters reportedly heard a barking sound and quickly shot the first humanoid form he saw. This happened to be one of the other Bigfoot hunters. The victim lived but all were arrested after they tried to cover up the shooting.

A ROCK

In 2013, Sharon Hill covered a story on her *Doubtful News* blog about a Utah man who found a 75-pound rock he believed was the fossilized head of a Bigfoot. The man knew it was a Bigfoot head — and not just a rock that looks a little like a face — because he himself had seen Bigfoot twice. His rock doesn't look a lot like a Bigfoot face. At best it looks like it was made by some high school kid who was trying to chisel out a miniature version of an Easter Island head and gave up 20% of the way through. A local newspaper sent pictures of the rock to a couple paleontologists. Both concluded it likely wasn't a fossil. The man had just found a weathered rock. The rock finder wasn't impressed with the conclusions of the experts. He felt they couldn't see it was a fossilized Bigfoot head because they had tunnel vision. In the words of Hill "One thing geologists do on a fairly regular basis is to inform people who think they found something awesome like a fossil or meteorite that it is not that thing, and there is a more mundane explanation. You know, often they don't believe us."

A TARGET SHOOTER

In 2018 a Montana man was setting up targets to practice his shooting when a Ford F-150 came roaring up and the driver fired multiple shots at him. The targeted target shooter was able to take cover behind a tree until the shooting stopped. When he confronted the pickup driver, he was told "I thought you were Bigfoot...if I see something that looks like Bigfoot, I just shoot at it." The pickup driver also cautioned the not-Bigfoot to wear an orange vest when he's out in nature.

Twenty Seven Unusual or Interesting Cryptids by State and Province

ALABAMA: HUGGIN' MOLLY

Huggin' Molly is reputed to be a seven-foot tall female ghost that haunts the streets of Abbeville, Alabama. It is so named because apparently the ghost likes to chase down people and hug them. That wouldn't be so bad if it were not for the claim this ghost woman also screams in your ear as she does it. The legend seems to be a local cautionary tale meant to scare children into returning home before dark.

ARIZONA: THUNDERBIRD

As the story goes, Arizona's *Tombstone Epitaph* newspaper is reputed to have published in the late 19th century a photo of a group of hunters standing proudly in front of their kill: a pterodactyl-like thunderbird. Many people will swear they've seen this photo but no one has ever been able to produce a copy.

ARKANSAS: FOUKE MONSTER

The Fouke monster is a Bigfoot variant known as a "skunk ape" due its reputed smell. It's described as being seven-feet tall and having red glowing eyes. The Fouke monster is the subject of the low-budget cult classic movie *The Legend of Boggy Creek* (1972).

CALIFORNIA: GIANT SPACE BRAINS

This was only seen once, in the early 1970s, but it's such a creepy story it seems to live on in the annals of cryptid mythology. As the story goes, in 1971, two friends in Palos Verdes, California reported seeing two giant blueish/grey brains floating in the middle of the road. Each had a single red eyeball poking out of their frontal lobes. The men, terrified, drove right on past the space brains. When they arrived home, they claimed their trip, which should normally take twenty minutes, had taken two hours. One of the witnesses underwent hypnosis about five years later and recalled the brains warning him something bad was going to happen to humanity in 1987. Nothing overly horrible happened to humanity as a whole in 1987 other than Starbucks opened its first store outside of Seattle.

COLORADO: SLIDE-ROCK BOLTER

Saturday Night Live introduced us to the "landshark" back in the 1970s. Colorado lumberjacks back in the 19th century gave the world the concept of a "mountain whale" known as a slide-rock bolter. More than a hundred years ago, lumberjacks felling trees near mountains told of seeing a whale-like creature that reputedly only lives at the top of mountains with steep slopes. The slide-rock bolter was said to wait patiently for unsuspecting tourists to wander by at the bottom of the mountain. It then, well, slides down and gobbles up the tourist. The creature could be a cautionary tale told to scare off tourists. However, it would seem one forest ranger took the legend seriously enough that he tried to bait the slide-rock bolter with an explosives-filled dummy. Something set the dummy off prematurely and leveled half the nearby town of Rico, Colorado.

CONNECTICUT: MELON HEADS

This cryptid should not be confused with 1990s punk band The Melon Heads. In Connecticut, melon heads are said to be a family of cannibals that live in the woods of the southwest. In some tellings, the cannibals are reputed to also be the product of inbreeding. Folklorists view the Melon Heads tale as a localized version of Scotland's Sawney Bean cannibal legend.

DELAWARE: SELBYVILLE SWAMP MONSTER

I'm not sure what's harder to believe. Delaware has its own Bigfoot-like cryptid or it has a swamp. The Selbyville swamp monster started as a hoax in 1964 by the editor of a small local newspaper named the *Delmarva News*. He wanted to increase circulation. So, he got a couple friends to put together a crude Bigfoot costume made out of a racoon coat. They took a photo and ran it in the pages of the *Delmarva News*. The trio kept up the hoax for a few months. One of the men dressed in costume would jump out at passing cars (don't try this at home!). They soon tired of the ploy. Despite rolling up the hoax, people report sightings of the Selbyville swamp monster today.

GEORGIA: ALTAMAHA-HA

Georgia has its own Nessie. It's known as altamaha-ha. According to native legends, altamaha-ha is a sturgeon-like cryptid that lives in the Altamaha River. The legend got a large boost in 2018 when it was reported a mysterious creature had washed ashore on a Wolf Island beach. Many speculated it was a dead altamaha-ha. The body was later found to be a hoax, made from the body of a shark and papier-mâché. A New York artist who calls himself Zardulu revealed the stuffed shark body was his latest attempt to merge performance art and viral memes. Zardulu is probably best known for creating the viral "Pizza Rat" video.

ILLINOIS: TUTTLE BOTTOMS MONSTER

Most states have their own version of Bigfoot. Illinois is no different. However, the Tuttle Bottoms monster, the Illinois Bigfoot, is distinguished by having a long anteater snout. Sightings started in southern Illinois in the early 1960s.

INDIANA: THE GREEN CLAWED BEAST

The green clawed beast is said to inhabit the Ohio River near Evansville, Indiana. It was sighted in 1955 by a pair of friends swimming in the river. One friend claimed to be attacked by the creature. Their description seems awfully close to that of the monster from *Creature from the Black Lagoon* which came out the year before. This a clear example of what *Monster Talk* host Blake Smith calls a scriptid.

KENTUCKY: THE HOPKINSVILLE GOBLINS

Around the same time the green clawed beast was being reported in Indiana, a family in Hopkinsville, Kentucky reported battling it out with small goblin-like creatures. The family described their home being attacked by creatures about four feet high with pointed ears and yellow glowing eyes. Evidence suggests some members of the family were intoxicated and might have been seeing great horned owls.

LOUISIANA: ROUGAROU

The rougarou is a were-creature that comes to Louisiana via French Canadian immigrants. The term is a Cajun derivative of "loup-garou", which is a North Eastern Quebec werewolf legend. "Rougarou" was actually trademarked by the owners of the New Orleans Hornets as a possible name for their NBA team.

The loup-garou is the subject of a popular card game in Quebec.
Photo Credit: Karl Mamer

MARYLAND: GOATMAN

The goatman is reputed to be a scientist who accidentally mutated himself in a crazed killer humanoid goat. He sports an axe and roams the wilds of Beltsville, Maryland. The goatman apparently attacks teens venturing out to popular "lover's lanes". This tale seems to have gotten started in the early 1970s and it's a variation on the classic "hooked killer" urban legend.

MICHIGAN: DOGMAN

The dogman of Michigan is a kind of cicada-like cryptid. The dogman only emerges on years ending in the number 7. Its first sighting allegedly took place in Wexford County, Michigan in 1887. The cryptid got a boost in 1987 when a Michigan DJ released a popular song called "The Legend". The song recounted what was purported to be the history of the dogman. In fact, the song was an April Fool's Day hoax. The DJ made up the history recounted in the song out of whole cloth. In 2007 filmmaker Mike Agrusa released what is known as "the Gable Film". Agrusa initially claimed it was an 8mm home movie from the 1970s he had acquired in an estate sale. It's about two minutes of people zipping around on snowmobiles and chopping wood. It ends with what is purported to be footage of the dogman rushing the camera holder. Agrusa later admitted the film was a hoax, but a well-engineered hoax. Agrusa used vintage film equipment and his own collection of snowmobiles from the era to heighten the film's authenticity. Though some cryptid experts asserted it couldn't possibly be an actor in a suit, Agrusa's dogman was exactly that: it was him scampering around on all fours wearing a ghillie suit.

MINNESOTA: ICEMAN

The Minnesota iceman was a hoaxed sideshow exhibit. It appeared to be a hominid frozen in a block of ice. It was billed in the 1960s and early 1970s variously as the fabled "missing link" or a frozen Bigfoot corpse. The creature has several provenances. The name comes from one version where the owner claims to have shot a Bigfoot-like creature in the forests of Minnesota. The Smithsonian Institute was called in at one point to investigate. Some people were worried it might be an actual human murder victim and this was someone's attempt to follow the maxim "the best way to hide is out in the open". Smithsonian investigators determined the Minnesota Iceman was most probably a latex dummy after discovering the owner had earlier contracted with a California prop shop to make an ape-like dummy.

NEBRASKA: WALGREN LAKE MONSTER

Between 1921 and 1923 several newspapers, including the illustrious *London Times*, began circulating a story of a monster in Nebraska's Walgren Lake. The monster supposedly resembled a giant alligator, but with horns. Its favorite meal was livestock that grazed too close to the lake. The source of every story was traceable back to a newspaperman and serial prankster named John G. Maher. Maher had successfully carried off four previous hoaxes, including burying a cement statue that convinced some archaeologists they had found a petrified human. Although the Walgren Lake monster was known to be a hoax by the 1950s, many still believe in it today. Its size and description have grown and changed over the decades.

NEW JERSEY: THE JERSEY DEVIL

The Jersey devil is one of America's oldest cryptids. Reports go back to the first half of the 18th century. The legend goes that in 1735 a witch known as Mother Leeds was living in the Pine Barrens of New Jersey and gave birth to her 13th child. That child was born a devil. The citizens of the Garden State have been reporting sightings of the creature ever since. It's said to have the head of a horse and the wings of a bat and terrorizes anyone who ventures into its forest. It's suggested the legend of the Devil was an attempt to smear the family of Mother Leeds. The Leeds were actually not a coven of witches, but a politically influential family given to a form of Christian esotericism. Their beliefs and politics came into conflict with the Quakers. Elements within the Quaker population sought to recast the Leeds family as, well, devils. The Leeds family crest featured a stylized Wyvern, which some scholars think became the basis for the legend.

NEW YORK: LAKE GEORGE MONSTER

In 1904, artist Harry Watrous (died 1940) fashioned a lake monster out of a ten-foot log. He attached a neck and head to the log. He floated it in Lake George, the favorite summer fishing hole of a friend and rival. Although Watrous had intended to fool his friend, he actually started a monster frenzy among the locals. Amused by this, he returned to the lake several more times with his log to generate more sightings. He kept up the hoax for a number of summers before giving up. Sightings of "Georgie" have continued long after Watrous abandoned his ruse.

OHIO: THE LOVELAND FROG

Since the mid-1950s, residents of Loveland, Ohio (a suburb of Cincinnati) have reported seeing a humanoid shaped frogman. The first recorded sighting is an unsourced story from 1955 that a travelling businessman (never named in the tale) saw three frogmen on the side of the road. The year 1955 seemed to be ripe for sightings of green cryptids. Indiana had its green clawed beast sighting and Kentucky had its Hopkinsville goblins. The Loveland frog would appear to be another "scriptid" influenced by 1954's *Creature from the Black Lagoon.* Unlike the one-off nature of the green clawed beast and the Hopkinsville goblins, the Loveland frog has had repeat sightings. The frogman popped up again in 1972. This time a police officer reported seeing the Loveland frog scurry across the road in front of his cruiser. Video of the frogman turned up in 2016. It was allegedly taken by a pair of teens wandering around while playing Pokémon Go.

SOUTH CAROLINA: LIZARD MAN

South Carolina's lizard man supposedly lives in the Scape Ore Swamp and has a habit of attacking and damaging cars. The first report emerged in 1988 when teenager Christopher Davis (died 2009) was driving home from his late-night job at a fast food restaurant. He reported his car was attacked by a seven-foot tall lizard man with red eyes and three fingers. According to one version of his tale (he told more than one) a lizard man leapt on top of his car and apparently did quite a number on the paint job. No pictures have ever been produced of these scratches, however. A local radio station offered a $1 million reward for the live capture of the creature. The area immediately became a tourist hotspot. A local airman reported a follow-on sighting, offering lizard scales and blood as evidence. He later confessed he was perpetrating a hoax in hopes of keeping the legend alive. Another incident of car damage was reported in 2008 and blamed on the lizard man. Evidence included blood traces. The blood turned out to be from a dog. In 2017 before a solar eclipse the South Carolina Emergency Management Division sent a tweet warning locals to keep an eye out for the lizard man. The tweet was meant to be humorous but not everyone interpreted it as such.

TEXAS: BLACK-EYED CHILDREN

Claims that black-eyed kids are wandering around Abilene, Texas, knocking on doors, and asking to come in appear to have popped up in 1996 on an email mailing list devoted to ghost stories. It emerged following a plot line in the *X-Files* about a mysterious black oil that infects people, leaves their eyes an inky black, and controls them. Stories of creepy otherworldly children are a common trope found in urban legends and the black-eyed children became a new variant. Despite everyone having a cell phone camera, no one has ever captured a picture.

VERMONT: NORTHFIELD PIGMAN

Vermonters, when raising a glass of maple syrup around a campfire, will tell the spooky tale of Sam Harris (but not *that* Sam Harris). This Sam Harris was a teenage boy that went missing on Devil's Night of 1951. It is said he was taken by the Devil and turned into a crazy pig killer. Harris dawns the head of slaughtered pigs and terrorizes the good people of Northfield as well as tourists visiting a waterfall known as "the Devil's Washbowl". The Pigman was reportedly seen at least twice in 1971. One sighting was by a farmer investigating strange noises out in his field. The other was by a group of teenagers out drinking.

VIRGINIA: THE BUNNY MAN

Fairfax County, Virginia is home to something known as Bunny Man Bridge. Under this bridge is a most unusual "bridge troll": a bunny man with an axe. The bunny man is reputed to have murdered numerous people in and around the bridge. It has become, in recent years, a popular Halloween destination for so-called legend trippers. This legend seems to have its genesis in an actual event. In the weeks before Halloween of 1970, a man wearing either a bunny costume or a white outfit topped with bunny ears approached a parked couple. The oddly attired man yelled at them they were on private property. The couple drove away quickly. After arriving home, they noticed a hatchet on the top of their car. They reported the incident to police. Interestingly, only the boyfriend reported seeing the man with bunny ears. The girlfriend told the police the man's costume resembled the white robes and the hood of a Klansman. There was a second sighting shortly before Halloween 1970. A security guard patrolling a construction ground was threatened by a man wearing a bunny costume and hefting a long-handled axe. The bunny man issued a similar warning about trespassing. Newspaper reports of the bunny man resulted in dozens of additional sightings and forever cemented the local legend which has grown over decades of telling.

Photo Credit: Karl Mamer

WISCONSIN: THE BEAST OF BRAY ROAD

The beast of Bray Road is a werewolf legend that got its start after a 1936 sighting. The man-sized werewolf is said to haunt Bray Road, which is near the farming community of Elkhorn, Wisconsin. People have reported sightings as recently as 2020. Half eaten bodies of deer and livestock are usually blamed on the beast of Bray Road.

ALBERTA, CANADA: MERMAN

When you think of mermaids and mermen, you don't normally think of looking in the landlocked Canadian Rockies for evidence. You will, however, find in Banff, Alberta a well-preserved merman on display at a shop called the Indian Trading Post. The store is over 100 years old. It began as a place where trappers could sell their furs. Today it's a kind of curio shop. As befitting a curio shop, it has what is reputed to be a merman. The merman appears to be half human half fish and fully desiccated. Taxidermists would recognize the creature as a "gaff". A gaff is a novel creature created by taxidermists by stitching together parts from different animals. Not all gaffs are intended as hoaxes, however. It's generally considered an art form.

ONTARIO, CANADA: THE TUNNEL MONSTER OF TORONTO

In August 1978 a resident of Toronto's Cabbagetown went out looking for his lost cat. Assuming it might have gotten lost in a sewage tunnel near his apartment building, he entered the tunnel with a flashlight in hand. He didn't find his cat. What he did find was a three-foot tall humanoid creature with grey fur. In English, it hissed at him "Get out". So, he did. The story was reported a few months later in the *Toronto Sun* newspaper. The incident might have been forgotten but it has been immortalized in the works of UFO author Commander X. Commander X claims the incident is proof there's an abandoned alien city under Toronto.

QUEBEC: CHAMP

Champ is Lake Champlain's version of Nessie. The lake spans the border between Vermont and Quebec. Some cryptozoologists claim Champ was first spotted by explorer Samuel de Champlain. However, the source of that claim is from a fake quote found in a 1970 issue of *Vermont Life*. The most famous piece of evidence for Champ's existence is a "blobsquatch" known as the "Mansi photograph". That photo appears to show the head and curved neck of a sea serpent. However, the water where the monster was allegedly spotted isn't much deeper than 14 feet. That is likely too shallow for a large sea monster to be submerged up to the neck. It has been suggested what was photographed was a partially submerged log with a hooked branch resembling a sea monster neck and head.

Top Cryptids by Country: A to Z

A cryptid might be best defined as an animal thought by some to exist but lacking the kind of evidence generally required by the scientific community to establish its factual reality. The greater the delta between the quality of evidence offered and the certainty of proponents about the creature's factual existence, the more the animal is in the "cryptid" category. So, a tentative suggestion by ornithologists that ivory-billed woodpeckers might not have gone extinct based on a few sightings by experienced birders would make that bird a poor inclusion on a list of cryptids. A bold assertion that fuzzy photos and easy-to-hoax footprints is absolute proof of the existence of Bigfoot would make it a cardinal example of a cryptid.

Here we explore some of the lesser known, and more surprising, cryptids in 25 different countries.

AFGHANISTAN

Yabalik-adam is said to be a wild man with yellow body hair that stalks Afghanistan's Pamir and Kunlun Mountains. It is said one was captured by hunters in 1912 and fed raw meat until authorities came and took it away.

BELGIUM

Maasie is a water monster that is reported to inhabit the Maas river in Belgium. It's described as looking like a crocodile. This has led many to speculate it actually is someone's discarded crocodile pet.

CAMEROON

Cameroon is home to one of the most famous cryptids after Bigfoot, Nessie, and the yeti: Mokele-mbembe. Mokele-mbembe is reputed to be a species of sauropod dinosaur that supposedly escaped extinction. Mokele-mbembe is a favorite of the young Earth creationist set who believe if they can capture one it is definitive proof of a young Earth.

DOMINICA

The soucouyant is a blood sucking hag alleged to prey upon sleeping people. While asleep, the soucouyant draws blood from the person's arms and legs, leaving tell-tale blue and black marks. Anyone unlucky enough to be fully drained by a soucouyant is doomed to not only die from blood loss but return from the dead as one.

ECUADOR

The entzaeia-yawá is an aquatic tiger that is said to inhabit the rivers of Ecuador. It's reputed to be slightly bigger than a jaguar with paws that resemble duck feet. Some believe it is a living species of saber-tooth tiger.

FRANCE

France is one of about half a dozen countries with phantom kangaroo sightings. Be skeptical if you see a kangaroo off in the distance but if you were west of Paris and you did happen to spot a phantom kangaroo, it might not actually be a product of your imagination and too much cheese. There is a population of wallabies living in the forests of Rambouillet. They have been there since a handful of red-necked wallabies escaped from a French zoo in 1970.

GHANA

Sasabonsam is a tree-dwelling vampire-like creature said to inhabit the jungles of Ghana. It's humanoid shaped with a human face and a long beard. It favors hanging from a branch upside down, clinging with iron hooks at the ends of its feet. There it waits to envelop unwary hunters with its short, leathery bat wings.

HUNGARY

Not many cryptids are ever said to be extinct. Hungary's fisher pig is a rare exception. It is featured in the legends of Hungary shepherds. It is reputed to have gone extinct near the end of the 19th century according to the writings of Hungary's famed horticultural writer Imre Somogyi (died 1947). Unlike regular feral pigs that graze, the fisher pig, as the name suggests, lived on a diet of aquatic life found in local swamps.

INDONESIA

The island of Java is supposedly home to a giant bat cryptid known as the ahool. The ahool is a child-sized creature with a 12-foot wingspan. By comparison the largest known bats (the so-called "flying foxes") have wingspans that max out just shy of 5 feet. The ahool is said to have occasionally dined on humans.

JAMAICA

The duppy is a malevolent spirit that can take many forms. In Jamaica, a duppy has been known to manifest itself as a three-legged horse. It comes out at night to terrify locals and kill people with its poisonous breath. Although having hooves and only three legs, it apparently is adept at climbing trees to avoid hunters.

KENYA

Kenya's nandi bear makes the Alaskan Kodiak look like a cuddly teddy bear. Artwork depicts it as towering over human victims. No fence can contain the nandi bear as it is said to be able to leap over fences or smash its way through. Physical evidence includes track sightings (although no one has occasioned to take a photograph or plaster cast). It is said four specimens were sent to various museums. Conveniently, all seem to have disappeared.

LIBERIA

Emela-ntouka is an elephant-sized creature said to have the tail of a crocodile and a single ivory horn arising from the end of its snout like a rhino. According to legend it is generally an herbivore but is said to attack and kill elephants and hippos that happen to cross its path. Its favorite method of taking these large animals down is spearing them in the belly with its horn, leaving them to bleed to death. Cryptozoologists of the creationist bent believe it is yet another example of a living dinosaur, hence proof to them of a young Earth.

MONGOLIA

Mongolia's Gobi Desert is home to the famous Mongolian death worm. Although one is tempted to instantly compare it to Frank Herbert's giant sandworms, the Mongolian death worm is said to be only about 2 feet long. Though this worm is "only" 2 feet long, don't trifle with it. It is said to be highly poisonous. Merely touching it can result in instant death.

NORWAY

Eighteenth and 19th century Norwegians living along the coast of Norway reported sightings of a cryptid known as a super-otter. Although referred to as a super-otter, Norwegians believed it was a whale-like sea serpent. It took the name because of a superficial resemblance to a giant otter. Some cryptozoologists have suggested Norwegians were spotting some unknown species of primitive whale that still has vestigial limbs.

OMAN

Oman's desert oasis of Bahla is said to be a hive of jinn activity. Jinn (or better known as genies to the West) are said to be supernatural spirits with something of a pre-programmed agnostic bent towards the reality of Allah. If you were looking to find ground zero for the jinn, Bahla is your go to place.

PARAGUAY

Curupira is said to be a man-like cryptid that walks on feet turned backwards. They use their backwards feet to confuse hunters who might be trying to track them. They seem to enjoy exacting great revenge on hunters that kill animals taking care of their young. Some cryptozoologists believe curupira sightings attests to the existence of a spider monkey that supposedly died out in the late Pleistocene.

QATAR

The Qatar ghul is an Arabic version of the ghoul. It is said to feast on human corpses. Upon consuming a dead body, the ghul takes on the form of the deceased.

RUSSIA

Despite the woolly mammoth having gone extinct over 4,000 years ago, people have been reporting live sightings of the behemoth through the modern age. The Internet lit up back in 2012 when video footage, shot "blobsquatch" style, showed what appeared to some as a woolly mammoth crossing a Siberian river. To many it looked like an out-of-focus bear with a fish hanging out of its mouth.

SURINAME

Suriname is reputed to be one of the stomping grounds of the hairy South American dwarf. The creature appears to be, in essence, a pygmy form of Bigfoot. Similar to Bigfoot, the cryptid has a wide range. Sightings have been reported as far north as Suriname and as far south as Tierra del Fuego.

TANZANIA

If you're in Tanzania and hear the sound of a rooster look up, not down. You could have a crown crested tree snake drop on you. Its venom is said to be instantly deadly to humans if bitten. The crown crested tree snake subsists on a diet of maggots and kills larger animals to "farm" new maggots.

URUGUAY

First spotted in Uruguay in the 19th century, the minhocão is reputed to be a subterranean serpent, over 150 feet long, that tunnels underground. It is said to tunnel with such ferocity that anything above ground it passes under suffers terrible damage. Witnesses describe it as having scales like an armadillo, causing some cryptozoologists to suspect it's a giant version of the species. Those who study folklore and legend ascribe the creature to a myth created to explain damage caused by earthquakes.

VENEZUELA

All new world monkeys have tails. That's kind of a rule. But Swiss anthropologist George Alexis Montandon begged to differ. He claimed he came across a photo of a strange, tailless ape taken by his friend Louis François Fernand Hector de Loys. De Loys was a geologist in the 1920s who spent time in Venezuela searching for oil. He reported discovering a strange tailless ape that walked upright and was considerably larger than any local monkey species. Naturally he shot it and took some photos. Other scientists were not impressed. They said the photo simply depicted a normal-sized spider monkey with its tail cut off. As for its unusual size, the single photo de Loy produced curiously had nothing in the photo by which the creature's size could be judged. Researchers have suggested the photo was a hoax. Montandon was keen on finding support for a pet crank theory of his that humans evolved in many different parts of the world from local species of monkey. Since humans in South America lack tails and no species of New World monkey is tailless, Montandon needed "de Loys monkey" as a kind of missing link.

YEMEN

Yemen is home to a rare cryptid that proved to be real. The Yemeni waran (a species of monitor lizard) was long considered to be legendary. It wasn't until the mid-1980s that the creature was discovered to be real. A zoologist was watching a 1985 documentary about Yemen and spotted living examples. It was then realized the British Museum actually had a few specimens of the lizard but scientists there had considered them a hoax.

ZAMBIA

The swamps of Zambia are said to be the home of a smooth-skinned alligator known as the nsanga. This fearsome scaleless gator is said to feast on hippos. Some cryptozoologists think it's a species of dinosaur that escaped extinction. Where the mokele-mbembe is said to be an extant species of sauropod, the Nsanga is thought to be an extant species related to the triceratops.

Nine Classic Loch Ness Monster Photos Explained

Nessie has been a long-time boon for Scottish tourism. Forbes pegs the monster's value at about £60 million a year for the local economy. A lot of interest in Nessie over the decades has been driven by anomalous photos published in books and featured on TV shows.

THE HUGH GREY PHOTO (1933)

In 1933 the newspaper *Scottish Daily Record* published a photo by Hugh Grey of what it purported to be the first ever photo of the Loch Ness monster. The photo is full of motion blur or possibly a double exposure. In any case, it requires a great deal of mental filtering to see a sea monster. It could be of a whale-like creature dipping its head into the water. Some have suggested it's a picture of Grey's dog mucking about in the water. However, Darren Naish, author of the *Tetrapod Zoology* blog, has argued convincingly that the photo is of a swan dipping its head into the water. The position of the neck and body shape match well.

THE SURGEON'S PHOTO (1934)

Although Grey's photo is the first photo of Nessie, the Surgeon's photo is the most iconic. It became known as the Surgeon's photo because it was ostensibly taken by London doctor Robert Kenneth Wilson. As the story goes, he and his friend Maurice Chambers were holidaying near the loch in 1934. The photo features what looks like a long, arched neck with a small plesiosaur-like head. It's the first photo to suggest perhaps Nessie was a small population of extant plesiosaurs. Experts on plesiosaurs doubt what's in the photo is a living example, however. Plesiosaurs simply weren't able to crane their necks in that fashion. A 1984 analysis of the photo suggested the "creature" in the closely cropped image was probably only three feet long. More definitive evidence of the Surgeon's photo being a hoax came after the death of Chambers. His personal papers revealed the photo was faked. A toy submarine with the head and neck attached was floated in the lake. Two photos were then taken. Chambers enlisted Wilson to be the "front man" in hopes his status as a doctor would lend credibility. It certainly did.

THE TAYLOR FILM (1938)

In 1938, South African tourist G. E. Taylor is reputed to have filmed something in the loch using early 16 mm color film stock. He gave the footage to science writer Maurice Burton. Burton included a still from the film in his 1961 book *The Elusive Monster*. There doesn't seem to be any surviving copy of the film and it's unknown if anyone has seen it other than Taylor and Burton. The photo still from the film only shows a couple black blobs. They could be anything.

THE STUART PHOTO (1951)

The Stuart Photograph (aka "The Humps) is a grainy black-and-white photo that appears to show three humps cresting the water. The photo was taken in 1951 by a government employee named Lachlan Stuart. In the 1970s, researchers were able to identify the photo's location and realized the patch of water was close to the shore and very shallow. In the 1980s, a man by the name of Richard Frere admitted he aided Stuart in creating the photo. He had supplied Stuart three bales of hay which were used to create the humps.

THE MACNAB PHOTO (1955)

In 1957 a book called *More Than a Legend* (by Constance Whyte) featured a photo of Nessie that is maybe only second to the Surgeon's photo in fame. This one features Urquhart Castle on the right side of the frame and what looks like the humps of a giant sea serpent cresting out of the water. The photograph was taken by bank manager Peter MacNab in 1955. According to MacNab he was framing up a photo of the castle when he noticed a disturbance in the water and captured two photos of Nessie. The published photo was a mystery until 1976 when a researcher by the name of Roy Mackal was able to obtain the original negatives. What was on the negatives wasn't what was published in Whyte's book. It was clear the published image had been doctored. It has been suggested MacNab took a photo of a boat's wake, painted out the boat, and re-photographed the retouched image.

THE RAYNOR FILM (1967)

Richard Raynor was 17 when he filmed something moving rapidly and creating a wake through the loch. Many consider it the most compelling film ever taken of Nessie. Raynor, who these days works as a tour boat captain on the loch, is skeptical of Nessie being the best explanation for his own film. He's suggested what he filmed might have been a group of swimming birds as he himself has witnessed birds producing a wake identical to what he filmed. Given the distance and the camera's resolution, the only thing it picked up was the wake itself.

THE FLIPPER PHOTO (1972)

The so-called Flipper Photo seemed to be smoking gun evidence of the reality of Nessie. It was taken in 1972 by a group called the Academy of Applied Science. The group used an underwater camera with a bright strobe light to capture what is undeniably a diamond-shaped flipper of some unknown animal. The photo generally circulated was claimed to be "computer enhanced". Interestingly, the so-called enhancement method has never been published and that is key. The original appears to be a mass of bubbles, although one region is suggestive of a flipper. No one since has ever been able to produce a computer enhancement routine that's been able to transform the original into such a crisp flipper photo. It would appear the flipper photo was artistically enhanced.

THE LOCH NESS MUPPET (1977)

It's probably no coincidence the clearest photo ever of Nessie is also the least believable. It is, in essence, a modern reboot of the Surgeon's photo but in color and taken from a three-quarters perspective. The *Daily Mirror* ran it on page 1. It's so obviously a static puppet of some kind (the creature makes no obvious wake in the photo) that skeptics have dubbed it the Loch Ness Muppet.

APPLE MAP NESSIE (2014)

The *Daily Mail* ran a screencap from an Apple Maps satellite photo that appears to show a strange creature in the loch. It was estimated at 100 feet long and featured two flippers. The only problem is it doesn't look anything like a sea serpent. Mick West, creator of *Metabunk*, quickly figured out it was a satellite image of a boat wake. One can just make out the boat itself in the original image if you play with the photos sharpness and contrast a bit.

Eight Mystery Carcasses

Common everyday animals can look pretty mysterious and terrifying if devoid of their fur. Likewise, animals that have decomposed at sea and then washed ashore can be hard for the average person to identify. They immediately get identified as some kind of unknown nightmare creature.

THE BEAST OF EXMOOR

In 2009 a rotting carcass was discovered on a beach west of the Exmoor moorlands. It appeared to have the teeth of a big cat. Local mystery mongers wondered if this was the carcass of the legendary Beast of Exmoor. The Beast of Exmoor is said to be an "alien big cat" that preys on local livestock. Sightings of the alien big cat started in the 1970s. A number of zoologists, including Darren Naish, were able to quickly identify the skull as belonging to a grey seal.

THE DEMON DOLPHIN OF SAKHALIN

In 2015 a creature washed ashore on the island of Sakhalin. It was characterized by a long beak and what was described as fur. Several photos were taken of the carcass but tides washed it back out to sea before tissue samples could be taken. The best bet was the body was a highly decomposed bottlenose whale. Some argued this was an unknown marine mammal creature as some of the photos appeared to show fur. Whales and dolphins do not have fur. The fur, however, was likely connective tissues exposed by decomposition.

HUMAN FEET

Since 2007, human feet have been washing ashore on British Columbia beaches. At least 21 feet have been found. Most were found in sneakers. It was initially suggested this was the work of an entirely bizarre serial killer. However, at least three were traced to people who had died of suicide or accidents at sea. It has been suggested that when a person dies at sea, decomposition causes the body to separate at the ankle. Modern sneakers offer both buoyancy and a level of protection from scavengers. Pacific currents eventually carry the floating feet to the coasts of British Columbia.

MONTAUK MONSTER

Montauk, New York is home to the Brookhaven National Laboratory. The facility supports research in high energy physics and biosciences. At least seven Nobel prizes have been awarded to scientists who did their research at Brookhaven. The lab is also the subject of many conspiracy theories. When an odd hairless creature washed up on the shores of Montauk, pictures of it quickly circulated around the net. Many tin foil hatters thought it might be some ghastly creature cooked up in the labs of Brookhaven. Zoologists identified the "monster" as a raccoon, which had likely died and lost its fur due to decomposition and wave action.

SAN DIEGO DEMONOID

As the name implies, the San Diego Demonoid washed ashore on a San Diego beach. The hairless remains featured terrifying fangs. *Vice* ran with the story in 2012 after a 19-year-old snowboarder sent in some photos which he thought might be a chupacabra or some kind of hell beast. Zoologist Darren Naish was able to identify the creature as an opossum based on its dentition.

SCOTTISH POLAR BEARS

In 2016, several carcasses with what looked like white fur were found washed ashore on the Scottish island of Colonsay. Amateurs speculated they were the remains of a family of polar bears that had died at sea. Marine biologists suggested the remains were minke whales, which were common to Scottish waters. What looked like fur were strands of blubber.

STRONSAY BEAST

The Stronsay beast might be the first publicized mystery carcass. It washed ashore on the island of Stronsay in Scotland in 1808. Scottish naturalist Patrick Neill (died 1851) believed it was an unknown sea serpent and gave it the scientific name *Halsydrus pontoppidani* (Pontoppidan's sea-snake). Neill was a gardening editor for the *Encyclopedia Britannica* and had no expertise in marine biology. Other contemporary scientists disagreed with Neill and suggested it was most likely a basking shark.

ZUIYO-MARU CARCASS

In 1977 the Japanese fishing trawler *Zuiyo-maru* pulled up something in its nets that looked like a decomposing plesiosaur. The crew believed they had caught an ocean-going version of Nessie and took a photo of it. That photo has been reproduced hundreds of times in articles and books about the supposed reality of sea monsters. Young earth creationists cite the creature as evidence of a young Earth. It sure does look like the creature had a long neck and dorsal fins. It didn't help that at the time three reputable Japanese marine biologists suggested to the media it could be a plesiosaur. Although DNA was not available at the time, skeptical scientists were able to use amino acids from the carcass to make a match with a basking shark.

Eight Anomalous Animals

Anomalous animals are animals that are generally thought to have gone extinct in more recent times but there is some evidence of an extant population. As well, they are potentially new species with a better degree of biological plausibility than traditional cryptids like sea monsters and living dinosaurs.

BLOOP

In 1997 the U.S. National Oceanic and Atmospheric Administration (NOAA) recorded a loud, ultra-low-frequency underwater sound. It was dubbed "Bloop". The source was an area of the Pacific Ocean over 1,000 miles west of the southern tip of South America. It was not a man-made sound nor was it from any natural source NOAA knew about at the time. The sound was so loud, however, it was suggestive of a creature far larger than a blue whale. Cryptozoologists and various mystery monger shows ran with this as evidence some mysterious leviathan is trolling the depths of the ocean. Cue headlines about stunned and baffled marine biologists. That picture, however, was inaccurate. The story of Bloop comes to us from a 2002 article in *New Scientist* magazine titled "Calls from the Deep". The article was about odd sounds NOAA has cataloged over the years. In addition to Bloop, it had other colorfully named mystery sounds like Train, Whistle, Upsweep, Slowdown, and Gregorian Chant. One of the NOAA personnel being interviewed by *New Scientist* named Christopher Fox suggested Bloop *might be* biological. And if biological, the creature had to be large. Subsequent writers, citing the short bit from the article about the sound, expanded the story of Bloop into a huge mystery. Follow-on stories sanded off Fox's provisional suggestions Bloop could be biological and spun it as the sound could *only* be biological in origin. Years later NOAA did manage to identify Bloop's source. It wasn't from a sea creature. It was an ice quake caused by a disintegrating iceberg.

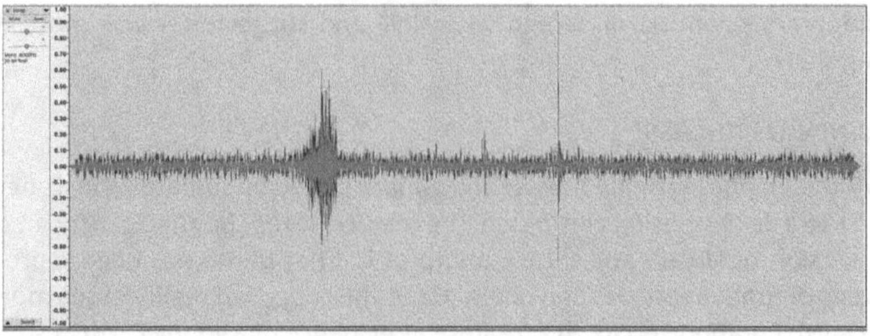

The Bloop sounds file downloaded from NOAA.

IVORY-BILLED WOODPECKER

Depending on who you ask, the American ivory-billed woodpecker is either extinct, probably extinct, or critically endangered. That said, the last sighting everyone agrees upon was in 1944 in Louisiana. Hunting and logging are thought to be responsible for the bird's demise. Birdwatchers have reported sightings of the ivory-billed woodpeckers over the decades. Most of these sightings have, however, been misidentifications of other species of woodpecker. Videos and photos of possible living examples have proven to be little better than blobsquatches.

ORANG PENDEK

The Orang Pendek is a creature from Indonesian folklore. Legends portray it as a three-foot tall hominid living in the mountain jungles of Sumatra. Over the last 100 years there have been sightings by locals and Dutch colonists. On its surface, the plausibility of Orang Pendek seems reasonable enough. And cryptozoologists have been really pulling for this one to be real. Cryptid hunters put a lot of stock in indigenous legends and eyewitness accounts. Unfortunately, solid evidence has eluded believers, despite multiple expeditions to find a colony of living Orang Pendeks. A total lack of physical evidence hasn't stopped some more reality challenged cryptid fans from suggesting maybe Orang Pendek isn't just an unknown member of the ape family but an extant population of *Homo floresiensis* (aka "hobbits").

NEW YORK SEWER GATORS

This is probably one of the most enduring urban legends about New York City. It has been kicking around since the 1930s. The story goes there are supposedly alligators living in the sewers, feasting on New York's uncheckable rat population. Rats are in such abundance that New York's sewer gators have grown to huge proportions. Gators ostensibly got a foothold in New York's sewer system because pet owners started flushing their gator pets down the toilet when the creatures outgrew their aquariums. Many in New York have a friend of a friend who knows a guy who works as a sewer worker and has confirmed the story. The story gets even better when you add in the city's public works department arms its sewer workers when they have to descend into the depths. Although there have been a few confirmed alligator sightings in and around New York City as a result of irresponsible people illegally keeping them as pets, the notion that gators are an invasive species that have found their perfect niche is purely the stuff of legend. The reality is gators are cold blooded and would not long survive a New York winter, even underground.

THYLACINE

The thylacine is sometimes better known as the Tasmanian tiger. It was a carnivorous marsupial found in Australia and on the island of Tasmania. Before mammals were introduced to the continent, the marsupial occupied a niche typical of wolves and predatory cats. The thylacine were wiped out by a combination of human encroachment, hunting, and being outcompeted by introduced species. The last living thylacine was kept in a Hobart zoo and died in 1936. After the death of this attraction, the zoo went looking for a replacement but could not find any more. It's assumed that captive thylacine was the last of its kind. Several extensive searches for surviving members of the species were conducted in the 1960s and 1970s but came up empty. After the thylacine was officially declared extinct in 1980, there have been a number of sightings by the hopeful but no indisputable physical evidence has been found. Some researchers suggest sightings in the 1980s might have been of actual surviving thylacine before it went really-totally-no-joking-this-time extinct in the 1990s.

ROTHSCHILD-NEUVILLE TUSKS

The Rothschild-Neuville tusks were a pair of tusks from an alleged unknown African species. The tusks were purchased in 1904 by Baron Maurice de Rothschild and zoologist Henri Neuville from an ivory dealer in Addis Ababa. Local hunters who were shown the tusks suggested they came from an aquatic creature that inhabits local lakes. Rothschild and Neuville published a paper about the tusks after three years of study. They concluded the tusks were most similar to elephant tusks but had enough significant differences that they likely came from an unknown species of elephant. The tusks were deposited in Paris' National Museum of Natural History but have subsequently disappeared. No similar tusks have ever been found. It is likely Rothschild and Neuville simply misunderstood what they were looking at. Those in the cryptozoological camp suggest these tusks are evidence of a living population of *Deinotherium* (a prehistoric ancestor of the elephant that went extinct about 7 million years ago).

UNICORN GOAT

In the early 1980s, Timothy Zell, a self-professed wizard, started touring the renaissance festival circuit with a real, live unicorn named Lancelot. Upon closer inspection Lancelot turned out to be a goat, not a horse. But, still, it did have a single horn growing out the center of its head. Zell was cagey about the origins of his unicorn goat but that may be because Zell was patenting his process for turning a two-horned goat into a one horned goat. Zell was granted patent #4429685 in 1984. It has since expired. Zell's process is based on a process discovered in the 1930s that you can take a kid's horn buds and transplant them to the middle of the goat's head. They will fuse and grow as a single horn. Zell eventually rented Lancelot out to Ringling Bros. who promoted the goat as a real unicorn. The circus got in some trouble in Florida. The circus was charged with breaking a law that prohibits exhibiting disfigured or malformed animals.

ZANZIBAR LEOPARD

The Zanzibar leopard was (or is ... we'll get to that) an African leopard that made its home on an island in the Zanzibar archipelago. Local legends associated the big cat with witchcraft and a campaign was launched in the 20th century to wipe it out. It appears the efforts of the locals were successful. No one had seen a living example since the early 1980s. Conservationists considered the Zanzibar leopard extinct by the 1990s. However, in 2018 a trail cam recorded what was thought to be a live Zanzibar leopard, raising the possibility there are still a few left in the wild.

Fifteen Scientists Interested in Cryptozoology

Cryptozoology attracts a wide variety of people, from the true believers to lovers of "Bigfoot skepticism". It's important to remember, cryptozoology covers a spectrum of belief. An interest in cryptozoology does not necessarily mean a belief in the reality of mysterious monsters. A number of scientists over the decades have tucked into the field. Here are some of the better-known scientists on that spectrum.

MICHAEL M. AMES

Michael M. Ames (died 2006) was a Professor of Anthropology at the University of British Columbia (UBC). He was a member of the Order of Canada, one of Canada's top cultural honors. Ames co-wrote a book in 1980 titled *Manlike Monsters on Trial: Early Records and Modern Evidence*. The book is a collection of scholarly essays on the possible existence of Bigfoot and the yeti. Most of the papers are from a 1978 conference UBC hosted on the subject.

DONALD BAIRD

Donald Baird (died 2011) was a paleontologist and curator of Princeton's Natural History Museum. Baird spent his life studying fossilized tracks and developed a technique for making casts of fossil tracks using latex. In his retirement he published an article titled "Sasquatch footprints: a proposed method of fabrication" that outlined how Bigfoot tracks, particularly those being cited as genuine in the cryptid literature, could be faked.

HENRY H. BAUER

Henry H. Bauer is a retired chemistry professor. He's published both skeptical and paranormal works. Many consider his 1985 book *Beyond Velikovsky: The History of a Public Controversy* one of the best debunkings of Velikovsky's *Worlds in Collision*. Bauer is also a big believer in UFOs and he's an ardent HIV denialist. In the 1980s he published a book titled *The Enigma of Loch Ness* where he argues for the reality of the Loch Ness monster.

JOHN BINDERNAGEL

John Bindernagel (died 2018) held a PhD in Biology from the University of Wisconsin-Madison. Bindernagel billed himself as the "Bigfoot Biologist" and spent many years in British Columbia researching what he thought was a real creature. He was the Bigfoot Field Researchers Organization's curator. He published two books of his research: *North America's Great Ape: The Sasquatch* (1998) and *The Discovery of the Sasquatch: Reconciling Culture, History and Science in the Discovery Process* (2010).

DAVID J. DAEGLING

David J. Daegling is a Professor of Anthropology at the University of Florida. In 1999 he conducted a study of the Patterson-Gimlin Bigfoot film. He concluded the Patterson-Gimlin film has too many issues (poor image quality, film artifacts, camera movements) to draw any confident conclusions about what is in the film. There's nothing that eliminates the subject being non-human but then there's nothing that eliminates the subject being a person in an ape suit.

BERNARD HEUVELMANS

Bernard Heuvelmans (died 2001) was a Belgian zoologist with an early interest in cryptozoology. Some consider him a founder of the field. In the late 1940s he became interested in the possibility a small group of dinosaurs had survived extinction. He primarily focused on sea monsters through his career but in 1955 he wrote a more general book on cryptids called *On the Track of Unknown Animals*. This book covered the possibility of sea monsters plus mokele-mbembe and the yeti. His 1955 book is also suggested as the first popular work to introduce the term "cryptozoology".

WOLF HENNER FAHRENBACH

Wolf Henner Fahrenbach was a zoologist at the Oregon Regional Primate Research Center until he retired. He has devoted his retirement years to proving the reality of Bigfoot. He's probably best known for his study of some 706 Bigfoot casts. He concluded based on his cast analysis that Bigfoot has an average foot length of 15.6 inches and an adult Bigfoot likely weighs 2,000 pounds. He's also investigated the sex life of Bigfoot and has suggested polyandry is not out of order for the species.

JANE GOODALL

Jane Goodall is considered one the leading experts on chimpanzees. She has been studying the primate for over 60 years. Goodall has suggested she has not ruled out the existence of Bigfoot and would like to believe the creature exists. However, she does find it odd that no one has ever found remains of the creature.

WILLIAM CHARLES OSMAN HILL

William Charles Osman Hill (died 1975) was a British primatologist. On his passing, Hill's collection of research specimens was turned over to the Royal College of Surgeons' Hunterian Museum in London. Part of his collection was what was reputed to be a finger of a yeti, part of the "Pangboche Hand". It was brought to him by explorer Peter Byrne. Byrne spun a wild tale of having "fingered" the finger from a Nepalese temple and spiriting it back to London with the help of Jimmy Stewart's wife. Byrne gave the finger to Hill to analyze. Hill concluded the finger was hominid and possibly that of a Neanderthal. The finger underwent DNA analysis in 2011 and was determined to be that of a modern human.

GROVER KRANTZ

Grover Krantz (died 2002) was a Professor of Anthropology at Washington State University. He was one of the first academic scientists to give serious attention to the study of Bigfoot. Krantz was initially skeptical of the Patterson-Gimlin film and of Bigfoot in general. However, in 1969 he examined the so-called "Cripplefoot" cast and thought he saw dermal ridges. He felt this was too much detail to be hoaxed and became convinced of Bigfoot's reality. He later reexamined the Patterson-Gimlin film and argued the creature's bent-knee gait was impossible for a man in a suit to mimic. In 2019, Krantz's cousin Laura Krantz created a limited run podcast called *Wild Thing* that looked at her cousin's work and the state of Bigfoot research today.

JEFFREY MELDRUM

Jeffrey Meldrum is an Associate Professor of Anatomy and Anthropology at Idaho State University. He's somewhat taken up Grover Krantz's baton. Meldrum is a popular guest on various TV shows that cover topics like Bigfoot. In 2011, Meldrum was lured to Siberia to investigate alleged Bigfoot tracks found in a cave near the city of Kemerovo. Meldrum discovered all the tracks were of a right foot and a little too perfect. Meldrum concluded he was an unwitting participant in a publicity stunt by the city to develop a tourist attraction.

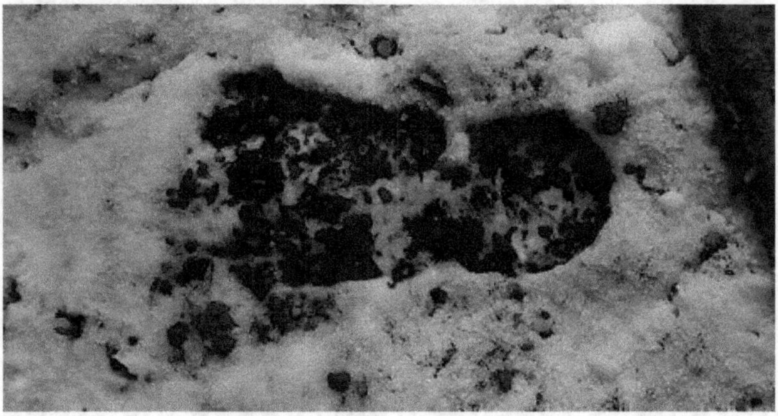

Photo Credit: Karl Mamer

DARREN NAISH

Darren Naish is a paleontologist and author of the long-running *Tetrapod Zoology* blog. He sometimes delves into the world of cryptozoology, bringing a skeptical, scientific perspective on sea monsters. He's written three skeptical books on cryptozoology between 2013 and 2019. His most recent is *Hunting Monsters: Cryptozoology and the Reality Behind the Myths*.

DONALD PROTHERO

Donald Prothero is a geologist and paleontologist. He was a Professor of Geology at Occidental College until 2012 and is now retired. Prothero co-wrote *Abominable Science: The Origin of Yeti, Nessie, and Other Cryptids* in 2013 with Daniel Loxton. It is probably the best and most comprehensive critique of modern cryptozoology ever published. Prothero is famous for appearing on Ben Stein's game show *Win Ben Stein's Money* and defeating the creationist. Prothero has quipped he won because Stein knows nothing about science.

IVAN T. SANDERSON

Ivan T. Sanderson (died 1973), along with Bernard Heuvelmans, is considered one of the founders of cryptozoology. He had a MA in Botany from Cambridge. In the 1930s he found work as a travel writer and gained some popularity writing about nature for magazines like the *Saturday Evening Post*. In 1961, he wrote an early popular book on hominid cryptids called *Abominable Snowmen: Legend Come to Life*. A review of the book by the journal *Science* described Sanderson's standards of evidence as "unbelievably low". Sanderson's credibility will always be called into question after he fell for a giant penguin hoax in 1948. Mysterious giant penguin footprints were found on beaches around Clearwater, Florida. Sanderson proclaimed the tracks as genuine and impossible to fake. They were later revealed to be faked by a pair of pranksters.

ESTEBAN SARMIENTO

Esteban Sarmiento is a primatologist and was for a time a research associate at the American Museum of Natural History. Sarmiento's analysis of the body proportions of the creature in the Patterson-Gimlin Bigfoot film suggests it is a man. He noted the plantar surface of the feet are light colored yet the palms are dark. This difference is unknown in mammalian biology.

Eight Vampire Legends from Around the World

BAOBHAN SITH (SCOTLAND)

The Baobhan sith are part vampire, part succubus. They hail from the highlands of Scotland. Baobhan sith appear as beautiful women with deer hooves. They prey upon victims at night and vanish before the sun rises.

BETAL (INDIA)

The Betal is a creature from ancient Indian folklore. The Betal is an undead creature that hangs upside down like a bat from trees in cemeteries.

LILITH (JUDAISM)

In Jewish mythology Lilith was Adam's original wife. She left Adam when she wasn't content to play simple housewife to Adam. Later Jewish mythology from Central Europe evolved her into a vampire-like creature of the night.

MANANANGGAL (PHILIPPINES)

The Manananggal are hideous female vampires. They have fangs and bat wings and, oddly, can detach their upper torsos from their hips and legs and fly off into the night in search of victims.

NACHZEHRER (GERMANY)

These German vampires are not created by being bitten but result after a suicide. The person taking their life emerges as a vampire-like creature hell bent on devouring his/her family.

SEKHMET (ANCIENT EGYPT)

Sekhmet was an incarnation of the Egyptian goddess Hathor. In one myth, Ra decided to wipe out humanity because humans were turning away from the gods. He sent Sekhmet to do his bidding. As Sekhmet moved across the face of the earth slaughtering people, she drank their blood.

SHTRIGA (ALBANIA)

The Shtriga of Albanian folklore have a taste for the blood of babies. When in human form, a Shtriga is said to appear as a disfigured woman. They can also take the form of a flying insect.

SOUCOUYANT (CARIBBEAN)

In places like Dominica, Haiti, and Grenada, there exists the legend of a blood sucking hag called a Soucouyant. During daylight, she appears as a hag. At night she transforms into a fiery being that enters homes and sucks the blood of sleeping people.

Seven Key Milestone Along the Way to the Modern Werewolf Myth

The modern-day werewolf legend has an ancient pedigree. Our modern-day conception of the werewolf bears little resemblance to original werewolf legends. Here are seven key milestones in the development of the modern werewolf myth.

THE FIRST LYCANTHROPE - 2100 BC

The *Epic of Gilgamesh*, the oldest surviving work of fiction, comes to us from around 2100 BC. One part of the epic describes a shepherd who was a bit over the top in his worship of the goddess Ishtar. As punishment for his stalkerish worshipfulness she turned him into a wolf.

WEREWOLVES APPEAR ONLY DURING CERTAIN CYCLES - 425 BC

The first suggestion werewolves transform based on certain cycles comes to us from a work by Greek historian Herodotus. In 425 BC he described a tribe in northern Ukraine called the Neuri. According to Herodotus, each member of the tribe can, once a year, turn themselves into a wolf for a few days.

LYCANTHROPY IS A CURSE - 380 BC

The idea that lycanthropy is a curse may have gotten its start in Plato's *Dialogues* from 380 BC. In it Plato quotes Socrates speaking about a tyrant king who was transformed into a wolf after eating human entrails. Ovid and other authors seem to have merged this with a legendary king called Lycaon. Lycaon, the king of Acadia, doubtful of Zeus' omniscience tried to test him by feeding him a dish made out of human flesh. Zeus quickly detected the origin of the mystery meat and turned Lycaon into a wolf as punishment.

LYCANTHROPY'S FIRST APPEARANCE- 1584 AD

The earliest known use of the term "lycanthropy" seems to be found in Reginald Scot's 1584 skeptical book on witches called *The Discoverie of Witchcraft*. He felt the very notion that humans could transform into wolves was absurd.

WEREWOLVES VULNERABLE TO SILVER - 1640 AD

German records from 1640 AD report the German city Greifswald was overrun with werewolves. The townsfolk first unsuccessfully tried to battle them with conventional weapons. They then collected silver buttons, cutlery, and buckles and melted them down into silver bullets. This seems to have done the trick as modern Greifswald is werewolf free to this day.

FULL MOONS AND THE WEREWOLF BITE - 1935

The notions that lycanthropy is passed on by a werewolf's bite and a full moon starts the transformation got their start as plot devices in the 1935 film *Werewolf of London* (this one did not star Lon Chaney Jr.).

VAMPIRES AND WEREWOLVES SHARE A UNIVERSE - 1943

The *Twilight* series has fixed in people's minds werewolves and vampires share the same universe as either allies or rivals. However, the first movie to include both a vampire and a werewolf was *Return of the Vampire* from 1943.

Seven "Were-d" Creatures

In mythology, therianthropy is the ability for humans to shapeshift into animal forms. The belief some humans have about the reality of this fantastical ability goes back as far as the paleolithic age. The best-known example is the legend of the werewolf. Here are six "were-d" creatures from legend you may not have heard about.

KUMIHO (KOREA)

The Kumiho was made famous by HBO's *Lovecraft Country*. The Kumiho is a fox with nine-tails that can shapeshift into the form of a beautiful woman. The Kumiho in woman form seduces men and then eats their heart or liver.

SKIN-WALKERS (NORTH AMERICA)

Some first nations people, notably the Navajo, have legends of people with the ability to don the pelts of animals and then take their forms.

WERECROCODILES (THAILAND)

A folktale from Thailand known as the *Krai Thong* tells the tale of Chalawan, the lord of the crocodiles. While in his cave lair he took the form of a human but outside of his lair he took on the form of a crocodile. Chalawan had a taste for human flesh and would prey on humans that ventured into the river near his lair.

WEREHYENA (MIDDLE EAST)

Legends of werehyenas are common among people of the Middle East, Near East, and the Horn of Africa. Unlike many were-creatures that are humans that turn into animal forms, werehyenas are hyenas that can turn into humans.

WEREJAGUAR (MESOAMERICA)

Olmec culture is replete with images of what archaeologists generally consider depictions of their chief deity, a werejaguar. There's some debate among archaeologists, however, whether these statues and images represent the culture's chief god or even a deity.

WERETIGERS (INDIA)

India has three primary types of weretiger. Some Indian tribes have legends of evil sorcerers who transform themselves into tigers by circling an anthill and chanting a spell. Other legends involve people, to use a *Game of Thrones* term, warging themselves into real tigers while they sleep. Unknowingly these sleepers cause their feline wargs to eat humans and livestock. Finally, a more benign weretiger involves people with the ability to take the form of a tiger in their dreams.

YUXA (TARTAR)

In Tartar mythology, a Yuxa is any snake that manages to live to 100 years old. Upon reaching that milestone, the snake gains the ability to turn itself into a beautiful woman or handsome man and create new offspring.

Twelve People on the Mothman Death List

Cryptozoologist Loren Coleman maintains something he calls "The Mothman Death List". This list contains the names of 100+ people who ostensibly had some connection to the Mothman and then succumbed to an apparent curse. The list includes not just people who reported Mothman encounters but people who partook in the making of *The Mothman Prophecies* movie, were associated with the author of *The Mothman Prophecies* book, wrote about the Mothman, or wrote about Mothman-like cryptids. Given the original Mothman incident and Keel's book research took place about a half century ago and the movie involved a huge number of people (over 300 people) more than 18 years ago, it's not surprising a fair number of people connected to any of these Texas sharpshooter targets have died. Here are some deaths Coleman has connected to the curse.

IVAN T. SANDERSON

Ivan T. Sanderson was a British biologist who fell hard for cryptozoology. He is probably best known for being fooled by the Giant Penguin Hoax of 1948. John Keel consulted with Sanderson during the writing of *The Mothman Prophecies*. Sanderson died of brain cancer in 1973.

FRED FEED

Fred Feed was a documentary maker and winner of a Peabody Award. He made a number of television documentaries in the 1960s. He was approached at one point by John Keel to do a documentary about the Mothman and other bits of high strangeness. Freed died of a heart attack at age 53 in 1974.

JIM KEITH

Jim Keith was a conspiracy author and former Scientologist. His book *The Octopus* tried to argue the suicide of Danny Casolaro (a reporter investigating the Inslaw case) was a state-sponsored murder. Keith is also reputed to be the mysterious Commander X, author of several trippy UFO books. Keith wrote about a connection between Mothman sightings and the Men in Black. Keith died in 1999 at the age of 49. He fell off a stage at Burning Man and injured his knee. Complications from the surgery resulted in a blood clot entering his lung and killing him.

AARON REBSAMEN

This inclusion seems in poor taste. Rebsamen was the teenage son of an artist who did the cover art for a Coleman book about the Mothman. In 2002, Rebsamen took his own life at the age of 14.

TED DEMME

Ted Demme was the nephew of director Jonathan Demme. He was also friends with Mark Pellington, director of *The Mothman Prophecies*. He died in 2002, the same year his friend's movie came out. It's not clear why the Mothman curse is picking off everyone BUT the director himself.

ROBERT STACK

Robert Stack died at the age of 84 in 2003. He hosted the show *Unsolved Mysteries* which did a piece about the Mothman in a 2002 episode.

JESSICA KAPLAN

Jessica Kaplan is listed as a scenic artist for *The Mothman* movie. Her name shows up on Coleman's Mothman Death List. However, the inclusion confuses her with screenwriter Jessica Kaplan who died in a plane crash in 2003. Screenwriter Kaplan had no connection with *The Mothman* movie. IMDB lists at least six Jessica Kaplans. All different people.

ROBERT SANDERS

Robert Sanders died of suicide in 2003. He was related to one of the people who died when Point Pleasant's Silver Bridge collapsed in 1967. Coleman seems to increase the mystery by noting Sanders died during an unusual "death flap" at the end of August 2003. "Flap" seems to be defined as three other people dying during the week. The other three "mysterious" victims include a person who died in a car accident, an 84-year-old man, and a 75-year-old woman. Hard to believe, eh?

JENNIFER BARRETT-PELLINGTON

Jennifer Barrett-Pellington was the wife of the film's director Mark Pellington. She died in 2004 at the age of 42 from complications due to a ruptured colon. As previously noted, some three hundred people are listed on the movie credits. If you also include their partners, you can effectively double the target pool.

JOHN KEEL

John Keel, of course, wrote *The Mothman Prophecies* book. He died at age 79 in 2009.

FRANK FRAZETTA

Frank Frazetta was an artist who brought to life many works of fantasy with his paintings and illustrations. He's probably best remembered for his *Conan* and *Lord of the Rings* art. He died at age 82 in 2010. He led a long, productive, successful life. But, hey, he illustrated the cover of an edition of Keel's *The Mothman Prophecies* book. So, I guess he could have lived a longer, more successful life if not for thumbing his nose at the deadly curse of the Mothman!

BOB ROACH

Bob Roach was a sculptor who created the Mothman statue in Point Pleasant. He died at the age of 81 in 2015. More shocking proof those who involve themselves with the Mothman are not immortal.

NOT DEAD YET

Loren Coleman is not only the originator of the list but he's written two books about the Mothman legend. Given so many before him have died for lesser entanglements, there can be no explanation possible as to why Coleman is still drawing breath.

Fifteen Things You Might Not Know about the Legend of Boggy Creek

The Legend of Boggy Creek, directed by Charles Pierce (died 2010), is a 1972 docudrama about the Fouke Monster, a Bigfoot-like creature reported near the town of Fouke, Arkansas. The film featured live interviews with locals who claimed to have seen the creature and dramatized reenactments of historical encounters. It was the first movie based on a cryptid legend to use the docudrama style. Many subsequent cryptid movies "aped" this style. The film is a favorite of many cryptid fans, be they true believers or skeptics.

ARTWORK

The Legend of Boggy Creek's iconic poster was done by Ralph McQuarrie. McQuarrie went on to achieve fame as an artist for *Star Wars*.

AUDIENCE

In much the same way films today are made as "direct to video" or "direct to streaming", many films in the 1970s were made for drive-in movie theaters. *The Legend of Boggy Creek* was purposely targeted to drive-in movie audiences.

BOX OFFICE

The film was made on a budget of $166,000 but grossed $25 million. Much of the funding came from a trucking company which was one of Pierce's clients. It was one of the top ten grossing movies of 1972.

COPYRIGHT STATUS

Pierce consigned the rights to his film to the major investor, L.W. Ledwell. In 2018, L.W. Ledwell's son Steve passed the copyright to Pierce's daughter Pamula Pierce Barcelou.

CONTROVERSIES

It has been reported the film's actors did not initially get paid and had to sue to get their wages. One of the film's more notable scene stealers was outdoorsman Julius E. "Smokey" Crabtree (died 2016). He sued Pierce for duress caused by the film. Crabtree alleged the film resulted in harassment by monster hunters.

DIRECTOR

Before making the film, director Charles Pierce owned an advertising company and made commercials for trucking companies and farm equipment makers. For a time Pierce played a character named Mayor Chuckles on an Arkansas children's TV show.

DISTRIBUTION

Pierce premiered his film at the Perot Theatre in Texarkana. The film was a local sensation and was picked up by film distributor Howco. Howco was previously involved in the release of B movies packaged as double features.

INFLUENCES

The makers of 1999's *The Blair Witch Project* have cited director Charles Pierce's pseudo-documentary style as an inspiration for their own film.

NARRATOR

The movie was narrated by Vern Stierman. He was the weatherman for a local NBC affiliate.

RERELEASE

For years, there was no official home video release. Several poorly mastered bootlegs were, however, available via VHS and DVD. In 2019 a restored version was released on DVD. Pierce's daughter had managed to find a good quality copy of the film that had been deposited with the British Film Institute.

SETTINGS

Pierce wanted to shoot a scene at a home where a witness claimed he was attacked by the Fouke monster. Pierce offered the homeowners $2,000 to film on their land.

The owners rejected the offer and demanded more money. Pierce passed and shot his scene at a look-alike home in Texarkana.

SPINOFFS

The film spawned three unofficial sequels: *Return to Boggy Creek* (1977), *Boggy Creek: The Legend is True* (2010), and *The Legacy of Boggy Creek* (2011). *Boggy Creek II: And the Legend Continues* (1985) is the only official sequel directed by Pierce. The official sequel was given the full *MST3K* treatment in season 10. In 1974, Julius E. "Smokey" Crabtree self-published *Smokey and The Fouke Monster*. It was a book about his participation in the film and the effects the film had on the local community.

THEME SONG

The film's theme song "Ballad of Boggy Creek" was written by Earl E. Smith and sung by the director Charles Pierce. Smith himself later went on to write the script for *Sudden Impact*, a 1983 Dirty Harry film.

TITLE

The film's working title was *Tracking the Fouke Monster*.

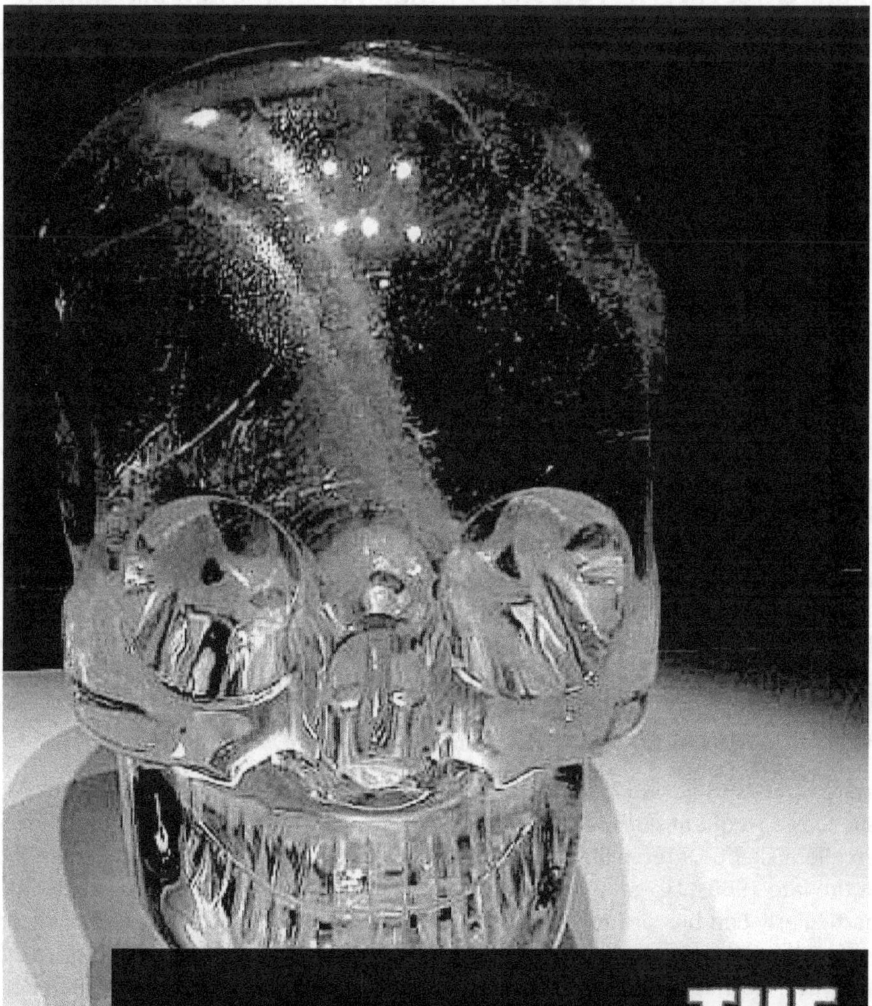
THE
PARANORMAL

Eight Astoundingly Wrong Psychic Predictions

EDGAR CAYCE PREDICTS POLE SHIFT AT THE START OF THE NEW MILLENNIUM

Edgar "the Sleeping Prophet" Cayce was asked to give a reading about what changes people might expect in the year 2000 or 2001. He predicted there would be a pole shift. This does happen, although the process takes about a thousand years. Geologists when comparing them to deep, geological time will intone pole shifts happen "quickly". But to a geologist, anything that happens in less than 25,000 years is "quickly".

JEAN DIXON PREDICTED THE RUSSIANS WOULD BEAT AMERICA TO THE MOON

Jean Dixon died in 1997. She was probably, for her time, the best-known celebrity psychic. At one point she was offering advice to Nancy Reagan. In terms of presidents, she seemed to have had a real thing for Nixon. She predicted Nixon would win the 1960 presidential election (Kennedy won). Later she predicted Nixon would go down in history as the best president America ever had. Charitably, the jury is still out on that. She seemed to be constantly predicting the end of the world. She predicted World War III would start in 1958. She thought there would be a second Holocaust in 1980. She thought the Christian messiah would return in 2000 and bring a new era of peace. Dixon, raised Catholic, seems to have gotten her Revelation chronology out of order because she also predicted Armageddon in 2020, when the Antichrist would rise and launch a war on humanity. Most amusingly, she predicted the Russians would beat America to the moon.

IRENE HUGHES FAILED TO UNDERSTAND SHORT-PERIOD COMETS

Irene Hughes (died 2012) was an American psychic who did the talk show circuit. She was a frequent guest on *The Merv Griffin Show*. Although Roger Ebert was an excellent skeptic later in life, he seemed to have fallen pretty hard for Hughes back in the late 1960s. He gave her key press in the *Chicago Sun-Times* for predicting a particularly bad blizzard for the Windy City in January of 1967 (which set a record with 23 inches of snow). In a 1967 *Chicago Sun-Times* article, he extolled her accuracy, particularly in the world of politics. In the article, Hughes made the amusing prediction that George Romney (father of Mitt Romney) would win the 1968 presidential election. Her powers of presidential prognostication seem to have abandoned her that year, never to return. For example, she saw Carter being reelected in 1980 after dumping Mondale and picking an African American running mate. She also foresaw Halley's comet returning in 1980, predicting it would bring with it massive earthquakes, tidal waves, and wars. If she had bothered to consult an encyclopedia (this was an era before Google), she might have noticed Halley's comet wasn't actually scheduled for a return until six years later. As history records,

Halley's comet returned in 1986, not 1980, and it was certainly not the cause of any upheavals. The comet was pretty much a dud.

SYLVIA BROWNE TELLS PARENTS THEIR CHILD IS DEAD

Sylvia Browne was a self-proclaimed psychic who enjoyed saying cruel and insensitive things to her paying fans. In 2002, Browne told the parents of Shawn Hornbeck that he was kidnapped by a dark-skinned man with dreadlocks and was dead. In reality he was alive, having been kidnapped by a white man with short hair. Not deterred, two years later she told the mother of Amanda Berry that her daughter was dead and "in water" (a typical vague guess of so-called psychic detectives). The grieving mother died a couple years later, going to her grave thinking her daughter was no longer alive. In reality Berry was alive and being held captive (along with two other women) by Ariel Castro. She managed to escape in 2013. Browne also predicted she would live to age 88. She thankfully did not last that long, having died at 77.

THE PSYCHIC TWINS PREDICTED A DULL 2000 ELECTION

Linda and Terry Jamison are identical twins who claim to be psychics. They're also not really hard on the eyes. They make for popular headline fodder in grocery store tabloids. They generate a lot of vague and/or obvious predictions phrased with enough wiggle room that if they miss by a year or three, well, they're still "right". Misses include predictions Pope John Paul II would die in 2004 and there would be a rash of attacks on federal buildings in Georgia in 2002. Their most amusing miss was their prediction the Bush/Gore 2000 contest for the presidency would be "a crashing bore". Not so much.

ANTHONY CARR PREDICTED PRINCE ANDREW'S RISING STAR

Some bill Anthony Carr as the world's most documented psychic. Who knows what that even means? What is true, Carr seems to generate at least 100 predictions every year. In 2018 he amusingly predicted Prince Andrew would rise in prominence, reinvigorate the royal family, and set himself up to replace Prince Charles as the next monarch. Carr completely missed Prince Andrew would certainly be in the headlines but for the wrong reasons, his royal duties would be removed, and public buildings all across the world named in his honor would have his name scraped off.

PSYCHIC NIKKI PREDICTED SEATTLE'S DOOM IN 2013

Psychic Nikki is a Canadian psychic who makes the rounds in Canadian media. In 2013 she predicted Seattle would be "in ruins after an earthquake". It never happened. Starbucks, Microsoft, and Amazon abide. The same year she also predicted New York would be hit by a snowstorm that would bury them in 25 feet of snow. That didn't happen either.

PSYCHICS RUNNING AT THE MOUTH OVER THE RUNAWAY BRIDE

In 2005, Jennifer Wilbanks faked an abduction to get out of a wedding with a Christian dude. Wilbanks was assumed to have been abducted while jogging. Possibly owing to the fact she was a pretty white woman about to marry a nice Christian man, her story went nationwide. Skeptical paranormal investigator/researcher Benjamin Radford documented the retinue of psychics who came forward to offer "help". Most claimed she was dead, offering typical nonspecific guidance like her body would be found by water or bushes. Of course, she turned up a few days later alive and well in Albuquerque, New Mexico.

THEY ALL MISSED 9/11

Every psychic failed specifically to warn us about the terrorist attacks on 9/11. New York-based "Intuitive Consultant" Maria Papertros in a list of predictions for 2001 seemingly missed this singular event in her own backyard. Embarrassingly she saw great things in 2001 for the airline industry, predicting airlines would be marketing new and innovative forms of air travel.

Seven Tricks of Cold Reading

Cold reading is a common technique employed by so-called psychics to convince people they know more about a person than they should. Since there's apparently no conventional way a psychic could have known there was someone in your family with a name that starts with the letter "D", that psychic must have gotten that information through paranormal means like mind reading or talking to the dead! Here are seven common but effective cold reading techniques employed by psychics on sitters.

START WITH THE OBVIOUS

Most people have a close relative or friend named "John" or a relative with a heart condition.

USE TERMS THAT SOUND SPECIFIC BUT LETS THE SITTER DEFINE THE SPECIFICITY

Most of us have lost a loved one recently. "Recently" can mean yesterday or even a year ago. A "loved one" itself doesn't imply the relationship, sex, or even species. A "loved one" could be a relative, a friend, or a pet.

TALK LIKE A HOROSCOPE

Horoscopes use something called the Barnum effect. The Barnum effect is a psychological phenomenon where people identify with personality traits that can apply to anyone. Leos are likely to agree personality traits associated with Cancers describe them if you simply tell them the traits are representative of people born under their sun sign. So, suggest things like "people think you're really happy but you have a secret pain." Most of us go through life trying to appear reasonably happy and engaged but most of us also do have some sadness in our lives.

Photo Credit: Karl Mamer

FEED THEIR OWN INFORMATION BACK TO THEM

A lot of people go to psychics because they just want someone to listen to their problems. We don't remember half of what we say to people. A good cold reader will remember certain specifics the sitter revealed earlier and then feed them back as if the information was revealed to them by spirits.

FRAME YES/NO QUESTIONS IN THE NEGATIVE

This ensures no matter what the sitter answers, you can spin it as a hit. For example, a cold reader might ask a question like "you haven't gotten a promotion recently, have you?" If the person answers yes, well, it's an astounding hit. If the person answers no, then you can spin it in an obvious direction: "yes, you're starting to feel undervalued at work". Who runs around screaming their overpaid and over appreciated at work?

GO WITH A STEREOTYPE

A guy with a ball cap and a mustache probably has a pickup truck or wants one. A middle-aged woman without a wedding ring is probably not having a lot of success in the dating world. A middle-aged married man would probably agree he doesn't get to spend enough time "with the boys".

NO MEANS IT'S PROBABLY SOMETHING IN YOUR FUTURE

If the middle-aged single woman answers "no", she has a great single life and meets a lot of great men, then simply reply that you're seeing storm clouds on the horizon or the spirits really, really want you to know things may take a turn for the worse soon. Our good luck rarely holds. When it goes into the toilet, a person may be inclined to think that the psychic saw it all coming. The psychic now has a return customer.

Nine Odd Places Claimed to Be Major Energy Vortices

In the 1920s a wealthy British businessman named Alfred Watkins thought he stumbled on a pattern no one else had discovered. Ancient structures and sacred sites in Britain seemed to line up. He theorized this was on purpose. He called these lines you could plot through various ancient structures "ley lines". His compatriots in mainstream archaeology weren't super impressed. There is no good reason one ancient society should give any consideration to lining up their holy site with the holy site of another culture unrelated to them and even unknown to them. The counterculture in the 1960s revived the concept and attributed spiritual energies to these lines. Over the decades New Agers have further imbued them with all kinds of pseudoscientific qualities. Eventually it occurred to them that since we live on a globe, these various lay lines intersect. In the view of New Agers, where two or more ley lines intersect, you find a "negative energy vortex" that acts as a kind of spiritual magnet and, like, anything can then happen, man. Below are nine reputed vortices found around the world in some unusual places:

1139 WEST FULTON MARKET, CHICAGO

Until recently this address used to host an art gallery and event space. It is also reputed to be ground zero of an energy vortex. Yes, there's a psychic energy vortex right in the heart of Chicago's West Loop district. Not as well-known as Sedona, certainly, but you'll find it listed in various tourist guides offering off-the-beaten path attractions. For a time, a sign pinpointed the exact location of the vortex: a loading dock behind the art gallery. The exact location was determined by a shaman. The sign is gone and in 2019 the building started undergoing redevelopment. A six-story office tower was slated to be built over the energy vortex.

AVEBURY, UK

Avebury is home to a stone circle constructed by denizens during the Neolithic period. There's no clear agreement among legit archaeologists as to why these ancients built such a structure. New Agers seem to think it's because the ancients knew a major negative energy vortex runs right under the place. So why not mark it with some large stones arranged in a circle?

AYERS ROCK, AUSTRALIA

Ayers Rock is known today by its Aboriginal name "Uluru". It's basically a giant 1,000-foot-high sandstone hill in Australia's Northwest Territories. Many might know it as a backdrop for that Meryl Streep movie *A Cry in the Dark*. New Agers believe it's a center of energy healing. The local Aborigine population views it as sacred and wishes people would stop trying to climb their holiest site and refashion their sacred stories to stroke the egos of rich white people.

BIMINI, BAHAMAS

Bimini is an island chain that's part of the Bahamas. Bimini was considered by Ponce de León as a possible location for the fountain of youth. In the late 1960s, divers discovered a natural stone feature off the coast of the north Bimini island that resembled a road or a wall. So-called psychic Edgar Cayce believed Bimini is a remnant of Atlantis. Cayce followers got pretty excited when they thought there was an ancient road discovered at the bottom of the sea. New Agers think a giant Atlantean power crystal is buried under the island and this screws up the navigational instruments of boats. Boaters have no problem with their instruments when navigating the waters of Bimini, however.

CALGARY, ALBERTA, CANADA

New Agers think a major vortex runs right under Calgary. If you're unfamiliar with Calgary, think of it as Canada's Houston and Dallas all rolled into one. People there love their crude oil, their beef industry, drinking large amounts of rye whiskey, and generally being cruel to animals during the annual Calgary Stampede. This is not what you'd call a particularly spiritually enlightened part of the world. What good is a spiritual vortex if it can't at least get people to be kinder to cows? New Agers are not fazed. This just means, unlike Sedona, the Calgary vortex is untapped and waiting to be exploited by crystal selling hucksters!

EASTER ISLAND, CHILE

Easter Island is home to those famous stone heads the misinformed think were made and moved by aliens, Atlanteans, or locals with psychic powers. It's no surprise the woo-inclined would view this island as some nexus of mysterious energies.

OKLO, GABON

This region of Gabon is reputed to be home to a vortex. It's possible New Agers selected this location because it's the only known place in the world where a natural nuclear fission reactor once formed. French uranium miners discovered a seam of uranium with a ratio of U235 and elements more commonly found after nuclear fission. It's theorized U235 was in high enough concentrations 1.7 billion years ago that a chain reaction was able to start. It's estimated this natural reactor ran for about 500 million years and produced 100 kW of power.

MARYHILL STONEHENGE

You'll find Maryhill, Washington, a small unincorporated community on the banks of the Columbia river, on lists compiled by "vortex hunters". I guess that's a thing. What the community lacks in population (it's home to about 58 people) it makes up for in multiple ways. It offers a beautiful view of the Columbia river gorge. It has an impressive art gallery. Finally, it's home to a Stonehenge replica dedicated to the memory of soldiers lost in World War I. Its creator Sam Hill believed the real Stonehenge was an elaborate site where warrior druids sacrificed their victims to a war god.

THE WAWEL CHAKRA

Wawel Castle
Photo Credit: Karl Mamer

Wawel hill is found in Krakow, Poland. In legend it was the home of a fire-breathing dragon. Today an impressive castle complex sits atop the hill. It's a key tourist attraction in the city. New agers believe a chapel wall found on the castle grounds is the location of an energy vortex and one of the earth's seven chakras. This power is not to be trifled with! It seems the castle's administrators have come to view New Agers wanting access to the chapel as a nuisance and have roped off the area.

Six Hauntings That Were Debunked but No One Cares

THE AMITYVILLE HORROR

It was pretty hard not to get through the 1970s without reading *The Amityville Horror*, a book based on a supposed haunted house in Amityville, New York. The book describes a series of increasingly horrific events that transpired within the house. As the book's popularity increased, more people began to look into the book's claims. Claims easy to verify proved not to coincide with reality. The nail in the coffin came when William Weber, the lawyer for Ronald DeFeo (the guy who murdered his family in the house), admitted to *People* magazine in 1979 he and the family partnered to cook up a good tale over a bottle of wine. The family ended up suing some detractors. Most of their allegations were thrown out of court. The few that made it into a venue of fact and evidence were dismissed, with the judge commenting "...it appears to me that to a large extent this book is a work of fiction." Despite all evidence to the contrary, people today still point to this as smoking gun evidence of the reality of ghosts and demons.

THE ANSON GHOST LIGHTS

One of the problems with ghostly happenings is they're not predictable or repeatable. The so-called ghost lights of Anson, Texas are another matter. On clear, dark nights, on a certain stretch of road (conveniently located near a cemetery), ghostly lights can regularly and reliably be seen drifting along the horizon. The prosaic explanation was people were merely seeing the lights of cars on a distant highway. Others wishing for a supernatural explanation doubted that explanation. The car light hypothesis was, however, confirmed back in 2011 when a psychology professor and his students, using iPhones and GPS, were able to systematically link car headlights to sightings of the lights. Still, almost a decade later, you will find many people and websites still promoting the so-called Anson Ghost Lights as genuine paranormal activity.

THE DEMON HOUSE

The media was quite a buzz in 2014 about a mysterious house in Gary, Indiana. A mother and her three children rented the home and within days started to claim demonic experiences. They reported a laundry list of phenomena well covered in various horror movies: one of the children was seen to levitate above her bed, the house was plagued by black flies, shadowy figures were observed, and one of the children was alleged to have been possessed. The family's doctor and children protection services investigated and believed their behavior was "delusional" and the young children were simply performing to meet their mother's expectations. All of this was a surprise to the property's landlord and numerous previous and subsequent tenants. None had reported any mysterious happenings. The story may not have gotten its legs if it had not been "authenticated" by the local police chief. Joe Nickell's investigation of the hunting demonstrated the police chief was a proponent of the supernatural. Paranormal researcher Zak Bagans boosted the house's reputation a few years later with his 2018 documentary *Demon House*.

MACKENZIE HOUSE

This stately Toronto house was home to William Lyon Mackenzie. Mackenzie led the Upper Canada Rebellion against British rule in the 18th century. It didn't go well. People in Mackenzie's house late at night reported hearing disembodied footsteps. In the 1960s, the house underwent a televised exorcism. Skeptic Joe Nickell investigated the house in the 1970s and discovered the disembodied footsteps were coming from a rather noisy exterior metal staircase a cleaning crew was using to access the building next door. Despite the documented source of the mysterious noises, the Mackenzie House today still gets promoted as Canada's most haunted house.

THE MOST EVIL HAUNTED HOUSE IN AMERICA

The History channel's *The UnXplained* ran an episode in 2019 that proclaimed the most evil haunted house in America was found in Eugene, Indiana. Remember this isn't the most haunted house in America. This is the most *evil* haunted house in America. There's a difference. We're told the home, known as Willows Weep, has an evil past. Numerous former owners were driven to suicide or died under mysterious circumstances. Very little of the offered story is documented. The home's current owners seem to be a group of ghost hunters who promote the house as haunted and sell admission to other ghost groups. That the house is haunted or evil is news to the home's previous residents. They not only dispute reports the house is haunted but take exception to ghost hunters trading on the genuine tragedy that befell their family inside the house. Their stepfather was diagnosed with cancer and couldn't afford the medical treatment so he killed himself in the house. They take exception to the ghost hunters trading on wholly inaccurate versions of their family's history.

SAN ANTONIO GHOST TRACKS

There is a haunted rail crossing in San Antonio, Texas. As the legend goes, a school bus stalled on the tracks. The driver wasn't able to get the bus restarted in time and a train crashed into the bus, killing numerous young children. The spirits of these young children stand guard over the rail crossing. If you park your car on the tracks (and presumably leave it in neutral) the ghost children will roll your car off the tracks. People have tested this and it seems like a car will indeed roll off the tracks. Ghost believers see this as proof. People who understand terrain and gravity point to the downward incline found at the top of the track crossing. Some over clever ghost believers have taken to putting baby powder on the back of their car to capture handprints. And, indeed, some do find handprints in the powder. Skeptics point out they're likely the prints of the person spreading the baby powder. The biggest problem with this story is the school bus disaster never happened. Or at least didn't happen in San Antonio. The actual disaster took place in Utah. Despite all this, ghost tour companies promote these tracks as a local "a-track-tion".

Eight Useless Ghost Hunting Devices

Ghost hunting groups like to load up with all kinds of so-called ghost hunting tools that help them, apparently, establish the presence of a real "live" ghost. The problem is there's no experimentally established baseline and they're not tested against controls to determine what constitutes a false positive. Why does a "ping" on a GBX-1000 actually indicate a ghost? What else could be setting it off? Ghost hunters, trying to seem science-y, never stop to do this actual hard work of establishing these things before they trundle off to a haunted house. Getting a reading on a device you can't personally explain doesn't mean it's being set off by a ghost.

EMF METERS

EMF meters are legit devices used to measure electromagnetic fields. They have a wide variety of uses but in general they need to be used in a careful and precise manner. They cannot be simply waved around like some kind of ghost tracker. No matter where you go we are bathed in electromagnetic fields and if not properly used they are prone to giving false positives. There is no good reason to think ghosts emit electromagnetic radiation other than for years ghost hunting groups have been asserting this as an established fact.

THERMAL CAMERAS

Ghost hunters assert so-called cold spots are evidence of a ghost's presence. The devices don't actually measure ambient air temperature but take surface readings. Not all surfaces will radiate the same amount of heat (for example, cement versus wood).

GHOST BOXES

Also known as spirit boxes, these are devices that quickly scan up and down the AM/FM dial. They record, basically, static and snippets of radio broadcasts the scanner catches as it spins up and down the dial. Ghost hunters claim you can ask the ghost box a question and somehow ghosts can manipulate these radio bands to communicate messages. It makes no sense why ghosts would have this ability. The reality is these devices generate a lot of random noise allowing the listener to bring audio pareidolia to bear.

ION DETECTORS

These things are maybe better used to detect radon gas in a basement. There is no good reason to suspect a ghost, and not radon gas, might be setting one off. Why ghosts should be giving off ionizing radiation isn't well explained.

REM PODS

REM pods are the opposite of EMF detectors. They generate random EMF. The theory goes if ghosts give off EMF maybe they're attracted to EMF. So, these things amount to a duck call whistle for spirits.

XBOX KINECT

The Kinect is an Xbox attachment used by games to track the motion of people playing the game. It uses a camera wired to software that recognizes human shapes and represents these shapes as wireframes. Ghost hunters found if you take the raw camera view and point it at an empty room, the camera sometimes displays stick figures where no people are standing or moving. Ghost hunters think the Kinect is, therefore, detecting and mapping wireframes to ghosts. The problem is in the ghost hunting world, there is no such thing as a false positive. Kinect does a pretty good job of recognizing human bodies because the software is tuned to err on the side of a false positive. It will happily lock onto anything in the room that roughly resembles a human form: a chair, a stand-up vacuum cleaner, and the like.

MOTION-ACTIVATED CAT TOYS

You can find these on Amazon for about $15. They're a plastic ball with LED lights and an accelerometer. If the ball detects it has been moved, it triggers the accelerometer and the fun begins for the cat. People have repackaged these as ghost detectors, with a considerable markup in price. The balls are purposely set to be very sensitive to any movement, like a bit of vibration on the floor. It makes it more fun for a cat. But it also helps set off the ball for ghost hunters pacing around the device waiting for ghosts to play with it.

STATIC ELECTRICITY DETECTORS

Ghosts, apparently, generate static electricity as they float around a room. Some ghost hunting equipment manufacturers offer $100+ static electricity detectors with LED lights that they claim are able to indicate the movement and direction of a ghost in the room. Controlled tests reveal they can't even follow a staticky sweater.

Ten Problems with *The Amityville Horror*

For many of a certain age, the book *The Amityville Horror* by Jay Anson offers the classic example of a haunted house. The book purports to document horrors experienced by a family that moved in shortly after the previous residents were murdered. Since the book's release, people have looked into many of the claims and found they didn't comport with reality.

THE WEATHER WAS WRONG

The book claims the homeowners saw cloven hoof prints in fresh snow. Records show it did not snow on the day indicated in the book.

THERE WAS NO POLICE VISIT

The book claims the homeowners were visited by the local police, yet police records show no officers were dispatched to the house.

VENGEFUL GHOST DAMAGE NEVER HAPPENED

Windows and locks ostensibly destroyed by the home's vengeful spirit were documented by the follow-on owners as being in original condition.

THERE WAS NO RED ROOM

The hidden "red room" was not hidden or ominous in any way. It was simply a small basement closet (so small a person couldn't stand in it) that would have been immediately obvious to any purchaser.

THE HOUSE WAS NOT BUILT ON SACRED TRIBAL LAND

The book alludes to the house being built on land sacred to the Shinnecock Indians. Historians and Shinnecock natives themselves have pointed out that's simply not true.

NO PRIESTS WERE HARMED IN THE MAKING OF A HOAX

The priest who supposedly was called to bless the home, told to "get out", and was later supposedly tormented by the ghosts never existed. He was made up for the book. The family did contact a priest named Father Pecoraro. However, he maintains he never visited the house but only spoke to the owners over the phone.

GREEN SLIME IS STILL ONLY FOUND IN THE MONSTER MANUAL

The book describes the family seeing green slime oozing from the walls. Everyone concerned (the family and author) has admitted this was an embellishment.

WAS THE FAMILY IN THE HOUSE FOR 28 DAYS?

The events in the book cover a span of 28 days. However, neighbors peg the family's occupation of the house at less than 10 days.

FLY INFESTATION

The book claims the house was plagued by swarms of flies, even in winter. This seems to have originated from the cleaning woman sent in to clean the house after the murders. She reported finding a lot of flies. That's not surprising given the house was splattered in blood.

NO PURCHASER SINCE HAS EXPERIENCED ANYTHING

Since the publication of the book, the house has been sold several times. No owner has reported any odd happenings. The home's biggest curses seem to be high property taxes and an endless stream of tourists looking for the house. To dissuade visitors, subsequent owners have managed to get the home's official address changed and replaced the iconic eye-like windows with simple square windows.

Nine Museums about Ghosts, the Occult, and the World Beyond

GHOST MUSEUM PENANG (PENANG ISLAND, MALAYSIA)

This museum is devoted to the lore of ghost legends across Asia and points beyond. Fans of Korean and Japanese horror cinema might find this museum ideal. The museum is low on woo and high on creepy displays and props.

ZAK BAGANS' THE HAUNTED MUSEUM (LAS VEGAS, NEVADA)

Zak Bagans is the host of a ghost hunting show on cable TV. He opened a "haunted" museum in 2017 in Las Vegas. One of the museum's gimmicks is that you have to sign a waiver that indemnifies the museum against negative interactions with the supposedly haunted and cursed exhibits found in Bagans' museum. Among the museum's objects are two copies of the "Crying Boy" painting. Reproductions of the painting were reputed to have started fires in UK homes that hung the painting. The museum also boasts a mirror that Bagans says was owned by Bela Lugosi and is cursed. Lugosi's family, however, has no memory of the actor ever owning the mirror.

THE WARRENS' OCCULT MUSEUM (MONROE, CONNECTICUT)

This museum was opened by self-described "demonologists" Ed and Lorraine Warren. The Warrens were involved in the evolution of the Amityville horror myth. They ran their museum out of the back of their home. Both Warrens have died in recent years and the museum is currently on pause. When it was open, attendees could look at various knickknacks the Warrens had collected over the years. They insisted many of the items displayed were cursed or infused with evil! The center piece is a ridiculous looking Raggedy Ann doll they named Annabelle. This doll and their claims it once killed a man became the basis for a wildly popular series of Hollywood movies.

MUSEUM OF WITCHCRAFT AND MAGIC (CORNWALL, UK)

The Museum of Witchcraft and Magic has been open for 70 years. The museum's aim is to help visitors understand the history of witchcraft and magic in Europe. Much of the museum is dedicated to folk magic traditions and the European witch trials. You'll find exhibits devoted to more modern esoterica like Paganism and Aleister Crowley.

MUSÉE DES VAMPIRES (PARIS, FRANCE)

The *Musée des Vampires* (AKA, the Museum of Vampires) was started by Jacques Sirgent in Paris. Sirgent, who apparently holds a job as a vampirologist, has collected a number of vampire artifacts over the years. He eventually opened his personal residence to the public so others could enjoy them. Much of the collection is memorabilia from vampire movies (posters, props, masks, etc.). Hammer House of Horror fans will be right at home. There are some historical objects, like a 19th century vampire protection kit, antique books about vampire lore, and a crossbow apparently designed to kill vampires.

SALEM WITCH MUSEUM (SALEM, MASSACHUSETTS)

It would probably come as no surprise Salem, Massachusetts hosts a museum dedicated to the Salem Witch Trials. The museum is broken into two sections. One is devoted to the 17th century moral panic that caused Christians in Salem to execute 19 innocent women and men. The other section is devoted to the history of witches in general. The museum's mission is to help visitors recognize the causes of witch hunts and how such witch hunts manifest in modern times (the internment of Japanese Americans, McCarthyism, and the like).

NATIONAL MUSEUM OF FUNERAL HISTORY (HOUSTON, TEXAS)

The National Museum of Funeral History addresses humankind's oldest terror. We're all going to end up in the ground. The museum tries to educate the public on this last ritual. You'll find displays detailing how cultures through space and time have dealt with death.

H.R. GIGER MUSEUM (GRUYERES, SWITZERLAND)

H.R. Giger (died 2014) is famous for giving movies like *Alien* their distinctive look. Fans of his amazingly creepy "biomechanical" style will find a huge collection of movie props designed by Giger and actual works of art created by the hand of the artist himself. You'll also find a bar in the museum all done up like a set of an alien ship.

HAW PAR VILLA (SINGAPORE)

Many describe Singapore's Haw Par Villa as a theme park. But it is more of an open-air museum and sculpture garden devoted to the Chinese afterlife. Want to know about the 10 levels of Chinese hell and what that might look like? Come on down to Haw Par Villa. Best of all, this attraction is totally free.

A level of Chinese hell as depicted at Haw Par Villa.
Photo Credit: Karl Mamer

Six Cursed or Haunted Items

There are many items that have come to be associated with some kind of curse. These cursed item legends are generally exercises in confirmation bias. Someone thinks he or she notices a pattern, like owners have had bad fates. The person then looks for outcomes that confirm the existence of the pattern and ignores the great body of outcomes that don't confirm the pattern.

THE ANGUISHED MAN

"The Anguished Man" is a painting by an unknown artist. It's owned by Sean Robinson. He claims he inherited it from his grandmother and it is haunted. Why? According to Robinson the artist used his own blood to make the painting and then promptly killed himself. So, of course it's haunted. Robinson posts YouTube videos of his painting backed by scary music. He takes pains to prove if you have a camera pointed at a painting 24/7, you'll eventually catch some light anomalies. Thing is he seems to believe these light anomalies are actually ghosts. The videos are as exciting as watching paint dry or in this case watching dry paint. Despite there being absolutely nothing here, paranormal TV shows have fallen over themselves trying to incorporate the painting into their show. And, naturally, there's a movie in the works.

THE CHAIR OF DEATH

The Chair of Death at one point was found in a pub in Thirsk, UK. The chair is reputed to be a chair owned by an 18th-century murderer who killed his father-in-law while he sat in the chair. The killer, on his way to the gallows, ostensibly put a curse on the chair. The chair got a reputation for being the "chair of death" during World War II. The pub where the chair was found was a favorite of Canadian air force personnel. A legend grew that airmen who sat in the chair never returned from their next mission. In the 1970s the chair was donated to the town's museum and hung high up on a wall, ostensibly to prevent people from sitting in it.

THE CRYING BOY PAINTING

In the mid-1980s the British tabloid press whipped up a considerable amount of paranoia over a mass-produced print of artist Giovanni Bragolin's "The Crying Boy" painting. As the story goes, fire departments were noticing a large number of house fires had one thing in common: they all found a copy of "The Crying Boy". Even more mysterious despite the homes suffering large amounts of fire damage, the painting remained unscathed. Panic set it among people who owned copies of the painting and joined organized burnings of "The Crying Boy". Researchers looking into this legend found the print's varnish acted as a fire retardant, hence aiding its survival.

THE DYBBUK BOX

The Dybbuk Box is a small wine cabinet that its owner claims is haunted. It takes its name from a malevolent spirit of Jewish legend known as a Dybbuk. The box is famous for being an early example of people trying to sell haunted or cursed items on eBay. The owner included a long story of tragedy that had ostensibly befallen him after purchasing the wine cabinet from an estate sale. He claims it was once owned by a Holocaust survivor and previous owners had similar bad fates. The box was sold and was flipped again on eBay by the new owner. It ended up in the hands of a director of an osteopathic medical museum who wrote a book about it. That owner eventually sold it to Zak Bagans' museum of haunted curios.

THE HOPE DIAMOND

The Hope diamond is probably the best-known item that carries with it a legend of a curse. It's a 45-carat blue diamond. People who maintain the diamond is cursed point to the bad fates that befell many of its owners. For example, the diamond was once in the possession of Louis XVI and he was guillotined! Of course, Louis XVI's gruesome end had a little more to do with politics and social unrest than a curse. The majority of the diamond's owners lived lifespans that were not unusually short. It's found today in the Smithsonian's National Museum of Natural History. The diamond has only been beneficial to the museum. It led to a large increase in attendance.

OSCAR THE THERAPY CAT

Oscar is a cat adopted by a Rhode Island nursing home. Workers began to notice Oscar preferred to nap next to patients a few hours before they expired. Oscar came to the world's attention after this observation was published in the *New England Journal of Medicine*. Due to the difficulty in controlling the meanderings of a cat, no controls were really applied to the observations. There's a very good chance this is a product of confirmation bias. People remember when Oscar slept next to a patient that died and forget when the patient didn't expire. Potential hypotheses of how a cat can predict impending death include the cat can smell subtle chemicals given off early in the dying process or dying people are usually still and Oscar simply finds their rooms peaceful. Naturally, patients in the nursing home couldn't help but fear Oscar was not just partial to sound sleepers but was actually a harbinger of death.

Seven Ancient and Historical Curses

CURSE OF THE BILLY GOAT

Fans of the Chicago Cubs attributed the team's decades long losing streak to the Curse of the Billy Goat. As the legend goes, the owner of the Billy Goat Tavern was ejected from a game during the 1945 World Series. The owner was reputed to then lay a curse upon the team. "Them Cubs, they ain't gonna win no more". Many attempts were made during the years to lift the curse. Most attempts seemed to involve goats in one way or another. The last attempt was in 2016 when a vegetarian restaurant urged Chicago citizens to go meat free. Well, good luck. Oddly 2018 was absent of any notable attempts to lift the curse. This is also the year the Cubs broke the losing streak by winning the World Series.

CURSE OF TIPPECANOE

According to those who believe in the Curse of Tippecanoe, any President of the USA elected to office in a year evenly divisible by 20 is destined to die in office. For example, William Henry Harrison was elected in 1840 and died in office. Abraham Lincoln was elected in 1860 and died in office. John F. Kennedy was elected in 1960 and died in office. No president since Kennedy has died in office. Curse proponents do point to the assassination attempt on Reagan (elected 1980) and a 2005 attempt on George W. Bush's life (Bush was elected 2000). The so-called curse was said to have begun with William Henry Harrison but the pattern wasn't actually noticed until the 1930s. There's no record of anyone ever uttering such a curse during the time of Harrison. A curse was retconned by the Ripley's franchise in 1948. It attributed this bad luck to Harrison's persecution of Native Americans during the Battle of Tippecanoe.

THE RING OF SILVIANUS

The Ring of Silvianus was found in the 18th century by a British farmer plowing his field. The ring was apparently owned by a Roman named Silvianus and lost during the Roman occupation of Britain. Although known as the Ring of Silvianus, the ring has another Roman's name engraved on it. The name found on the ring is Seniciane. In the 19th century an excavation turned up a tablet with a curse, penned by Silvianus. Silvianus' curse was directed at a person named Seniciane who he believed stole his ring. In 1929 archaeologist Mortimer Wheeler thought there might be a connection between the ring with Seniciane's name on it and the lost ring curse authored by Silvianus. None other than J. R. R. Tolkien was called in to establish if there was a connection between the ring and the curse. Many think Tolkien's inspiration for Sauron's ring comes from his work on this artifact.

KING CASIMIR IV'S CURSE

In 1973 Polish researchers received approval by Cardinal Karol Wojtyła (who became Pope John Paul II) to open the tomb of King Casimir IV. Reportedly within days of opening the tomb, four of the twelve researchers present died. Many drew parallels to the supposed King Tut's curse. One of the researchers who didn't pass away found the *Aspergillus flavus* fungus present in the tomb. There was speculation the researchers succumbed to the fungus' toxins. However, no autopsies were performed to confirm this as the cause of death.

KING TUT'S CURSE

The discovery of Tutankhamun's tomb in 1922 by Howard Carter was a media sensation. Like any big media event, cranks will always try to grab some of the spotlight. The media of the day started to give press to armchair "experts" who knowingly spoke of the pharaoh's curse. Anyone who entered a pharaoh's tomb would surely die. Lord Carnarvon, Carter's financier, died of a blood infection six weeks following a visit he made to the tomb. Carnarvon's death seemed to confirm the reality of the curse. Naturally, given loose criteria of who was involved with the tomb's excavation, it's not unlikely you'll find people who died before their time.

Carter, first in the tomb, lived a full life, as did most of the 25 people who helped excavate Tut's resting place.

TUESDAYS

In the same way people avoid staying on the 13th floor of a hotel, Greeks avoid trying to do anything of importance on a Tuesday. The day seems cursed. Greeks will point to bad things that have happened on Tuesday. For example, Constantinople (then under control of the Byzantine Empire) was sacked by Roman Catholic crusaders on a Tuesday (April 13, 1204 to be exact). The city fell for good to the Ottoman empire on another Tuesday (May 29, 1453).

THE SCOTTISH PLAY

Maybe the second thing you learn about *Macbeth* in English class is that the members of the company staging it never refer to the play by its title but always call it "the Scottish play". Those who do are said to bring about a curse that will doom them or doom the whole production. The origins of this curse are murky. It certainly doesn't go back as far as the 17th century. Some believe a 19th-century critic Max Beebohm seeded the idea in a review. He wrote the play was notorious for a chain of bad luck that followed those who staged it. Believers in the curse will point to things like a riot that broke out after a staging of the play in 1849 or a close call Laurence Olivier (died 1989) had with a falling stage weight in 1937. Like many curses, the best explanation is the law of large numbers and confirmation bias. Thousands of people every year across the planet are staging *Macbeth*. Someone associated with some production is likely to have some tragedy befall them during or shortly after the play's staging.

Twenty Professions with Unusual Superstitions

You will find professions where success (or survival) has a large luck component. Consequently, many practitioners react by developing certain superstitious behaviors. Many will openly admit there's no rational basis but at the same time they might quip "but why tempt fate?"

ASTRONAUTS

The first human in space Yuri Gagarin realized he needed to go before his historic launch. Halfway to the launch pad he got out of the bus transporting him and he took a leak on one of the bus's tires. Astronauts who fly out of Russia's Baikonur Cosmodrome all repeat this act. Astronauts who launch from the Kennedy Space Center all consume a breakfast of steak, eggs, and cake. It doesn't matter what time the launch is, that is their final Earth meal until they return. More recently, as a nod to Douglas Adams, some astronauts have started to take a towel with them into space.

BASKETBALL PLAYERS

Many basketball players swear by the "hot hands" phenomenon. That is, a player is more likely to make a basket if the previous basket attempt was successful. A 1985 paper by psychologists Thomas Gilovich, Amos Tversky, and Robert Vallone examined this belief and found statistics did not support it. Since that 1985 paper, there have been about a dozen follow-on papers investigating the hot hands phenomenon. Some confirmed the 1985 findings, others have suggested the phenomenon might be real.

COLLEGE FOOTBALL PLAYERS

There's a general feeling among college football players that winning the Heisman Trophy will invariably doom your team from winning the next bowl game it plays. The curse hit winners for a stretch between 2003 and 2008. Five of six winners went on to lead their teams to defeat that year. Statistically, over the award's history, teams with Heisman winners have not underperformed.

FARMERS

Some farmers believe anything planted by a pregnant woman will guarantee a bumper crop.

FIREFIGHTERS

Many firefighters develop superstitions about clothing. Boots and suspenders have to be put on in a certain order or certain items need to be hung up last after a fire call. Some firefighters develop rituals about their helmet, notably they believe you should never clean it out after a call.

FLIGHT CREW

Like hoteliers and elevator makers, some airlines and flight crews have an aversion to the number 13. Following the crash of Continental Flight 1713 in 1987, Continental Airlines ensured there were no gate 13s at its terminals or row 13s on its planes.

JAPANESE OLYMPIC ORGANIZERS

With the postponement of the 2020 summer games in Tokyo due to the Covid epidemic, Japanese Olympic organizers started to murmur about a 40-year cycle of bad luck for hosting the Olympic games. Japan was supposed to host the games in 1940 but World War II canceled those games. Forty years later Moscow was supposed to host the games in 1980 but there was a massive boycott by Western nations (the Soviet Union still went ahead with the games). And then in 2020 a world-wide pandemic postponed those games.

JAPANESE RESEARCH SCIENTISTS

According to a 2017 article in the journal *Nature* many research labs in Japan hold an annual memorial service for animals they've used in research and then later had to put down. They do so to express gratitude for the animals' unwitting service.

KNITTERS

There's a belief among some knitters that if you knit a sweater and give it to a significant other, that will bring about the end of the relationship.

MARINES

U.S. Marines on patrol tend to discard a brand of candy called Charms if they find them in their ration kit. Marines have come to believe eating them will cause a rainstorm. Marines also apparently avoid eating apricots. This superstition started during the Vietnam War. Marines started to notice fellow casualties had cans of apricots on them. Word spread and soldiers decided it was better not to eat or even carry around apricots just to be safe.

MIT HACKERS

The online *Jargon File* (a dictionary of hacker terms compiled by Eric S. Raymond) recounts the story of a magical switch in an MIT computer lab. The switch was crudely affixed to the outside of one of the lab's PDP-10 minicomputers. The switch had two settings: Magic and More Magic. A hacker visiting the lab noted the switch was set to More Magic. Any computer scientist worth his/her own silicon knows you don't throw a switch in a lab if you don't know what it does. However, this switch was obviously not connected to anything. The hacker flipped the switch to its lower magical state. The computer it was attached to promptly crashed. The hacker repeated the test a couple more times in the presence of witnesses and each time the setting was switched from More Magic to Magic the computer promptly crashed. Lab members wisely stopped fiddling with the switch and left it set for More Magic.

MUSICIANS

Many up-and-coming young musicians fear becoming part of the "27 club" so named for the number of high-profile musicians who died at age 27. This includes Jimi Hendrix, Janis Joplin, Jim Morrison, Kurt Cobain, Kristen Pfaff (bass guitarist for Hole), B. Boon (lead singer of the Minutemen), and Amy Winehouse. Statistically, musicians are no more likely to die at age 27.

NASCAR DRIVERS

NASCAR was fraught with accidents and deaths in its early days. A few superstitions emerged among drivers. One was a ban on selling or consuming peanuts in the shell. Apparently, NASCAR drivers noticed peanut shells were frequently found in cars involved in deadly crashes. In 1964 NASCAR driver Joe Weatherly was killed during a race. As legend has it, two $50 bills were found in his pocket. Since then trying to spend a $50 bill at a NASCAR event is unwelcome.

NAVY SAILORS

U.S. navy sailors believe you should never wash your coffee mug while at sea. It's okay to rinse it out but washing it out with soap and water will bring bad luck from the coffee god.

A navy man's mug after a tour at sea.
Photo Credit: Sean Stone

POLICE OFFICERS

Many police officers have an aversion to commenting to another officer "it's quiet" or "it's a slow night". Upon uttering that, the shift will end up being the exact opposite.

PHYSICISTS

Among some physicists there is a belief that some fellow researchers exude bad luck. Their mere presence can cause technical glitches. This superstition is known as the Pauli Effect, named after Wolfgang Pauli and a play on his Pauli Exclusion Principle. The Pauli Effect is stated as "Wolfgang Pauli and a functioning device may not occupy the same room". The effect is said to be directly related to a physicist's talent. The more talented, the stronger the effect.

PROPERTY DEVELOPERS

Many property developers are subtly building the number eight into luxury developments to appeal to well-heeled Chinese buyers. Eight is a lucky number in China and much of East Asia. Developers will ensure a property has eight rooms on each floor, eight-meter ceilings, and prices incorporate one or more eights.

REALTORS

If a house isn't selling or you aren't getting desirable offers, realtors of a religious bent might suggest burying a statue of St. Joseph in your front yard. The trick is one needs to bury the statue upside down with St. Joseph's face pointed in the direction of the house that needs to be sold. When the house does sell, the former owner is required to unbury the statue and give it a place of honor in the new house. The trick here is most houses will sell eventually and the seller almost by definition will be getting a price he or she is happy with.

SAILORS

Women were considered bad luck and banned from sailing ships of old. Authorities cite at least two rationales for this ban. One, a woman aboard a ship of men would distract them from their duties and, of course, a woman aboard would anger the sea gods. To ensure calm waters, boats featured topless women carvings on the bow. Sailors believed this would shame the sea.

SOCCER PLAYERS

Many European soccer players get a feeling of dread if they find out Mick Jagger is attending their home game. Jagger is reputed to jinx a team and guarantee its loss.

Six Internal Debates in Woo

At times it seems like woo is one big tent. Everyone's ideas are accepted, even if contradictory, as long as they are at odds with the "official story". However, it's not always as collegiate in the woo tent. Here are six major debates among different branches of woo.

9/11 TRUTHERS: MADE IT HAPPEN VS LET IT HAPPEN

Conspiracy mongers who doubt the "official" story of 9/11 generally agree on one central point. George W. Bush needed a reason to go to war and rob Americans of various freedoms. There they part ways. Nine-eleven conspiracy mongers then break down into two basic camps. There's the "made it happen" crowd. These people believe George W. Bush, in league with a constellation of operatives (CIA, Mossad, the Illuminati, etc.), organized an attack on the World Trade Center and the Pentagon. Their rallying cry is "9/11 was an inside job". Some in this camp think the planes were empty. Others think there were no planes. The planes people witnessed and filmed were some kind of complex hologram. In either case, explosives were planted some time before the attack and that is what brought the twin towers and Building 7 down. The other camp believes George W. Bush knew Osama Bin Laden was planning an attack but had the American military and other security elements stand down.

AIDS DENIAL: HIV EXISTS OR DOESN'T EXIST

AIDS denialists separate themselves into two major warring camps. Some believe the HIV virus exists and was created in a genetics lab for some nefarious purpose (population reduction or to line the pockets of Big Pharma). Spike Lee shared that belief in a 1992 interview he gave to *Rolling Stone*. Another camp believes HIV exists but it's actually a harmless virus. AIDS is a result of an immoral lifestyle, drug abuse, or poor nutrition. Molecular biologist Peter Duesberg is probably the leading proponent of the idea that HIV exists but it has no or little role in AIDS. On the other side is the UK magazine *Continuum* that claimed there was no evidence HIV even existed. In 1996 the magazine started offering a cash reward to anyone who could produce "one scientific paper establishing actual isolation of HIV". It seems to have missed HIV had been isolated a decade earlier. The magazine folded in 2001 after yet another editor died of an AIDS-related disease.

BIGFOOT: SHOOT VERSUS DON'T SHOOT

If you see a Bigfoot, should you shoot it? Some Bigfoot researchers suggest the only way to 100% convince people of Bigfoot's existence is to bring in a body. It's a damn shame one of them has to die but forward science! Some on the other side argue Bigfoot is a human scion and killing a Bigfoot is tantamount to murdering your cousin. In the don't shoot camp are also those who think Bigfoot is on the order of a more distantly related ape (like a gorilla) yet argue the creature's population is so vanishingly small that killing even one could quickly put the species on the road to extinction.

CREATIONISM: YOUNG EARTH VERSUS OLD EARTH

As skeptics we probably encounter so many young Earth creationists, we might miss there is a branch of creationism called old Earth creationism. Young Earth creationists generally believe the earth and the entire universe is no more than 10,000 years old. Somehow between 35,000 and 50,000 species were saved 5,000 years ago on a boat and within a few thousand years they migrated and evolved into millions of species. Yet, young Earth creationist will tell you that this kind of evolution couldn't possibly have happened over millions of years. Old Earth creationists accept a longer time frame. Some subscribe to gap creationism. According to *Genesis 1:1* their god created the universe and Earth and then millions or billions of years passed and then their god created humans, plants and animals. From there, *Genesis 1:2* kicks in. The rest of *Genesis* then plays out over the next six or ten thousand years. This explains why radiometric dating keeps showing the earth is old, really old. It explains why we are seeing light from stars millions of light years away. There is also a branch of old Earth creationists that subscribe to day-age creationism. Each day of creation is a kind of "god day". What would seem like a day to their god is actually millions or billions of years to humans.

FLAT EARTH: IS AUSTRALIA REAL?

Australia presents a problem to flat Earth believers. If you were to fly from Santiago, Chile to Sydney, Australia on a flat Earth, the most direct route would take you over Dallas. No one has ever observed this. Oddly it's all ocean. In a similar vein, the most direct flight from Johannesburg to Sydney would take one over mostly land (Africa, the Middle East, Asia). Again, no one on such a flight has ever observed this. That is, of course, because in real life planes follow a great circle and that takes passengers on routes that mostly overfly ocean. There are three camps that try to explain why no one has ever observed flying over Dallas or China on these routes. Some flat Earthers believe in a conspiracy of airline pilots who fly the round Earth route between Santiago and Sydney. On a flat Earth this would mean doubling the flight time. One now may wonder how flight times match round Earth flight times? Turns out pilots just burn a lot of gas flying the plane at supersonic speeds to match the flight time of a subsonic flight between the destinations. Some flat Earthers believe pilots catch super-fast tail winds that have never been demonstrated to exist to make up for time. The last camp takes a more parsimonious approach. Australia simply doesn't exist. People who think they're visiting Australia are actually visiting a much closer island that's nowhere near where the real Australia should be.

UFOLOGY: NUTS AND BOLTS VS EVERYONE ELSE

The early days of Ufology were ruled by people advocating the "nuts and bolts" hypothesis. That is UFOs were physical craft being piloted by space aliens from another world. The National Investigations Committee on Aerial Phenomena

(NICAP) primarily represented this wing. It was composed largely of ex-military and people with an aeronautics bent. On the other side of the fence were the Contactees, led by George Adamski. These people espoused a modified version of Theosophy, claiming to be in contact with space aliens and receiving messages of universal peace and brotherhood. In the late 1980s and early 1990s there emerged on BBSes and the early Internet what became known as the "dark side hypothesis". Aliens were not our friends. They are kidnapping people and conducting ghastly medical experiments on humans. Our own governments know about this but are looking the other way. There was quite a bit of conflict between traditional nuts-and-bolts proponents and those espousing the dark side hypothesis during the early days of online culture. Finally, there's the interdimensional hypothesis put forward by Jacques Vallée. Vallée believes there's some kind of interdimensional consciousness that has been manifesting itself to humanity for generations. Sightings of gods, angels, ghosts, fairies, jinn, and UFOs are merely human interpretations of this force's machinations.

DESCARTES
MATH &
LOGIC

Twelve Logical Fallacies Used by the Woo Crowd

AD HOMINEM

Instead of attacking the argument, one attacks the person. The person is no good, therefore his/her theories or facts can be immediately dismissed without need for critical examination. Darwin's character is frequently a target of creationists in books and in online articles. Darwin was a sadistic hunter. Darwin was a racist. Darwin suffered from mental illness. Someone *this* bad couldn't possibly have published a theory worth considering.

APPEAL TO ANTIQUITY

Something must be true or is probably true because it's old. This is a favorite fallacy of the alt med crowd. Typically, it's the first line in their brochure. Acupuncture has been used for thousands of years, therefore you can rest assured it works. The assumption here is if it didn't work, people would have figured that out a thousand years ago and simply stopped using it. As James Randi liked to say, such beliefs are unsinkable rubber duckies. Nothing in alt med ever falls out of favor. Trepanation is alive today. The appeal to antiquity also plays to anti-modernism biases. It usually tees up or is teed up by the Appeal to Nature fallacy (see below).

APPEAL TO AUTHORITY

An appeal to authority asserts a statement is true simply because the person making the claim is an authority figure. Certainly, heeding the advice of your doctor or car mechanic is sound practice. One could, however, demonstrate the evidential basis of doctor's orders. In contrast, if the argument for the validity of a claim rests solely on the supposed authority of the claimant, then one commits the appeal to authority fallacy. Fans of homeopathy frequently commit this fallacy. They may point to the fact the UK's royal family is a big proponent of homeopathy. If these very rich and ostensibly well-educated people are using it, it must be effective.

APPEAL TO IGNORANCE

This fallacy is committed when the person believes the proposition is true simply because the person believes there's no evidence to counter it. Alternatively, the person can't think of an alternative explanation, so therefore his explanation wins by default. Religious people will argue since you can't prove their god does not exist, therefore it does exist. Likewise, they will argue since no one has an explanation for how the universe began without a god, therefore god is the only logical explanation. Naturally, there are explanations of how a universe can come from nothing. As well, simply saying "we don't know yet" is a perfectly logical position to take.

APPEAL TO NATURE

Another alt med favorite. Something is necessarily better because it is natural. What is natural is problematic. Is a pill made in a factory natural because it contains St. John's Wort? Of course, there are a lot of things in nature that, if ingested, will kill you. Anthrax is a wonderful natural substance. That doesn't mean it's harmless.

FALSE EQUIVALENCY

The idea here is that since two things share some common traits they must be equal. This is a favorite legal tactic of creationists. They claim evolution is just a religion and since religions can't be taught in science class, evolution should not be taught. Arguably science and religion share a few common traits. You have to take it on faith you can reasonably learn about nature through observation and the laws of science are universal. Since religions also require faith, the creationist will conclude, science is religion. Deciding two things are equal based on shared traits only gets you so far. Dogs and cats share a lot of common traits. They have fur. They eat meat. They have tails. They birth litters. Therefore, a dog is a cat. If you're trying to understand if two things are equivalent, you also need to understand how they're different. Religious beliefs, for example, are rarely subject to revision as new data emerges.

FALSE DILEMMA

Also known as a false dichotomy, this fallacy involves the proposition there are only a limited number of possible explanations. Since the arguer has eliminated the undesirable possibilities by some means, the last explanation standing must be the correct one. C.S. Lewis probably made this fallacy famous in various apologetic writings arguing for the divinity of Jesus. Jesus was either divine, a liar, or a madman. Since liars wouldn't stick to their lies under a threat of execution and Jesus seems too wise, well spoken, and rational to be a madman, therefore Jesus must have been honest about his divinity. Arguments like this typically ignore there are other possibilities. For example, in the case of Lewis' trilemma, it's possible Jesus is simply a mythological character or Jesus never claimed he was divine. That claim was later attributed to him by the authors of the gospels decades after the fact.

MOVING THE GOALPOST

A person agrees that a certain amount or kind of evidence is required to demonstrate the truth of a proposition. When the requested evidence is provided, the person moves the goalpost and demands more evidence. Moving the goalpost is a favorite tactic of denialists. For example, HIV denialists will argue AIDS has never fulfilled Koch's Postulates therefore one cannot say with confidence HIV causes AIDS. When shown a scientific paper that demonstrates HIV fulfills Koch's

Postulates, they will then move on to some other demand, like a test for HIV that never produces false positives. Vaccine denialists are also famous for moving the goalpost to maintain their belief vaccines cause autism. Their original claim was if you simply removed the thimerosal from childhood vaccines, the autism rate would plunge. It did not. They then moved the goalpost to claiming aluminum in vaccines causes autism. When that was shown not to be true, they moved to unspecific claims about toxins in our vaccines.

POST-HOC ERGO PROPTER HOC

This is Latin for "after this, therefore because of this". If Y happens after X, X must be the cause. X could be the cause or X might be necessary but not sufficient to cause Y or X might just be spuriously related. For example, jails have very few rich white people. Does that mean rich white people are more moral? Or perhaps they can just afford better lawyers and stay out of jail? Post-hoc reasoning is the trademark of the anti-vaccine crowd. Children seem to be diagnosed with autism around the time they get their MMR shot. Therefore, the MMR vaccine must cause autism. The reality is the earliest autism can be detected is also around the time children get their MMR vaccine. As one doctor put it, a parent brought her child in for his MMR vaccine. While in the waiting room, the child had a seizure. The doctor noted if the child had the seizure *after* receiving the vaccine, the mother would have forever blamed the vaccine and the doctor.

SHIFTING THE BURDEN OF PROOF

This fallacy is nearly identical to the Appeal to Ignorance fallacy. A person asserts the truth of a proposition and then expects the skeptic to marshal evidence as to why the proposition is wrong or why the evidence better supports another possibility. The *McLean v. Arkansas Board of Education* court case in the 1980s partially hinged on so-called creation science's attempt to shift the burden of proof. The state was trying to introduce creation science into the biology curriculum. In their view evolutionary scientists had not met the burden of proof to establish evolution as the best explanation for the origin of species, therefore creationism wins by default. A judge saw right through that and ruled for the plaintiffs.

SPECIAL PLEADING

This fallacy most commonly takes the form of a person insisting their pet belief is exempt from some standard of evidence or an ability to be detected or measured. Usually special pleading is offered as a post hoc explanation for why the expected evidence is absent. Alt med types rely on a lot of special pleading to explain why rigorous tests of homeopathy fail to demonstrate results. Homeopathy treatments are so highly specialized to the individual that there is no general test of homeopathy's effectiveness. Parapsychologists long ago reached for the special pleading fallacy to explain why skeptics, following the same experimental design, can't replicate their positive results for psychic abilities. As it turns out there's a phenomenon called "psi missing" and this is caused by the presence of so-called "goats", that is anyone skeptical of the existence of the paranormal.

STRAW MAN

This fallacy ignores the actual argument and re-casts it as a weaker argument. This straw man argument is then easier to knock down. Many skeptical positions are based on a nuanced argument that woo believers will try to oversimplify and create a straw man in the process. For example, skeptics will point out to UFO believers that eyewitness testimony is weak evidence. UFO believers will quickly retort "so you're saying thousands of people all across the globe are lying?" Or a skeptic might say "I think the chances another alien civilization is visiting Earth is low." A UFO believer would straw man that argument by countering "So you don't think there's any other life in the rest of the universe?"

Fifteen Cognitive Biases That F' up Skeptics and Woo Proponents

The human brain takes a number of mental shortcuts just to get us through the day. Many of these shortcuts manifest themselves as biases. Here is a list of biases that affect our rational judgment.

ANCHORING

Judgments about the value of one thing can be affected by anchoring the thing to another similar item. For example, what might seem like a pricey hotel room may seem value-priced if anchored to a more expensive room rate. One can see the anchoring effect in woo. For example, you anchor a cancer sufferer with a complex, unpleasant, and uncertain science-based treatment and then offer them a simple treatment like a fruit-based diet that comes with the promise of a 100% cure rate.

AGENT DETECTION

This is the bias to perceive some kind of intelligence or life force behind some random or natural occurrence. As the saying goes, if you mistake some rustling leaves for a tiger, you look foolish. If you mistake a tiger for rustling leaves, you are dead. There's some data (Douglas, Karen & Sutton, Robbie & Callan, Mitchell & Dawtry, Rael & Harvey, Annelie. 2016. "Someone is pulling the strings: Hypersensitive agency detection and belief in conspiracy theories") to suggest people who fall deeply for conspiracy theories have an over-tuned agent detection bias.

AVAILABILITY BIAS

When we try to assess the risk or benefit of a certain course of action, we tend to give undo weight to recent examples. A recent plane crash in the news tends to make one overestimate the risks of flying. Much modern political discourse is informed by the availability bias. Although crime rates have steadily fallen since the 1970s, the availability bias is used by politicians to keep the electorate in a state of fear of crime.

BACKFIRE EFFECT

A 2010 study suggested trying to debunk a certain belief tends to reinforce the belief. The idea here is people tend to remember the most salient details (i.e., the woo) and end up forgetting the more subtle explanations as to why the story was wrong. Attempts to replicate the 2010 study have failed to find a backfire effect, however. That said, one should probably err on the side of making the debunking as exciting as the woo itself.

BANDWAGON EFFECT

This is the tendency to make decisions based on popular ideas instead of objective facts. The lure of being on the winning side can be a powerful motivator and short-circuit our inclination to check facts or seek evidence for a claim. Your Facebook peers reposting memes about the alleged similarities between Horus and Jesus might cause you to repost the meme without checking if there is any legitimate scholarly research on the alleged connection.

BARNUM EFFECT

This is the tendency to believe vague statements made to us are actually highly accurate and insightful readings. Horoscopes and psychic cold reading techniques are the classic examples of the Barnum effect in action. While many might be skeptical of horoscopes, it's easy to miss personality tests like the Myers-Briggs likewise operate on the principle of the Barnum effect.

DECLINISM

This is the belief the past was better and the future can only be worse. Every decade seems to have people raging things are going to hell in a handbasket and the past is seen as some golden era. Many white middle-aged males tend to assert the 1950s was some golden era of prosperity. They ignore the 1950s was not the best time for people of color, women, the LGTBQ community, and the physically challenged.

ENDOWMENT EFFECT

This is the tendency to demand more to part with something we own than what we might have been willing to pay for it originally. For example, a person might have willingly bought a concert ticket for $100 but would require considerably more money to sell the ticket. Knowledge and experience you worked hard to acquire might be hard to abandon even in the face of new evidence.

GAMBLER'S FALLACY

Who of us hasn't watched someone play a losing slot machine and then jumped on it thinking it's now due for a big payout? The Gambler's Fallacy is the belief an unlucky or lucky streak affects independent future events.

ILLUSION OF CONTROL

This is a tendency to overestimate the control you have in a generally chaotic world. We tend to attribute our success to our own unmatched abilities and fail to recognize the role luck plays in our success. This bias may be at play with psychics. Many psychics fool themselves into thinking random hits are the result of some ability.

IN-GROUP BIAS

This is the tendency to treat in-group members more favorably. We, as skeptics, are more likely to forgive fellow skeptics for errors and ethical transgressions than woo proponents who commit the same transgressions.

THE JUST WORLD HYPOTHESIS

This is the tendency to view positive outcomes for you as right and proper as you are a good person. Negative outcomes for others may be simply bad people getting what they deserve. Victim blaming is a classic example. The bias can lead us to seeking reasons for the status quo instead of changing the status quo.

MASLOW'S HAMMER

This is the notion if the only tool you know how to use is a hammer, all problems look like a nail to you. This is the tendency to try and fit a complex phenomenon into our area of expertise. For example, engineers tend to over represent the scientifically trained in the world of creationism. Many see biology as a complex machine that looks designed from an engineering standpoint. As skeptics we tend to wade into debates over matters of taste or opinion and try to bring the tools of skepticism to inform people of the correct answer.

RESTRAINT BIAS

This is our tendency to overestimate our self-control. We would like to think given certain temptations we would make the right choice. For example, if one were a scientist one would think you would never fudge your data.

SURVIVORSHIP BIAS

This is the tendency to view surviving members of a group as representative of the wider group. For example, we see a successful restaurant and think we can do that too, not understanding 95% of restaurants fail. When, say, a celebrity is asked the secret of his or her success, the celebrity typically offers "never give up!" Multi-level marketing scams use the survivorship bias well to their benefit.

Twelve Aphorisms Skeptics Should Live by

Murphy's Law is probably the best example of an aphorism: Anything that can go wrong, will go wrong. They're short, clever observations about reality that are usually phrased as rules or laws. Here are twelve that skeptics should employ when venturing out into the wide world of woo.

BENFORD'S LAW OF CONTROVERSY

"Passion is inversely proportional to the amount of real information available." This was coined by sci-fi author Gregory Benford in 1980. Wade carefully into debates full of passion.

CLARKE'S FOURTH LAW

"For every expert there is an equal and opposite expert." This was formulated by sci-fi author Arthur C. Clarke (died 2008). Most are familiar with his third law "Any sufficiently advanced technology is indistinguishable from magic." That's a good one to remember too but his fourth law is best remembered by lay people trying to debate highly technical arguments from cranks. For example, in 9/11 debates, there's always a seemingly well-credentialed expert that conspiracy theorists will cite.

ELLISON'S LAW

"The two most common elements in the universe are hydrogen and stupidity." This is courtesy of sci-fi author Harlan Ellison (died 2018). Never be surprised you encounter so much stupidity out in the world.

HANLON'S RAZOR

"Never attribute to malice that which can be explained by stupidity." This is an excellent first approach to debunking conspiracy theories. Is what people are alleging just the product of stupidity? Why was America caught off guard on 9/11? Hanlon's Razor suggests we should first look to error and incompetence.

HITCHENS'S RAZOR

"What can be asserted without evidence can also be dismissed without evidence." This was first phrased by Christopher Hitchens (died 2011). If the person making a claim can't be bothered to marshal any cogent evidence for the claim, it's better to simply dismiss the claim.

KRANZBERG'S FIRST LAW OF TECHNOLOGY

"Technology is neither good nor bad; nor is it neutral." Historian Melvin Kranzberg (died 1995) wrote a lot about the history of technology and came up with six "laws of technology". This one reminds the skeptic that technology's impact on society is complex. It is good to remember this when engaging naturalistic woo types who might argue against things like nuclear energy, Wi-Fi, or mRNA vaccines. Don't try to simplify the ramifications of complex modern technology down to an argument like "a hammer can be used to build a house or bash in a skull".

LEM'S LAW

"No one reads; if someone does read, he doesn't understand; if he understands, he immediately forgets." This adage comes to us from sci-fi author Stanislaw Lem (died 2006). Are you surprised the woo proponent didn't read your source and offers arguments for things already debunked by your source?

NEWTON'S FLAMING LASER SWORD

"That which cannot be settled by experiment is not worth debating." This was suggested by Australian mathematician Michael Alder. Many scientists take a dim view of philosophy, viewing it as mere word play and have a tendency to take an extreme Newtonian view on how to pursue knowledge. It's a check digit against over-applying Occam's Razor and the scientific method to all aspects of the human experience.

SAGAN'S STANDARD

"Extraordinary claims require extraordinary evidence." Sagan's Standard reminds us if someone's claim violates a well-established theory, the evidential bar is high. Carl Sagan (died 1996) first introduced this aphorism on *Cosmos* (1980). CSICOP co-founder Marcello Truzzi (died 2003) coined a version about two years before Sagan that goes "an extraordinary claim requires extraordinary proof." Thomas Jefferson (died 1826) might have introduced the earliest formation that went "A thousand phenomena present themselves daily which we cannot explain, but where facts are suggested, bearing no analogy with the laws of nature as yet known to us, their verity needs proofs proportioned to their difficulty."

Statue of Thomas Jefferson in Paris.
Photo Credit: Karl Mamer

SAYYRE'S LAW

"In any dispute the intensity of feeling is inversely proportional to the value of the stakes at issue." Sayre's law was formulated by economics historian Charles Issawi (died 2000). He coined it after reading a 1973 quote in the *Wall Street Journal* by political scientist Wallace Stanley Sayre (died 1972). Sayre stated, "Academic politics is the most vicious and bitter form of politics, because the stakes are so low." Many debates with woo can spiral out of control over arguments with amazingly low stakes. Convincing a crank the moon isn't hollow probably has very little real world consequences.

STURGEON'S LAW

"Ninety percent of everything is crap." This comes to us from sci-fi author Theodore Sturgeon. Philosopher Daniel Dennet reminded readers in his book *Intuition Pumps and Other Tools for Thinking* to approach any debate with Sturgeon's Law in mind. Don't waste time debating the crap. Ask yourself what are the best arguments? Deal with those.

TWYNMAN'S LAW

"Any statistic that appears interesting is almost certainly a mistake." Tony Twynman (died 2014) measured radio and TV audience data in the UK and this was his check digit. It's an excellent reminder when you hear about things like faster-than-light neutrinos or some possible alien signal captured by a radio telescope.

Nine Mistakes Skeptics Make Arguing with Woo People

IGNORING HYMAN'S CATEGORICAL IMPERATIVE

CSICOP founding member James Alcock coined this term for Ray Hyman's warning "Do not try to explain something until you are sure there is something to be explained." One will often find when you start to access primary sources about a supposed paranormal event, there's no evidence the event even happened. The best argument against a paranormal claim is to show it didn't happen in the first place or there's a complete lack of contemporary attestations.

TRYING TO USE FACTS TO ARGUE A PERSON OUT OF A BELIEF THE PERSON DIDN'T COME TO BELIEVE THROUGH FACTS

If you stop to think about many of your own cherished beliefs, they're not all built on a bedrock of fact and logic. They've always just felt right. It's something a parent or teacher told you decades ago, and you've made it part of your life's philosophy.

NOT HOLDING THEM TO THE BURDEN OF PROOF

You can get lost in trying to prove why a proposition is wrong, especially if you're trying to present sources the woo proponent won't view as legitimate in the first place. Require the woo proponent to establish his or her case. Ask questions. "Why do you think that?" "What is your evidence?" "Why do you think that's a good source?" "Why is your explanation better than simply admitting we don't know and may never know?"

LETTING YOURSELF GET GISH GALLOPED AND SIDETRACKED

Woo proponents like to pile it on high and deep. You'll find once you demolish one claim, the woo proponent will then bring up another claim and then another one and then another one. You'll spend all your time going down rabbit holes. Establish straight out you are happy to examine the woo proponent's three best pieces of evidence. You might be surprised at how often a woo proponent can't articulate even one. If they can't, argue at your own peril. If they can, deal with them in order of importance and don't let the woo proponent sidetrack the argument.

NOT SETTING A GOAL POST FROM THE START

Before you try to supply evidence, ask what the person you're arguing with would agree is valid evidence, a valid source, and what would it take for the person to accept his/her belief was falsified. If you don't, you'll waste a lot of time having the woo proponent handwave away peer reviewed research because "they're being paid to lie by the drug companies!"

NOT SETTING YOUR OWN GOAL POST AND EXIT CONDITIONS

What is your goal for arguing? What do you hope to win? Arguing with woo proponents can sometimes be a great way to test your own assumptions and learn how we know what we know. Other times you may know you'll never convince the woo proponent but you have an audience of fence sitters that can be compelled by fact and reason. When should you extricate yourself from an argument you will never ultimately win? Do you think you will get the woo opponent to admit his/her entire worldview is based on nothing but lies? Sometimes a good time to take the off ramp is when the woo proponent returns to arguments that were already debunked.

QUESTIONING MOTIVATIONS FOR A BELIEF, NOT EVIDENCE FOR A BELIEF

Penn Jillette once quipped the difference between a cynic and a skeptic is the latter questions the evidence that led you to a belief while the former questions your motivations for holding that belief.

YOU RAGE YOU LOSE

It's pretty easy to lose your cool with someone who handwaves away unimpeachable evidence and airtight logic. If you're in the argument to help the fence sitters come to a better decision, being the first one to start raging and using the F word will only lose you credibility with the fence sitters.

USING THEIR TERMS

Don't use the woo debater's term. It's medicine, not Western medicine or allopathic medicine. It's a lifesaving and medically necessary abortion, not a partial birth abortion. It's conservative judicial activism, not originalism. It's evolutionary science, not evolutionism.

Seven Interesting Yet Spurious Correlations

Tyler Vigen is the creator of a site called *Spurious Correlations*. From several large and disparate data sets he finds very strong correlations that are, on the face of it, obviously lacking in any kind of causal mechanism. His site offers a powerful and amusing reminder that simply because two things seem highly correlated it doesn't mean there is any relationship. Here are seven strong and strongly amusing spurious correlations from his site:

1. Age of Academy Award Best Supporting Actress Winner vs Divorce Rate in Alaska (correlation: .738)
2. Suicide by crashing of motor vehicle vs German passenger cars sold in the US (correlation: .936)
3. Cost of bananas vs US spending on science and technology (correlation: .940)
4. Total revenue generated by arcades vs Computer science doctorates awarded in the US (correlation: .985)
5. Divorce rate in Maine vs Per capita consumption of margarine (correlation: .993)
6. Apple phone sales vs People who died by falling down stairs (correlation: .995)
7. Money spent on pets vs Lawyers in California (correlation: .998)

Three Math Theorems that Attract the Most Cranks

Math is hard. Math also leads to some real-world conclusions we find disturbing. If X spending continues at Y pace, I'll have no more money in my bank account. This asteroid has a high percentage of hitting Earth and ending all life as we know it. We are going to have to spend billions on a space program to deflect it. We can respond to the potential reality math has revealed by doing something difficult (cutting back on consumption or spending billions to save the planet). Or we can ignore the problem. There's also a secret third option beloved by cranks. Show the math is simply wrong. Many math cranks seem to be religiously motivated. They think the theorem says something about reality that's contradictory to the reality they assume their god has promised them.

SQUARING THE CIRCLE

Ancient geometers spent considerable time trying to figure out how one could fit the area of a circle into a square of the same area. What they didn't realize is there is no square root of Pi. You can find squares that are close to the area of a given circle but never exactly equal. This was proven an impossible task when it was established in 1882 that Pi was a transcendental number, that is a number with no root. That there's a proof of this doesn't seem to square with math cranks. Circle square-ers seem to be given to self-publishing books detailing their methods for squaring circles. This mostly involves a very mathematically laborious way of rounding Pi to 3.2.

CANTOR CRANKS

Georg Cantor in the 19th century established there are actually some infinities that are larger than other infinities. In essence, the set of all real numbers (any number that can be written on a number line like 2, 1/8th, -0.75, etc.) is infinite in size. So is the set of all-natural numbers (1, 2, 3, 4, and so on). If you try to map members of the natural number set to members of the real number set, Cantor was able to show there would always be numbers in the real number set that you can't map to natural numbers. It's like building Ikea furniture. You follow the instructions and then at the end discover you've got extra stuff. Extra stuff in the real number set means this infinite set is bigger. When Cantor published his proof, he met quite a lot of resistance in the world of math. But like any well-tested mathematical proof, people came around. However, the idea that one infinity could be larger than another is deeply unsettling to math cranks. It's as if dictionary definitions aren't absolutely accurate all the time. Typically, the way to attack Cantor's idea of larger infinities is to simply assert there is no infinity in real life. However, many Cantor Cranks are of a religious bent. They can't deny the infinite as they were promised at least one by their god. So, most Cantor Cranks attack the problem by simply creating a straw man version of Cantor's proof and then knocking that down.

FIFTH POSTULATE PROVERS

Euclidean geometry is based on five postulates. You have to accept them as true if you want to do Euclidean geometry. You can't prove any of them. If you don't want to take them as true, then shove off. Go do some other kind of geometry, hippy. Fortunately, these postulates are not difficult to accept as true. The first four seem pretty hard to deny. All right angles are equal to each other. You can draw a straight line between any two points. A line can be extended to any length you desire. Any circle can be described by defining a point in the middle and a straight, finite line going out. Real basic, obvious stuff. But the last one, not so much. It can't be stated in a simple sentence. This is where your grade school grammar lessons on how to construct compound sentences comes in handy. Basically, if two straight lines intersect with another straight line and both sides of those intersections are right angles or at least their angles add up to 180 degrees, well, the lines are parallel and will never meet. But God forbid if that ain't the case. If the interior angles are less than two right angles, then, hold your britches, those lines will intersect as sure as God made little green apples! However, if you take that fifth postulate and flip it around, say parallel lines meet some place way, way out there, or throw it away entirely then you're living, literally, in a whole different universe. That you can modify Euclid's fifth postulate and the math still works has unsettled generations of cranks who don't like the idea that what you're measuring with your ruler might not be 100% true but just *mostly* true. The best way to return to a safe, absolute universe is to prove the fifth postulate using the other four. Smart people worked the problem for about twenty centuries before giving up and learning to live with non-Euclidean geometry. A hallmark of any crank is, however, they think they've seen something really simple that people have overlooked for hundreds or even thousands of years. Most crank "proofs" of the fifth postulate end up being long winded proofs the author unknowingly accepts the fifth postulate as a true.

Fourteen Criteria on the Crackpot Index

There have been a number of attempts over the years to come up with a way to mathematically describe cranks that haunt the inboxes and message boards of math and physics professionals. RAND Corporation employee Fred J. Gruenberger developed an index in 1962 called "A Measure for Crackpots". His scale ranged from 0 to 100. Higher the number, less of a crackpot the person in question is. In the early 1990s, mathematical physicist John C. Baez developed a list of criteria for his Crackpot Index. He flipped Gruenberger's scale. Higher the index number, the bigger the crackpot. Below are some of the more amusing criteria on Baez's index and their point values.

1. The person repeats claims that have already been corrected (5 points for each statement)
2. A word is in all caps without good reason (5 points for each word)
3. Feynman is name dropped (5 points)
4. The person offers prize money if others can find flaws in his/her theory (10 points)
5. The person needlessly invents new terms (10 points for each new term)
6. The person intimates relativity or quantum mechanics is "fundamentally misguided" (10 points)
7. Established theories are hand waved away as being "only a theory" (10 points)
8. The person compares him/herself to Newton (20 points for each mention)
9. The person has named the theory after him/herself (20 points)
10. A critic is accused of being a "hidebound reactionary" or a "self-appointed defender of the orthodoxy" (20 points)
11. Claiming Einstein himself was working on the theory in his final days (30 points)
12. The theory comes, in part, from information supplied by space aliens (30 points)
13. The person alleges they are being persecuted like Galileo (40 points)
14. The person offers a groundbreaking theory that offers no predictions (50 points)

Eight Examples of Mathematical Folklore

EINSTEIN'S CHAUFFEUR DELIVERED A RELATIVITY LECTURE

The story goes Princeton hired Einstein a chauffeur who drove him around to an uncounted number of lectures the great scientist was giving on his Theory of Relativity. Just before one lecture, Einstein became ill. The resourceful driver said "I've sat in on your lecture so many times now I know it word for word. Why don't I give it?" Einstein agreed. Sure enough the driver delivered the lecture perfectly. When the Q&A portion of the lecture started, an attendee stood up with what he was sure was a gotcha question for Einstein. Without batting an eye, the driver chortled "that question is so simple I'm going to let my driver sitting in the back of the room answer it." One thing that makes this legend hard to believe is by the time Einstein was on the academic lecture circuit he was probably one of the most recognizable people in the world. It's doubtful Einstein could have just hung out in the back of the room and not be noticed.

FERMAT'S LAST THEOREM MEMES

In 1837, Fermat proposed the formula $a^n+b^n=c^n$ has no solution if *a*, *b*, and *c* are positive integers and *n* is any value greater than 2. He scribbled this in the margins of a book he was reading. He also added he had thought of a "marvelous" proof but the margins of the book weren't enough to contain it. Fermat died before he ever found a sheet of paper big enough to contain his proof. For more than a century and a half, mathematicians struggled finding one. Many over those years had wished Fermat had a Denny's paper placemat in front of him instead of a math text. Most mathematicians eventually concluded Fermat probably didn't have an actual proof. When Andrew Wiles successfully proved it in 1994 the proof ran about 100 pages. Fermat was certainly right about the inability of a book margin to contain a proper proof. Since 1837, there have been a number of jokes and legends related to Fermat's claim of having solved his conjecture but not writing it down. As one story goes, a math professor would make the claim before a talk that part of his talk was going to include a proof of Fermat's last theorem. When the professor failed to deliver, he would quip "oh, I just put that in there in case I died before giving my talk".

GALILEO DROPPED WEIGHTS FROM THE TOP OF THE LEANING TOWER OF PISA

It's often said that Galileo (died 1642) dropped two spheres of different weight from the top of the Leaning Tower of Pisa to demonstrate objects fall at the same rate regardless of their mass. The problem is there is nothing in Galileo's own writings to indicate he had ever conducted this experiment. Most historians believe Galileo suggested it as a thought experiment but never carried it out. The legend is found in a Galileo biography written by his student Vincenzo Viviani. Viviani's account of when Galileo was supposed to have conducted the experiment predates Galileo's formulation of his law of falling bodies.

THE MATHEMATICIAN WHO DID HIS GREATEST WORK THE NIGHT BEFORE HE DIED IN A DUEL

Évariste Galois (died 1832) was a budding math genius. His most famous contribution to math was something called Galois theory, which shows certain equations can be solved using radicals. Galois was also something of a hot head and political activist. He took part in an 1831 Bastille Day protest and came loaded for bear. He brought with him several pistols, a rifle, and a dagger. He ended up in prison. Prison didn't seem to cool his jets. Shortly after being released from prison he ended up agreeing to a duel. Who he dueled or the exact reason why is lost to history. Some suggest it was over a woman. The night before the duel, Galois had a feeling like he was not going to be the victor. He spent much of the night writing final letters, including what is described as his "mathematical testament". From this a legend was born that Galois' most important mathematical insights were put to paper on the eve of his ill-fated duel. In reality, his final math was an outline of some work he had been doing. On the morning of May 30, 1832, he took a bullet to his abdomen. Apparently, all involved (including Galois' "second") fled the scene, leaving Galois to die. He was eventually found by a farmer and brought to a hospital where he died the following day. His death has spawned at least one conspiracy theory. Because we know so little about the reasons for the duel, the other participants, and their odd refusal to obey the forms of a duel (your second should help you off the field of honor), some contemporary writers have posited Galois was the victim of a cop conspiracy. Somehow the police used Galois' hotheaded nature against him and set him up in a situation where he could not refuse a duel that he was sure to lose.

NEWTON'S LAWS OF GRAVITY WERE INSPIRED BY AN APPLE FALLING ON HIS HEAD

The story about Isaac Newton (died 1727) developing his laws of universal gravitation after being bonked on the head by an apple is a near perfect metaphor for how our brain sometimes generate insights after some kind of sudden, novel stimulus. However, the story is a mutated version of an anecdote Newton shared with his biographer William Stukeley almost four decades after he published his *Principia*. Stukeley reports around 1665 the scientist had witnessed an apple falling from a tree and was given to wonder why the apple fell straight down, and not sideways. This inspired him to understand the nature of gravity.

WAS NEWTON'S "STANDING ON THE SHOULDERS OF GIANTS" QUOTE A SICK BURN?

For many years Newton was in a dispute with scientist Robert Hooke (died 1703) over criticism he received from Hooke about his work on optics. Now, Hooke was no schlub. He coined the term "cell" and made major contributions to the worlds of physics, astronomy, microscopy, and paleontology. Before Newton and Hooke had a falling out, Hooke wrote Newton some kind of fanboy letter in 1676. Newton famously wrote back that Hooke actually helped his research. Newton penned "You have added much several ways...". Newton finished his letter with his famous line "If I have seen further it is by standing on the shoulder of giants." Historians have read a lot into this line. Some have seen it as Newton being humble. Many discount that as Newton claimed at times he was chosen by the Christian god to interpret scripture. He rarely comes across as having a single humble bone in his body. Others think maybe Newton was trying to insult Hooke. Hooke was apparently a slight man and possibly of stunted height due to a childhood illness. So, for Newton to suggest Hooke was a "giant" was balls out sarcasm on Newton's part. Others have cast doubt on this interpretation. When Hooke wrote his letter, the scientists were still on good terms. As well, Hooke may not have been all that short for a man of his time.

SOURCE OF THE COFFEE QUOTE

There's an often-repeated quote in Math culture that goes "A mathematician is a device for turning coffee into theorems". Hungarian mathematician Paul Erdős (died 1996) was famous for drinking vast quantities of coffee while he worked, so it was initially assumed the quip came from him. Erdős' capacity to knock back cups and cups of Joe and other "stimulants" began to worry his friends and colleagues. He was offered $500 if he could stop for one month. He accepted the challenge and managed to go the month cold turkey. He pocketed the $500 but complained during his abstinence "Before, when I looked at a piece of blank paper my mind was filled with ideas. Now all I see is a blank piece of paper." Erdős himself claimed the coffee quote was not his but came from fellow Hungarian, coffee lover, and mathematician Alfréd Rényi (died 1970). It was suggested the quip might have been inspired by German wordplay. The German word *Satz* can mean either "theorem" or "coffee residue". However, both Rényi and Erdős spoke Hungarian so it might be unlikely they were fond of making puns in German

"THIS IS NOT MATHEMATICS, IT'S THEOLOGY!"

Mathematician Paul Gordan (died 1912) was considered the "king of invariant theory". This is a branch of abstract algebra that deals with mathematical functions that do not change (that is, they are invariant). Gordan spent twenty years working on something called the finite basis theorem. It so defined his career that many simply called it Gordan's Problem. In 1888 mathematician David Hilbert (died 1943) came along and solved Gordan's Problem with a few tidy and innovative pages of math. Hilbert used a new method of proof he invented called an existence proof. In essence, the trick is to first show there are only two possible cases. If you can then show one case is impossible, the other case is then logically and necessarily true. Upon submitting his proof to Gordan, legend has it Gordan rejected the proof exclaiming "This is not mathematics, it's theology!" It's a great line and you don't have to know much about math to feel Gordan's hurt resulting from seeing 20 years of work wrapped up in a lazy afternoon by some young upstart. The problem is Gordan may not have uttered it and if he did, no one is sure when or why he made the statement. The quote is attributed to Gordan by an associate who was eulogizing the mathematician a quarter century after Hilbert's proof. Gordan and Hilbert were most definitely not rivals but worked closely together. Gordan encouraged Hilbert to work on his problem. Gordan went on to adopt Hilbert's methods. Some suggest if Gordan did utter the phrase, it might have been in praise.

Six and a Half Examples of Using Benford's Law to Detect Chicanery

Benford's Law is the observation that in large, real-world data sets the leading number of any member of the set is not random but follows a declining curve. That is, more members of the data set will start with the number 1 and 2 than 8 and 9. So if you were to look at a list of counties in the USA, you might think their population numbers would be all over the map in terms of the leading number. So, you'd see population numbers with leading numbers like **9**,569, **7**54,987, **3**2,441, and so on. What you find is about 30% of the population numbers will have a leading number of 1, 18% of the population numbers will have a 2, 12% will have a 3, and so on. Humans have certain biases as to what data sets should look like and what randomness should look like and can be pretty bad at fudging a random dataset. Benford's Law can be used to suggest such fraudulently created data sets.

CHECK FRAUD

Many financial institutions will put a hold on any check over $1,000. By Benford's Law, you should not see a spike of checks being written for amounts ranging between $700 and $999.

COVID REPORTING

During the early days of the Covid epidemic, a number of countries were accused of underreporting their case numbers. A number of researchers applied Benford's Law to the reporting datasets offered by countries around the world. Curiously, datasets that failed Benford's Law tended to come from nations run by dictators. Nations like Tajikistan, Russia, Turkey, Myanmar, and Brazil offered datasets that didn't jibe with Benford's Law.

DETECTING PONZI SCHEMES

A number of researchers have retrospectively run gains and losses from Bernie Madoff's Ponzi scheme against Benford's Law. Madoff (died 2021) ran his scheme between 1990 and 2008. Researchers found the first eight years of Madoff's data appears to have violated Benford's Law. It is possible, Madoff got wise to how Benford's Law was increasingly being used to detect accounting fraud and ensured his reporting comported with expectations.

EMPLOYEE EXPENSES

Many corporations apply lesser scrutiny to employee expenses that are below a threshold. For example, any individual travel expense over $30 might require approval and increased documentation. Benford's Law has been used to smoke out employees submitting an excess of fraudulent expense claims that would normally be paid out.

TAX FRAUD

For a number of years, the IRS has been applying Benford's Law to tax records. Numbers that defy Benford's Law are considered a red flag and increase the chances the person or businesses will be audited.

UNUSUAL NETWORK TRAFFIC

The amount of data people are downloading follows Benford's Law. The law has been used to detect outgoing employees slowly siphoning off company data they intended to use in a new competitive business.

THE HALF: ELECTION FRAUD

Many have suggested various elections around the world, from Iranian to US elections, have been fraudulent because results have defied Benford's Law. Studies applying Benford's Law to datasets of known fair elections as well as datasets of simulated fair and fraudulent elections have revealed voting patterns don't reliably follow a Benford-like curve.

Six People Who Beat the Gambling Industry at Their Own Game

EDWARD O. THORP: FATHER OF CARD COUNTING

In the late 1950s, mathematician Edward O. Thorp used an IBM 704 computer to analyze the game of blackjack. Based on his analysis he developed a method of card counting. Card counting is a way of keeping track of cards that have already been played from the deck which can then inform you about the likelihood of winning cards coming out of the shoe. With a $10,000 stake from a Nevada professional gambler, Thorp put his method to the test and more than doubled the stake in a weekend. Word quickly spread of this new technique and casinos responded by ensuring dealers shuffle the deck more, particularly after a player has placed a large bet. Thorp, with the help of Claude Shannon (himself the father of information science), developed a small wearable computer that allowed the user to increase his/her chances of winning roulette.

RICHARD JARECK: TOOK CASINOS FOR A SPIN

Richard Jareck (died 2018) was a medical doctor and researcher at Germany's Heidelberg University. He realized many roulette wheels eventually develop biases through wear and tear. In the 1960s, he toured various casinos in Europe and over the course of a month he would patiently observe particular wheels. He had to observe about 10,000 spins to determine if a wheel had a bias. If it did he would then place appropriate bets. Given roulette wheels in Europe have only a 2.7% edge for the house, knowing the bias was enough for Jareck to turn the odds in his favor. His rang up winnings of about $2 million before European casinos started to ban him. He famously quipped to *The New York Times* in 1969 "if casino managers don't like to lose, they should sell vegetables."

Photo Credit: Karl Mamer

DANIEL CORRIVEAU: KENO CRACKER

One day in 1994, Daniel Corriveau popped into a provincially run casino in Montreal and hit three consecutive jackpots on a computer Keno machine. That is to say he correctly "guessed" 19 out of 20 numbers three times in a row. No one in the history of keno had ever hit 19 correct numbers even once. Three times in a row? You have about a 1 in 6 quintillion chance of pulling that off. The haul was $450,000, which seems pretty low given the odds. Still, not bad for a few minutes of work. Unfortunately, casino security wouldn't let him out the door with his winnings. They rightly computed hitting three jackpots in a row was something that shouldn't ever happen once in the lifetime of the universe. They suspected fraud. Did he rig the machine? Was he working with a confederate on the inside? Corriveau explained he was a computer programmer and he had used a branch of chaos theory to predict the correct numbers. When that story got out, the press ran with it. A few people in the world of computer security and cryptography were half worried and half skeptical. If true, Corriveau's methods could crack any cypher on the planet. But such an astounding breakthrough doesn't usually come from an unknown freelance computer consultant with a computer no more powerful than a 286 at his disposal. Corriveau, hoping to get his money through legal means, revealed his trick. He realized the casino wasn't a 24-hour operation. He figured they turned off the Keno computer each night and then turned it back on at the same time in the morning. He understood enough about how computers generate pseudo random numbers from a time-based seed that he could then predict which numbers were likely to be generated at a particular time. The provincial casino corporation eventually had to admit nothing Corriveau did was cheating or against house rules so they eventually paid him his money. Curiously, to this day, you will still find stories about Corriveau's win that claim he won using a breakthrough in chaos theory.

JERRY SELBEE: WAS ROLLING IN IT

In the early 2000s, Michigan introduced a new lottery draw called Winfall. Like many lotteries, you win the jackpot if you pick all six numbers but win smaller prizes if you correctly pick between two and five of the numbers. Like many lotteries, if the grand prize isn't won, it gets rolled forward for the next draw. The grand prize keeps growing and growing. However, most lotteries don't up the payout for the smaller prizes. Those are fixed. Part of Winfall's gimmick was the grand prize would cap out at $5 million and if there were no winners after each week $2 million would be distributed (the lottery used the phrase "rolldown") among the smaller prizes to increase their payout. Jerry Selbee was a retired mathematician and realized at some point the "rolldown" would make it profitable to buy large numbers of tickets and collect the winnings on the smaller prizes. For example, if you have a 1 in 4 chance of winning and you buy 4 tickets, you're statically likely to win. However, it's a bad investment to spend $4 on a guaranteed win of $2. But if the rolldown pays out at $5, the $4 bet is in your favor. Selbee and his wife kept a close eye on the Winfall lottery's rolldown. When the payouts were in their favor, he and his wife fanned out across Michigan to buy thousands of dollars in tickets. Eventually the Michigan lottery corporation figured the math out for themselves and pulled the game. But before then the couple netted about $8 million.

MOHAN SRIVASTAVA: DISCOVERED A SCRATCH TICKET TELL

Mohan Srivastava is a Canadian statistician. In 2003, he noticed an Ontario lottery scratch game called Tic Tac Toe had a series of numbers embedded on the "teaser" portion of the ticket. When these numbers lined up in a certain way, Srivastava realized the ticket had a 95% chance of being a winner. However, it didn't suggest this size of the prize. Being a statistician, he realized he couldn't play this "tell" to early retirement. The scratch game had pretty low payouts. Srivastava determined it probably wasn't worth his time. He did feel, however, it was his civic duty to inform the lottery corporation about this tell. If he figured it out, others would too. A series of phone calls to the lottery corporation kept resulting in the brush off. No doubt lottery corporations get cranks calls all the time from people alleging they've discovered some secret to beating the lottery and demanding money to keep silent. After a few futile attempts, he went another route. Srivastava bought a stack of scratch tickets. One pile he indicated 95% were winners. The other pile he labelled losers. He couriered the tickets to the lottery corporation. Within two hours of the package being delivered, security officials with the corporation returned his call.

JOHN KANE: HITS IT BIG AT VIDEO POKER

John Kane played a lot of video poker over the years. According to his lawyer "He's played more than anyone else in the United States... I'm not exaggerating or embellishing..." Lawyer? We'll get to that. In 2009, he discovered a bug on a particular brand of video poker machine. Through a complex series of button presses he could get the machine to repay any win a second time but at a 10x rate. He let another friend in on the action and they quickly started playing the machines in casinos across Nevada. After one particularly improbable series of wins, Kane was arrested. Casino security discovered the bug Kane was exploiting. The federal government charged Kane under a computer fraud law that makes it illegal to exceed any authorized access one is given to a computer. Kane got before a judge in 2011. The court found in favor of Kane. By virtue of the fact the machine tacitly authorized the person to access the game information and Kane could access an old hand suggested to a judge Kane had not broken the law.

Nine Cases of Pseudoscience and Fraud in Computer Science

One can be forgiven if one were to think computer science is a field remarkably free of pseudoscientists looking to use and abuse the field. Woo in computer science and computers, while rare, is out there.

BADBIOS

One of the best ways to ensure your computer never gets infected by a virus is to air gap it. That is, don't connect it to anything else, like another computer or the

Internet. In 2013, computer security researcher Dragos Ruiu announced his discovery of a novel computer virus he dubbed "badBIOS". He believed once this virus took residence in a computer, it replicated itself by emitting ultrasound signals through the computer's speakers. Air gapped computers would pick up these signals via their microphones and start running code. Ruiu reported it had infected computers in his lab. He shared data which he thought proved the existence of the virus. However, other computer experts saw nothing unusual in his data. Computer experts have not ruled out the possibility a clever virus could infect other computers via sound waves but the reality of such computer viruses has yet to be proven.

DENNIS MONTGOMERY 1.0

Dennis Montgomery was the CTO of a video software company called eTreppid Technologies. Montgomery had no formal training in computers or any scientific discipline. He did have what some may describe as a gambling problem and a prior arrest for passing bad checks. Somehow, he was able to talk his way into a CTO job with eTreppid Technologies. More amazingly, in 2004, he was able to convince the Pentagon his company's software could analyze Al Jazeera broadcasts and detect messages being sent to terrorists. Obviously, his software could not do what he claimed. Whether the FBI ran any kind of background check on Montgomery before the Pentagon awarded him a $30 million no-bid contract is unknown. What is known, however, at least one Orange alert was issued by the Bush government in response to so-called intelligence supplied by Montgomery. This resulted in Air France having to recall flights. France was not impressed and started an investigation into Montgomery's allegations. French intelligence provided the Bush administration with their findings. Bush officials quickly understood they had been had. Montgomery seems to have escaped any kind of prosecution due to an illegal search by the FBI and government fears of security secrets having to be revealed in a public trial.

DENNIS MONTGOMERY 2.0

Dennis Montgomery was *persona non grata* during the Obama years. He popped up in 2014 as an expert witness for Sheriff Joe Arpaio who was defending himself against a racial profiling lawsuit. Montgomery told the court he had a piece of software called Hammer that had detected evidence of a conspiracy against Arpaio. Montgomery's data was handed to national security experts. It was determined Montgomery's data was "junk". He re-emerged in the wake of the US 2020 presidential election. He claimed to be at the center of Trump's stolen election conspiracy theory. Somehow his original Hammer software became the name of a supercomputer. Layered on top of this Hammer supercomputer was a new mythical piece of software called "Scorecard". Montgomery claimed he had invented both as tools to be used by intelligence services but it was abused by those services to alter election results. The US election was, of course, not stolen and there is zero evidence any computer or software was used to alter votes.

GIGATEXT

In the 1980s a French-speaking driver in Saskatchewan got a speeding ticket and challenged the legality of the ticket on the grounds it was only in English. The case went all the way to the Canadian Supreme Court. The court ruled Saskatchewan had to provide the text of its laws in both English and French. The court also gave Saskatchewan a deadline. Any law not translated after the deadline would be unconstitutional. It looked like English-to-French translators were going to get a lot of work out of the Saskatchewan government. Guy Monpetit, a businessperson with big ties to Canadian Conservative Party politicians, saw an opportunity with Saskatchewan's then Conservative government. In 1988, he sold Saskatchewan on the idea of doing its legal translation using a universal translator machine his company Gigatext had developed. He got the provincial government to invest $4 million in Gigatext. Monpetit assured the government his company could translate all its legal text using an astounding new AI translation process invented by his partner Dr. Douglas A. Young. Young was a professor in the Department of Computer Science at the University of Manitoba. Young had an interesting idea that words and phrases in all languages could be represented by a common set of "interlingua" symbols. You map all your English and French words and all possible nuances onto those symbols. When you need to translate blocks of texts from English to French, the mapping takes care of the translation and nuance. That was the theory anyway. Young's demonstration system, at best, could only successfully translate about 50% of sample text. Success was defined as text that was reasonable enough that a French editor had something cogent to smooth out. As a first batch, Gigatext was given 45 statues to translate. When the government came to check on the progress a few months later, Gigatext hadn't translated a single one. It also discovered Monpetit had spent the $4 million he was given. He had spent most of the money on a private jet and "loans" to other political buddies. As well, the government discovered Monpetit had billed them $2.9 million for old LISP computers that had, at best, a value of about $39,000. Certainly, machine translation isn't horrible today, but in the late 1980s a Babel Fish was a most impossible device.

HANXIN CHIP

In 2003, Chen Jin announced he had developed the first made-in-China digital signal processing chip. He dubbed it the Hanxin. DSP chips are a key component in mobile phones. In the early 2000s China was looking to move up the value chain in terms of designing and manufacturing cell phones. Chen's Hanxin chip was a good candidate to get China there. Chen was hailed as something of a national hero and given government financing. He was made the head of his own research group. This team released several more versions of the Hanxin chip until an anonymous whistleblower spilled the beans. Chen had not actually developed his own chip. He had merely purchased American chips, sandpapered off the chip's serial numbers, and repackaged it as his own design.

PIXELON

Pixelon was a dot com company that claimed in the late 1990s to have breakthrough technology that would allow users to stream high quality video over the Internet. Although that's like a duh thing today, in the late 1990s to suggest you could push full-screen, broadcast quality video through nascent home broadband infrastructure was a rash claim. Even more rash, the company spent $16 million (literally half the money it raised from venture capitalists) on one company party in Las Vegas. The company hired The Who, KISS, the Brian Setzer Orchestra, and half a dozen other top name acts to perform. Although that seems like a total waste of money, it totally wasn't. Nope, not one bit. See, the party was going to be broadcast live over the Internet as a demonstration of the company's video technology. The company did no such thing, however. It streamed the bash over Microsoft's own Windows media player. The company's ruse quickly unraveled after the disastrous and misleading launch party. Investors, a little too late, started to question what they had actually invested in. They should also have asked who they invested in. The founder, who went by the name Michael Fenne, turned out to be David Kim Stanley. Stanley had previously been convicted in 1989 of running a stock scam that targeted the parishioners of his father's church. Stanley's sentence was deferred to give him a chance to pay restitution to his earlier victims. After a few payments he vanished and then remade himself as Michael Fenne and started Pixelon. Stanley was re-arrested in 2000 after his company went right into the toilet.

POLYBIUS

There's an urban legend about an arcade machine called Polybius that was created by the CIA or some other nefarious government organization to conduct mind control experiments on players. Several copies of the game were installed around the Portland, Oregon area in 1981. It had disturbing and addictive effects on players, causing players to experience hallucinations and sometimes driving them to commit acts of violence. Men-in-black-type figures were said to periodically service the machines (ostensibly downloading some kind of data the machines were keeping). The legend seems to have gotten its start around 2000, possibly as a ruse by the founder of the coinop.org website to drive traffic to his page.

Teaser screen of a retconned Windows version.
Photo Credit: Software author unknown

NASA BELIEVES SANSKRIT IS THE BEST PROGRAMMING LANGUAGE FOR AI

There's an odd urban legend that gets passed around by, I guess, fans of Sanskrit that NASA has determined it's the perfect language for programming, notably for programming AI applications. The language is so amazingly useful that NASA makes learning Sanskrit compulsory for many of its high-tech workers. However, none of that is actually true. This legend got started by someone misreading a 1985 paper by NASA researcher Rick Briggs titled "Knowledge Representation in Sanskrit and Artificial Intelligence". A challenge back in 1985 was getting computers to understand natural language. Many languages have loads of ambiguity. If I use the word "bank" within a sentence do I mean a financial institution, the act of depositing my money in a financial institution, or counting on some outcome ("I'm banking on you coming to my party...")? Briggs argued Sanskrit's more rigorous rules of grammar made it a good candidate as a testing language for computers. In other words, if you want to start tackling the problem of natural language processing, you could create your own artificial language that lacks ambiguity and has a perfectly predictable grammar structure or you could not reinvent the wheel and just use Sanskrit.

SOFTRAM

RAM was very expensive in the 1990s. Windows 3.1, Windows 95, and follow-on Windows bloatware suddenly increased the need for additional RAM on computers. A company called Syncronys (died 1998) offered a software package called SoftRAM that would double your onboard RAM via a proprietary compression scheme and make your computer super speedy. The $30 price tag seemed a deal compared to spending $300 to add an extra 4 megabytes of RAM to your 386 system. Syncronys sold hundreds of thousands of copies. It was one of the top selling software packages in 1995. The software came with a very impressive looking dashboard, full of cool gauges, that showed how much additional RAM your computer was seeing after the tool compressed data in memory. It was a slick offering. When researchers ran it through tests, the software did nothing it claimed it could do. The dashboard's gauges were spewing meaningless numbers. What SoftRAM actually did was resold a developer tool provided by Microsoft to bump up your default pagefile RAM disk. The programmers at SoftRAM, while great at UX, were incompetent in almost everything else. Researchers found SoftRAM shipped with a developer flag turned on that actually made your computer run slower. The FTC eventually forced SoftRAM to stop selling its cash cow. The company went bankrupt a few years later.

ON THIS SITE
IN 1897 NOTHING
HAPPENED.
PSEUDOHISTORY

Five Astounding Historical Events that Never Happened

THE EXODUS

There's zero archeological evidence tens of thousands of Jews left Egypt, wandered through the deserts of the Sinai for decades, and displaced the indigenous inhabitants of Canaan. Even more problematic, Canaan itself was under Egyptian control during the supposed time of the Exodus. News that the runaway slaves had been found would have gotten back to Pharaoh.

THE DONATION OF CONSTANTINE

For centuries popes asserted their political authority over Rome, the Papal states, Constantinople, and various parts of the former Roman empire based on an imperial decree in the 4th century by Constantine the Great. The decree is known as the Donation of Constantine. The emperor is said to have issued the transfer of authority after Pope Sylvester I converted Constantine to Christianity and cured him of leprosy. The problem is no one had seen a copy of this imperial decree until the 8th century (four centuries after it was supposedly issued). It wasn't until the 15th century that Catholic historians began to realize there were too many anachronisms in the supposed 4th century decree to be genuine. For example, it used terms like "Judea" and "Byzantia" which were not used in the 4th century. It's generally believed the document was forged in the 8th century as a check against the possibility the Byzantine Empire, which was circling the drain at that point, would start making claims on Rome and parts of Western Europe.

MARTIN LUTHER NAILED 95 THESES TO A CHURCH DOOR

Martin Luther (died 1546) did have a grudge list of 95 theses. However, it's most likely he simply mailed them. There's no historical record of him ever nailing them to any kind of door. The legend of him nailing them to a church door first appears in the historical record some three decades after the fact.

WITCHES WERE BURNED IN SALEM

Many women in 17th century Salem were certainly accused of witchcraft and executed quite horribly. However, most met their fate at the end of a hangman's noose. At least one was crushed by heavy rocks. None were burned at the stake. This was, however, a punishment meted out to women accused of witchcraft in Western Europe.

MASS HYSTERIA DURING THE ORIGINAL BROADCAST OF ORSON WELLES' WAR OF THE WORLDS

Rating services of the time indicate only about 2% of the listening audience was actually turned into Welles' broadcast. It's generally accepted the next day newspapers, who were in stiff competition with radio during the Great Depression for limited advertising dollars, wildly exaggerated the impact of Welles' broadcast.

Seven Historical Figures Who Never Existed

ST. CHRISTOPHER

Many nervous neophyte car drivers have relied on a St. Christopher medal hanging from a rearview mirror to help them get from point a to point b. St. Christopher is well known among many Catholics as the patron saint of travelers. He was supposedly a martyr killed by either the Roman emperor Decius (249-251 AD) or Maximums II Daia (308-313 AD). However, there are no contemporary records of his existence or martyrdom. This is actually not that unusual as many of Catholicism's best-known saints have no historical evidence of their existence. St. Christopher first appears in the historical record in the 6th century. As legend has it, he helped a small boy cross a dangerous river. After having barely made it to the other side, the boy revealed himself to be the Christ Child and promptly vanished. Owing to the questionable nature of his existence, the Catholic Church in 1970 removed him from their calendar of veneration days, ostensibly to make way for newer saints with better historical pedigrees. Although removed from this calendar, St. Christopher is still considered a saint. The Catholic church leaves it up to local dioceses as to whether or not he should be venerated in their liturgical calendar.

POPE JOAN

In the 13th century it became widely believed that a woman had surreptitiously entered the priesthood disguised as a man and got elected pope. Legends of her papacy appeared in some writings by Dominican friars. She ruled as Pope John VIII for a brief period in 855 AD following the death of Leo IV and before the ascension of Benedict III. Her ruse was, however, uncovered when she became pregnant and went into labor in the middle of a papal procession. As legend has it horrified onlookers descended on her and stoned her to death upon discovery. Others have it she died in childbirth. Outside of the writings of a few Dominicans, there is no record of a Pope John VIII taking the Lateran throne at that time (there was a real John VIII but he ruled between 872 and 882 AD). It's also well documented the gap between the rule of Leo IV and Benedict III was a matter of weeks at best. There was no time for her to be elected, take the throne, and do the various things attributed to her rule. Many consider the legend an example of religious satire written by a few disgruntled religious workers. As is the rule for satire, it's not good satire unless someone believes it's true.

PRESTER JOHN

Prester John was a supposed Christian priest-king who ruled over a mysterious, magical, and powerful Christian nation in the Far East. Many believed he was a scion of one of the original magi that supposedly bestowed gifts upon the infant Jesus. Prester John's capital was so big it was said it would take a person four days on foot to completely circle. Legends placed him variously in the Far East, India, and later Ethiopia. The legend gained traction with the people of Europe in 1165 when a hoaxed letter supposedly written by Prester John began circulating among Europe's religious authorities. Pope Alexander III so believed in the authenticity of the letter he wrote a reply to Prester John and sent it with an envoy making a pilgrimage to the Holy Land. As various crusades in the Holy Land began to fail, Prester John was cast into the role of savior. He became known as the "hammer of the infidel". Several doomed crusaders held out hope Prester John and his holy warriors would roll up at the last minute like the 7th Cavalry and save the day. No such help ever came.

NED LUDD

The Luddites were a 19th century labor movement famous for destroying their automated knitting machines. There is some debate over their motivation. Some argue Luddites destroyed their machines in an ultimately fruitless effort to prevent job loss through automation. Other historians argue working conditions in weaving factories were so horrible, the only way workers could get their bosses' attention was by destroying their machines. Whatever the case, the group ostensibly took their name in honor of an 18th century weaver named Ned Ludd. As the story goes, Ludd, who was a young man, didn't always exhibit the best on-the-job motivation. In an effort to inspire him, his boss would whip him. After one particularly vicious whipping, Ludd sought some career advice from his father. The father, himself a retired weaver, told him to "square your needles." Ludd took this to mean "it's payback time". Ludd took a hammer to his boss's knitting equipment. Ned's example apparently inspired 19th century Luddites. When arrested and urged to give up their leaders, Luddites claimed they were following the orders of "General Ludd". Manifestoes by the Luddites were also frequently signed by Ned Ludd. Luddites had widespread sympathy among the British public and for a time seemed unstoppable. The movement came to a rapid end when the government passed a shoot-to-kill order in response to Luddite protests.

RANI PADMINI

Padmini was the Helen of Troy of Hindu legend. She was supposedly a Sri Lankan princess of such amazing beauty men went to war over her. There are a number of accounts of the trials men of the subcontinent went through to win her hand. In one version, she was a kind of Mulan figure. She spent her days training as a sword fighter and became one of the finest warriors in the land. Only the man who could defeat her in combat would win the right to marry her. One of the more amusing legends of Padmini involves her owning a talking parrot. The talking parrot was given to singing the praises of Padmini and could talk about nothing else. Her father, the king, grew so tired of the parrot's endless prattle that he had it banished from his kingdom. The parrot didn't go quietly and would hype up her beauty to anyone it encountered in flight. Word got around and she became desired by two Hindu kings and one Muslim king. The Hindu king Ratan Singh got to her first. However, on the journey to bring his new bride back to his kingdom the other Hindu King, Devpal, captured them and imprisoned them in his fortress. Singh and Devpal agreed to fight a duel for Padmini. The duel proved fatal to both Singh and Devpal. The Muslim king Alauddin Khalji took advantage of the chaos and brought his army to lay siege to Devpal's fortress. As Alauddin's men breached the fortress's defenses, Padmini, to protect her honor, killed herself by performing a ritual act of self-immolation. Padmini's story was generally considered a legend until 19th century British writer James Tod wrote local histories of the subcontinent and presented Padmini as a historical figure. Padmini quickly became a symbol for a growing Hindu nationalist movement. Her self-immolation was seen as an example for Hindus. It exemplified the level of personal sacrifice required to win dominance over their Muslim rivals.

TOKYO ROSE

Although many modern movies and TV shows have depicted Tokyo Rose as a single woman with a seductive voice taunting and teasing lonely American GIs over an open radio frequency, there was no single woman named Tokyo Rose. There were certainly a number of Asian women, fluent in English, who broadcast Japanese propaganda. None however went by the name Tokyo Rose. This was a general appellation given by troops in the South Pacific to any female broadcaster of Japanese propaganda. The legend of a singular woman broadcaster named Tokyo Rose got its genesis when confused reporters Henry Brundidge and Clark Lee in 1945 offered $2,000 (about $25,000 in today's money) for the real Tokyo Rose to step forward and give an interview. LA-born Iva Ikuko Toguri D'Aquino took the bait. She had been trapped in Japan during the war years and was pressured into participating in propaganda broadcasts. Following the interview, the reporters reneged on the deal and turned her transcript over to occupying powers. She was arrested but released a year later. Although she was a known radio personality who went by the on-air name Orphan Annie and performed in musical and comedy skits, there was no evidence of her having done anything more treasonous than read scripts handed to her. In fact, many male allied prisoners of war were forced to do the same. There was also evidence Aquino used her access to prisoners of war to smuggle them extra food. After her release she tried to return to the USA but this time radio host Walter Winchell agitated for her arrest. She was re-arrested and forced to stand trial in San Francisco. There she was successfully prosecuted and found guilty on one count of treason. She was sentenced to ten years in jail. She served six and was released on parole. On President Ford's last day in office he granted Aquino a full pardon.

ANNA SPRENGEL

Anna Sprengel is reputedly the founder of a 19th century British secret society called The Golden Dawn (more formally, the Hermetic Order of the Golden Dawn). The Golden Dawn was devoted to the study of metaphysics and the paranormal. The Golden Dawn had profound effects on the Wiccan revival of the 20th century and other occult movements. The first Golden Dawn temple was founded by William Wynn Westcott in the UK. Westcott, a former Mason, wanted to create a rival secret society and sought an ancient backstory. He claimed he had come across a mysterious coded letter. Upon decoding it, it revealed the address of a German countess named Anna Sprengel who was a priestess of an ancient Rosicrucian order known as The Golden Dawn. After an exchange of letters, she authorized Westcott to found a branch in the UK. There's no evidence there ever existed a German countess by the name of Anna Sprengel. Westcott's supposed letters from Sprengel were written in rather poor German and full of errors non-native speakers typically commit. It's unlikely a highly educated German countess would be so unfamiliar with her native tongue.

Four People Who Fooled Others into Believing They Were Someone Else

DANG TAN NGOC

Dang Tan Ngoc was a 76-year-old Vietnamese citizen with a history of trying to convince American military authorities he was an MIA American soldier "left behind" after the Vietnam War. In 2006, Ngoc claimed he was Green Beret Master Sgt. John Hartley Robertson. Robertson was recorded KIA when his helicopter went down over Laos in 1968. Fellow combatants deemed the crash not survivable. US military authorities were dubious of Ngoc's story. Ngoc spoke no English and didn't remember the names of his children. He couldn't get one biographical detail correct. Ngoc was fingerprinted and a DNA sample was taken. His prints did not match Robertson's and his DNA did not match a reference sample given by a relative. It should have ended there. However, in 2013 Ngoc came to the attention of Canadian filmmaker Michael Jorgensen. Jorgensen believed the circumstantial evidence supported Ngoc's story. Jorgensen made a movie called *Unclaimed*. The filmmaker somehow convinced the sister of Robertson that Ngoc was her long-deceased brother. After the movie premiered, the family raised funds for a DNA test and discovered, to no one's surprise, Ngoc was not Robertson.

ANNA ANDERSON

Many people have claimed to be Anastasia, the daughter of Russia's Tsar Nicholas II. Anastasia, along with the rest of the Romanov royal family, were executed by communist revolutionaries in 1918. Anna Anderson (died 1984) is probably the best-known claimant. Anderson was challenged for most of her adult life with mental illness. The idea she was a surviving Romanov seems to have been suggested to her while she was in a German asylum following a suicide attempt. A fellow patient believed she bore a strong resemblance to Anastasia's older sister Tatiana. Word got out among Russian émigrés in Germany that a surviving Romanov daughter was residing at the asylum. After Anderson's release, members of the Russian émigré community decided she wasn't actually Tatiana but was more likely the younger Anastasia. For the rest of her life she was supported by an ever-changing cast of people who seemed to want to parlay her supposed claim on the Tsar's assets into a payday for themselves. It's not clear if Anna Anderson herself ever fully believed she was Anastasia. She was, however, party to a three-decade long lawsuit in German courts to get herself declared Anastasia and the rightful heir to the Romanov's remaining European assets.

MM and WD

In 1997, the UK medical journal *The Lancet* published an interesting study titled "Clinical findings in three cases of zombification". The study examined three individuals in Haiti (identified by initials only) who were alleged to be zombies, that is a deceased person returned from the grave. The reported cases had similar stories. A relative died, was buried, and then a few days later was seen wandering zombie-style in public. The families took the individuals in, believing they were zombies returned from the dead. DNA fingerprinting was used to determine if the person identified as a zombie was indeed their deceased relative. In the cases of MM and WD, DNA fingerprinting eliminated that possibility. In both cases, the individual thought to be their revitalized relative was simply a homeless, mentally ill stranger. Due to cultural beliefs and personal grief, the families fooled themselves into believing these strangers were their lost family members and their mental illness was a symptom of the zombification process.

The Seven Root Races of Helena Blavatsky

Helen Blavatsky founded a quasi-religious movement called Theosophy in the late 19th century. At its core is a belief that the real history of humanity has been hidden from us. Blavatsky claimed she had tapped into this hidden history of humankind after discovering there existed a secret brotherhood of "ascended masters". These masters have attained supernatural powers by reviving religions of the ancients. A part of Theosophy's historical mythology is before the rise of humankind there existed progenitor races ("root races") that arose on lost continents. Theosophy's influence on modern spiritualism (the New Age movement), Ufology, and conspiracy thinking cannot be understated.

FIRST ROOT RACE: POLARIANS

Earth's alpha race supposedly emerged around the North Pole when the planet was still cooling and before mountains appeared. Blavatsky's Polarians did not possess bodies but were ethereal spirits that reproduced like amoeba and lacked sentience.

SECOND ROOT RACE: HYPERBOREANS

Blavatsky borrowed a Greek legend about a land in the north that was home to a race of giants called Hyperborea. There she placed her second root race to emerge on Earth. According to Blavatsky Hyperborea encompassed the North Pole, Scandinavia, Northern Canada, and Northern Asia. It's not apparent Blavatsky's Hyperboreans were giants but she described their skin color as yellow.

THIRD ROOT RACE: LEMURIAN

Theosophist mythology pins the rise of the third root race about thirty-four million years ago, arising on the so-called lost continent of Lemuria. Although Blavatsky doesn't go into great detail about what the third root race looked like, she claimed they reproduced by laying eggs. Richard Sharp Shaver in one of his *Shaver Mysteries* stories picked up on this and described it as a world populated by lizard men. In Blavatsky's view, the egg-laying Lemurians didn't die off. They evolved to reproduce like humans and eventually evolved into Africans.

FOURTH ROOT RACE: ATLANTEAN

Yup. You knew Atlantis had to figure into this somehow. Atlanteans evolved from the Lemurian root race and came to populate the continent of Atlantis about 1 million years ago. Atlanteans flew around in airships powered by a mystical energy called vril. Apparently, they abused this vril power and their civilization was destroyed. Surviving Atlanteans dispersed and became the founders of Middle Eastern, South and Central American, Southern European, and Southeast Asian civilizations. You know, all the people who did great stuff but just weren't quite as evolved (physically and spiritually) as good ol' white people.

FIFTH ROOT RACE: ARYAN

Good ol' white people were the last of the historical root races to evolve in Blavatsky's mythology. They started in India and migrated to Greece and the rest of Western Europe. Jews ("Semitic people") seem to be Aryans in Blavatsky's writings but she believed they became spiritually degenerate.

SIXTH ROOT RACE: CALIFORNIANS

The sixth root race has yet to evolve but oddly Theosophists believe they themselves will give birth to the sixth root race in Baja, California via a eugenic program run by the group. Don't start trying to corner Baja property now, however. The *Kwisatz Haderach* isn't slated to emerge until the 28th century.

SEVENTH ROOT RACES: TBD

The seventh root race will emerge in 28 million years on an all new continent that will emerge from the Pacific Ocean. Theosophists don't seem to have a lot to say about the seventh and possibly final evolution of intelligent life on Earth. In general, they will be super spiritual beings that eliminate sin.

Twelve Mythical Lands and Lost Continents

AVALON

Avalon ("the isle of fruit trees" in Cornish) first appears in the works of Georffrey of Monmouth's pseudo historical *Historia Regum Britanniae* from 1136. It is reputed to be a mystical island where King Arthur's sword Excalibur was forged. It is also reputed to be King Arthur's final resting place, having been taken there after he was mortally wounded. Avalon is characterized as a land of plenty. So much food grew naturally that there was no need to farm. Early writers placed Avalon on a hill near Glastonbury called Glastonbury Tor. This hill is not an island today but it was surrounded once by a swamp making it islandish for a time. The problem with this being the reputed location of Avalon is Georffrey of Monmouth indicated one needed a voyage at sea to get to the island. Arguing in its favor is a late 12th century claim that monks found the actual tomb of King Arthur and Queen Guinevere after their abbey burned down. Historians generally chalk this claim up to a case of pious fraud. The monks invented the legend in hopes of attracting pilgrims and soliciting donations to rebuild.

ATLANTIS

Plato used Atlantis as an allegory for a city state and civilization that just couldn't play nice. It was a warring empire that tried to best tiny little Athens but failed because Athenians were far too moral of a people. The people of Atlantis eventually fell out of favor with the gods and their island nation sank into the Atlantic Ocean. New Age types have in modern times transformed Atlantis into a utopia that carried its culture and technology to the four corners of the globe.

EL DORADO

The Spanish legend of El Dorado was originally about a king named El Hombre Dorado or "The Golden Man". The legend, from the 16th century, tells of how this king draped himself with gold and then drowned in Colombia's Lake Guatavita. By the 17th century, the story of El Hombre Dorado and various legends about lost cities in the Andes got conflated and firmly cemented into a legend about a lost city of gold. The "hombre" bit got dropped. More modern interpretations further spin the legend into a city literally made of gold. Gold is pretty malleable and heavy. It's not exactly an ideal building material. Like any mythical treasure, there's always one or more maps out there enticing explorers. Season 1 of the Travel Channel's *Expedition Unknown* had Josh Gates setting out to find the Peruvian version of El Dorado based on his reading of a four-hundred-year-old document found in the Vatican archives. It will come as no surprise Gates did not find a lost city made of gold.

GARDEN OF EDEN

Many who believe in the literal interpretation of the Bible have pondered over the location of the Garden of Eden. *Genesis* describes the Garden of Eden as being watered by a river that parts into four separate waterways as it flows out of the garden. The Bible lists the rivers as the Tigris, Euphrates, Pishon, and Gihon. *Genesis* lists a few other geographic locations like the lands of Havilah and Cush. Cush is thought to be a reference to Kash, which was what the Egyptians called Lower Nubia. The best guess as to what Havilah is referring to is possibly a region in southwestern Arabia. If Eden is supposed to be located somewhere between Sudan, Iraq, and Saudi Arabia, that is a pretty big footprint and doesn't narrow things down. Many fundamentalist Christians deal with "the river problem" by suggesting Eden is in the area bracketed by the Tigris and Euphrates and the Pishon and Gihon rivers were wiped out by Noah's flood. The Mormons are famous for ignoring the clues in *Genesis* and locating the Garden of Eden in Jackson Country, Missouri. The first president of Boston University William F. Warren (died 1929) proposed in his 1885 book *Paradise Found* that the Garden of Eden was located at the North Pole on a lost continent. According to *Paradise Found*, this lost continent sounds like it was a crowded place. Not only was it home to the Garden of Eden but it was home to Atlantis, Avalon, and Hyperborea. Brinsley Le Poer Trench (died 1995) was the 8th Earl of Clancarty and a Ufologist. He proposed in his 1960 book *The Sky People* that *Genesis* really describes a network of canals and that points straight to Mars. Adam and Eve were really Martians and their flight out of Eden was really them escaping a disaster on Mars and flying to Earth.

HELL

Christians generally view Hell as some kind of other worldly place where bad people go. Some, however, believe Hell is a physical location one can find or there's some kind of discoverable nexus to Hell. In 1989 the Christian Trinity Broadcasting Network picked up a story written by a Finnish Pentecostal about Russian geologists digging a hole to Hell in Siberia. Apparently after digging a 9-mile hole straight down, they lowered microphones into the hole and picked up the sounds of people screaming in pain. The Russians did drill a 7.5-mile hole on the Kola Peninsula (which is not in Siberia). This is known as the Kola Superdeep Borehole. However, microphones were never lowered into the hole to listen for sounds of the eternally tormented.

HYPERBOREA

Hyperborea is a fabled land in the north. It first appears in the works of Herodotus in about 450 BC. Although later writers have fashioned it as an arctic abode, Greeks didn't place it that far north. Greek writers generally placed it in Thrace. Later sources moved it increasingly north, eventually believing it could be found in France or Britain. Modern day believers in the reality of Hyperborea claim it is a land near the arctic circle.

IRAM OF THE PILLARS

Iram (sometimes cited as Irem) is a lost city cited in the *Koran*. Middle Eastern writers have built upon its mention in the *Koran* and created embellished tales of Iram being a treasure city that was destroyed by Allah. There's debate among historians and archaeologists if the city is merely legendary or based on a real city from a region mentioned in the Bible called Aram. The lost city first entered Western thought via the *Arabian Nights* collection of Middle Eastern folk tales. Iram moved into more esoteric thought when H.P. Lovecraft adopted it as home base for his Cthulhu cult. You'll find it mentioned in his stories "The Nameless City" and "The Call of Cthulhu". Some mystery monger writers like to frame Iram as the "Atlantis of the sand". It's a realm where the Nephilim or Biblical giants lived before the rise of humans.

LEMURIA

Lemuria was a proposed lost continent located in the Indian ocean. It offered an explanation as to why fossils in Madagascar are more closely related to fossils found in India than Africa. In essence it was a proposed land bridge between South Asia and Madagascar. The theory was first put forward in 1864 by zoologist Philip Sclater (died 1913). It was abandoned when continental drift was established. Many esoteric writers of the 19th century jumped on Lemuria, assuming it was the Indian ocean's version of Atlantis complete with a highly advanced civilization. In 1899 author Frederick Spencer Oliver wrote *A Dweller on Two Planets*. In it he claimed survivors of whatever disaster sank Lemuria found refuges under Mount Shasta. Lemuria was further fixed in the minds of the woo crowd when *Amazing Stories* published Richard Sharpe Shaver's highly influential "I Remember Lemuria" short story in 1945.

MU

It would seem every ocean needs a lost continent and the Pacific is not shortchanged on this account. Mu is reputed to be a continent that once existed in the Pacific. Unlike Lemuria that had, for a time, a plausible theoretical existence, Mu was made up out of whole cloth by archaeologist Augustus Le Plongeon (died 1908). In the mid-19th century, he began writing about his work translating sacred Mayan texts and discovering references to the lost continent. It was home to an advanced civilization before it disappeared beneath the waves. According to Le Plongeon, one survivor, Queen Moo, made it to Egypt and founded the Egyptian civilization. Evidence? None. But just as the so-called Bimini road is considered a remnant of Atlantis, the Yonaguni monument rock formation off the coast of Japan is considered evidence of Mu. Of course, both the Bimini road and the Yonaguni monument are entirely natural underwater features that have been co-opted by pseudo archaeologists who trade in notions of hyperdiffusionism.

RAMA'S BRIDGE

There's an Indian legend about a divine prince named Rama. His wife was kidnapped by a demon king and spirited away to the island of Sri Lanka. Rama raised a mighty army of monkeys to help him rescue his wife. The only problem was it's not easy to invade an island nation in force unless you have a fleet of ships. Rama didn't have a fleet. He had something better: enchanted rocks that float on water. Rama's monkey battalions tossed these magical rocks into the Indian ocean and built a bridge, allowing Rama to defeat the demon king and rescue his beloved. Between India and Sri Lanka there is a shallow strait that features a string of limestone shoals. These shoals are collectively known as Adam's Bridge. That name comes from an Islamic legend that Adam crossed from India to Sri Lanka after being booted from the Garden of Eden. Orthodox Hindus believe Adam's bridge is not a chain of natural shoals but betray signs of being artificial, hence proving the truth of the legend. The Indian government has made many proposals over the years to dredge the shallow waters around Adam's bridge and open it to shipping. However, proposals are always met with fierce opposition from orthodox Hindus who believe Adam's bridge is a holy site. The shoals have been dated to about 1.7 million years old, which places them not only well before when the Rama tale was supposed to have taken place but well before modern humans even existed. That doesn't dissuade pseudo-historian Michael Cremo (author of *Forbidden Archaeology*) from simply rolling the timeline back and treating these shoals as evidence of his pet theory that human civilization is millions if not billions of years old.

SHANGRI-LA

Shangri-La is a creation of author James Hilton. He wrote about it in his 1993 novel *Lost Horizon*. Hilton describes it as a lush valley hidden deep in the Kunlun mountain range. There the people of Shangri-La live in harmony, never age, and never experience disease. Although Shangri-La is a creation stemming from a Western fetish for the ancient mysterious wisdom of the Far East, many people believe it to be a genuine lost Utopia. Hilton himself admits he was inspired by reading an 1850 travelogue written by two French priests who visited Tibet. There are a few regions that claim to be the home to Shangri-La in an attempt to attract tourist dollars. Tibet, naturally, claims to be home to this lost Utopia. The Sichuan and Yunna provinces have also called dibs. Yunnan went as far as to rename one of its cities Shangri-La City in 2001 to cement its claim. Season 2 of the Travel Channel's *Expedition Unknown* has Josh Gates embark on a two-episode quest to find the real Shangri-La in Nepal. By the end of the second episode, Gates proclaimed Nepal's Mustang region the home of the fabled Shangri-La.

THULE

Similar to Hyperborea, Thule comes to us from Greek and Roman legend. It was considered the most northerly region of the world. Pytheas claimed to have traveled there in the 4th century BC. Many ancient writers have interpreted Thule to mean either North Scotland or the Norwegian island of Smøla. The legend figured prominently in the works of Nazi occultists who believed Thule was the origin of their Aryan race. Nazis based their interpretation of Thule on a hoaxed manuscript from 1867 called the *Oera Linda Book*. *Oera Linda Book* was likely created by Cornelis Over de Linden (died 1874). He claimed it was passed down to him from his grandfather. The manuscript describes an ancient matriarchal culture devoted to the worship of Frya. The book seems to have been written in an attempt to upend the notion that Greek civilization influenced much of European culture. The *Oera Linda Book* claims this mythical "Frisian civilization" passed their alphabet onto the Greeks and Phoenicians. The work was almost immediately rejected as a hoax due to a large number of obvious anachronisms. It was largely forgotten until a German translation made its way into Nazi circles.

Four Civilizations Alleged to Have Discovered America Before Columbus

"America had often been discovered before Columbus." — Oscar Wilde

Columbus didn't discover the Americas, although it is fair to say his voyages opened the Americas up for permanent colonization/conquest by Europeans. It is widely accepted the Americas were populated by Paleo-Indian hunter gatherers who crossed from Asia via Beringia (commonly referred to as the Bering land bridge) tens of thousands of years ago. There are people who, for various nationalistic motivation, assert some other nation or culture "discovered" America before Columbus.

CHINESE

Gavin Menzies, a former Royal Navy officer, popularized the notion the Chinese discovered the Americas (plus pretty much everything else) several decades before Columbus. He revealed all this in his best-selling pseudo-historical book *1421: The Year China Discovered the World*. What we know is admiral and explorer Zheng He conducted several wide-ranging expeditions between 1405 and 1433. The admiral's trips were impressive. Zheng got as far as Kenya. His voyages generally followed the coasts. However, there's no evidence Zheng managed any distant blue water voyages. Menzies asserts Zheng didn't stop at Kenya but made it all the way to South America. There he founded the Peruvian civilization. But wait. He still didn't stop there. He charted the Americas in full plus the North Pole. Menzies' evidence for all of this is weak. He claims Mongolian remains have been found in South America. He doesn't grasp when archaeologists of yesteryear classified remains as "Mongoloid" they didn't mean it was of Asian origin. It's a reference to the old Mongoloid/Caucasoid/Negroid skeletal classification system. The system tried to classify remains by morphology not by origin. So, classifying something as mongoloid implied, for example, the skull was rounder. Menzies seems to believe Central Americans had no knowledge or love for jade until it was introduced by Chinese explorers. This is not the case. Menzies also points out a lot of South and Central American art and carvings depict people with slit-like eyes. See, just like Chinese people! Menzies doesn't seem to be aware these works of art depicting people with slit-like eyes predate Zheng's supposed visit by several centuries.

ANCIENT HEBREWS

Mormons maintain the Americas were populated by a couple waves of ancient Hebrews, starting with a people called the Jaredites way back about 2,300 BC. There is zero archaeological evidence for this, although some Mormons work very hard to shoehorn the accomplishments of Central American civilizations like the Olmecs into their Tolkienesque pseudohistory. There have been a couple hoaxes over the years in support of this notion. Someone carved the Ten Commandments into an 80-ton stone in Los Lunas, New Mexico. The stone's inscription was written in what the inscriber thought was ancient Hebrew. This has come to be known as the Decalogue stone. The stone was first discovered in 1933 by archaeologist Frank Hibben. Hibben had a penchant for hyperdiffusionism, although that wasn't particularly unusual for the time. The writing on the stone betrays someone with a beginner's grasp of relatively modern Hebrew, not someone with a native grasp of ancient Hebrew. The text reveals errors and changes to the traditional Hebrew decalogue that a native speaker of ancient Hebrew would not have made.

TEMPLARS

The idea that the Templars fleeing persecution in the early 14th century ended up in America (along with their treasure and looted holy relics) is a fairly modern one, spurred on by goofy History channel shows and treasure hunting cranks not content to merely hunt for mythological pirate treasure. How you get the Knights Templar from Europe to North America requires making a lot of stuff up. In 1456, about a century and a half after the Templars ceased being a going concern, William Sinclair built the Rosslyn Chapel in Scotland. Although the Templars did build some actual sites in Scotland, Rosslyn Chapel is not one. In the 1980s, speculation by *The Holy Blood and the Holy Grail* crew started a cottage industry of cranks trying to tie the chapel's many beautiful and intricate carvings to Templar imagery. Among those carvings are what some allege are representations of corn and cactuses unique to the Americas. Given the chapel was built decades before Columbus made it to the Americas, how could Sinclair have known about such plants unless he or someone he knew had visited the Americas? Pseudohistorians reason since he was a Templar because "carvings" (he wasn't a Templar), he must have helped Templars discover North America in their bid to escape persecution. The reality is almost all the plants depicted in the chapel are simply highly stylized nature images not meant to represent actual plant species. The carvings are ultimately an armchair botanist's Rorschach test.

Photo Credit: Karl Mamer

VIKINGS

The Vikings did make it to the Americas before Columbus. There is undeniable archeological evidence Vikings made it to Newfoundland as early as 990 AD. They settled in L'Anse aux Meadows for a short time. It is suggested a lack of game did not provide enough food for a sustainable settlement. There is no evidence Vikings expanded their settlements beyond L'Anse aux Meadows. However, there are people, usually brimming with Nordic pride, that really, really want you to believe Vikings made it into the heartland of America and established longer term settlements. Most notably, in the late 19th century, Olof Ohman (died 1935) magically discovered what became known as the Kensington Runestone. Runes carved into the stone ostensibly described the explorations of a band of Vikings in the 14th century. The problem is, the language on the runestone is written in 19th century Swedish, not what a 14th century Viking would be speaking or writing. Some Rhode Islanders also think Newport was the Viking version of Plymouth Rock. They argue a tower in Newport (known to be built in the 17th century) is actually a tower built by Viking settlers between the 11th and 14th centuries. This was first claimed in an 1837 book by Danish archaeologist Carl Rafn. The problem is, not only is it well documented Benedict Arnold built the tower on his property but excavations indicate the tower was built upon earlier settlements that are well dated after the time the Viking supposedly built the tower. It would be entirely impossible for the Vikings to build an earlier structure overtop later features.

Eight Odd Examples of Historical Revisionism

History can be uncomfortable for some. It forces people to confront their own historical myths, politically inconvenient truths, and less than complimentary views of themselves. Holocaust denial is probably the best example of historical revisionism. Here are some lesser known attempts to rewrite some aspects of history.

JESUS WAS A STORY MADE UP BY THE ROMANS TO PACIFY THE POOR

The New Testament certainly has a pro-Roman bent in parts. Paul's writings are clearly written to address Romans. Most Biblical scholars accept the Bible was tweaked to help early Christians avoid Roman persecution. There is also a small but legitimate group of Biblical scholars called mythicists. They argue there was no historical Jesus. Their view is the New Testament is an evolved mythology. Most mythicists stick to the realm of possibilities. Joseph Atwill is a different story altogether. In 2005 Atwill, a software developer, self-published the book *Caesar's Messiah: The Roman Conspiracy to Invent Jesus*. He took the pro-Roman apologetics found in the New Testament one step further. The whole Jesus story as told in the four gospels wasn't piecemeal evolved in the way, say, normal myths evolve but was a complete fabrication written by scholars working at the behest of Vespasian and his successors (Titus and Domitian). The idea was to create a new, more pacifistic religion for rebelling Jewish populations. The warrior Messiah that the Jews were expecting was actually a peacenik Messiah. Atwill seems to think the gospels are a coded retelling of Josephus' *The Jewish War*.

THE CLEAN WEHRMACHT MYTH

The Clean Wehrmacht Myth refers to a belief by some that the regular German army had no role in the Holocaust or other major atrocities of World War II. The SS and its leaders were solely responsible. The Clean Wehrmacht Myth got its start during the early days of the Cold War. West Germany had to reconstitute its army to help defend against the Soviet threat. Naturally, much of the officer corps in a 1950s era West German army would have been the same leadership that waged war against their NATO allies. The solution was to portray the SS as a "freakish aberration" and a criminal gang, while the conventional Germany army traced its roots to the noble Prussian school and had fought a clean, professional war. The Clean Wehrmacht Myth held considerable sway in Germany until the mid-1990s when popular historical works made it clear the Wehrmacht was involved and fully complicit in atrocious war crimes like helping the SS massacre Jewish populations.

MUSLIMS NEVER INVADED SPAIN

In 711 AD, North African Muslims known as the Moors crossed the Straits of Gibraltar and invaded the Iberian Peninsula, which would be modern day Spain and Portugal. There's an awful lot of Muslim architecture in Spain that argues for an extended Muslim occupation of the peninsula. This bit of history proved inconvenient for the Franco regime. Spanish historian Ignacio Olagüe (died 1974) ingratiated himself to the Franco regime by publishing a 1969 work titled *The Islamic Revolution in the Wes*t (the book was also published under the title *The Arabs Never Invaded Spain*). In it, he tried to cast doubt on the ability of the Moors to mount an invasion from North Africa. Ostensibly they weren't organized enough in 711 AD. Problem is, if you accept this, you need to explain all that Muslim architecture and art in Spain. Olagüe claimed it was produced by a group of wannabees. He argues nontrinitarian Christian sects rejected Catholicism and adopted the religion, language, and culture of Islam as a kind of FU to the Pope.

ARGENTINA SUNK THE HMS INVINCIBLE

Despite the fact the *HMS Invincible* remained active throughout the Falklands War, to this day some Argentines believe the British carrier was either sunk or severely damaged during the conflict. One does have to deal with the inconvenient truth that there is actual video of the *HMS Invincible* returning to the UK in good condition. Revisionists counter that a lookalike carrier was hastily assembled after the sinking. The myth seems to have emerged from a war time report by Argentinian pilots who claimed direct hits with an Exocet missile and a pair of 500-lbs. bombs on what they thought was the carrier. The British did report an attack at the place and time claimed by the Argentinians but the carrier was 30 miles away.

SLAVES WILLINGLY FOUGHT FOR THE SOUTH

The notion that Black men and women willingly enlisted in the Confederate army is part of the "Lost Cause" revisionist movement. The Civil War, revisionists argue, was really a noble defense of state rights and a war to keep aggressive Northern states from bringing their money-grubbing ways to the genteel, Christian people of the South. By way of proof of how much all concerned believed the South was worth fighting to preserve, Black slaves and freedmen willing joined the fight! The evidence of this is pretty thin. Revisionists like to point to a handful of photos of Black men posing in the garb of Confederate soldiers. Actual historians point out these photos were just cruel jokes, photos staged by their masters. Black slaves were pressed into service to build defenses and officers who were slave owners took one or two of their slaves with them to perform menial tasks. But it's not like these slaves had a choice.

THE IRISH WERE SLAVES TOO AND WERE TREATED WORSE

Revisionists like to play both sides of the coin. They will argue a certain established historical event (like the Holocaust or slavery) didn't happen. At the same time, they will argue even if it did happen, another group actually had it far worse. Yet you don't hear them complaining about it or demanding reparations! Slavery apologists who argue Black slaves were well taken care of and lived better lives than their descendants today will point to the plight of the Irish slaves in America. Claims are made, usually via Internet memes, that Irish slaves were about 10 times cheaper to buy than a Black slave, the vast majority of slaves in America were Irish not Black, you could legally kill your Irish slave, and Irish slaves being whipped or starved was a common sight in early America. And, yet, none of that is reported today in our history textbooks. It's true the first Irish settlers in America were not particularly welcome or well off. Many came as indentured servants. In essence, they signed really lopsided contracts where they would pay off the cost of their voyager to the new world with work. Slavery apologists touting the Irish Slave myth don't seem to make a distinction between indentured servitude and slavery. The children of Black slaves, for example, were property to be sold off. The children of Irish indentured servers were quite free. The myth took root in 1993 with the publication of *They Were White and They Were Slaves: The Untold History of The Enslavement of Whites In Early America*. The book was self-published by Holocaust denier Michael A. Hoffman II. Irish author Sean O'Callaghan piled on in 2000 with his book *To Hell Or Barbados: The Ethnic Cleansing of Ireland*. He invented lascivious new details like Irish women slaves were purposely bred with Black slaves. From there online racists have taken up the myth by posting and reposting a clutch of Irish slave memes.

OTHER LOSSES: EISENHOWER'S GENOCIDE OF GERMAN POWS

Other Losses is a 1989 book by James Bacque. Bacque, who had no training as a historian, made the startling suggestion Gen. Dwight D. Eisenhower orchestrated a post-war genocide of roughly 1 million German prisoners of war. The story goes that Eisenhower so hated the German people, and German soldiers in particular, that he created a new category for German POWs called "Disarmed Enemy Forces" (DEF). This allowed him to circumvent the Geneva Convention and starve prisoners to death by keeping food and medicine from millions of these reclassified German prisoners. Those liquidated by starvation were recorded as "Other Losses" by American POW camp commanders. Bacque's evidence for this genocide comes from his misunderstanding of Allied POW camp statistics and its use of the term "Other Losses". As Bacque read it, the Other Losses numbers were staggering, totaling close to a million. Bacque assumed it meant dead German POWs. By way of confirmation Bacque relied on the recollections of one "expert" witness, a 91-year-old colonel who told Bacque the figure indeed meant deaths. When journalists interviewed Bacque's expert, he denied telling the author "Other Losses" meant deaths. He claimed Bacque took advantage of his old age to get him to agree to put his name to words he never spoke. In reality, the Other Losses figure simply meant POWs in American camps who were transferred to British or French camps and hundreds of thousands of teenage boys and old men who were quickly released after surrender so they could go back to working on farms and caring for their families. Further the DEF categorization itself was not created by Eisenhower but by the agreement of allied forces.

EASTERN EUROPEAN JEWS ARE NOT JEWS

Parapsychologist Arthur Koestler (died 1983) thought he was helping when he published his book *The Thirteenth Tribe* in 1976. Koestler tried to prove that Ashkenazi Jews, the largest ethnic group of Jews, are not actually Jews but descendants of a nomadic Turkish tribe known as the Khazars. Somehow these Khazars tribes adopted Jewish ways, migrated west after the collapse of their society, and masqueraded as European Jews. In Koestler's mind, once he revealed the majority of European Jews are not actually blood related to the "Christ-killers" of the Bible, antisemitism would magically disappear. Koestler's work only provided additional talking points for white supremacists, particularly Identity Christians who insist they're actually the real Jews. It is certainly true some Jews fleeing Roman persecution found refuge among the Khazars. But genetic studies of Ashkenazi Jews firmly establish their lineage back to the Middle East.

Six Alternate Historical Timelines

There are some people that like to wade into the very deep end of the historical revisionism pool. They're not content to reframe some inconvenient historic event. They want to revise whole swaths of recorded history and eliminate centuries, if not millennia, from the conventional timeline.

THE PHANTOM TIME HYPOTHESIS

In the 1980s, Velikovsky fanboy Heribert Illig began publishing ideas about what would come to be known as the Phantom Time hypothesis. Illig noticed what he thought were some historical anomalies. First, we have a rich written history until about 614 AD, after that the historical record in many places in Western Europe gets kind of, oh, dark. For the next 297 years there's a kind of, oh, dark age. Written history then starts to perk up again about 911 AD. Illig also noticed when the Catholic Church changed from the Julian to Gregorian calendar, it only added ten days to the calendar. In Illig's estimation the Church should have added thirteen days, not ten. Since the Julian calendar went off kilter by a day for every century, the Church should have corrected for 13 centuries not 10 centuries. Therefore 300 years must have been added to the calendar at some point. Since, in Illig's mind, we have 300 years of almost zero written history between 614 AD and 911 AD, that's the perfect candidate for those phantom years. Why would someone do this? Illig believed the Holy Roman Emperor Otto III wanted his rule to coincide with the turning of the first millennium in 1000 AD. The motivation, as Illig saw it, was ascending the throne on the cusp of the new millennium would legitimize his rule. In Illig's mind Otto III didn't ascend the throne in 996 AD, he actually took the throne some 297 years before that. Otto III, understanding he wouldn't live three more centuries, conspired with Pope Sylvester II and Byzantine Emperor Constantine VII to add some 297 years to the calendar. This way they could ALL claim to rule during this portentous time period. Real historians point out that while in some places written history is pretty spotty between the 7th and 10th centuries, we do have some written history from that time period. Like, Charlemagne was a pretty big deal for a couple decades at the start of the 9th century. Illig's counter is that Charlemagne never existed and his history was made up by the Church to help paper over this time gap.

THE NEW CHRONOLOGY

Russian mathematician and pseudo-historian Anatoly Fomenko looked at Illig's claim that three centuries of European history were made up and said, "hold my beer". Fomenko asserts that written history before 1600 AD is mostly a fabrication written by crafty Jesuits who were conspiring with the House of Romanov. Fomenko's idea became known as the "New Chronology". Actual historians will tell you written history gets us back about 5,000 years to the time of the Sumerians. But Fomenko isn't buying. Written history starts in 800 AD. The histories of Greece and Rome — including their art, sculptures, and literature — are all recent forgeries or misinterpretations of actual historical events that really took place in and around the former Soviet empire. For example, Jesus was actually Byzantine Emperor Andronikos I Komnenos who ruled for a few years near the end of the 12th century. The Temple of Solomon has been confused by historians for the Hagia Sophia in Turkey. Historians have also managed to interpret the same event as being different events separated by centuries. For example, Fomenko thinks historians have conflated the Trojan War with the Crusades. The weaknesses of Fomenko's ideas are similar to those of Illig's. You have to ignore a lot of known written history. Fomenko's Jesuit conspiracy compiles the problem. There's zero evidence that thousands of Jesuit scribes and artists were occupied with the enormous task of forging ancient documents in a variety of dead languages, carving statues, minting ancient coins, and casting weapons of war.

ROHL'S NEW CHRONOLOGY

David Rohl is a British Egyptologist. In the mid-1990s, Rohl proposed a major revision to Egyptian chronology, knocking off about 350 years of history. For example, Rameses II ruled between 1279 BC and 1213 BC. Rohl, however, pegs his reign starting in 943 BC and ending in 877 BC. Rohl's re-dating is also referred to as the New Chronology, although it has no relation to the vast conspiracy-based New Chronology of Fomenko. Rohl does not seem to propose any kind of conspiracy to rewrite history. Rohl's New Chronology does come with some problems as a lot of dating of Near Eastern history is built ladder-like on the dating of Egyptian history. If you move Egyptian history three and a half centuries up the timeline, then you need to revise the timelines of Hittite, Assyrian, and Biblical history. Some people like Rohl's New Chronology because he has tried to sync up legendary Biblical figures and locations with people and places actually established in history. For example, Rohl links King David to a ruler named Dadua which is referenced in a series of clay tablets known as the Amarna letters. He pegs the Tower of Babel as being a ziggurat found in the ancient city of Eridu.

Egyptian statues found at the British Museum.
Photo Credit: Karl Mamer

Not surprisingly, Rohl's New Chronology has been rejected by mainstream Egyptologists although not quite as voraciously as alternative timelines offered by full bore cranks. Admittedly, there is some "play" in the standard Egyptian timeline, although most historians view shifting it up more than three centuries is going beyond the evidence.

AGES IN CHAOS

Immanuel Velikovsky (died 1979) followed up his *Worlds in Collision* with 1952's *Ages in Chaos*. That book, plus four other follow-on works, is another attempt to revise the ancient Egyptian timeline. Velikovsky tried to knock off about 500 years of Egyptian and Near Eastern history. Velikovsky's technique is similar to that of Rohl and Fomenko (though technically Velikovsky got there a few decades before either). You can bring amazing economy to the timeline if you assert historical figures that lived centuries apart are really the same person. Velikovsky claims the pharaohs of the 26th dynasty are basically a historical copy-paste of the 19th dynasty pharaohs. Velikovsky took the additional step of trying to merge known historical figures with people of Biblical legend. For example, he merged the 15th century BC Hatshepsut woman pharaoh with the legendary figure of the Queen of Sheba who, if she existed, lived around the 10th century BC. Velikovsky merges the Hyksos people (who ruled Egypt during the 16th and 17th centuries BC) with the Biblical legend of the Amalekite people. Suffice it to say, just as astronomers rejected *Worlds in Collision* as pure crankery, historians have likewise taken a dim view of *Ages in Chaos*.

THE ORION CORRELATION THEORY

As far back as 1973, it was suggested that the three main pyramids of Giza appear to align with the three stars that make up the belt in the Orion constellation. That is two of the belt stars are in line and one, the star Mintaka, hooks south a bit. Now, if you look at the three main pyramids of Giza and rotate them, you get a similar, but not exact, pattern. Robert Bauval, trained as an engineer (drink!), ran with this and published a book in 1994 called *The Orion Mystery*. Pseudo-historian Graham Hancock also went hard on this theory in his 1995 book *Fingerprints of the Gods*. Orion is an impressive constellation and the belt is a beautiful sight. So if you want to argue the Egyptians were inspired to lay out three of their greatest pyramids in a kind of backward upside down representation of Orion's belt, well, sure. It's as good of an idea as any. However, you don't get a super great match for how the stars looked in 2500 BC when the main pyramids were built. The Egyptians, if they wanted an exact match for Orion's belt, were fully capable of doing it. To solve for this, Bauval and Hancock reach for a vision by psychic Edgar Cayce that Atlantis was destroyed in 10,500 BC. They suggest encoded in great works of ancient architecture, like the pyramids, is this 10,500 BC date. Hence, if you roll back stellar procession to 10,500 BC, the main pyramids of Giza are an exact match! Well, not quite. Again, you can't get around that fact you have to flip the pyramids to match the Orion alignment. And actual astronomers who have run the stars back to 10,500 BC don't find such a great match.

CENTURIES OF DARKNESS

When people think of the Dark Ages, they usually think of the centuries of cultural decline that ostensibly followed the collapse of the Roman empire. Historians have been more charitable in recent decades to what Europeans accomplished during that period. The Dark Ages weren't quite as dark as popularly imagined. However, the term itself remains useful to refer to periods where written history is thin. Such a dark age emerged after the collapse of the Bronze Age around 1100 BC and picked up again about 350 years later in 750 BC. During this time intercultural trade collapsed along with record keeping. The building of monuments and walls diminished. Although there are many theories for this collapse, there is no generally agreed upon cause. One way out of dealing with a three and a half century long knowledge hole is to claim those 350 years never happened. Peter James put forward this idea in his 1991 book *Centuries of Darkness*. Like others before him, James tries to rewrite standard Egyptian chronology. His main line of attack is arguing some pharaonic dynasties weren't actually consecutive but overlapped. Hence, you can shave off three centuries and severely reduce the Dark Ages of 1100 BC. Over the decades, James has been no more successful than his colleagues in trying to radically change the Egyptian chronology ladder. Established scholars of the period have not adopted these new timelines. They can't quite get around the fact we can radiometrically date archeological objects to the time periods that supposedly didn't exist.

Eight Things *The Da Vinci Code* Got Wrong

The Da Vinci Code is a fine enough thriller but it's not a work of history. Unfortunately, readers have assumed Dan Brown's novel weaves well-researched, unimpeachable historical facts in with his narrative. They can't be blamed, I suppose, as author Dan Brown has asserted 99% of the novel is true. Here are facts presented that are simply not true:

MARY MAGDALENE WAS NOT FROM THE TRIBE OF BENJAMIN

The book asserts Mary Magdalene was from the tribe of Benjamin. There's no evidence this was the case. "Magdalene" is thought to refer to being born in the town of Magdala. Magdala was a town in northern Israel, while the tribe of Benjamin was from the south.

MARY MAGDALENE IS PICTURED IN THE LAST SUPPER

Brown claims Da Vinci swapped out St. John with a depiction of Mary Magdalene. His evidence is John looks pretty feminine. There are a number of issues with this. Da Vinci and artists of his time tended to depict young men in a feminine way. Da Vinci's other depictions of St. John show him in a similar fashion. As well, we have

Da Vinci's original notes on the last supper and he clearly labels the figure as St. John.

WALT DISNEY WAS A BIG PROMOTER OF THE LOST GODDESS

In Chapter 61, Brown drops this stunning revelation. Walt Disney wasn't a guy who set feminism back decades with numerous films that depicted women as helpless and facile. Turns out he was secretly trying to tell the world about the lost goddess through his films. Okay. The evidence here is pretty thin. In the late 1990s, someone had published a paper suggesting Ariel in *The Little Mermaid* was a Mary Magdalene archetype. Why? Because in one sing-song scene she finds the painting *The Penitent Magdalene* by Georges de la Tour in a sunken ship. If that isn't proof enough, need I remind you Ariel has red hair? Legend is the Magdalene was a redhead. Mic drop time! The reality is la Tour was known for his vibrant paintings of candlelight and the animators wanted a lost painting to go with a lyric Ariel sings wondering what fire is. The red hair was a deliberate choice as Disney didn't want Ariel to be a copy of the blonde-haired mermaid from *Splash*.

POPE CLEMENT V HAD THE TEMPLARS BURNED ALIVE

It is true many Templars were burned at the stake. However, Pope Clement V had nothing to do with it. France's King Philip IV was their persecutor. Brown further asserts Clement V had their ashes tossed into the Tiber river. Brown seems unaware Clement V had the Holy See moved to Avignon at the time. The Tiber river is in Rome.

TEMPLARS BUILT THE ROSSLYN CHAPEL

The Rosslyn chapel in Scotland looms large in Brown's mythology. He suggests the chapel was built by Templars. It is well established, however, the chapel was built more than 125 years after the end of the Templars.

PARIS WAS FOUNDED BY THE MEROVINGIAN KINGS

The book claims the Merovingian line (which the book maintains represents the bloodline of Jesus) founded Paris. This is not true. Paris was a going concern as far back as 51 BC. It is true the first Merovingian king Clovis I conquered Paris and made it his capital in the 5th century.

THERE IS NO ROSE LINE

Brown conflates two lines in France. One is the Paris Meridian. The other is a brass strip found in the Church of Saint-Sulpice in Paris. The Paris Meridian was an alternate prime meridian for a time. The brass strip Brown identifies as part of the Paris Meridian and Rose Line is an astronomical device used to help the church determine the advent of the Easter Equinox. This brass strip is actually 100 meters west of the Paris Meridian. The only known reference to a "Rose Line" is in a hoaxed document put out by the fake Priory of Sion society.

THERE ARE A SMALL NUMBER OF PEOPLE TODAY THAT CARRY ON JESUS' BLOODLINE

If Jesus did live to pass on his DNA 2,000 years ago, it is highly unlikely only a small group of people would carry on this bloodline. The reality is millions of people today could trace their lineage directly back to Jesus.

Six Locations of Supposed Ancient Nuclear Wars

Many religious texts, like the Bible and the Bhagavad-Gita, describe mythical conflagrations that, to some, sound like a nuclear war. Instead of accepting ancient people can create pretty fantastical myths, some pseudohistorians propose these religious myths describe incidents of actual ancient nuclear warfare.

SODOM AND GOMORRAH

The Bible has this to say about the destruction of Sodom and Gomorrah. "Then the Lord rained on Sodom and Gomorrah brimstone and fire from the Lord out of heaven..." Many have suggested over the years the Bible described what amounted to a nuclear attack. The first person with a semblance of scientific credentials to suggest this was Soviet-era mathematician Matest Agrest (died 2005). Agrest certainly had a legit scientific career, publishing 100 papers on topics ranging from math to astronomy. However, in 1959 he got fixated on what he called "Cosmonauts of Antiquity" and later the "paleocontact hypothesis". In a 1959 paper he put forward the idea Sodom and Gomorrah were destroyed by an atomic bomb. The bit about Lot's wife turning into a pillar of salt is viewed by Agrest as a poetic reference to Lot's wife being vaporized by the blast and leaving a shadow on the ground of the salty desert. By way of proof, he seems to make a lot out of tektites, a glassy silica formed by meteor impacts. Some have been found to be radioactive. In Agrest's mind these could only have been formed by nuclear explosions. Unfortunately, dating of these tektites seems to argue against his evidence. Tektites are millions of years old. If Sodom and Gomorrah were real, the cities are thought to have existed around 2600 BC. There's something of a gap there that can't be reconciled.

MOHENJO-DARO

Mohenjo-Daro is an amazingly well-preserved ancient city. It dates back to 2500 BC. It's one of humankind's earliest major cities. Very early excavations that were poorly documented suggested to archaeologists the city had been hit by a disaster and people had died suddenly and in place. Later review indicated the bodies were actually purposely buried and at different times. Believers in ancient nuclear wars, however, pass around tales of bodies found strewn across streets and huddled together before the end. They paint a very Pompeii-like scene. Since there are no known volcanoes near Mohenjo-Daro, the only possible explanation for this is an atomic blast and people killed suddenly by radiation! So, what about radiation? Proponents claim the area is surrounded by a layer of radioactive dust. Some unnamed Soviet-era researcher is also frequently cited as having found skeletons with some of the highest radiation levels ever found. It will come as no surprise these stories that get copy and pasted across the Internet never come with any attribution to an actual scientific source.

JODHPUR

If you're to believe a post on Rense.com from fall of 2000, a housing development just outside of Jodhpur, India sits atop a layer of radioactive ash from an ancient nuclear war. To add credibility to the story, the findings were confirmed by a Western archaeologist named Francis Taylor. We're told findings were consistent with what appears to be a nuclear attack described in an ancient Indian Sanskrit legend called the Mahabharata. The legend describes a "projectile charged with all the power in the Universe" and a "column of smoke and flame as bright as 10,000 suns". The problem is there is no archaeologist by the name of Francis Taylor and the citation from the Mahabharata was made up out of whole cloth.

LONAR CRATER

Lonar Crater is another candidate site offered by cranks as the location of an ancient nuclear attack. Lonar Crater, found about 310 miles east of Mumbai, is an impressive water-filled crater. It's estimated to have been the result of a meteor impact between 52,000 and 570,000 years ago. The leading proponent of Lonar being the remnants of a crater blasted by an ancient atomic weapon is *Ancient Aliens* go-to guy David Childress. In Childress' mind, since geologists have not been able to find any remnants of the meteor, the crater is unexplained. Therefore, the best fit is some hyper advanced civilization that didn't exist 50,000 years ago set off an atomic bomb. Despite not finding any meteor remnants, the crater is a nearly textbook example of what results when a small piece of the universe impacts a planet like Earth. It bears no signature of being blasted out of the surrounding basalt by a nuclear bomb.

THE LIBYAN DESERT

The Libyan desert is strewn with something called "desert glass". Desert glass was highly prized by the pharaohs of Egypt and can be found in ancient jewelry. Geologists generally consider such glass found in Libya as well as Australia to be the result of a meteor impact or possibly a lightning strike. According to the elaborate mythology of Zecharia Sitchin, Libya was apparently home to a spaceport that was nuked by rival ancient aliens. Since we found sand fused into glass after the Trinity tests in New Mexico, in the mind of Sitchin and fellow travelers, Libyan desert glass was a result of warring space aliens.

MARS

Physicist John Brandenburg analyzed gamma ray data from the Mars Odyssey orbiter and found an unusual concentration of radioactive elements in an area on Mars. At a 2011 scientific conference he introduced the idea that a natural nuclear reactor might have formed on the red planet and produced the radioactive elements. Given at least one such natural nuclear reactor has been found on Earth, his idea was treated as an interesting possibility. If only Brandenburg had stopped there. He went on to suggest a high ratio of radioactive xenon-129 in the atmosphere was evidence that a massive nuclear explosion had taken place on Mars. Brandenburg estimated the explosion amounted to the equivalent of a one million megaton bomb. Brandenburg believed this then explained what happened to Mars' water and atmosphere. The scientific community found that idea less plausible. Again, he should have stopped there. But no. He eventually concluded the explosion wasn't accidental. It was the result of some vengeful space aliens discovering a primitive Martian civilization and nuking the planet from orbit. Oh, yeah, he also believes the face on Mars is not natural. That is a lot to hang on the presence of a higher than expected concentration of xenon-129 in the Martian atmosphere. Actual planetary scientists, which Brandenburg is not, understand there are many natural sources of Mars' atmospheric xenon-129.

Eight Possible Explanations for Flood Mythology

Cultures around the world and throughout history have their own legends about deluges. Biblical literalists see this as proof of a global flood. Others, not so much. Here are several proposed hypotheses that explain the origin of flood mythologies found in various cultures. They are organized from least plausible to most plausible.

SATURN WENT NOVA

Immanuel Velikovsky attempted to explain a number of events in the Bible by appealing to a most ludicrous form of planetary mechanics in his 1950 book *Worlds in Collision*. It has been described by some as "planetary billiards". In Velikovsky's view, a conjunction between Saturn and Jupiter caused Saturn to go nova and expel a huge cloud of hydrogen. When the earth passed through this ejecta, a hand wavy electrical process caused the hydrogen to combine with oxygen in Earth's atmosphere producing rain. A lot. Of. Rain. No legit planetary scientist thinks this is even possible.

A COMET

There's a legitimate modern theory that much of Earth's water came from collisions with bodies of ice in space during the early years of our planet's formation. The natural philosopher William Whiston (died 1752) sort of got this right around the edges when he proposed in his 1696 treatise *A New Theory of Earth* that Noah's flood was caused by Earth passing through the tail of an icy comet. It seems this theory got some support from Isaac Newton. Newton, himself, was no stranger to trying to explain Biblical events through science of the day.

TOLLMANN'S BOLIDE HYPOTHESIS

Tollmann's bolide hypothesis suggests around 7640 BC the earth was pummeled by several large asteroids leading to the Holocene extinctions. As well, asteroids that hit the ocean are hypothesized to have caused giant waves that would have caused large coastal floods around the world. Hence, different cultures that have flood mythologies might all be passing down tales of the effects of Tollmann's hypothesized asteroid strikes. Scientists across multiple fields of study (geology, paleoclimatology, astronomy) have found little evidence to support Tollmann's idea, however. Evidence offered by Tollmann in support can be explained by more mundane processes.

BLACK SEA DELUGE HYPOTHESIS

Noah's Flood: The New Scientific Discoveries about the Event that Changed History is a 1998 book that suggests glacial melt resulted in a rapid expansion of the Black Sea around 6500 BC. Flood narratives from the region (like those found in the *Epic of Gilgamesh* and *Genesis*) may have been influenced by the devastation wrought by the Black Sea flood. Although an intriguing hypothesis, it's not one popularly held by geologists today.

FLOODING CAUSED BY THE END OF THE LAST GLACIAL PERIOD

It has been suggested flooding caused by glacial melt that started about 18,000 years ago has been preserved in many flood myths. For example, many parts of Mesopotamia now under water were once fertile land occupied by humans until about 8,000 years ago. Melting glacial waters created for a time a lake in North America that covered large parts of Manitoba, Ontario, Northern Minnesota, and Eastern North Dakota. Indigenous Americans may have preserved tales of this massive lake in their flood myths.

IT WAS HOW ANCIENTS EXPLAINED SHELL FOSSILS ABOVE SEA LEVEL

The ancient Greeks, Romans, and Egyptians noticed shell and fish fossils could be found well above sea level. Greek scholars suggested one or more global floods to explain the existence of these fossils.

EPIC OF GILGAMESH

The *Epic of Gilgamesh,* which contains within it a flood narrative, was written between 2100 BC and 1200 BC. This predates *Genesis* by several centuries. The earlier flood narrative in *Epic of Gilgamesh* so closely matches the *Genesis* flood myth that the vast majority of historical scholars agree the *Genesis* version is a simple retelling, likely adapted after or during the Babylonian exile. The *Epic of Gilgamesh* version, found on tablet XI, tells the tale of gods displeased by humanity and seeking to kill them with a global flood. One god saves a righteous man named Utnapishtim by commanding him to build a boat and load it up with his family and animals, both domestic and wild. Utnapishtim sends out birds to determine if the waters had receded. Utnapishtim's boat comes to rest on a mountain top (Mount Nimush).

SYNCRETISM

Almost all cultures have had to deal with devastating floods and it's not uncommon to blame natural disasters on a vengeful god. Flood narratives tend to likewise act as cautionary tales: be prepared for future disasters, know what stuff to save. As Christian missionaries moved across the Americas and Asia, they encountered local flood tales and, in an effort to convert the locals, engaged in what is known as syncretism. This is taking local deities and creation myths and trying to align them with the Christian equivalents to show the locals they're already halfway to being good members of the missionary's faith. Hawaiian and native American flood legends that existed purely as oral narratives have been shown to have changed substantially after Christian missionaries set up shop.

Seven Problems with *Chariots of the Gods*

Although not the first book to suggest we have been visited by aliens in the distant past, Erich von Däniken's 1968 *Chariots of the Gods* has done the most to popularize the claim. As one might expect, the book is not without problems.

THE NAZCA LINES ARE ALIEN LANDING STRIPS

The lines are carved into rather soft soil. If they were runways, you would not want to land any kind of aircraft or spaceship on it.

THE NAZCA LINES OFFER PARKING AREAS FOR SPACESHIPS

In *Chariots of the Gods,* von Däniken includes a closely cropped photo of a bird image on the Nazca plain. The crop is so tight you have no idea what the image even is and von Däniken doesn't bother to tell you. He notes the image resembles an aircraft parking area found in modern airports. Odd, no? It will come as no surprise von Däniken made no effort to check his work. The area depicted is actually the bird's knee joint and it's only a few feet across. Whatever von Däniken thinks its function is (beyond representing the knee joint of a bird), it can't possibly be a holding area for flying craft.

DID ANCIENT PERUVIANS MOVE A 20,000 TON BLOCK?

Von Däniken makes a pretty big deal about a supposed 20,000-ton block found near Sacsayhuaman, Peru and leaves the reader to wonder how ancient humans could have moved something so huge. How indeed? It wasn't until at least 2008 humans were able to create a crane (China's Taisun crane) that could lift 20,000 tons. So how did the ancients do it? They didn't. This stone doesn't seem to exist anywhere other than inside the pages of *The Chariots of the Gods* and a few web pages that simply repeat his disinformation. Archaeologists have well documented the area and have never made any reference to this 20,000-ton block.

ANCIENT PEOPLE CAME UP WITH GOOD CALENDARS

Throughout *Chariots of the Gods*, von Däniken seems incredulous that people could have come up with reasonably accurate calendars. How could, for example, the Mayans have figured out the solar year was 365 days long (which it's not) unless told by aliens? He seems unable to grasp that ancient peoples, over generations of observing the seasons and movements of celestial bodies, have figured out reasonably accurate calendars so they knew when to plant crops.

Photo Credit: Karl Mamer

PAKAL'S SARCOPHAGUS LID

Pakal was a Mayan ruler from the 7th century. Arguably few would know anything about him today if von Däniken hadn't made his sarcophagus lid a centerpiece of his ancient aliens mythology. Von Däniken asserts it depicts a man in a space capsule working controls. Flames are emanating from beneath the space capsule. What it really shows is Pakal sitting under the splendor of the "world tree". In Mayan culture the world tree grew out of the underworld, up through the world of humans, and into heaven. His pose is a little odd to people unfamiliar with how Mayans depicted newborn babies but it's instantly recognizable to scholars as a symbol of his rebirth. And what von Däniken says is flames or exhaust is really a two-headed snake helping open the heavens up to Pakal.

THE MYSTERIES OF EASTER ISLAND WERE SOLVED YEARS BEFORE CHARIOT OF THE GODS

Von Däniken can't for the life of him figure out how a small population of about 2,000 Rapa Nui people could move such large stone statues. Although he quotes some historical work done on the island by Thor Heyerdahl (died 2002), he doesn't seem to be aware Heyerdahl offered a plausible explanation in his 1957 book *Aku-Aku*. Heyerdahl found a six-man team could move a statue into place in about 18 days by simply laying it flat on a simple sledge and dragging it.

THE COMPLETE LACK ANYTHING ALIEN

It's pretty hard to find a single paragraph in *Chariots of the Gods* that isn't simply an elaborate question. Why would people do this? How could people do this? Doesn't this resemble a 1960s era astronaut? He offers a lot of questions he can't seem to personally answer but what von Däniken never actually offers is a single item of truly alien manufacturer. Not a single accidentally waylaid alien tool or even a bit of alien garbage found in a pit. No king's tomb features a shiny alien object given as a gift. Zip. *Nada*.

Fourteen Out of Place Artifacts and How They Really Got There

Out of Place Artifacts (or OOPArts as some skeptics like to call them) are artifacts that were supposedly found *in situ* but can't be explained by so-called conventional science. These OOPArts are variously held out as clear evidence:

1. The world is not as old as scientists claim!
2. The world is actually a lot older than scientists claim!
3. Space aliens have visited us in our ancient past!
4. Atlantis was real!

Despite assertions science can't explain OOPArts, it's not difficult to find websites and papers featuring scientists actually explaining how the OOPArt isn't so OOPy.

ACÁMBARO FIGURES

Waldemar Julsrud was a German immigrant in Mexico. He was living near a city called Acámbaro. In the 1940s he reported finding some interesting clay figures resembling dinosaurs. Thinking he had stumbled on a stash of ancient artifacts, he paid locals to search for more. He paid a peso per figure. In all told, locals in and around Acámbaro "found" approximately 33,000 figures. The figures represented not just dinosaurs but humans riding dinosaurs and doing battle with them. Young Earth creationists believe the Acámbaro figures are proof positive humans and dinosaurs lived together. It escapes them that the locals simply made them in return for Julsrud's ready money. Archaeologists who have examined the figures are, of course, suspicious. There's no way to actually date them scientifically but they defy anything that's ever been found before. You may find one or two ancient cultural objects in reasonably good condition, but you don't find 30,000 small, intricate clay objects, ostensibly buried for thousands of years, unearthed in pristine, unbroken condition.

ANTIKYTHERA MECHANISM

The Antikythera Mechanism is an ancient Greek bronze clockwork device that calculates the motion of the moon and planets. It is the oldest known example of an orrery. It was discovered in 1901 by sponge divers investigating a shipwreck off the coast of a Greek island named Antikythera. It has been dated to about 150 BC. Pseudohistorians, notably those pushing hyperdiffusionism, claim ancients could not have created what is, in essence, a sophisticated computer! Therefore, Atlantis! Although this is the first example ever found, scientists long suspected the ancient Greeks had such devices as they were described in the writings of Cicero. Modern day analyses of the device suggest this wasn't the first copy built. There's no hint of trial and error in its build. It's thought we've never encountered a similar device because they would be rare in the ancient world. Such a device would be incredibly expensive to purchase. Being made from bronze would mean broken versions wouldn't be idly tossed in the trash. The ancients tended to melt down and reuse bronze when they could.

BAGHDAD BATTERY

Although scholars assume the so-called Baghdad battery is really a terracotta pot meant to store a sacred scroll, there are those that insist the copper and iron bits found with it suggest it is really an ancient battery. From there, claims range from the semi reasonable (ancients discovered it produced a tingly effect and used it in religious ceremonies) to the semi unreasonable (ancients used it to electroplate objects) to the whacked (it was technology passed down to ancient Persians from space aliens or Atlanteans).

CRETACEOUS FINGER

The Creation Evidence Museum of Texas has on display a Cretaceous stone that looks a bit like a finger. If you squint really hard at an indentation at the end of the tube-shaped bit of rock, you can see what looks like a nail bed. Hence, it's a finger! One needs to ignore for a moment no other fossilized human body part has ever been found in a Cretaceous rock. The finger to real scientists is indistinguishable from tubes created in Cretaceous rock by an ancient form of burrowing shrimp.

DAVENPORT TABLETS

As European settlers moved west into the heartland of America, they discovered huge mounds, remnants of large structures created by indigenous peoples. European settlers, however, couldn't imagine "Indians" were capable of building such sophisticated structures. Many latched onto the idea some sort of European civilization had gotten there first. To the mind of settlers, they were not actually stealing land from indigenous people. They were just taking back what was left to them by their mysterious European forefathers. Figuring into this myth were the Davenport tablets. They were found in an indigenous burial mound near Davenport, Iowa in 1877. The tablets were sheets of slate and carved with zodiac symbols and what was reputed to be Phoenician script. Their authenticity was almost immediately called into question. The script and zodiac symbols appeared to be copies of similar depictions from the 1872 edition of *Webster's Dictionary*. As well, the carvings on the slate had the hallmarks of being worked recently by modern tools. It has been long suspected the tablets were planted as a mean-spirited prank on a local archaeologist who was finding solid evidence that these ancient structures were indeed the work of indigenous people. Despite the tablets being rejected almost immediately by contemporaries, hyperdiffusionists still trumpet the Davenport tablets as evidence some progenitor civilization (e.g., people from Atlantis!) brought "civilization" to the Americas and Africa.

ELTANIN ANTENNA

In 1964, a US research ship named the *USNS Eltanin* photographed the ocean's floor off Cape Horn. One of the photos featured something that looked like a radio or TV antenna. The object was dubbed the Eltanin Antenna. Ufologist and ancient astronaut proponent Brad Steiger (died 2018) did a lot to popularize the notion this was a photo of an actual antenna set up in an inconspicuous place by space aliens. Scientists familiar with life in the deep ocean, however, recognize it as a form of carnivorous sponge called *Cladorhiza concrescens*.

The Eltanin Antenna
Photo Credit: US Government

ICA STONES

These stones, supposedly found in Peru's Ica province, are claimed to be ancient rock carvings that depict dinosaurs living contemporaneously with humans. They're held up as proof positive by young Earth creationists that the world is only 10,000 years old and dinosaurs survived the flood. If you happen to be of the ancient astronaut bent, you'll find Ica stones depicting space aliens. Although the Ica province was the subject of a number of legit archeological expeditions in the 19th and 20th centuries, no expedition ever discovered anything resembling the Ica stones. It wasn't until the 1960s that these stones began turning up when a couple collectors started happily paying locals real money for each stone they supposedly "found".

KLERKSDORP SPHERES

Ancient aliens types love the Klerksdorp Sphere. They were found in pyrophyllite taken from mines near Klerksdorp, South Africa. They are spheres with interesting equatorial grooves that give the appearance of tooling. Proponents of ancient aliens make wild, unsubstantiated claims about their perfection, balance, and hardness. Claims include NASA testing indicates the spheres could only be made in zero-G and they are made from an unknown alloy with uncanny hardness. Oddly, no one can seem to produce a copy of this NASA study. These stories about the Klerksdorp spheres can be traced back to a bogus article printed in an issue of the *Weekly World News* published in the 1980s. Vedic creationist Michael Cremo seems to have viewed the *Weekly World News* as a legitimate source and gave the story wider circulation in his 1993 book/screed-against-mainstream-science *Forbidden archaeology*.

LONDON HAMMER

How did a 19th century hammer get encased in solid rock supposedly millions of years old? If you listen to creationists, our dating methods are woefully inaccurate. If you listen to Hare Krishnas, humans have been alive for millions of years and littering the planet with broken hammers. Scientists have a rather ready explanation. It's a 19th century hammer someone dropped and was encased by a fast hardening sediment. It's a process similar to a phenomenon known as a petrifying well. If you leave an object in well water with high mineral content, it will take on a stony exterior after a few months.

MEISTER PRINT

In 1868 fossil hunter William Meister found what he thought was a fossilized sandal print in the 28-million-year-old Wheeler rock formation in Utah. Intriguingly the fossilized sandal seems to have crushed a pair of small trilobites. Obviously, humans and trilobites were not contemporaries. The fossil found its way into the hands of Mormon creationist Melvin Cook (died 2000) who asserted it was evidence of a young Earth. Cook was a chemist and explosives expert by trade. Real geologists who examined the Meister Print recognized it as a result of natural spalling, that is rocks breaking off of larger formations. The Wheeler formation has numerous examples. Meister simply found one that randomly resembled a human sandal. Real archaeologists understand that there are a lot of things that can resemble prints. A sample of one is rarely accepted as evidence. A trackway is typically required before something that looks like a print is determined to be an actual print.

MICHIGAN RELICS

The Michigan relics are a collection of artifacts supposedly uncovered by James Scotford in 1890. As the story goes, he was out in Montcalm County, Michigan digging a posthole for a sign when he uncovered some 797 objects that appear to be of ancient Near Eastern origin. Among the relics were a clay cup, tablets, ornaments, tools, and a tobacco pipe. Some were adorned with images depicting Biblical scenes like Noah's ark. Most were covered in markings that resembled cuneiform and Egyptian hieroglyphs. Michigan Secretary of State Daniel Soper purchased the artifacts and planned to turn them into a tourist attraction. The relics were suspect from the start. The cuneiform characters were simply stamped into the clay in random order. They betrayed no natural language. The relics were easily destroyed by water. It seems impossible they could have survived centuries in damp soil. Soper is generally considered the perpetrator of the fraud. He had served as Secretary of State for only about a year before he was forced to resign after being accused of embezzling government funds. He had also been caught planting Native American artifacts and then pretending to discover them. Although no real scientist was fooled, the Michigan relics did pique the interest of Mormons who saw this as possible evidence that ancient Hebrews had traveled to the American Midwest. The Mormon church acquired them and moved them to their own museum in Salt Lake City. In 2001 the church had BYU archaeologist Richard Stamps examine the relics to definitively establish their authenticity. Stamps concluded, as had so many others, they were not genuine. The Mormon church donated the items to the Michigan History Museum in Lansing.

PALUXY RIVER MAN TRACKS

Texas' Paluxy river is a rich source of dinosaur trackways. One appears to be from a herd of sauropods being stalked by a theropod. Creationists claim you can also find examples of human footprints preserved next to dinosaur prints. Therefore, the earth is young and therefore Jesus! Many of the prints identified as human are simply too big to have been made by a man. This is not a problem for creationists as they must have been left by giants! The reality is many tracks over time become eroded and infilled and can vaguely resemble a human print. What creationist offer as human tracks looks nothing like human footprints to actual trained paleontologists. Some of the tracks that appear more human, both in size and shape, were simply carved by the faithful during the Great Depression in the hopes of kickstarting a tourism industry.

SPARK PLUG IN A 500,000-YEAR-OLD GEODE

This "ancient" spark plug is known to some as the Coso artifact. The story goes some prospectors in 1961 were collecting geodes. After cracking one of them open, they found a spark plug in its middle. The prospectors claimed they had the geode dated and its age was reported to be 500,000 years old. However, the prospectors never explained their dating method or what lab carried out the dating. Photos of the spark plug indicate it's a Champion brand spark plug from the 1920s. Similar to the London Hammer, it's simply a modern artifact that got encased in a fast hardening sediment. People who hold this OOPArt to be evidence of ancient aliens or Atlantis don't really explain why such advanced civilizations never advanced beyond Model T-era spark plugs.

STEGOSAURUS AT TA PROHM

Creationists claim a Cambodian temple known as Ta Prohm has a carving that depicts a stegosaurus. You have to first ignore that the temple has carvings of numerous mythical creatures. You also have to ignore that the head is all wrong and the tail does not feature characteristic spikes. Oh, you also have to squint really hard to interpret the background foliage as dermal plates. Most real scientists assume the carver was trying to depict a boar or possibly a pangolin.

Five Theories about Amelia Earhart's Fate

In 1937, Amelia Earhart was attempting to become the first female pilot to circumnavigate the globe. On July 2 of that year, Earhart and her navigator Fred Noonan departed New Guinea and disappeared. There has been much speculation over the decades about her fate. Some of it is reasonable. Some of it is crazy. Here are five theories about what became of this most famous aviator.

EARHART RAN OUT OF GAS AND DITCHED IN THE OCEAN

Earhart refueled in New Guinea and hypothetically had enough gas to get her to her next stop which was Howland island. This was the longest leg of her flight and was just at the edges of her plane's maximum range. It is known Earhart reported she was running low on fuel about 200 miles from Howland island and was unable to receive a critical radio direction signal from a nearby ship that was positioned to guide her to Howland. The strength of her last radio contacts suggested she was never very far from Howland. It's quite reasonable that she and Noonan missed the island. It is speculated she had to ditch her plane in the ocean after exhausting all fuel reserves. Earhart and Noonan drowned after ditching. Historians generally agree this is the most reasonable and consistent theory.

EARHART MADE IT TO GARDNER ISLAND

This one is still within the realm of general reality. Having realized she missed Howard island, Earhart turned south to look for the nearby Gardner island, which was uninhabited. Earhart was able to land. Tragically they spent their remaining days on Gardner and died. Search and rescue planes did overfly the island. The searchers reported seeing signs of recent habitation but nothing indicating Earhart and Noonan were on Gardner. Ground searches later in the year turned up no indication Earhart and Noonan made it to the island. In 1940, a skeleton of a 5' 4" human was found along with a sextant box. Detailed examination of the skeleton indicated it was that of a male and was actually too short to be Earhart who was known to be about four inches taller. Since 1988 a group called The International Group for Historic Aircraft Recovery (TIGHAR) has mounted several searches of Gardner island and its surrounding waters. The searches have turned up a number of artifacts like aluminum and Plexiglas that could have come from Earhart's plane but nothing found has been definitive.

EARHART AND NOONAN WERE EXECUTED BY THE JAPANESE

CBS news anchor Fred Goerner (died 1994) published a 1966 book titled *The Search for Amelia Earhart*. The book was on *The New York Times* best-seller list for half a year. It was the first published work to suggest Earhart and Noonan landed in Japanese controlled Saipan, were captured, and then executed. The theory was given greater exposure in 1990 by an episode of NBC's *Unsolved Mysteries*. The show included an interview with a woman in Saipan who claimed to have witnessed their execution. However, her story has never been corroborated by any other witness. The History channel picked up on this theory in a 2017 documentary called *Amelia Earhart: The Lost Evidence*. It moved the location of their capture from Saipan to the Marshall Islands. The documentary's big reveal was a supposed photo of Earhart and Noonan in the custody of Japanese soldiers. There are a couple people in the grainy photo that could be the aviators. However, it was discovered the photo was taken about two years before their 1937 disappearance.

EARHART WAS TOKYO ROSE

At the end of World War II, there was a certain mania for identifying Tokyo Rose, that is the women who participated in Japanese broadcasts to allied forces under that general moniker. A rumor was started at war's end that one of the broadcasters was Earhart. Ostensibly she survived her capture and was forced to broadcast English-language propaganda. Earhart's husband investigated this rumor. He listened to hours of broadcasts but was unable to identify his wife's voice.

EARHART FAKED HER DEATH AND ASSUMED A NEW IDENTITY

Former US Air Force pilot Joseph Gervais (died 2005) got it into his mind that a friend of a friend of Earhart he met at a party named Irene Bolam (died 1982) was actually Earhart. What transpired was years of what would be termed stalking today. Gervais delved into every aspect of Bolam's past in a fevered attempt to find proof Bolam was not who she claimed to be. He gave his research to author Joe Klaas (died 2016) who published Gervais' allegations in a 1970 book titled *Amelia Earhart Lives*. In a contemporary review of Klaas' book, *Time* magazine described Gervais' claims as being built on "a slithering foundation of fanciful codes, anagrams, leading but unanswered questions, and hints". Upon publication of the book, Bolam lawyered up, produced an affidavit documenting in fine detail why she could not be Earhart, and then sued the book's publisher McGraw-Hill for $1.5 million. The publisher settled out of court and ceased selling the book.

Four Legendary Treasures that Probably Don't Exist

There's a certain class of people that spend extraordinary amounts of time and real treasure trying to find mythical treasures. Despite others having failed, and sometimes dying, these treasure hunters assume things will turn out differently for them. After all, they have the right map this time. They've read the clues correctly. They have the right equipment to find the treasure. And, most importantly, they're willing to dig one foot deeper than the previous guy. Regardless of their imagined "edge", they never do find the treasure.

THE OAK ISLAND MONEY PIT

Since the 18th century, people have been trying to find buried treasure on Canada's Oak Island. It all started when a group of boys in 1795 thought they had found evidence someone had buried a treasure in what became known as the money pit. Theories about what is hidden on the island range from the semi plausible (pirate booty) to the bizarre (the Holy Grail). Since 1795, increasingly well-equipped groups of excavators have tried to dig up the so-called money pit. Holes quickly fill with sea water. Treasure hunters believe this is caused by boobytrapped flood shafts. They offer this as proof of both the treasure's reality and its value. Who would go to such great lengths to protect something if it were not priceless? Local geologists contend the flooding is a result of natural sinkholes. The most recent attempt to find the treasure is by a pair of Michigan brothers and is the subject of a popular History channel show called *The Curse of Oak Island*. Despite bringing to bear 21st century excavation technology, these new searchers have been unable to find the fabled treasure after more than *seven years of digging*. It's increasingly doubtful anyone from the 18th century (or earlier) had the technology to bury a treasure that well.

THE LOST DUTCHMAN'S GOLD MINE

The so-called Lost Dutchman's Gold Mine is probably one of the most famous of several lost gold mine stories popular in America's West. Others in this category have names that read like Hardy Boy novels: The Lost Padre Mine (California), the Wheelbarrow Mine (Idaho), and the Lost Bucket Mine (Oregon). Since 1892, treasure hunters have been searching for a forgotten gold mine in the aptly named Superstition Mountains of Arizona. As the legend goes, a German expat named Jacob Waltz started telling anyone in Phoenix, Arizona who would listen about how he stumbled across a mine featuring a rich vein of gold. Unfortunately, after being attacked by a band of Apache he was forced to flee for his life and he was never able to return to the mine. His story was lent credence by a box of high-quality gold ore he had in his possession. Before he passed away, he shared the location with a friendly baker named Julia Thomas. Thomas herself tried to find the mine but failed. Instead of taking another whack at it, she found it much more profitable to sell maps to treasure hunters who were starting to hear about the *Deutsch* man's lost mine. Others since Thomas have found the surest fortune is simply selling maps and books to the gullible. Despite people having spent over 125 years searching for the lost mine, no one has found it. Many have died in the search. The difficulty of the local terrain, in combination with Arizona's hot summers, have taken the lives of many ill-prepared and misinformed searchers.

THE NAZI GOLD TRAIN

There's never a good outcome when self-described "independent historians" venture out with a shovel. Such is the case for the seventy-year quest by numerous "independent historians" who think this time they've cracked the secret resting place of the fabled Nazi gold train. As the legend goes, the Nazis, in the dying days of World War II, filled a train full of gold and other plundered valuables. They rolled it into a secret tunnel someplace in Lower Silesia (which is part of modern Poland) and then buried the entrance. Like many legends, there's just enough facts here to make the story plausible to some. The Germans indeed used trains to haul war booty from Eastern Europe back to Berlin. The Germans also began constructing a complex of tunnels in Lower Silesia but never managed to finish any of them by the end of the war. This was known as Project Riese. The project's goal is a bit hazy as plans were lost but historians believe the aim was to move critical war production, and up to 20,000 workers, underground. Also, by moving further east, this put the complex out of the range of American and British bombers. To some, it would seem logical if you wanted to hide your war booty, the unused tunnels of Project Riese would be a good place to stick it. The problem is, beyond this rank speculation, there's zero documentation the Nazis ever moved large amounts of treasure to any of these tunnels. Between 2015 and 2016, two "independent historians", Piotr Koper of Poland and Andreas Richter of Germany, thought they had found evidence of a buried train using ground penetrating radar. They told reporters they were 95% sure they had finally located the Nazi gold train. Unfortunately, they didn't seem to know how to use their equipment. Polish geologists responded that their ground penetrating radar showed no such thing. While they may have found a tunnel, possibly a natural tunnel, there was no indication there was a train. The local Polish authorities, buoyed by the promise of future tourism dollars, ignored the experts and gave Koper and Richter permission to dig. After a week of digging, not only did they not find the treasure, or a train, or tracks, they didn't even find a tunnel.

THE LOST TEMPLAR TREASURE FLEET

The Templars certainly accumulated a large amount of wealth before their order was rather violently ended in the early 14th century by greedy kings who wanted that wealth for themselves. A recent legend has emerged in the world of pseudohistory that several Templars got word of the coming crackdown and loaded up their fleet with much of the organization's gold and booty from the Holy Land (which may or may not include the Holy Grail, the Ark of the Covenant, and the body of Jesus). The fleet set sail just in the nick of time and went someplace to hide the treasure. Much of this legend found its genesis in Baigent, Leigh and Lincoln's *The Holy Blood and the Holy Grail* (1982) and Baigent and Leigh's follow up book *The Temple and the Lodge* (1989). In their 1989 work Baigent and Leigh speculate the Templars found safe harbor in Scotland. There they used some of their treasure to form the Freemasons. Author Andrew Sinclair extended the legend in his 1992 book *The Sword and the Grail*. In that he claims the Templars didn't stay long in Scotland but kept sailing west until they managed to discover America about a century before Columbus. After stopping off in Nova Scotia to presumably bury their booty on Oak Island, Sinclair asserts they headed down to New England and settled in Rhode Island. There they built the Old Stone Mill tower in Newport. That particular claim is problematic as it's well documented the tower was built a couple centuries later by Benedict Arnold. There is, of course, no evidence any Europeans, other than Vikings, made it to North America before Columbus. The only record of any Templars making off with treasure is found in the testimony of Jean de Châlons. De Châlons, under the pain of torture, testified he heard one fellow Templar managed to bolt with another Templar's treasure. In the same testimony, de Châlons reported he heard several other Templars had managed to escape in a fleet of 18 oared vessels. These oared vessels, designed for the coastal waters of the Mediterranean, have, in the retelling by pseudohistorians, become ocean going sailing ships loaded with treasure. Nowhere in de Châlons testimony does he indicate any treasure was aboard. Even if that were the case, it is doubtful Templars were capable of rowing themselves from Scotland to Nova Scotia and then down to Rhode Island in shallow bottom coastal vessels.

Eight "Treasures" Found on Oak Island

So-called reality shows like *The Curse of Oak Island, Finding Bigfoot, Chasing UFOs,* and a legion of ghost hunting shows go on for years. The most ardent followers of these shows always seem to miss the fact that the hosts never find what the shows are claiming to be searching for. This is despite years of looking by so-called experts with excellent financing and the latest equipment. No one ever concludes "after a well-financed, extensive search, we have to admit the thing we're looking for probably doesn't exist". The cast of *The Curse of Oak Island*, led by the Laginas brothers, has spent the better part of a decade looking for a fabled treasure buried on a rather small island off the coast of Nova Scotia. They've brought to bear excavating technology unavailable to the ancient people who supposedly buried the treasure. They've yet to find a treasure that experts say simply doesn't exist. But they've found, well, stuff.

A SPANISH COPPER COIN

After a whole season of digging, the Laginas brothers found a Spanish copper coin dated to the 17th century. You can pick one of these up on eBay for about $20. Spanish money was actually the coinage of choice in Nova Scotia as late as 1817. The province traded a lot with the Caribbean. As well, many Nova Scotians found work as privateers raiding Spanish shipping in the West Indies.

18TH CENTURY BRITISH COINS

In season two, team members sweeping the island with metal detectors found two 18th century King George II copper coins. Such coins were usually used to pay the wages of British soldiers. It would not be unusual for Oak Island to host a contingent of British troops en route to the mainland.

Photo Credit: Karl Mamer

A "TEMPLAR" COIN

In season two, a couple episodes after finding the King George II copper coins, the crew found a piece of tin they alleged was a Templar coin. The bit of tin, even after a good cleaning, is so eroded it's not clear what it is. People knowledgeable about the real Templars suggest the group wouldn't have made an early and dangerous transatlantic crossing to hide a stash of tin coins. Most likely what was found was a medallion of some sort, possibly something that adorned a bottle of wine produced by a monastic order.

A SOUVENIR SWORD

Season three trotted out J. Hutton Pulitzer who was offering a Roman sword that was allegedly found off the coast of Oak Island in the 1940s. How did a sword from about 300 AD make it to North America? A scientific analysis of the sword's metal indicated the sword was actually made in the late 19th century. It turns out the sword is an old souvenir, something people bought during a trip to Italy back in the day.

A LEAD CROSS

In a season five episode, the brothers discovered a lead cross. The episode intimated it might not even be a lead cross but a symbol of the Phoenician goddess Tanit. This, of course, let the show introduce a whole new theory that ancient Phoenician buried the Ark of the Covenant on Oak Island. The season five episode tilted back to its favorite Templar theory when the cross's metal was dated to the 12th or 13th century. Basically, anything from the 12th or 13th century instantly means Templars! Of course, people just don't throw out lead. They melt it down for generations and make other stuff out of it, like crosses. Lead crosses are found from time to time by excavations across Nova Scotia. They were popular items with Acadians who inhabited the area.

A BROOCH

Near the end of season five, the team found what might be the only semi-valuable thing on the island: a small silver brooch set with a red garnet. The piece was dated to the 16th century. This led the show to wildly speculate that what was really buried on the island was not the Ark of the Covenant, Templar treasure, Sir Francis Drake's gold horde, the lost treasure of Captain Kidd, or the works of Shakespeare written in the hand of Francis Bacon but treasure stashed on the island by Marie Antoinette's maids.

A DECORATIVE LOCK PLATE

The team's dreams of finding treasure chests is buoyed by the discovery of a lock plate, something that might have been on a treasure chest. Or someone's luggage.

A SILVER BUTTON

Season seven opened with the grim news that the Laginas brothers' partner Dan Blankenship had died. But good news quickly followed. A metal detecting foray turned up a silver button. A conservator estimated the button dates from a time just before the American revolution.

RELIGION &
PHILOSOPHY

Twenty Seven Times the World Was Supposed to End

There always seems to be a significant portion of the population that believes they are living in the end times. Predictions about the end of the world come and go with regularity. Here is a list of failed predictions.

500 AD

Hippolytus of Rome (died 235 AD) was an important early Christian theologian. A lot of his interpretations of the *Book of Revelation* have influenced modern Christian eschatology. He is generally identified as the first person to predict the second coming based on mathematical calculations. He reasoned in the 3rd century that the second coming would come 6,000 years after the birth of Adam. Why 6,000 years? Well, his god made the earth in six days and everyone knew a biblical day equals 1,000 years. So, the world had a shelf life of another 6,000 years. And since everyone knew Adam was born in 5,500 BC, the second coming would be in 500 AD. It didn't happen.

801 AD

Spanish theologian Beatus of Liébana (died 798 AD) in his 776 AD book *Commentary on the Book of the Apocalypse* is another doomsayer to take up the math challenge for predicting the second coming. Like Hippolytus, Beatus subscribed to Earth's 6,000 year best before date. In the study of eschatology this 6,000 year "times up" is known as the sabbatical millennium. Since Beatus lived a couple hundred years after the world was supposed to end in 500 AD, he took another look at the calcs. He factored in a few things like the Holy Spirit's jersey number and came up with a revised date of 801 AD. Beatus of Liébana croaked a few years short of the mark. If he had lived, he would have discovered the end didn't happen.

1000 AD

In the run up to the Year 2000, it was conventional wisdom that much of Christendom was approaching a state of near panic that the world was going to end January 1, 1000. The claims were, however, overblown. Your average European was a peasant toiling in the fields. They had little concept of the actual calendar year. Even the church itself seemed not overly concerned by the Year 1000. Surveys of Papal bulls written between 970 AD and 1000 AD reveal no mention of an imminent apocalypse. However, there is evidence religious leaders in England were warning of the imminent rise of the Antichrist in the years running up to Y1K. The Bishop of London, known by the singular name Wulfstan (died 1002 AD), estimated Satan's 1,000-year time in the clink was coming to an end about 1000 AD or a few years after. It didn't happen.

1284 AD

Here a pope did go on record. Pope Innocent III didn't like Islam very much. He called for the Fourth Crusade. He also figured the founding of Islam was tied to the end of the world. Innocent III calculated the world would surely end 666 years after the founding of Islam, which put the end of the world at 1284 AD. Innocent III died in 1216 so he didn't live long enough to see that it didn't happen.

1524-1526 AD

Thomas Müntzer (died 1525) was a protege of Martin Luther in the late 15th and early 16th centuries. Whereas Luther cozied up to the wealthy and powerful, Müntzer sought his power base among the German peasant class. Müntzer subscribed to theologian Joachim of Fiore's idea that history plays itself out in three ages which parallel the Trinity. The first age is the time of the Old Testament and the father god. The second age is the time of the New Testament. The Second Age began with the advent of Jesus and closed out in 1260 AD (chosen because *Revelation* spoke of two prophets who will scuttle about for 1,260 days and, well, 1,260 days surely meant the year 1260). The third age is the age of the Holy Ghost and the end times. Müntzer believed he was living in this third age of Christian history and had a messianic role to play in the final chapter. He figured that chapter was destined to end on page 1524 AD or thereabouts. His reading of the Bible, however, indicated the world wasn't going to end itself. It needed a guy like him to start it. In Müntzer's view he was the "sickle of God". He organized a group of German peasants to rise up against the oppressive ruling class. They took as their symbol a rainbow flag. While Müntzer believed he had both destiny and a god on his side, he seems to have forgotten the oppressive ruling class had forts and cannons. His Apocalyptic Tripwire Force planned to attack the town of Frankenhausen. His peasants hesitantly turned out on the field of battle, lined up against government troops who brought real weapons like those aforementioned cannons. The peasants were hesitant to attack, despite Müntzer's assurance he could protect them because he had the miraculous ability to catch cannonballs in his shirt sleeves. Things might have ended peacefully after the government offered a parley, but the peasants noticed a rainbow had descended over the town. They took this as a sign from their god that victory was assured and spoiled for a fight. The government obliged. Müntzer's peasant army was quickly slaughtered by cannon fire. History doesn't record Müntzer as having caught even one cannonball. Müntzer survived the slaughter of his followers long enough to be captured, tortured, and then executed. His world came to an end. The world didn't.

1666 AD

People obsess over the meaning of *Revelation*'s "number of the beast". Most versions of *Revelation* cite the number as 666 although the oldest known version of the book cites it as 616. An extremist group of Puritans in England called the Fifth

Monarchists were hung up on the 666 variation and were convinced things would wrap up in 1666 AD. They tried to stage a coup in London in 1661 AD. Their plan, if you can call it that, was to take the city in the name of "King Jesus". They failed, as did their prediction about the end of the world in 1666 AD. That said, 1666 AD was not the best year for Londoners. In 1666 AD London was not only in the middle of a massive plague but in September of that year was the Great Fire of London.

1673 AD

William Aspinwall (died 1662) was an early American colonist, arriving in Boston in 1630. There his religious nuttery helped divide the settlers. America's welcome mat was withdrawn for him. He returned to London in 1662 and joined the Fifth Monarchists to stir it up again. Always the contrarian, he seemed to doubt the accuracy of the 1666 AD prediction. He more reasonably concluded the end of the world was coming in 1673 AD because *Revelation 11* speaks about two witnesses that prophesy for 1,260 days. And since a day is a year, he simply added 1,260 years to 413 AD to get 1673 AD. Err why 413 AD? Well, that's the year he proclaimed Christendom had entered a spiritual Babylonian captivity. History records 1673 AD as being a pretty dull year. The Netherlands and Denmark signed a defense treaty. The Dutch navy did briefly seize New York City and creatively renamed it New Orange. Curiously history does not record the world ending.

1697, 1716, 1717, 1727 AD

The American Puritans were quite hung up on the end times. None more so than Puritan Minister Cotton Mather. Mather was a central figure in the Salem Witch Trials. He helped set up courts and encouraged prosecuting women and men for the crime of witchcraft. Thinking witches were real and people needed to die for the "crime" was not the end to his nuttiness. Mather is on record for setting at least four dates for the "big one". He first calculated the end would come 1697. When that passed without lights out, he looked at his numbers again and decided he really meant 1716. Mather reportedly spent most of 1716 "in an agony of anticipation". When nothing happened, he offered 1717 as his next guess, then finally 1727. He would have probably kept on rolling the date forward if he had not died in 1728.

1795 AD

Richard Brothers (died 1824) was a British naval Lieutenant and one of the founders of a pseudo-historical claim called British Israelism. This is a notion that some collection of citizens of the UK and parts of Canada are descendants of the ten lost tribes of Israel. Brothers believed it was his job to uncover these "hidden Jews" and help them return to Israel before his god destroyed the world's nations and made Brothers, conveniently, "Ruler of the World". You may wonder, reasonably, what qualified Brothers for the job of running an entire planet? Easy. Birthright. Brothers suggested he was a descendant of King David. In 1791 he asserted his god was itching to do this ASAP but was holding off as a favor to Brothers. However, that solid was temporary as Brothers claimed the world was going to end November 19, 1795 come hell or high water. Turns out Brothers was not crowned ruler of the world. Because he kept writing letters to politicians proclaiming the imminent death of the monarchy, he was arrested for treason and found insane. While in an asylum, he was able to publish pamphlets and gained an influential following. However, his following mostly abandoned him when November 19, 1795 passed without the end of the world.

1814 AD

Joanna Southcott was born in England in 1750. Near the end of the 18th century, she became convinced she was a prophet and the woman of the apocalypse as described in Chapter 12 of the *Book of Revelation*. She also became convinced (at age 63) she was pregnant. Given she was a virgin, she concluded she was pregnant with the messiah and would give birth on October 19, 1814. Those concerned they might not have a place in heaven after the second coming could buy a holy seal (a piece of cardboard with some verse) from her that absolutely guaranteed the purchaser was one of the 144,000 elect. She attracted quite a following. About 100,000 people described themselves as adherents to the Southcottian Movement. She did not, in fact, give birth to anything on October 19, 1814. She did, in fact, die a bit more than two months after her due date. Her followers, however, were not fooled. They were certain she was simply in a trance. They hung on to her corpse for about a month after her death before they realized people in trances don't rot. Unable to deny she was what John Cleese would term "a dead parrot" they reluctantly agreed to an autopsy. Yep, she was not only dead but there was no evidence she was ever pregnant. The Southcottian Movement did not die off with news of her death or revelation she was never pregnant. Her followers assumed the dragon of *Revelation* had snatched away the baby messiah. The group splintered into a core of "purists" (the Southcottian Society) and a breakaway group called the Panacea Society. The Panacea Society continued as a going concern until 2012.

1836 AD

Some regard the Lutheran clergyman Johann Albrecht Bengel (died 1752) as one of the greatest biblical scholars of the 18th century. He is known for publishing a Greek translation of the New Testament and an influential commentary on the work. Bengel liked to dabble in math and tried to use it to calculate the end times. Bengel thought he had figured out there were two timelines operating within the Bible. Normal time and "prophetic time". Calculating prophetic time was really easy. Just take the number of the beast (666) and divide by the 42 months the beast is allowed to exercise authority (*Revelation 13:5*). That gives you 15.86 years. So, feed that back into the Apocalypse-O-Tron 9000 with 96 AD (the year the *Book of Revelation* was written) as your starting point along with some other milestones from *Revelation* and the *Book of Daniel* and, boom, you get 1836 AD. June 18th to be exact. Given Bengel was born in 1687, he knew sure as shit he wasn't going to be around to be proven wrong. Not many people in the year 1836 AD would have known about Bengel's end of the world claim if it were not for two more widely read contemporaries of Bengel: John Wesley (died 1791) and Johann Heinrich Jung-Stilling (died 1817). Wesley co-founded Methodism in England. Johann Heinrich Jung-Stilling was a German author who attracted a cult-like following of Mennonites in the late 18th century. Both promoted Bengel's idea that the world would end in 1836. While Wesley preached the end date from the pulpit, Jung-Stilling worked it into a popular mystical novel he wrote called *Das Heimweh*. In this novel, Jung-Stilling's protagonist Christian von Ostenheim escapes the coming 1836 apocalypse by leading a small band of Mennonites to a mystical land in Central Asia called "Solyma". Solyma is an unspoiled land the Antichrist cannot reach. Mennonites, thrilled they were central characters in a novel, adopted the book as a second Bible. As 1836 drew near, thousands of Mennonites figured this mystical land of Solyma was actually the Russian Caucasus and moved there to sit out the end times that never came.

1844 AD: THE GREAT DISAPPOINTMENT

The Great Disappointment is probably one of the best-known failed end-of-the-world predictions. Like others before him, Baptist preacher William Miller (died 1849) dipped into the *Book of Daniel* and *Revelation*, used the day/year principle, and concluded Jesus would return without a doubt on March 21, 1843. Or maybe March 21, 1844. Either or, on one of those days Jesus was destined to come out of the clouds and cleanse the world with fire. Miraculously the elect would be spared. It didn't happen. Miller's followers (numbering in the hundreds of thousands) had bet the farm, some literally, on the world ending. Most of his followers seemed to accept the prophecy business wasn't beyond a math error here or there. When the world continued as normal after March 21, 1844, his followers settled for a wait of a few more months. They set their hopes on a third calculation. This time, totally guaranteed, they would see the world cleansed by fire on October 22, 1844 (which coincided with Yom Kippur). This one wasn't actually set by Miller but somewhat foisted on him by his followers who, at this point, had no Plan B if the world didn't end. They had given up jobs, family, and businesses. Through all this, the general public mocked Miller's followers ("Millerites") quite openly. Miller's critics were proven right. The world didn't end. Thousands of Millerites were reported to be seen openly weeping and in a state of great despair. Undaunted, many decided Jesus had returned but in a spiritual sense and went on to form the Seventh Day Adventist church. Although Miller's prediction failed and he was roundly mocked by other Christians of the day, Miller's major contribution to Christian eschatology was the idea that true Christians would be taken up to heaven and not have to experience the horrors predicted in *Revelation*.

1881 AD

Mother Shipton was reputed to be a soothsayer that lived in the 16th century. Some sources suggest she died in 1561 AD. There is no contemporary documentation that attests to her existence. There is no written mention of her until nearly a century after her ostensible death. In 1641 AD a book called *The Propheceyes of Mother Shipton* was published. It was alleged to be a collection of Shipton's visions of the future, written in a Nostradamus-like cryptic verse. It was noted for a number of stunningly accurate postdictions, notably the death of Cardinal Wolsey in 1519 AD. Oddly no post-1641 AD vision proved to be accurate. Over the centuries a small cottage industry of authors published their own bound collection of newly found Shipton prophecies. Like the original published in 1641 AD, they were notable for stunningly accurate postdictions but Shipton's powers of prophecy always seemed to abandon her after the collection's publication date. In 1862 AD, Charles Hindley, a bookseller in Brighton, UK, republished a book of Shipton prophecies collected by Richard Head in 1684 AD. Head himself republished *The Propheceyes of Mother Shipton* with additional material. Hindley, in kind, tacked on prophecies of his own. One prediction he wrote up for the book had this couplet "The world to an end shall come / In eighteen hundred and eighty one". Hindley might have been a good bookbinder but he was no poet. Many people not understanding the book was filled with postdictions took quite seriously the supposed prophetic powers of Mother Shipton and hunkered down for the end in 1881 AD. The panic seizing a significant portion of the UK population caused many newspapers of the day to publish skeptical takedowns of Hindley's hoax.

1914 AD

The Jehovah's Witnesses' precursor The Watchtower Society made a number of end-of-the-world predictions. The cult's founder Charles Taze Russell (died 1916) believed an invisible Jesus had returned in 1874 AD and set a best before date for October 1914. Russell predicted Israel would be restored and multiple nations would attack it. This would culminate in the destruction of all world governments. When that didn't exactly happen, Russell did point out World War I had started and that was certainly a sign of the end times. He kicked the apocalypse can down the road a few years and confidently predicted Christendom would get its ass handed to it in 1918.

1934 AD

Walter Marks (died 1951) was an Australian parliamentarian. As a well-to-do lawyer and championship yachtsman, nothing about him screamed "nutbar". Things changed, however, when Marks read some kook lit written by a Christadelphian lecturer named Charles P. Wauchope. In 1921, Wauchope published a couple books based on end-of-the-world lectures he was giving to his fellow Christadelphian congregants. His view was basically Palestine was about to go to shit and the British

Empire was tasked with charging in there and winning the battle of Armageddon. Armageddon was slated for 1934. It's not certain how Marks happened upon Wauchope's writings but one thing is clear: They both ran in the same social circles in Oz. Wauchope was chairman of Australia's legendary Balfours commercial baked goods empire. In the middle of a November 1921 debate in parliament on the state of Anglo-Japanese relations, and fears that Japan was a looming threat, Marks decided he would let everyone in on a little secret. There was never going to be a war in the Pacific involving the Japanese. He knew because of Wauchope's shocking predictions. Armageddon was slated for 1934 and the Japanese and Chinese were going to ally with Anglo forces in the Battle of Armageddon. Seeing what he described as "smiles" on the faces of his fellow elected officials, Marks urged skeptical members of Parliament to read Wauchope's works. If they only read what he had read, they would surely agree with him: Armageddon was slated for 1934. Marks suggested he could talk for hours on the topic but he was mercifully silenced by the Speaker who pointed out he had exhausted his allotted time. Turns out the world didn't come to an end in 1934 and history records there was, in fact, a war in the Pacific.

1935 AD

Wilbur Glenn Voliva (died 1942) was quite a piece of work. Not only was he an influential evangelical preacher, an evolution denier, and a flat Earther but he also owned a radio station. He was actually the first evangelical preacher to own his own radio station. He used it to broadcast his flat Earth rants. In 1906 he was elected "General Overseer" (I'm not making that term up) of the Christian Catholic Apostolic Church. Since the church owned a company called Zion Industries and Zion Industries owned 100% of the land in its company town of Zion, Illinois, Voliva was effectively the town's mayor, chief of police, and head of the school board. Being in charge of the police he made it very difficult for other churches to set up shop in town. Any attempt to do so was met by overwhelming police harassment. And being in charge of Zion's schools, he required them to teach flat Earth. Voliva managed to amass quite a fortune running these endeavors. His worth was estimated to be north of $5 million in 1927 dollars. That would be close to $100 million today. He spent part of his fortune on, ironically, a round-the-world tour in 1931. Visiting some 19 countries, he not only preached flat Earth nuttery, but he started claiming the world would end in 1935. Shortly after announcing the end would come in 1935, he seems to have revised the date to 1940. He toggled back to 1935 when, in spring of the same year, he was ousted as Zion's General Overseer and a legit secular mayor won a general plebiscite. The ouster was motivated by the flagging fortunes of Zion Industries and the mysterious disappearance of company funds. Predicting the end after an election defeat wasn't just sour grapes. Voliva and his few remaining followers truly did hunker down on September 28, 1935 expecting a death toll so large he predicted it would form a 160-mile-long river of blood. It didn't happen. In what can only be described as the greatest act of turning the knife after an election victory, the newly elected government added a globe Earth to Zion's license plate stickers. This effectively forced Voliva to advertise the reality of a round Earth on the back of his car for the last few remaining years of his life.

1936 AD

Herbert W. Armstrong (died 1986) was another early pioneer in the world of religious radio broadcasting. He was also a big proponent of British Israelism (the notion that the real lost tribes of Israel are hidden among the populations of the UK, Canada, and the USA). He is probably best known for his *The World Tomorrow* broadcast (which later evolved into a syndicated TV show of the same name) and his *The Plain Truth* magazine. Pickup boxes offering free copies of *The Plain Truth* were a common sight for decades at most major bus stops in North America. Armstrong started as an advertising man until he lost his job in the Great Depression. He eventually concluded he was the last apostle of Jesus, formed the World Wide Church of God, and proceeded to tell the world we were living in the last days. Terrible things were always just on the horizon. For example, Armstrong predicted in 1965 the Pope was going to bring Hitler back to life. Unless you believe the History channel, Hitler is still quite dead. His first prediction, given in 1934, was for 1936. That didn't happen. In 1940 he revised the date to 1943. By the 1950s he seems to have learned to stop setting dates only a few years in the future and giving his contemporaries more ammunition against him. In 1953 he suggested a range of dates between 1972 and 1982. In 1956 he set 1975 as the most likely year World War III would begin. He predicted by 1975 society would be dominated by something he called "push-button leisure". The worst part of your day will be the personal helicopter ride you enjoy to work. This easy life won't last long. It will start a new Babel and Armstrong's god was going to punish us for it through starvation and World War III.

1954 AD

Dorothy Martin was the leader of a UFO cult called the Seekers. She had previously studied Dianetics before joining the Contactee movement. Martin told her followers, via automatic writing, she was warned by aliens from the planet Clarion that the earth was going to be hit by a global disaster on December 21, 1954. Faithful Seekers were told not to worry. On December 17 (4 PM to be exact), they would be rescued by UFOs. A group of sociologists (Leon Festinger, Henry Riecken, and Stanley Schachter) got wind of the group's doomsday prophecy and sought to investigate the psychological phenomenon of cognitive dissonance. Festinger wondered how would the cultists deal with the world not ending at the appointed time? They gained the trust of Martin and were given access to the group's vigils. Disturbingly, many of the cultists quit jobs, left their families, and gave away money and personal possessions they were certain they wouldn't need on Clarion. The group assembled in Martin's backyard on December 17. Martin instructed them to remove any pieces of metal they had on their person as they would not be able to board the rescue saucers. Apparently, this also included metal in women's bra clasps. The saucers didn't come at 4 PM. The group hung around the backyard (this was Chicago in December so it couldn't have been the most pleasant of waits) until 5:30 PM before retreating to the warmth of Martin's living room. The group tried to figure out what went wrong and concluded December 17 was a practice run. The saucer men wanted to see if they were actually committed to their departure. Martin had the Seekers reconvene the next day but again the saucer men were no shows. They were sure cutting it close. After all, the world was ending in just three more days. Martin once again assured her group they were coming. In fact, they had just sent her a message from Clarion that they would not disappoint Martin. The group hung on until the prophesied day of the disaster. Again, the rescue saucers didn't come. Group members were reduced to tears. And then it happened. Martin's automatic writing kicked in with a message from the saucer people. Turns out Martin's group had proven themselves so faithful and devoted (and willing to go braless) that their god decided to stop the disaster and save the entire world. In 1956, Leon Festinger et. al. went on to publish these events in a landmark work called *When Prophecy Fails*. Martin got out of the eschatology biz, recast herself as Sister Thedra, and formed a group of New Age Channelers called the Order of Sananda and Sanat Kumara. Martin died in 1992.

1975 AD

The Jehovah's Witnesses, not having learned from their days as The Watchtower Society, jumped back into the End Times prediction game. In the 1960s and early 1970s from the pulpit and various publications they began to warn their followers something big was coming in 1975. Could it be, I dunno, Christ's return, doom for the unbelievers, or maybe the Battle of Armageddon? The math worked. Their god's rest day began in 4026 BC. Applying the old sabbatical millennium trick, that got them to 1975 AD. J-dub members who thought about leaving the church were warned they'd lose their lives in the impending Armageddon. Cult members organized under the rallying cry of "stay alive until '75". The year came and went without incident. The church, with egg on its face, went into full denial. It never said 100% the world was going to end in 1975. It just thought maybe something *could* happen in 1975. I mean, stuff happens. Why not some stuff happening in 1975? Any church member that thought the world was going to end and did stuff like canceling insurance policies and pulling their kids out of school had no one to blame but themselves. They read way too much into all the apocalyptic and doomsday language the church was using for over a decade. Silly, faithful.

1982 AD

John Gribbin is an astronomer who should have known better when he published *The Jupiter Effect* in 1974. He and co-author Stephen Plagemann (died 1998) claimed a long expected "alignment" of the planets would cause disasters on Earth. Their argument was that in 1982 basically all of the solar system's planets (which included Pluto back then) would be on the same side of the sun. This isn't something that happens in the lifetime of many people. It happens roughly every 200 years. The fact nothing happened 200 years ago should have been a big tip off their proposed "effect" was complete bunk. Now, any astronomer worth his finely ground lenses should have been able to calculate the tidal force effect on Earth of all those planets would be negligible. Gribbin and Plagemann understood this but believed these tidal forces would cause increased sunspot activity which would create increased solar winds. Increased solar winds would slow Earth's rotation and that would trigger earthquakes. The book was something of a best seller and as 1982 drew near people increasingly grew worried. In the popular interpretation, people spoke of the planets actually lining up in a straight line and this overwhelming gravitational pull would destroy Earth. Gribbin's and Plagemann's fellow scientists described their book's central thesis a fantasy. Charles Richter (as in the Richter Scale guy) called their idea "pure astrology in disguise". Even if every planet was on the same side of the sun, it would have a negligible effect on the sun's surface. Their idea that solar wind could affect Earth's rate of spin had no basis in real science. Despite overwhelming criticism from their peers, Gribbin and Plagemann doubled down a bit in 1982 offering the 1980 eruption of Mount St. Helens as evidence of the Jupiter effect. Of course, the planets weren't even close to being aligned on one side of the sun in 1980. In the late 1990s, Gribbin saw the light and retracted his Jupiter effect claims. Gribbin went on to write a number of popular legit books about quantum mechanics, black holes, and climate change.

1988 AD

Writers of books about Bigfoot, Nessie, UFOs, and the Bermuda Triangle did very well for themselves in the 1970s. A market for nonfiction woo was arguably created by Hal Lindsey and his book about the looming apocalypse called *The Late, Great Planet Earth*. Published in 1970, within three years it sold over 3 million copies and by the mid-1980s over 16 million copies were in print. In the book, Lindsey argued the founding of the modern state of Israel in 1948 started a 40-year apocalyptic countdown. In Lindsey's reading of the Bible, with the founding of Israel there could only be one more generation before Jesus returned. His book tried to map a lot of the crap going on in the 1970s onto Biblical predictions of what the run up to the end times should look like. Like many before him, he ignored every age is full of misery and it's super easy to map world events onto your own revenge fantasy reading of the Bible. Although Lindsey avoided setting specific dates, he told at least one audience in the 1970s Jesus could return by 1977. His 40-year countdown had history coming to an end by 1988 at the latest. In a sequel, published in 1980, Lindsey doubled down and suggested the decade of the 1980s could be the last.

1990 AD

In the late 1980s Elizabeth Clare Prophet (died 2009) moved her Church Universal and Triumphant HQ to a ranch in Montana. There she had her followers construct a massive bomb shelter out of used oil tanks. Prophet warned her followers there was going to be a nuke war between America and Russia on March 12, 1990 and they'd need those bomb shelters to survive. She knew this because she channeled various ascendant masters, which included the regular cast like Jesus and Buddha, plus a few curious ones like Merlin and Sir Lancelot. Many people who knew Prophet but had not drunk her Kool-aide came to the conclusion she was certifiable. Those who had drunk her Kool-aide were happy to toil away on a slapped together bomb shelter that was intended to house hundreds of followers for months on end. When the appointed day arrived, her followers filled the shelter to capacity. March 12, 1990 passed without incident. Prophet, like most doomsayers, had a solid explanation. Turns out March 12 wasn't actually the end of the world. It was all just a drill. But it did mark the start of a 12-year cycle in which the Four Horsemen of the Apocalypse would stalk the planet. They are still stalking.

2004/2014 AD

Our old friend Joanna Southcott, who predicted she would give birth to the Messiah in 1814, left some writings that modern day followers interpreted as meaning the world would come to an end in either 2004 or 2014. Fortunately, Southcott did not leave them unprepared. Several years before her death she left a sealed box she said contained a collection of prophecies that could be opened by her followers to help the UK through a coming time of crisis (very Hari Seldon!). However, her orders stipulated the box couldn't be opened by just anyone. It had to be opened in front of every bishop of the Church of England. During the Crimean War and then the First World War her followers urged the church to open the box. The church steadfastly refused to waste time on what was little more than a cult's publicity stunt. Today there seems to be three different groups that claim to have Southcott's box. Psychic researcher Harry Price (died 1948) claimed he had the box. He managed to get one reluctant Church of England bishop to witness its opening. The box apparently contained nothing prophetic. Its contents included some random papers, dice, a lottery ticket, and a pistol. Others claimed the box was passed on to Annie Stitt, head of the Southcottian Society. She gave the box to the British Museum in 1966. It is said the museum opened that box and found a collection of inconsequential papers. The museum itself disputes this account and has said it has no record of ever receiving such a box. The Southcottian splinter group the Panacea Society claims they have the real box and have it hidden away for safekeeping. For a number of years in the 1960s and 1970s, the society ran regular ads in the *Daily Mail* and *Sunday Express* urging the Church of England to send in the bishops to open the box. The Panacea Society itself folded in 2012, having seen several Southcottian end-of-the-world predictions pass without incident. Today all that remains is a charitable trust set up in the group's name and a museum dedicated to the society in Bedford, UK.

2011

Harold Camping (died 2013) was the owner of a Christian radio network. In 2005 he started warning his listeners that the world was going to end May 21, 2011 with the second coming of Jesus. This wasn't the first time he predicted the end of the world. He had previously predicted the end would come September 6, 1994. Camping's 2011 date came from his calculation that Noah's flood was in 4990 BC. To that he added 7,000 years based on a quote from *Genesis 7:4*: "Seven days from now I will send rain on the earth". As we've seen these Biblical literalists always interpret a day to mean 1,000 years. In 2010, followers of Camping's prediction kicked into overdrive. Many self-financed billboard campaigns proclaiming the end on May 21, 2011. Several followers ran up huge credit card debits they assumed they wouldn't need to pay back upon the coming of their lord. As most sane people predicted, May 21 came and went without fire or brimstone. Camping refused to comment for several days. He eventually returned to the microphone and assured his misled followers May 21 was, in fact, judgment day just like he predicted. Thing was, no one noticed because it happened in secret. A small math error had him confuse this secret spiritual judgment day with the actual physical end of the world. Not to worry, that would come October 21. That didn't happen either. Camping's radio network quickly removed from its online archive any episode where Camping predicted the end of the world. Camping's own world came to an end on December 15, 2013 when he fell at home and died from complications.

Photo Credit: Karl Mamer

How the World Ends According to Nine Different Religions

BUDDHISM

Buddhists generally have no actual beginning or end. Time is an endless cycle. Chinese Buddhism did, however, adopt aspects of local apocalyptic beliefs. Chinese Buddhism speaks of the return of the Buddha (known as the "Maitreya") at a time when all knowledge of Buddhism is lost. The lost years will be marked by the collapse of society and a whole lot of sexual depravity. The Maitreya will restore Buddhism and bring about a virtuous world.

HINDUISM

Hinduism buys us some time. Hinduism teaches time is a cycle, each lasting upwards of 8 billion years. Each cycle culminates in a decline and then a restoration. After about 311 trillion years of these cycles, the universe will collapse into a singularity and then be reborn. Cyclical declines are characterized by the arrival of a Vishnu avatar named Kalki. Astride a white horse, he will lead an army to restore righteousness and begin the next cycle.

ISLAM

Similar to Christianity, many Muslim sects believe the end will involve a false messiah named Al-Masih ad-Dajjal engaged in a battle with Jesus. After Jesus puts down the false messiah, the sun will rise in the west and the dead will be reincarnated. All will then face the Day of Judgment.

JEHOVAH'S WITNESSES

Jehovah's Witnesses believe we are actually living in the end times. It all kicked off in 1914 and will end in a great tribulation. The great tribulation has been "imminent" since 1995. But when it does finally come, Jehovah will use the UN to destroy all false religions. The saved, that is the J-dubs, will live out their lives forever on an Eden-like Earth.

JUDAISM

According to Judaism, the temple in Jerusalem will be restored and a messiah from the House of David will be appointed king. This messiah will lead the world into an era of peace. The dead will then be resurrected. All will then face judgment.

MORMONISM

Mormons believe wars and earthquakes will increase before Jesus returns. Jesus will raise the dead in two phases. The righteous (the best of Mormons) will be raised and given the reward of eternal life in a business class heaven. The rest will then be resurrected and sorted into either a premium economy heaven or an economy heaven. Those who were good but just weren't hip to the Mormon thing will go to premium economy heaven. Non-Christians and generally bad people (liars, whoremongers, and sorcerers) will get thrown back into coach.

RASTAFARIANISM

Rastafarians believe Ethiopian monarch Haile Selassie (died 1975) will return from the dead, judge the living, and bring Rastafarians and the descendants of those taken during the slave trade to Mount Zion in Africa. There everyone will live in peace and harmony.

SIKHISM

Sikhs don't view the universe as having a beginning and end, at least one that can be defined and comprehended by the human mind. Sikhism urges followers to not worry about issues ultimately beyond them and concentrate on just being good in the day to day.

TAOISM

According to Taoism's Divine Incantations Scripture, the coming of a messianic figure named Li Hong will signal the end of the world. Before the end, Li Hong will rescue the virtuous.

ZOROASTRIANISM

Zoroastrians have one of the oldest end times mythologies. There will be a battle between good and evil. A savior will emerge and resurrect the dead. Judgment day is pretty hardcore. Lava will flood the earth. The good will be unharmed but the wicked will burn.

The Four "Best" Arguments for God

ARGUMENT FROM DESIGN

This is sometimes referred to as the teleological argument. A watch implies a watchmaker. Hence a biological organism or something that looks designed like the eye must have a designer. To a theist, it seems improbable that random processes could assemble even one complex protein, let alone something like an eye. To suggest the eye could be a product of chance is, to theists, to suggest if I toss all my daughter's Lego pieces in the air at once, they could conceivably land in such a way as to self-assemble into Buckingham Palace. The error here is theists assume organisms and biological features only come about by chance. Evolution has demonstrated complex biological systems and organs emerge from random mutation plus selection. Selection is not random. Douglas Adams poked fun at this argument with his puddle analogy. The puddle is amazed at how well the hole fits it and assumes the hole was created specifically to contain it.

ONTOLOGICAL ARGUMENT

The Ontological Argument was first laid down by Anselm of Canterbury in the 11th century. It goes like this: You can imagine the greatest of all possible beings (i.e., a god). A being that only exists in your mind and not reality isn't the greatest possible being. That is a god that lacks greatness in every other aspect except being real isn't truly the greatest. Therefore, since we can imagine the greatest of all possible beings and that requires a reality component, a god must exist. The problem with this argument, besides it makes almost no sense, is you have to accept a perfect god can exist. That's just not a given. If you propose "reality" is a property of an unknown entity, that is a property to be determined. It's not a property you can just bolt on and make necessary.

PRIME MOVER ARGUMENT

This one goes all the way back to Aristotle. It's also known as the First Cause argument. Motion is caused by a mover. The universe is in motion. There must be a universe mover. That is a god. The most obvious problem with this argument is you have to accept there is at least one god as an exception to this rule. If one assumes there has to be a prime mover, the argument doesn't necessitate it has to be a god of any sort. As well, why does everything have to have a cause? There is no rule that says nature has to conform to human logic. Causation is a human concept. Science has a long history of showing the universe doesn't conform to our simple understanding of how it works.

FINE TUNING

The argument goes there are a range of universal constants (the strength of gravity, the strength of the strong and weak forces, etc.) that, in conjunction, all seem fine-tuned for life. If just one was a little stronger or weaker, life could not possibly exist. For example, if gravity were a little weaker, stars and planets would have never formed. If the strong nuclear force was a bit stronger, stars in the early universe would have quickly burned through all their available hydrogen. Theists will then argue the odds all of these being set exactly are so improbable that a god had to have had a hand in determining the constants. Fine tuning sounds compelling with its myriad of constants that had to be set just right. However, it's a science-y "god of the gaps" argument. Since no one can explain how a range of constants are all what they are, therefore, a god. It's hard to say how the constants came about when we don't have a good understanding of how the universe came about before the Big Bang. Naturalistic explanations include the possibility of a multiverse. Across an infinite number of universes, we just happen to live in one with constants that support life as we know it. Alternatively, the universe may have been cyclical. It can't fully expand and quickly collapses back on itself until the constants are within some relationship with one another. The argument is also a false dichotomy: the constants are either random or require a god to set them. There may well be a deeper pre-Bang physics we don't understand and these values are simply a necessary byproduct.

The Six Deadliest Cults

Cults seem to offer the worst facets of religion and somehow lure hundreds or thousands of people who are just dying to be right.

AUM SHINRIKYO (BODY COUNT: 12 INNOCENT SUBWAY COMMUTERS)

This Japanese cult viewed themselves as the embodiment of Isaac Asimov's Foundation. They viewed themselves as a group of psychic scientists who would continue civilization after the apocalypse. The problem with apocalypses is they don't ever come in your lifetime. Some cults figure it's best to help it along by committing horrible acts of violence. Aum Shinrikyo in 1995 attacked the Tokyo Subway system with home-brewed sarin gas. Their attack killed 12 commuters and injured over 1,000 others.

HEAVEN'S GATE (BODY COUNT: 39 ADHERENTS)

In 1997 this UFO cult, led by Marshall Applewhite, topped themselves by ODing on a mixture of apples sauce, Phenobarbitals, and vodka. For good measure they tied plastic bags around their heads before lying down for their awaited journey to a UFO they thought was following comet Hale-Bopp. Applewhite, a former music teacher, was one of the last four to die. When cops were tipped off of their mass suicide by an anonymous caller, they found the dead cultists all dressed in matching track suits and Nike "Decades" sneakers. They also found many of the members had castrated themselves several months before their suicides. Apparently, this was an attempt to suppress their sexual desires. Many have wondered about the cult's choice of Nikes for their so-called "away mission" uniform. The choice was not symbolic. The group wanted matching sneakers and were able to buy the Nikes in bulk for a reasonable price.

ORDER OF THE SOLAR TEMPLE (BODY COUNT: 74 INCLUDING ONE INFANT)

Nothing good comes when a neo-Nazi decides the second coming of Jesus is right around the corner. The Order of the Solar Temple was founded by neo-Nazi Luc

Jouret in the late 1970s. He believed his cult was charged with the duty of uniting Christianity and Islam. After knocking that off, they were to prepare the world for Jesus to return as a solar god-king. It turned out not to be that easy to get Christian and Muslim leaders to listen when you're a neo-Nazi who openly believes in UFOs and New Age mumbo jumbo. Amazingly, however, Jouret did manage to attract a large number of rich and influential members of European and Quebec society. Members tended to be older, accomplished, and affluent. The cult included Patrick Vuarnet (son of fashion mogul Jean Vuarnet), Robert Ostiguy (a mayor of a Quebec city), Edith Bonlieu (an Olympic downhill skier), Michel Tabachnik (a former conductor of the Berlin Philharmonic), Jocelyn Grandmaison (a reporter for a Quebec newspaper), and several influential employees of Quebec Hydro (which basically controls the power supply of America's East Coast). The cult split its time between a so-called temple in Quebec and a Swiss chalet. For reasons not super clear, in 1994 Quebec cult members decided a couple within their fold had recently given birth to the Antichrist and they stabbed the three-month-old baby, mother, and father to death with wooden stakes. This precipitated a chain of events that led 53 cult members to kill themselves over the space of several days. Cult members killed themselves by drinking poison, shooting themselves, and burning themselves alive in their Swiss chalet. Those unwilling to join with space Jesus were hunted down and murdered before the mass suicide. Surviving cult members didn't actually survive for long. Sixteen more topped themselves in 1995 in France with guns and self-immolation. That still didn't bring about the return of space Jesus. So another batch tried again in 1997, resulting in the death of five people.

BRANCH DAVIDIANS (BODY COUNT: 86 INCLUDING FOUR ATF AGENTS)

ATF agents raided the Waco-area compound of the Branch Davidians in 1993 acting on reports the cult had stockpiled several illegal weapons and explosives. This led to a 45-minute shootout that ended the lives of four agents and five Branch Davidians. Apparently at least two of the cult members were "mercy" killed by their own after sustaining injuries in the shootout. The shootout led their leader David Koresh to conclude this was the beginning of the apocalypse he told his followers was coming Real Soon Now. A controversial follow-up raid led by the FBI resulted in Koresh ordering his followers to burn their compound down around themselves. Only about seven of the cultists made it out of the building. The 76 who either remained or weren't able to get out were variously killed by the fire and rubble from the collapsing building. A few were shot by other cult members. David Koresh, himself, apparently had assistants shoot him.

PEOPLES TEMPLE (BODY COUNT: 913 MEN, WOMEN, AND CHILDREN)

In the 1970s, Reverend Jim Jones spirited away his followers to a "paradise" in the middle of a Guyanese jungle. Jones' paradise proved less than. A wellness check by a US congressman and a news crew set Jones off. He had the politician and news crew gunned down. He urged his followers to drink poisoned Flavor Aid (not Kool-Aid). Those who refused were simply shot.

MOVEMENT FOR THE RESTORATION OF THE TEN COMMANDMENTS OF GOD (BODY COUNT: 924, ROUGHLY HALF WERE CHILDREN)

This end-times cult was formed in Uganda in the late 1980s by Credonia Mwerinde (a former prostitute who said she had regular conversations with the Virgin Mary) and Joseph Kibweteere (a failed Ugandan politician and lapsed Catholic). Early on they enlisted several defrocked Catholic priests and former nuns. The cult believed the only way to survive the coming apocalypse, slated for January 1, 2000, was to adhere strictly to the Ten Commandments. The cult had over 1,000 adherents. As the end of the millennium approached, members were encouraged to sell off their worldly goods, donate the proceeds to the cult, and prepare for the second coming at the very stroke of Y2K. Most obeyed and when the world didn't end January 1, cult members — now without jobs, livestock, or any worldly goods — started to turn on Mwerinde and Kibweteere. Like so many cult leaders who confidently name a date for the end of the world and the day passes without incident, the founders blamed it on a math error. They claimed they miscalculated and the real apocalypse would happen in just a few more months on March 17, 2000. Hang in there. Those who held true gathered in a church on that day to await this totally guaranteed end. Their end came, but not in the way they expected. Kibweteere with the aid of several high-ranking cult members had the church rigged with explosives and chemicals to accelerate fires. They barred the doors and windows and burned everyone in the church alive. Estimates of the dead are a little inexact but the figure is generally put at 530 people. Several hundred cult members who didn't RSVP for the March 17 "celebration" were apparently killed off a few days before by poison, stabbing, or having their heads bashed in. Kibweteere died in the events, possibly one of the few to die willingly. Mwerinde appears to have skipped off and gone into hiding. There's a decades old outstanding warrant for her arrest.

Ten Odd and Disturbing Cults

CHILDREN OF GOD

The Children of God cult was founded in 1968 by David Berg (died 1994). Ostensibly an evangelical sect hell bent on converting the unsaved, the cult was really a sex ring involving women and children. An early tip off the cult was dangerous to young people might have been Berg's frequent denunciations of laws against pedophilia. The group not only promoted the sexualization of children but Berg abused the children of cult members. New cult members were recruited by a "sacrament" called flirty fishing. Cult women were required to prostitute themselves as "hookers for Jesus". The tactic added some 19,000 members. Women that refused were reminded their bodies didn't belong to them. Despite the evidence of child abuse and women being coerced into prostitution, Berg never served a day in jail.

FALUN GONG

Falun Gong means "Dharma Wheel Practice". It was founded in the 1990s by Li Hongzhi in China. It's a Buddhist offshoot that holds as dogma we're living in the last days and space aliens are intervening in human development. Li thinks space aliens gave us modern technology and airplanes to undermine morality. He also swears he can levitate and walk through walls. Falun Gong members have been persecuted in China and that has given them considerable cred in the West. The group has two high-profile outreach arms: the Shen Yun dance troupe and *The Epoch Times* newspaper. The Shen Yun group's colorful dance performances weave in the group's apocalyptic mythos and anti-evolution pseudoscience. *The Epoch Times* was for a spell a reasonably apolitical free pickup until Donald Trump was elected in 2016. After 2016 it started to promote Trump as a heaven-sent savior as well as various Covid pandemic conspiracies.

LORD'S RESISTANCE ARMY

Photo Credit: Karl Mamer

The Lord's Resistance Army is a cult-like rebel group that began waging a civil war against the Ugandan government in 1987. If the group doesn't sound familiar, you may remember "Kony 2012". Kony 2012 was an attempt by filmmaker Jason Russell to manufacture a grassroots movement to urge the U.S. to involve itself in the Ugandan civil war. Russell's apparent goal was to bring the leader of this cult army Joseph Kony to justice. Russell had a long track record as a fundamentalist Christian and Uganda at the time was trying to make being gay a crime punishable by death. It was the goal of fundamentalists to nurture nations willing to commit genocide against gay people. The reality is Kony was not much of a threat by the 2010s. He taught his cultists that smearing holy oil on their chests made them bullet proof. This defensive tactic proved ineffective, explaining why his army had dwindled to about a dozen members by the time the Kony 2012 movement kicked off.

NXIVM

NXIVM was founded by Keith Raniere in 1998. Ostensibly it was a group where members could help each other. Its teachings were partially based on the works of Ayn Rand and Tony Robbins. Sigh. It became popular with affluent people and B-list celebrities. The group had particular appeal to women. Participants included members of the Bronfman family, actress Allison Mack, and *Battlestar Galactica*'s Nicki Clyne. It quickly became a sex cult and pyramid scheme. As members moved up the pyramid, they were given different colored sashes to identify their rank. Those at the bottom of the pyramid were referred to as "slaves". Naturally, founder Raniere was at the top of the pyramid. Female members had to provide naked pictures of themselves as "collateral" as well as sexual services to Raniere. Women were also branded with Raniere's initials around their pelvic area. The group was broken up in 2018 when Raniere was arrested for sex trafficking. He was convicted in 2019 and sentenced to six years in jail.

RAËLISM

Raëlism is a UFO sex cult founded by Claude Vorilhon. Vorilhon worked as a sports reporter until he was supposedly contacted by space aliens in 1973. The aliens told him he was the latest in a line of some 40 prophets who tried and failed to properly deliver their message. You'd think super advanced aliens would get it right after ten or twenty kicks at the can. But no. He was given the name Raël and charged with not fucking up like Buddha, Moses, and Jesus before him. Vorilhon preaches a strange mix of free love and the potential of human cloning. They may be one of the only hippy-ish cult movements that fervently support GMO research. The Raëlian HQ has moved around over the years. For a time in the mid-1980s it was located just outside the Vatican when Vorilhon proclaimed himself to be the true pope. In 1992, Vorilhon relocated the HQ to Valcourt, Quebec.

RAJNEESHPURAM

Rajneeshpuram was a ranch commune founded in Oregon in 1981. It was the home of a cult founded by Shri Rajneesh (aka Osho). Rajneesh (died 1990) owned, at one point, over 90 Rolls Royce cars. Apparently giving Rajneesh yet one more Rolls was considered an act of devotion. Being a cult leader is good work if you can get it. The cult was not well liked by the locals. In response the cult worked hard to use local laws against the townsfolk. In one instance the cult brought hundreds of homeless people into their commune in an attempt to influence a local election. When that didn't work, they kicked the homeless people out and let the state of Oregon pay for their care and relocation. The cult also launched America's first biowarfare attack in 1984. Rajneesh's "enforcer" Ma Anand Sheela and other members poisoned several restaurant salad bars with *Salmonella*. The intent was to keep a significant number of locals away from the polls and ensure the cult's candidates won a county election. They managed to sicken over 700 people. Luckily no one died. Rajneesh was eventually kicked out of the USA for immigration improprieties. The cult's many hijinks were covered in a 2018 Netflix series called *Wild Wild Country*.

THAT GERMAN CROSSBOW SEX CULT

If this cult had a name, the members took it to the grave with them. Back in 2019, German police found five members of the cult dead. Three were killed by a crossbow. Two poisoned themselves. As best the police could determine, the cult consisted of Torsten Weiss and four women followers. Weiss owned a shop that sold replica swords. He also gave Medieval sword fighting lessons. Weiss and the four women members were involved in some sort of BDSM master/slave relationship. Women participants were believed to have been forced to sever ties with their family. They believed by killing themselves they would be reborn in a higher realm.

TRUE RUSSIAN ORTHODOX CHURCH

In the words of Tywin Lannister, "any man who must say 'I am the king' is no true king." I suppose that goes for cults. If you have to put "True" before your cult's name, truth is likely the least your cult has to offer. The so-called True Russian Orthodox Church was founded by Pyotr Kuznetsov. He felt the mainline Russian Orthodox Church had wandered off the path as, among other things, the church allowed members to eat processed foods. Kuznetsov would have none of that. In 2007, Kuznetsov was arrested by Russian authorities and charged with setting up a violent religious organization. In response, church members retreated to a cave to await the end of the world. They brought with them food and water along with a large number of gas canisters. The cult told authorities they'd ignite the fuel and kill themselves if the police tried to arrest them. The mainline Russian Orthodox Church sent in a crack squad of monks hoping to lure the cultists out of the cave. Unbelievably, that failed. The group did emerge when two of their members died in the cave and the smell of the putrefying corpses drove them out.

WORLD MISSION SOCIETY CHURCH OF GOD

World Mission Society Church of God (aka the WeLoveU Foundation) was founded in South Korea in 1964 by Ahn Sahng-hong. Ahn claimed to be the Second Coming of Jesus. His cult is roughly based on Seventh Day Adventist teachings. The cult attracted a fairly wide following in the 1970s and 1980s. Members had their "come to Jesus" moment when their belief in Ahn's divinity was challenged by his sudden death from a heart attack. Technically, immortal messiahs aren't supposed to die, at least not in such routine ways. Jang Gil-ja quickly stepped up. She is variously identified as Ahn's wife or Anh's "spiritual wife". It's not clear if she is the legal widow. In any regard, Jang quickly took over the reins of the church. How do you replace the Son of God? Easy, call yourself the Mother of God. Despite the title, Jang should not be confused as the Second Coming of Mary. Jang's Korean followers have stepped up efforts to recruit abroad, notably on college and university campuses. Former members report the cult follows typical recruitment tactics: take in lonely people, love bomb them, and demand large amounts of their savings and income. According to former cult members, in the run up to 2012 the cult preached the end of the world. Members were urged to give up their education and careers and prepare for the end. Post 2012, the cult denies they ever suggested that.

Five Outrageous Creationist Quote Mines

Creationists seem to believe science works by assertion. If you can compile enough supposed quotes by actual scientists saying that evolution can't possibly work, well, you've demolished evolution. Since the vast majority of scientists accept the fact of evolution, some creative editing of the quote is called for. This is known as "quote mining". By taking quotes out of context, removing the troubling affirmative parts with ellipses, creationists try to make it seem like the quoted scientist is forced to admit evolution doesn't work, the fossil record fails to show anything, or a god is actually a viable solution to the "problem" of evolution.

QUOTE MINE

"Astronomer Carl Sagan candidly acknowledged in his book *Cosmos*: 'The fossil evidence could be consistent with the idea of a Great Designer.'" Cite in *Life - How Did It Get Here? By Evolution or by Creation?* (1985). Published by the Watch Tower Bible and Tract Society.

Reality

Did Carl Sagan actually admit the fossil evidence supports the creationist "model"? No. Sagan's actual passage reads as follows with the quote mine highlighted in italics:

"*The fossil evidence could be consistent with the idea of a Great Designer*; perhaps some species are destroyed when the Designer becomes dissatisfied with them, and new experiments are attempted on an improved design. But this notion is a little disconcerting. Each plant and animal is exquisitely made; should not a supremely competent Designer have been able to make the intended variety from the start? The fossil record implies trial and error, an inability to anticipate the future, features inconsistent with an efficient Great Designer..."

This is an excellent example of how creationists simply don't read beyond the sentence or passage they think says what they hope it says or they simply fail to understand a basic literary technique used by authors of popular science books. Authors frequently introduce topics by leading with what seem like initial obvious explanations, objections, or insurmountable problems. They then work out how the problem was actually solved. A third possibility is creationists do understand what the scientist is saying but still dishonestly misquote under the belief a bit of pious fraud never hurt anyone.

QUOTE MINE

"It is entirely in line with the accidental nature of mutations that extensive tests have agreed in showing the vast majority of them detrimental to the organism in its job of surviving and reproducing, just as changes accidentally introduced into any artificial mechanism are predominantly harmful to its useful operation...good ones are so rare that we can consider them all bad." Cited in the book *Evolution Unmasked* by Laurence D. Smart.

Reality

The quote is largely from a 1955 article in the *Bulletin of the Atomic Scientist* by geneticist H.J. Muller. Creationists like to cite it in support of their notion that almost all mutations are negative and therefore it's nearly impossible to have enough beneficial mutations for anything to evolve. The problem with the quote mine is twofold. One, the final sentence isn't in Muller's original 1955 article. It's pasted in from a quip he made to *Time* magazine in 1946 about the dangers that may result to people living near a nuclear power plant if there's a radiation leak. Despite what the comic books suggest, you're not likely to get any beneficial mutations being exposed to hard radiation. Two, while some believed back then most mutations were detrimental, modern research has demonstrated the vast majority of mutations are neutral. Muller's views from six decades ago don't reflect current findings about mutations.

QUOTE MINE

"The answer is that the chances that the universe should be life-permitting are so infinitesimal as to be incomprehensible and incalculable. For example, Stephen Hawking has estimated that if the rate of the universe's expansion one second after the Big Bang had been smaller by even one part in a hundred thousand million million, the universe would have re-collapsed into a hot fireball." From a speech by creationist William Lane Craig.

Reality

Craig is referring to a passage in Hawking's *A Brief History of Time*. Similar to Sagan's quote, Hawking first sets up the problem:

"If the rate of expansion one second after the big bang had been smaller by even one part in a hundred thousand million million, the universe would have recollapsed before it ever reached its present size."

To which he then solves by suggesting:

"...the rate of expansion of the universe would automatically become very close to the critical rate determined by the energy density of the universe. This could then explain why the rate of expansion is still so close to the critical rate, without having to assume that the initial rate of expansion of the universe was very carefully chosen."

The Fine-Tuning Argument is a creationist favorite as of late. There are a number of constants in science that if they were just a little bit different would result in life as we know it being impossible. Why an all-powerful god couldn't make life happen anyway in a universe that is just vacuum or amorphous fields of particles, is rarely, if ever, addressed. The goal is simply to make it seem like life is so astoundingly improbable that even the world's best scientists have to throw up their hands at that point and go "yeah, we really don't have an answer." There are many ideas why these constants are what they are. Creationists just edit them out.

QUOTE MINE

"Charles Darwin acknowledged the inadequacy of evolution when he wrote, 'To suppose that the eye, with all its inimitable contrivances for adjusting the focus to different distances, for admitting different amounts of light, and for the correction of spherical and chromatic aberration, could have been formed by natural selection, seems, I freely confess, absurd in the highest possible degree.' (Darwin 1872)" Cited in *The Collapse of Evolution* (1996) by Scott Huse. This Darwin quote is also found in numerous creationist tracts and websites.

Reality

Yet again creationist authors fail to read past the quote they think demolishes evolution. In *Origin of Species* Darwin sets up the problem and then continues on for three pages on how a complex eye can evolve stepwise. Darwin gives numerous examples of extant animals with viable eyes that would be considered intermediate stages.

QUOTE MINE

"All paleontologists know that the fossil record contains precious little in the way of intermediate forms; transitions between major groups are characteristically abrupt." A quote by Steven Jay Gould cited on numerous creationist websites like Genesispark.com.

Reality

Gould was arguing for his theory of punctuated equilibrium. Gould argued that forms were stable for a long period of time and responded only to abrupt changes in the environment when they would evolve more rapidly. The odds these transitional forms would get fossilized and discovered was, in Gould's mind, rare. Gould believed transitional forms were rare but not absent.

The Net Worth of Ten Religious Leaders

CREFLO DOLLAR ($27 MILLION)

Creflo Dollar was born Michael Smith. He heads the Creflo Dollar Ministerial Association out of Georgia. *Forbes* estimates his wealth at $27 million. Smith is a prosperity gospel preacher and leads by example. He is known to own two Rolls Royce cars and several homes worth north of $1 million each. His church itself is not exactly poor either. It reportedly rakes in about $70 million a year in donations. His church was able to build a $20 million venue without using any bank financing. Smith probably came to many people's attention when in 2015 this vastly rich pastor called on his followers to up their giving so he could buy himself a $65 million Gulfstream jet.

BENNY HINN ($42 MILLION)

Benny Hinn's faith healing tricks have been well exposed by Randi over the years. Unfortunately, there is a never-ending supply of desperate people ready to hand over whatever little money they have left in hopes Hinn will heal them. According to *Money, Inc.* Hinn has managed to amass a net worth of at least $42 million.

JOEL OSTEEN ($40-$60 MILLION)

Joel Osteen is probably the poster boy for the so-called prosperity gospel brand of god slingers. He runs the Houston-based Lakewood Church, which is basically a former basketball arena. Estimates of his worth range between $40-$60 million. He is known to live in a home with an estimated value of $10 million. Osteen was widely criticized for refusing to take in people made homeless by Hurricane Harvey in 2017. This vastly rich pastor also applied for, and got, $4.4 million in federal Covid relief funds in 2020.

PAT ROBERTSON ($100 MILLION)

Failed presidential candidate and televangelist Pat Robertson's net worth has a wide range of estimates. The highest estimate is $1 billion but most authorities seem to agree $100 million is a safe guess. Robertson has business interests in

North America, the UK, Asia, and Africa. Holdings in Africa are reported to be a diamond mine in Zaire and a gold mine in Liberia.

THOMAS DEXTER "T. D". JAKES ($147 MILLION - $150 MILLION)

The net worth of pastor T. D. Jakes is at least $147 million. *Forbes* has him pegged at $150 million. Jakes founded Potter's House Church in Dallas, Texas. Jakes has rubbed shoulders with both George W. Bush and Barack Obama.

DAVID OYEDEPO ($150 MILLION)

David Oyedepo heads the evangelical Living Faith Church Worldwide denomination. His main church is based out of Nigeria, although he has at least 300 churches in Africa, the UK, and the USA. *Forbes* pegs his net worth at $150 million. He owns a private university in Nigeria, a publishing company, four jets, and has homes in the USA and the UK.

QUEEN ELIZABETH II ($350 - $500 MILLION)

Queen Elizabeth II is the official head of the Church of England. It's a bit hard to determine her exact personal wealth as much of what the Queen owns is ultimately owned by the state. For example, the Queen couldn't sell the crown jewels or Buckingham Palace. *Forbes* estimated her personal wealth at $350 million in 2020. Other groups have put her worth as high as half a billion.

KENNETH COPELAND ($300 - $760 MILLION)

Kenneth Copeland is by far the richest of the prosperity gospel televangelists. The *Christian Post* puts his net worth at $300 million. *Newsweek* pegs it at $760 million. He heads the humbly named Kenneth Copeland Ministries. Copeland likes to assure followers if you donate money to him, his god will give the donor a "hundredfold" return. Copeland and his ministry own a fleet of private jets and an airport. His defense for owning multiple jets is he refuses to "get in a long tube with a bunch of demons." It's unsure if he meant actual demons or people who can only afford to fly coach. During the Covid pandemic, Copeland urged his followers to keep paying their tithes even if they lost their jobs.

EDIR MACEDO ($1.1 BILLION)

Edir Macedo is a wealthy Brazilian businessperson and founder of the Universal Church of the Kingdom of God (UCKG). Macedo owns Brazil's second largest TV network and several dozen radio stations. *Forbes* estimates his worth at $1.1 billion. Macedo has styled himself as the bishop of his church. The UCKG claims to have 12 million followers worldwide. The church requires members to tithe 10% of their pre-tax income and opposes higher education for women.

PATRIARCH KIRILL ($4 BILLION)

Patriarch Kirill (nee Vladimir Mikhailovich Gundyayev) took over the reins of the Russian Orthodox Church in 2009 with the blessing of Vladimir Putin. According to *Forbes* in times past Kirill worked clandestinely as a KGB officer. Kirill is widely seen as a Putin puppet. In return for literally blessing Putin's socially conservative agenda and not speaking out against his moral transgressions, the church enjoys lucrative privileges. The Russian Orthodox church, for example, has the right to import foreign alcohol and tobacco duty free and resell them. The *Moscow News*, Russia's oldest English language newspaper, estimates Kirill's own private worth at $4 billion. Evidence of Kirill's wealth came to light when this ostensibly simple man of the cloth was seen sporting a $30,000 Swiss Breguet watch. The church tried to hide evidence of his expensive watch by Photoshopping it out of official images but in at least one case they forgot to Photoshop out the watch's reflection on a table.

AGA KHAN IV ($1 BILLION-$13 BILLION)

Karim al-Husayn Shah is the current Aga Khan. He is the spiritual leader to about 15 million Nizaris Muslims. *Forbes* estimates his wealth at $13 billion although other sources put it at $1 billion. Shah breeds race horses and owns stud farms in Ireland and France. He's been through a couple divorces. One wife received north of $20 million in a divorce settlement. Another got a payout of over $50 million in 2014. The latter was considered the most expensive divorce settlement in history until Jeff Bezos split with his wife in 2019. Bezos' wife received a $38 billion payout (and has since donated about $6 billion of it).

Six For-Profit Interests Owned by the Mormons

Mormons are required to give 10% of their pre-tax income to the Church of Jesus Christ of Latter-Day Saints. When Catholics toss money into the collection plate on Sunday, the vast majority of that money stays local. Mormon tithes go straight back to the Mormon HQ in Salt Lake City. It's estimated the church takes in about $7 billion annually from these tithes. The question becomes, what does the church do with that money? Officially the Mormons have an all-volunteer clergy, so all that tithed money has to go someplace. Much of it gets invested into secular businesses and property. Religions like the LDS are usually not required to disclose their finances. However, in 2019 a whistleblower revealed the church had amassed about $100 billion in assets. Below are some church assets sources have uncovered.

RANCH LAND AND ORCHARDS

According to Bloomberg the church owns over 1 million acres of land devoted to agriculture. The church owns ranches, farms, and orchards in the USA, Brazil, Argentina, and Mexico.

A MEGAMALL

In 2012 the church opened City Creek Center, a $1.5 billion megamall across from their Salt Lake City temple. The mall closes on Sundays.

HAWAII RESERVES

As the name implies, Hawaii Reserves is an LDS-owned company in Hawaii. It manages water treatment companies, a shopping mall, and a cemetery. It also owns several residential properties.

BENEFICIAL FINANCIAL GROUP

The Beneficial Financial Group is a Utah-based insurance company. It has about $3 billion in assets and revenues of over half a billion dollars a year.

BONNEVILLE INTERNATIONAL

Bonneville International is a Mormon-owned broadcasting company. It owns thirteen radio stations in markets like Seattle, Phoenix, and San Francisco.

SHARES IN TECH COMPANIES

The Mormon church has about $6 billion invested in Microsoft, Apple, Google, Amazon, and Intel stocks. Other publicly traded companies it has sizable investments in include Johnson & Johnson, Visa, and JPMorgan Chase. True to Mormon ethics, they have no investments in tobacco companies or Coca Cola.

Nine Antisemitic Traditionalist Catholic Groups According to the Southern Poverty Law Center

Vatican II was the Catholic Church's attempt to pull itself out of the dark ages. It allowed priests to forego Latin and say mass in the local language. Women were able to participate more in church affairs. All this happened back in 1965. To this day there are Catholic groups called "Traditionalists" that believe Vatican II precipitated moral decay in the church and society in general. Particularly galling to many traditionalist Catholics was Vatican II's insistence that Judaism isn't to blame for Jesus' crucifixion. It's no small coincidence that "traditionalist Catholic" and "rabid antisemite" tend to go adjective-hand-in-adjective-hand. Many of these groups are tracked by the Southern Poverty Law Center. Here are nine groups from a report they published in 2007.

CATHOLIC APOLOGETICS INTERNATIONAL (PENNSYLVANIA)

Catholic Apologetics International was founded in 1993 by Robert Sungenis. The Southern Poverty Law Center describes Sungenis as "one of the most rabid and open antisemites in the entire radical traditionalist movement". He's also a high-profile advocate of geocentrism. Sungenis and his group believe the Antichrist will "come from Jewry".

CATHOLIC COUNTERPOINT (PENNSYLVANIA)

Catholic Counterpoint is a traditionalist Catholic publisher. The group publishes and distributes aging videos of former Dominican priest John O'Connor (died 2006). O'Connor was a high-profile Catholic antisemite who traded in all the usual tropes about Jews controlling the world. The Catholic church fired him in 1991. O'Connor's best-known work is "Synagogue of Satan". In it he alleged Jews control the USA through the Federal Reserve and were the cause of both World Wars. You can get a copy from the group's site for $3.

CATHOLIC FAMILY MINISTRIES (NIAGARA FALLS, NY)

Catholic Family Ministries publishes the *Catholic Family News*. The paper was edited for several years by John Vennari until he died in 2017. Vennari's editorial position was that Jews are the eternal enemy and the Vatican is actually a puppet organ of Jews who infiltrated the church back in the 18th century. For a few years *National Review* magazine's Joseph Sobran (died 2010) had a column in *Catholic Family News*. Sobran was notorious for aligning himself with Holocaust deniers, giving talks at their conferences, and writing articles for the antisemitic *Journal of Historical Review*.

FIDELITY PRESS (INDIANA)

Fidelity Press's best-known publication is a magazine called *Culture Wars*. It's edited by E. Michael Jones. The magazine seems obsessed with imagined clashes between Christianity and Judaism. It offers "enlightening" articles like "Are the Jews Our Friends?", "The Judaism of Hitler", "Jewish Privilege", and a history of "Jewish Science". The magazine's creepiness caused at least one review editor to quit, objecting to the magazine's stance that Jewish organizations were responsible for the removal of prayer in school and the magazine was "becoming a journal of psycho-sexual conspiracy theory."

INTERNATIONAL FATIMA ROSARY CRUSADE (CONSTABLE, NY)

The International Fatima Rosary Crusade was founded in 1977 by a Canadian Catholic priest named Nicholas Gruner (died 2015). For years he published a magazine called *The Fatima Crusader*. Gruner was obsessed with the Third Secret of Fatima, the Vatican's refusal to release the text, and speculation about the secret's text. When the Vatican did release it in 2000, it did not confirm Gruner's speculation that it would be about evil Jews and a communist conspiracy. He spent the remainder of his life asserting the real secret had yet to be revealed. Gruner, like most traditionalist Catholics, saw Jewish people behind various plots. Amusingly Gruner's enemy number one was his rival publisher E. Michael Jones who he accused of being "secretly a Jew". Gruner liked to associate with Holocaust denial groups. He would sell his wares at their conferences. Gruner was stripped of his priesthood in 2001.

LEGION OF ST. LOUIS (NORFOLK, VA)

The Legion of St. Louis reprints social and economic works by the likes of Hilaire Belloc (died 1953), Rupert J. Ederer (died 2013), and G.K. Chesterton (died 1936) under its IHS Press imprint. Maybe it's a coincidence, but all those authors have been accused of antisemitic writings. Its catalog is rather more mild as compared to the group's offerings back in 2002. Back then it was selling more notorious works like Henry Ford's *The International Jew* and Holocaust denier Michael A. Hoffman II's *Strange Gods of Judaism*.

OMNI CHRISTIAN BOOK CLUB (PALMDALE, CA)

Omni Christian Book Club publishes a number of questionable works including a charming chapbook called *The Great Jewish Masque*. The book opens with the line "Jews, like other childish people, enjoy pretending, and when pretending brings them the tribute due to the character assumed, they revel in it." It only gets more toe-curling in its antisemitism from there. Although the Legion of St. Louis had the rare wisdom to pull Ford's *The International Jew* from its store, as of this writing Omni is still happily selling it ($44.95 for a four volume set).

THE REMNANT PRESS (FOREST LAKE, MN)

The Remnant Press was founded in 1967 in response to Vatican II, notably the church's abandonment of the idea Jews are responsible for the execution of Jesus. It publishes a newspaper called *The Remnant*. In its "About Us" section the paper proudly trumpets its opposition to "the Pope's sell-out to the Jews". It was one of the few publications to defend Mel Gibson's antisemitic rant when he was arrested in 2006 for a DUI. These days the paper seems less overtly concerned with the Catholic church's more enlightened approach to Judaism and more focused on spreading Covid craziness and whatever lies Donald Trump has been telling.

SLAVES OF THE IMMACULATE HEART OF MARY (RICHMOND, N.H.)

The Slaves of the Immaculate Heart of Mary was founded by a defrocked antisemitic priest named Leonard Feeney (died 1978). In terms of antisemitic priests, Feeney is probably second to Charles Coughlin (died 1979) in repute. Feeney preached Judaism was "working for the overthrow of that society by every means at its disposal." After Feeney kicked the bucket, the Richmond group continued to issue antisemitic crap. In 2004, New Hampshire's bishop called the group out for being blatantly antisemitic. In 2019, the church advised the group to stop representing themselves as a Catholic organization. It took them that long?

Eleven Pieces of Bad Biblical Advice

TIPS FOR SLAVE OWNERS

"Anyone who beats their male or female slave with a rod must be punished if the slave dies as a direct result, but they are not to be punished if the slave recovers after a day or two, since the slave is their property." - *Exodus 21:20-21 20*

The Bible offers a fair amount of advice on how to treat a slave but it is rather silent on the morality of owning another human being. So, apparently, it's okay to own and torture your slave. Just don't kill the person.

AVOID BLOOD TRANSFUSIONS

"I will set my face against any Israelite or any foreigner residing among them who eats blood, and I will cut them off from the people." - *Leviticus 17:10*

Back in 1945 Jehovah's Witnesses introduced a doctrine that forbade members from getting life-saving blood transfusions based on this passage. Numerous children have died because their parents refused this simple, lifesaving procedure.

OLD FASHION ADVICE

"Do not wear clothing woven of two kinds of material." - *Leviticus 19:19 19*

I'm not sure desert dwellers have ever had to buy a modern winter coat. This is one of those passages that literalists have to work hard to interpret. Apparently, the Bible wasn't actually saying don't wear clothing woven of two different materials. But it was just trying to reinforce the notion God's chosen shouldn't mix with others. So, interracial marriages are bad. See?

HANDICAPPED, DWARVES, PEOPLE WITH DAMAGED TESTICLES NEED NOT APPLY

"'For the generations to come none of your descendants who has a defect may come near to offer the food of his God. No man who has any defect may come near: no man who is blind or lame, disfigured or deformed; no man with a crippled foot or hand, or who is a hunchback or a dwarf, or who has any eye defect, or who has festering or running sores or damaged testicles." - *Leviticus 21:17-20*

The Bible offers very sensitive advice on how to treat people with physical challenges. Basically, only physically perfect people can become priests.

ACTUALLY, THAT'S A WAR CRIME

"The Lord said to Moses, 'Take vengeance on the Midianites for the Israelites. After that, you will be gathered to your people.' ... They fought against Midian, as the Lord commanded Moses, and killed every man ... The Israelites captured the Midianite women and children and took all the Midianite herds, flocks and goods as plunder. ...Moses was angry with the officers of the army — the commanders of thousands and commanders of hundreds — who returned from the battle. 'Have you allowed all the women to live?' he asked them. 'They were the ones who followed Balaam's advice and enticed the Israelites to be unfaithful to the Lord in the Peor incident, so that a plague struck the Lord's people. Now kill all the boys. And kill every woman who has slept with a man, but save for yourselves every girl who has never slept with a man.'" - *Numbers 31:1-18*

You'd think a good god would respond to the slaughter of children and noncombatants and the enslavement of young girls with some kind of consternation. *Numbers* merely related their god high-fiving Moses for a job well done and advising him on how to divide the spoils of war.

NOT GREAT PARENTING ADVICE

"If someone has a stubborn and rebellious son who does not obey his father and mother and will not listen to them when they discipline him, his father and mother shall take hold of him and bring him to the elders at the gate of his town...Then all the men of his town are to stone him to death. You must purge the evil from among you." - *Deuteronomy 21:18-21*

Look, parenting is hard, especially when kids hit their teenage years. But advising harried parents to enlist strangers to kill your kid with rocks is excessive.

THE BEST KIND OF CHRISTIAN IS AN UNMARRIED CHRISTIAN

"An unmarried man is concerned about the Lord's affairs — how he can please the Lord. But a married man is concerned about the affairs of this world...and his interests are divided...I am saying this for your own good, not to restrict you, but that you may live in a right way in undivided devotion to the Lord." - *1 Corinthians 7:32-35*

The Bible seems to suggest married people aren't great Christians.

GO AHEAD, THAT SNAKE WON'T KILL YOU

"They will pick up serpents, and if they drink any deadly poison, it will not hurt them; they will lay hands on the sick, and they will recover." - *Mark 16:18*

In America's Southeast, you will find Pentecostals and Charismatic cults that interpret that passage to indicate the truly faithful need to pick up live, venomous snakes and manhandle them during their worship services. The practice has helped over 100 snake handlers join Club Darwin.

DON'T TAKE YOUR FRUSTRATIONS OUT ON THE TREES

"Early in the morning, as Jesus was on his way back to the city, he was hungry. Seeing a fig tree by the road, he went up to it but found nothing on it except leaves. Then he said to it, 'May you never bear fruit again!' Immediately the tree withered." - *Matthew 21:18-22*

This story appears in *Mark*, *Matthew*, and *Luke*. *Mark* makes it clear the tree was not in season. An omniscient god, let alone a desert dweller capable of surviving 40 days in the wilderness, should probably have known when fig trees are likely to produce fruit. So, one should not expect trees to produce fruit out of season and one should definitely not kill off a tree for not producing fruit out of season.

MUD AND SPIT AREN'T GOOD FOR EYE CONDITIONS

"When Jesus had said this, He spit on the ground, made some mud, and applied it to the man's eyes. Then He told him, 'Go, wash in the Pool of Siloam'. So the man went and washed, and came back seeing." - *John 9:6-7*

If you're blind or experiencing seeing problems, rubbing mud in your eyes has no known therapeutic effect.

BAD MARRIAGE ADVICE

"Wives, submit yourselves to your own husbands as you do to the LORD. For the husband is the head of the wife as Christ is the head of the church, his body, of which he is the Savior. Now as the church submits to Christ, so also wives should submit to their husbands in everything." - *Ephesians 5:22-24*

Whoever wrote that was either not married or was really pissed off at his wife that day. Most married men will tell you their wives usually have a better idea about almost everything.

Ten Christian Heresies

AUDIANISM

Audianism teaches the Christian god is a physical being like a human. It takes as literal truth a passage in *Genesis* that states humans were created in the Christian god's image. Audianism emerged among Christians in Syria in the 4th century.

CATHARISM

Catharism was popular in Southern Europe in the 12th century. Proponents believed the world was lorded over by a good god (as represented in the New Testament) and an evil god (as represented in the Old Testament). Catharism neatly solved the Problem of Evil dilemma. Why does a good god allow so much evil? Proponents of Catharism answer this with the proposition the good god can't intervene in matters of the evil god.

FRATICELLI

The Fraticelli were 13th century followers of St. Francis of Assisi and believed the church had to follow St. Francis' example of poverty to the extreme. The Fraticelli preached that the church's wealth invalidated its right to issue holy orders to priests. That bothered the Pope. What seems to have bothered the Pope more, however, was the Fraticelli's extreme poverty required them to constantly beg for funds. They were competing with the church's own donor pool.

ICONOCLASM

Iconoclasm is the belief the veneration of religious icons and the production of religious artwork in general is irreligious and must be stopped. From iconoclasm we got the term iconoclast which today generally means a person who challenges widely held beliefs. Iconoclasm emerged multiple times throughout Europe's history, usually resulting in mobs destroying religious art.

MARCIONISM

Marcionism was a 2nd century attempt to reconcile the reality that the god depicted in the New Testament doesn't seem to be anything like the god depicted in the Old Testament. Marcionism teaches the god of the Old Testament was an inferior god given rule until the supreme god sent a purely spiritual non-human Jesus to Earth to reveal the truth and clean up the Old Testament god's shoddy work.

OPHITES

Ophites are Christians who revered the serpent in the garden of Eden. In the view of Ophites, the god of the Old Testament was separate from the god of the New Testament. The Old Testament god was a control freak and the serpent helped humanity free itself from the god's machinations.

PELAGIANISM

Pelagianism was a 5th century heresy that taught humanity was free of original sin. While Adam's fall set a bad example for humans, it didn't sully our essential nature.

Humans have free will and can achieve perfection without the need for divine grace.

SABELLIANISM

Sabellianism is a 3rd century heresy. Believers rejected the Trinitarian view that the father god, Jesus, and the Holy Spirit are three distinct but co-equal persons made of the same god substance. Sabellianism posited there is a single deity that manifests a Jesus and holy spirit aspects depending on the role this god is required to play. As a creative force, it manifests itself as a Father god. For redemption, it manifests as Jesus. For sanctification, it manifests itself as the Holy Spirit.

TRITHEISM

Tritheism takes Trinitarianism one step further. The Father god, Jesus, and the Holy Spirit are not just three independent personas made of the same divine substance but three completely independent deities. Tritheism was popular between the 3rd and 7th centuries.

WALDENSIANS

The Waldensian movement was founded in the 12th century by Peter Waldo. The Waldensian presaged many of the Protestant reforms of the 16th century. Waldensians translated the New Testament and mass into local languages. They believed in a lay clergy. They rejected the veneration of relics and the efficacy of holy water. Waldensians barely survived attempts to wipe them out in the 13th and 14th centuries. The group eventually joined with the Protestants following the reformation. The movement survived as a Protestant church into the 20th century.

Thirteen Fundamentalist Christians Divorced 20 Times and Married 32 Times

"So they are no longer two but one flesh. What therefore God has joined together, let not man separate." — *Matthew 19:6*

"Everyone who divorces his wife and marries another commits adultery, and he who marries a woman divorced from her husband commits adultery." — *Luke 16:18*

1. Jim Bakker (Charismatic TV preacher) - divorced and married twice
2. Kenneth Copeland (vastly rich Charismatic TV preacher) - divorced twice and married three times
3. Bill Donohue (President of the Catholic League) - divorced and married once
4. Kathie Lee Gifford (TV personality, graduate Oral Roberts University) - divorced and married twice

5. Newt Gingrich (family values politician) - divorced twice and married three times
6. John Hagee (mega church pastor) - divorced and married twice
7. Benny Hinn (faith healer) - divorced and married twice
8. Victoria Jackson (former SNL actor turned Christian comedian and activist) - divorced and married twice
9. Hal Lindsey (Christian author) - divorced three times and married four times
10. Chuck Norris (karate actor, author of Christian books) - divorced and married twice
11. John Osteen (TV preacher, father of Joel Osteen) - divorced and married twice
12. Robert Tilton (prosperity gospel preacher) - divorced twice and married three times
13. Mike Warnke (Christian evangelist author who made up stories of his Satanic past) - divorced three times and married four times

Ten Major Religious Scandals

CANADIAN RESIDENTIAL SCHOOL SYSTEM

The Canadian residential school system was a partnership between the Canadian government and various Christian churches across Canada. The schools were set up in the late 19th century with the explicit intent to remove indigenous children from their own cultures and place them in boarding schools where they would be indoctrinated in Canadian and Christian culture. The practice, viewed by many today as a form of cultural genocide, didn't end until at least the 1980s. Conditions in the schools were disgraceful and upwards of 6,000 children died under the so-called care of their Christian guardians. Many of the dead were unceremoniously buried in unmarked graves. It wasn't until the 1990s and 2000s that the Canadian federal government and many Canadian churches began to recognize their role and began offering apologies.

JERRY FALWELL JR'S BAD 2020

Jerry Falwell Jr. took over the running of Liberty University on the passing of his father. In 2020, Falwell brazenly published photos of himself with his pants unzipped, displaying his underwear, holding what appeared to be some kind of alcoholic drink, and his arm around a young woman who also had her pants unzipped. It didn't escape anyone's notice that students had been expelled from his university for less salacious activities. He also tweeted a picture of a Covid mask he designed featuring a white man in blackface standing next to a Klansman. Despite all of this he remained the head of an institute of higher learning. His downfall came, however, when later in 2020 allegations emerged that Falwell, his wife, and a pool attendant had entered into a three-way relationship with Falwell as the cuckold.

THE JIM BAKKER AFFAIR

Jim Bakker's PTL Club was on pretty thin ice with the FCC and the IRS in the late 1970s and early 1980s when it was alleged Bakker's ministry was misusing donations. The ice broke when it was reported Bakker had paid church secretary Jessica Hahn nearly $300,000 for her to keep secret about what was characterized as a brief affair. Darker versions of what went on allege Hahn was drugged and raped by Bakker. Bakker was forced to resign from the PTL Club and turned the leadership over to Jerry Falwell. Bakker's troubles were not over yet. A *Charlotte Observer* investigation alleged Bakker was overselling lifetime memberships to his Heritage USA complex, which included a yearly three-night stay at the complex's hotel. The hotel only had 500 rooms and no capacity to accommodate the tens of thousands of lifetime members sold on the deal. In 1989 Bakker was sentenced to 45 years in federal prison for fraud. Bakker ended up serving only eight years.

JIMMY SWAGGART DEFROCKED

Televangelist Jimmy Swaggart was booted from the Assemblies of God in 1988 when he was implicated in dalliances with female prostitutes. Unable to keep his cock in his frock, he was caught again with a prostitute in 1991. Swaggart still thumps the Bible today but his flock is a fraction of what it was before he took to the airwaves to famously intone "I have sinned against you!"

METZITZAH B'PEH AND TRANSMISSION OF GENITAL HERPES

A small segment of New York's ultra-Orthodox Jewish community practices something called *metzitzah b'peh*. During a circumcision ritual, after removing the foreskin, the rabbi basically applies his mouth to the baby's penis and sucks away the blood. There's a certain lack of basic sanitary practice in the ritual. That became apparent from a CDC report released in the early 2010s. According to the report, between 2000 and 2011 at least eleven babies had contracted genital herpes from the ritual and two had died. Despite calls to ban the practice or have rabbis modify the ritual by using a straw, nothing seems to have been done. Cases of babies contacting genital herpes from the practice have been reported as late as 2020.

MOUNTAIN MEADOWS MASSACRE

The Mountain Meadows Massacre was an attack ordered by the Mormon church on a wagon train of settlers passing through Utah in 1857. Some 120 men, women, and children were murdered by Mormons near Mountain Meadows, Utah. Their corpses were hastily buried by their Mormon assassins. All indications are the massacre was planned and ordered at the highest levels of the Mormon church, including Brigham Young. The Mormon church today denies such culpability. Only four low level participants were ever brought to justice.

ORAL ROBERTS GETS A DEATH THREAT FROM GOD

In January of 1987 televangelist Oral Roberts warned his followers that unless they donated $8 million to his Oral Roberts University by March of the same year, his god would "call him home". Most interpreted this to mean his god was going to kill him unless people forked up the dough. Although that kind of sick appeal is absolutely par for the course today in the world of televangelism, many viewers back then were disturbed by Roberts' claim. Several stations dropped his show. It looked like Roberts was going to come up short by the March deadline. However, a Florida dog track owner stepped in at the last minute and donated $1.3 million, putting Roberts' $8 million goal over the top and ostensibly saving his life.

SEXUAL ABUSE IN THE CATHOLIC CHURCH

Reports that priests were up to more than blessing crackers and cups of red wine started to emerge in the late 1980s and blew up into a scandal of nightmare globe-spanning proportions. In one study, more than 4,000 Catholic priests have been plausibly accused of the sexual abuse of minors. Lawsuits and out-of-court settlements have cost the church over $2.6 billion.

THE VATICAN BANK SCANDAL

The Vatican has its own bank which is curiously named The Institute for the Works of Religion. In 1982 it was at the center of the collapse of the Italian bank Banco Ambrosiano. The Vatican bank was its largest shareholder. Bishop Paul Marcinkus was not only the head of the Vatican bank but he was also a director of Banco Ambrosiano. In 1981 Italian police began investigating the bank for a wide range of illegal activities, including funding Nicaragua's Somoza dictatorship right alongside its Sandinista opponents. An investigation revealed the bank couldn't account for over $3 billion in funds. The head of the bank, Robert Calvi, was arrested but fled prosecution. He was later found hanging from the Blackfriars Bridge in London. It is assumed he was murdered. Italy issued an arrest warrant for Bishop Marcinkus. The Bishop refused to leave the grounds of the Vatican and avoided prosecution until the warrant was dismissed a few years later.

THE WHITE SALAMANDER LETTER

The White Salamander Letter was a forgery by Mark Hofmann. He claimed to have discovered it in 1984. The letter suggested Mormon founder Joseph Smith consulted with a demonic white salamander instead of an angel to uncover the *Book of Mormon*. Mormon leaders, alarmed that the letter might have given the church's enemies additional talking points, entered into negotiations with Hofmann to purchase the letter. The deal never transpired. Instead the Mormon leadership went public with the letter and its contents. Church apologists tried to spin the white salamander as being a symbolic depiction of an angel. The letter started the unravelling of Hofmann's forgery business and revealed church leaders, supposedly imbued with the godly ability of "discernment", were no better at telling truth from lies.

Eight Moral Panics

A moral panic is a widespread social fear not well founded on fact and logic. It is spread by the media, word of mouth, and opportunistic politicians. Typically, a disadvantaged group (known as a "folk devil") or a certain activity is held to blame for a wide variety of social ills.

THE WAR ON WHITE SLAVERY

In the late 19th century, as America urbanized, young women started to move off the farm and into cities looking for work. A woman's ability to date and socialize was, for the first time, not under the direct control of parents. By the 1910s, people began to fear white Christian women were being coerced by swarthy foreigners and African Americans into lives of prostitution. The friendly neighborhood ice cream parlor became a place where a young woman might be lured into a life of moral turpitude by male patrons and foreign owners. Politicians and religious workers started to refer to this as "white slavery". The US passed the Mann act (aka the White-Slave Traffic Act) in 1910 and made it illegal to transport a woman across state lines "for the purpose of prostitution or debauchery, or for any other immoral purpose". Police forces took an expansive view as to what an "immoral purpose" constituted. More often than not, the law was used to harass African American men travelling with their white wives or girlfriends.

THE RED SCARE

The Russian Revolution in 1917 and the growing labor movement touched off a fear of communist activities in America. The Red Scare panic was exacerbated by a new wave of post-World War I migration from Europe and fears these new arrivals were bringing the communist ethic to America's shores. In 1919, responding to this panic, President Hoover's Attorney General A. Mitchell Palmer requested funding from Congress to conduct raids against suspected communists. He warned the politicians he knew armed radicals were planning to overthrow the government on May Day 1920 and time was of the essence. Although Palmer only got a fraction of the money he requested, it was enough for him to conduct a number of sweeping raids that rounded up basically anyone and everyone at the wrong place at the wrong time. Although 3,000 people were arrested, the raids turned up a paucity of weapons. Four pistols were found along with some "bombs" which turned out to be inert iron balls.

THE SUICIDE SONG

"Gloomy Sunday" is a 1933 song written by Hungarian composer Rezso Seress. The song was something of a hit in 1936 on UK and US radio stations. It resonated with many suffering from the Great Depression. Somehow Hungary's notoriously high suicide rate became connected to this song. It became known as the "Suicide Song". Stories quickly spread of people topping themselves after listening to it. One story told of an Italian errand boy who heard a beggar humming the tune. The boy gave the beggar all his money and then leapt off a bridge. US radio stations pulled it from rotation and the BBC banned the song in all forms, which included a Billie Holiday cover. The BBC didn't actually remove its ban on "Gloomy Sunday" until 2002.

HOMOSEXUAL RECRUITMENT

In the 1970s, many jurisdictions began to remove laws that made it illegal to be gay. As well, several cities began passing ordinances that made it illegal to discriminate against gay people. In 1977, Miami-Dade County, Florida amended its local set of human rights laws to outlaw discrimination based on sexual orientation. Anita Bryant famously led a group called Save Our Children in opposition to the amendment. Bryant at the time was best known as the spokesperson for the Florida Citrus Commission and a crooner who had a handful of songs that charted in the 1960s. Bryant and her group raged the antidiscrimination ordinance actually discriminated against the right of Christians to discriminate against gay people. How can you teach your children the absolute truth of Biblical morality otherwise? Surveys at the time revealed most voters felt gay people were harmless. Bryant went on a tear to change this perception and convince the public gay people were anything but. They were a hidden threat to the children of America. Bryant sought to convince voters antidiscrimination laws encouraged and legalized gay people's ability to recruit children into the "homosexual lifestyle". Her theory was since gay people can't have children (they can... just not with each other) their only method of reproducing more gay people was luring innocent children. Many fears about gay people and Christian right talking points got their genesis in Bryant's campaign. For example, Bryant and her group tied being gay to pedophilia and bestiality. In one instance, Bryant claimed droughts were also a result of letting a lot of gay people live freely in your community. Bryant allied herself with Jerry Falwell who went on to spread Bryant's talking points. In a speech, she quipped "God made Adam and Eve, not Adam and Bruce". Falwell picked up her phrase and tweaked it to "Adam and Steve". Bryant's efforts unfortunately met with success. Voters widely repealed Miami-Dade County's amendment. Bryant and Save Our Children went on to get such ordinances repealed in places like Minneapolis and Eugene, Oregon. Although Bryant was successful in rolling back gay rights legislation and starting a gay moral panic, Bryant's career went into the dumpster. She became the butt of jokes on late night television and secular record companies ceased releasing her music.

THE DISCO RIOT

For a few years in the late 1970s, disco music was at the top of the charts. Disco was largely made by Black, Latino, and gay artists. For the first time white, middle America was exposed to these cultures in a ubiquitous fashion. Radio stations across the country were abandoning their traditional formats and switching to disco. Disco acts dominated the Grammy Awards. Many white middle Americans at this time were transitioning from being strong supporters of blue-collar Democrats to Reagan conservatives. In 1979, the Chicago White Sox encouraged fans to bring disco records to a July 2 doubleheader with the Detroit Tigers. Organizers promised the records would be piled in the middle of the field and blown up at the end of game one. Over 75,000 people turned out for the game and tried to cram into a ballpark that was designed for about 45,000 fans. Not all records were collected. Consequently, the first game was constantly being interrupted by attendees frisbeeing records at players on the field. As promised at the end of game one, a pile of records was assembled on the diamond and blown up. This caused about 5,000 of the fans, many drunk and high, to charge onto the field and tear up the grass, destroy batting cages, and start fires. Despite appeals over the loudspeaker to return to their seats, the anti-disco rioters continued. Police in riot gear were called in to disperse the crowd. It was clear to Black employees that fans weren't just bringing disco LPs to burn but the works of Black artists (like Marvin Gaye and Stevie Wonder) in general. Nile Rogers likened the event to a Nazi book burning. Black park employees felt personally threatened by these rioters and heard the N-word used quite liberally. Many disco artists reported after the Disco Riot of 1979, record companies and concert promoters stopped returning their phone calls.

SATANIC RITUAL ABUSE

The Satanic Panic and tales of Satanic ritual abuse have their genesis in the 1980 book *Michelle Remembers*. The book was written by Canadian psychiatrist Lawrence Pazder (died 2004). It tells the supposed true story of his patient Michelle Smith and the abuse she suffered at the hands of her Satanic mother (who had died more than a decade before the book's release). Pazder used hypnosis on Smith to supposedly recover repressed memories of being tortured, witnessing the sacrifices of adults and infants, and the summoning of Satan himself. Despite allegations of horrific physical abuse, Smith had no marks on her. Pazder, who was a devout Catholic, suggested maybe Jesus and the Virgin Mary helped erase her scars. People who fact checked Pazder's claims could not verify anything. For example, as a child Smith had allegedly been isolated for nearly three months as part of some Satanic ritual. School records indicated no such absence. The book was a best seller and made Pazder hundreds of thousands of dollars. He later divorced his first wife and married Smith. This seems a breach of professional ethics and the Catholic beliefs he was so fervently trying to protect. The book's topic of hidden Satanists in positions of trust and power over children came at a time when women were entering the workforce in large numbers and putting their kids in daycare. It's no surprise a number of daycare workers were accused of being Satanists and abusing children in their care in a fashion similar to details published in *Michelle Remembers.* The USA's McMartin preschool trial and Canada's Martensville Daycare Satanic sex scandal were two high profile examples of the insanity Pazder's book touched off. Pazder himself turned up at the McMartin trial as an expert witness.

DUNGEONS AND DRAGONS

During the Satanic panic of the 1980s, Dungeons and Dragons (D&D) was singled out for its own moral panic thanks to the efforts of Pat Pulling (died 1997). In 1982, Pulling founded BADD (Bothered About Dungeons and Dragons) after her son committed suicide. The son used a loaded pistol he found in Mrs. Pulling's bedroom to kill himself. As Pulling would tell it on the *Geraldo* show, up until the day of his death her son was a normal, intelligent, well-adjusted high schooler who suddenly, and without warning, committed suicide. Why? D&D, of course. During an after-school game the day before, his Dungeon Master ostensibly put a curse on his character. Pulling sued her son's school for allowing this harmful game to be played. Through BADD, Pulling assembled a motley collection of "experts" with questionable academic and professional credentials to talk about the dangers of Satan's game and testify at trials. For most of the 1980s, Pulling milked her tragedy, the tragedy of others, and the crimes of a handful of murderers who had an association with D&D. About the only skeptical audience she ever faced was the FBI. It found no rational basis for any of the ideas and methodologies in $300-a-head seminars she was giving to local law enforcement agencies around the country. The Game Manufacturing Association (GAMA) eventually took action to quell this unhinged moral panic. It commissioned sci-fi author Michael Stackpole to research the increasingly wild claims made by Pulling and BADD's "experts". Stackpole collated his research into his 1990 "Pulling Report". It detailed numerous inconsistencies with Pulling's story. Pulling's son, for example, was not the balanced child she portrayed him to be. The son spent nights in their backyard howling at the moon like a werewolf and slaughtered over a dozen rabbits the Pullings were raising. He also disemboweled a neighbor's cat. Stackpole crunched Pulling's numbers relating to kids who had committed suicide while involved with D&D. When compared to the national suicide rate, it was easy to see D&D players actually had a vastly lower rate of suicide. Stackpole talked to some of the parents of suicide victims Pulling identified. Most offered that D&D was not the cause. In fact, many felt the sense of belonging the departed found within the D&D community helped stave off the inevitable for a few precious months or years. Stackpole's publication of "The Pulling Report" had a devastating effect on Pulling's place in the media's golden rolodex and the fortunes of BADD. In less than a year, Pulling gave up her position as head of the group. She went on to sell real estate and died of cancer a few years later.

You too might be a Satanist if you have a well-loved copy of this on your bookshelf
Photo Credit: Karl Mamer

WAR ON CHRISTMAS

Not everyone who lives in North America is Christian and celebrates Christmas. As groups like Jews, Hindus, and atheists grew in economic power, corporations adjusted their messaging to recognize the solstice period is important and sacred in different ways to these groups. Some Christians have been less than receptive to such changes, believing efforts at inclusiveness are part of a war on their sacred beliefs. Henry Ford fired the first shot in the War on Christmas in the 1920s. In an antisemitic rant he published in his *The Dearborn Independent* newspaper, he claimed Jews were conspiring to pull religiously themed Christmas cards from the store shelves. In the late 1950s, the John Birch Society warned a UN-Communist alliance was seeking to encourage department stores to sell Christmas cards emblazoned with the UN logo. In the early 2000s, former Fox talking head Bill O'Reilly informed his dim-witted viewers the uniform of the anti-Christmas warrior was now being worn by Liberals. Again, the battleground seemed to be department stores. The anti-Christmas forces were pressuring Walmart to use the phrase "Happy Holidays" as opposed to "Merry Christmas". Broadcaster and Holocaust "questioner" Pat Buchanan upped the rhetoric. He likened efforts to be more inclusive to "hate crimes against Christianity". In 2015, department stores got some relief when Starbucks' annual holiday cup became the latest symbol in the left-wing conspiracy to extinguish Christmas. Since 1997, at the start of the holiday season (which seems to be October) Starbucks switches to a Christmas-themed cup. Though usually wordless, the cups typically displayed traditional Christmas imagery like snowflakes and evergreen trees. For the 2015 holiday season, they introduced a plain red cup. This didn't sit right with YouTube evangelist (that's a thing now) Joshua Feuerstein. He claimed in a widely shared video that removing all Christmas symbols was evidence Starbucks hates Jesus. He urged his followers to go to their local Starbucks, order, and when asked what name the barista should use to give the name "Merry Christmas". See, it would force Starbucks employees to use the banned words. It might have escaped Feuerstein that this protest was turning his followers into patrons and requiring them to buy Starbucks' more expensive (and higher margin) drinks. There is no IQ test, apparently, to become a YouTube preacher.

Do not taunt Happy Fun Ball Holiday Tree.
Photo Credit: Karl Mamer

Pre-Christian Origins of Six Christmas Traditions

THE DATE

From the outset, Christians did not celebrate Christmas or even the ostensible birth of Jesus. Early theologians seemed to mock so-called pagans for celebrating birthdays. As Christianity took hold in Rome and other parts of Europe, local winter solstice celebrations and customs were overlaid with Christian celebrations of the birth of Jesus. The date was fixed in the 4th century. The rationale for December 25 (versus December 21) as the exact date is lost to history. Some assume the date was chosen to comport with an ancient Jewish tradition that great men are born and die on the same date. Jesus' death was believed to be March 25 and December 25th is 9 months (a gestation) later. Some suggest December 25 was chosen to entirely replace the Roman celebration of solstice on December 25.

GIFT GIVING

Gift giving was a common practice during the Roman mid-winter festival of Saturnalia.

LIGHTS

The lighting of bonfires and candles in the run up to the winter solstice was a pre-Christian tradition across Europe and points beyond. Christmas lights themselves were first strung up in 1882 by Edward H. Johnson, vice president of the Edison Electric Light Company. Many viewed it as a publicity stunt as such lights were wildly expensive and beyond the reach of people of ordinary means.

MISTLETOE

Mistletoe has long been considered a symbol of male fertility by pre-Christian cultures. The Celts viewed mistletoe's white berries as symbolic of semen. Romans hung mistletoe above their doors during Saturnalia. Early Christians hung it up as protection from witches. It wasn't until the 18th century that it became associated with Christmas and took on a rather date-rapey tradition: any woman standing under mistletoes was fair game to be kissed by a man. Refusing was said to bring bad luck onto the woman.

TREES

Many religions that Christianity supplanted had long viewed the evergreen tree as a symbol of eternal life and life's perseverance in dark times. Pre-Christian Scandinavians believed they could scare away evil spirits before the New Year by setting up evergreen trees in their homes and barns.

WREATHES

Evergreen wreaths were a common symbol used by the Romans to celebrate their mid-winter festival of Saturnalia.

CONSPIRACIES

Eleven Impossibly Stupid Conspiracy Theories

THE DENVER AIRPORT CONSPIRACY

It's not unreasonable to think most airports are purposely built in such a way as to advertise the complete incompetence of government. It's another matter entirely to think an airport was constructed by the forces of the New World Order and/or the Illuminati to advertise their nefarious plan. Such is the claim of people who subscribe to the Denver Airport conspiracy. Much of the conspiracy theory revolves around people trying to point out various "tells" the forces of the New World Order have incorporated into the design. For example, it is claimed the runways are laid out in the form of a Nazi swastika (they're not). There's signage supposedly in an alien language (nope, it's Navajo). And then there's the artwork that is a coded reference to the New World Order's plan for world domination. Most of it is by muralist Leo Tanguma whose art is of the same genre that made Diego Rivera famous. To someone who feels a Filet-o-fish is exotic, Tanguma's vivid imagery can be bewildering. Oh, there is also a wild-eyed belief that the airport's baggage tunnels are really an underground transit system used by the New World Order to shuttle Chinese troops around the nation.

DINOSAURS ARE A HOAX

Dinosaur fossils have been studied since the 17th century. However, it wasn't until the 19th century that paleontologists started to organize these finds under the term "dinosaur".

Total fake!
Photo Credit: Karl Mamer

That term was coined by Sir Richard Owen (died 1892). There seems to be a branch of cranks that believe Owen just didn't invent the term but invented dinosaurs out of whole cloth. The line of reasoning is a little hard to follow. It's claimed Owen was a fierce critic of Darwin's theory of evolution and invented dinosaurs as either "gotcha" evidence evolution was impossible or he simply wanted part of the attention Darwin was getting. The conspiracy also alleges Owen discovered manufacturing dinosaur bones was an incredibly lucrative enterprise, a gravy train that continues in the world of paleontology today. The evidence that dinosaurs are a hoax rests on the belief that large dinosaurs couldn't get enough food to maintain their bulk. Therefore, it's impossible for them to have existed. Of course, scientists have studied the food requirements of large dinosaurs like the brachiosaurus. Research indicates food was plentiful for large dinosaurs and they could easily get the required calories to stay alive.

IRAQI DINAR INVESTMENT CONSPIRACY

Since 2007, people have been chasing a never ending get rich scheme that involves the Iraqi dinar. The Iraqi dinar trades for a fraction of a fraction of a penny. It hasn't changed much in years. For about $1,000 you can get over a million dinar. There are people who think the dinar is "soon" to take off in value. The dinar is slated to be on par with the US dollar. People who have been sinking money they can't afford into the scheme believe they'll be millionaires, if not billionaires, when some magical bit of monetary wizardry takes place. The only people getting rich are so-called "gurus" taking outsized commissions on sales of dinar to the gullible. These gurus feed some of the dumbest investors in the USA false tales of leaked government information and wild-eyed conspiracies that explain why they'll have to wait another day before they can cash in. The dinar investment conspiracy got a shot in the arm when Donald Trump was elected president. People were sure Trump was on the verge of forcing an upward revaluation of the dinar. They're still waiting.

OBAMA VISITED MARS AS A YOUTH

Vancouver, Washington lawyer Andrew D. Basiago is a member of the exopolitics movement. In 2009 he submitted his own photo analysis of NASA Mars images to *National Geographic*. He concluded it offered clear evidence of human and animal-like life on Mars. Not so much. *National Geographic* declined to publish his findings. He came roaring back a few years later with a story about his participation as a youth in a DARPA program called Project Pegasus. According to Basiago, starting in the 1970s, DARPA was teleporting people to Mars. He went on to claim President Obama, going by the name Barry Soetoro, was a fellow participant in this DARPA program. The Obama White House, in 2012, officially denied this story. Tommy Vietor, spokesperson for the National Security Council, did intimate, however, Obama had a fondness for Marvin the Martian cartoons.

OREO COOKIES ARE A PLOT BY CRYPTIC TEMPLARS AND SATAN

Have you ever really *looked* at an Oreo cookie? The chocolate wafers in the USA have an interesting textured pattern. Just above the Oreo word is a very small depiction of the Cross Loraine. This cross is also part of Nabisco's logo itself. It was chosen by the company's founder Adolphus Green who believed the symbol represented purity. In the early 2000s, netizens started passing around an urban legend that the symbol was really a depiction of the Templar Leviathan Cross, which was later adopted by Church of Satan's founder Anton LaVey.

THE MATTRESS BUSINESS IS A MONEY LAUNDERING FRONT

It's hard to tell if this one is a Poe or started as a Poe (ostensibly on Reddit) and now is just believed by stupid people. But basically, there seems to be a lot of mattress stores. More so than some people think is necessary. You tend to buy a mattress once every quarter century. Are there really that many people buying new mattresses every year that so many mattress stores can stay in business? If one were to think "no way", the only logical explanation is then they're simply fronts for money laundering. The reality is mattresses aren't cheap. At $1,000 a pop, you're doing pretty well if you're selling about 100 a month. That's $1.2 million a year. And that's about four times what the average small business grosses.

NUCLEAR WEAPONS ARE A HOAX

Back in 2013 there emerged a conspiracy theory that the most frightening weapons ever created by humans are actually all a lie. Nuclear weapons were impossible back in 1945 and they're impossible today. The hoax has been maintained for generations to siphon off billions of dollars by unscrupulous scientists. You may wonder, of course, what levelled Hiroshima and Nagasaki? Turns out it was just conventional fire bombing. The leading proponent, and possible originator of this conspiracy theory, is Norwegian Anders Björkman. He's a naval engineer (drink!) by trade. I don't know much about naval engineers but I'm pretty sure they don't study a lot of nuclear physics. So his insight into how these top secret weapons can or can't work may be limited. His view seems to be the myth of the atomic age started in 1942. Stalin and FDR agreed on a disinformation campaign to let their enemies think they had these weapons. On Björkman's website (which seems to double as his more legit marine consulting business called Heiwa Co.) he offers details about his €1 million prize for anyone who "can scientifically demonstrate that an a-bomb works". By "demonstrate" he means building a very small version of an atom bomb and, well, setting it off. Thankfully, no one has tried to claim this prize.

PENCILGATE

A number of UK voters participating in the 2014 Scottish referendum and the EU membership referendum became convinced the government provided golf pencils at voting stations so MI5 could erase and change votes. Proponents of this theory called it Pencilgate and urged voters to bring their own pen to voting stations to cast their ballot. One would think MI5 would have a more sophisticated method of vote rigging than the labor-intensive process of manually erasing thousands of votes.

WHAT BIG CAVITY DOESN'T WANT YOU TO KNOW ABOUT MERCURY FILLINGS

Many dentists fill cavities with a dental amalgam that uses a mixture of mercury, silver, and zinc. Dentists have been using the mercury-based amalgam for at least 150 years. Studies show it's safe, easy to work with, and is both cheaper and longer lasting than alternative resin compounds. However, there's a group of scare mongers that hear "mercury" (be it in your mouth or in your vaccines) and think the stuff is put there to make people sick. Mercury fillings have been blamed for the same constellation of symptoms that vaccines, Wi-Fi, and wind turbines have been blamed for. So, basically Chronic Fatigue Syndrome. A number of years ago a video known as the "smoking tooth video" started to pop up on YouTube. It purported to show a tooth filled with mercury amalgam off gassing the heavy metal. The video tends to get passed around by so-called "holistic" dental clinics. Naturally, they offer a filling removal service (which is not likely covered by your dental insurance). They'll also roll you over to some doctor associates to get chelation therapy. The problem with this smoking tooth video is mercury's boiling point is about 672 degrees Fahrenheit. With the exception of Alex Jones, it doesn't get anywhere near that hot in a human's mouth. Likewise, mercury gas is heavier than air. How can it be rising? It's impossible for the depicted gas to be mercury. These smoking tooth videos are actually just capturing water vapor.

WAYFAIR IS A FRONT FOR CHILD SEX TRAFFICKING

You thought Pizzagate was dumb? In 2020 Qanon believers, probably too deep into the mythology to ever surface back to normality, said "hold our beer". Behold: Wayfair is a front for child sex trafficking. How does that work exactly? Wayfair is an online retailer that sells a broad range of household goods. Some of the stuff isn't super cheap. Most people would go "I'm not paying $10K for a set of storage cabinets" and hit the resale shops looking for cheaper alternatives. But those ever crafty Qanon believers put two and two together and came up with 9.789 billion. See, some of these high-priced products have human names. And some of those names match the first name of youngsters on missing child registries. Since Wayfair knows there's no chance normal people are going to part with $10K for a metal storage cabinet, rich creeps (aka Democrats) can confidently order up an actual child sex slave by purchasing a "storage shelf" named Annabel. Sadly, no Qanon believer grasps they simply rewrote the Monty Python "Chest of Drawers" skit.

VACCINES HAVE 5G TRACKING DEVICES

Right around the time Covid started infecting the world, many telecommunication companies started deploying 5G technology. Naturally, the cranks quickly associated the erecting of 5G towers with the pandemic. As news began to spread that several vaccines for Covid were in the offing, the 5G-Covid link transmogrified into stories about the vaccines being a ruse to inject 5G tracking devices into the arms of the sheeple. Oh, and this was all being financed by Bill Gates. Many communications experts have noted antennas don't really work that way. An antenna no bigger than an imperceptibly small chip, injected subcutaneously, cannot possibly transmit any kind of usable signal. Further, communications experts pointed out everyone already HAS a tracking device on them. It's called a cell phone. In response to this cold dose of reality, someone released on Twitter the ultimate mic drop: an actual schematic of the very 5G chip these "experts" claimed could not be made. It didn't take long for people to identify what the schematic actually represented: a guitar FX pedal.

Eight Organizations Said to Run the World

FREEMASONRY

Freemasonry's history is murky. It's generally thought it grew out of guilds run by actual masons during the Middle Ages. In an era before ID badges, the only way guild members could prove their membership and their experience level was to learn a series of codewords and secret signs. Masons found a lot of work building grand cathedrals all around the UK and Europe.

Photo Credit: Karl Mamer

Masons seemed to have an "in" with the all-powerful Catholic Church and maybe God Himself. Wealthy people who never did a lick of work wanted a piece of that. Masons started offering "speculative" memberships to the wealthy. Protection was another big motivation for opening their doors to the rich. Guilds in the UK were being busted up by the government of the day. Letting in influential members of society was likely viewed as insurance. It didn't really help. Trade guilds were outlawed but the "speculative" aspect of masonry continued. Freemasons themselves are actually unsure of their own history so they cleverly invented one. They tie themselves to a mason (Hiram Abiff) who designed and built Solomon's temple. What's ultimately well-known is freemasonry became the pastime of the well-connected in the UK and Europe. As the British empire colonized various parts of the world, freemasons followed. Obviously, freemasonry took hold in America and many founding fathers were freemasons.

THE ILLUMINATI

The Illuminati was founded in 1776 by Adam Weishaupt. It was founded in Bavaria with the intent to encourage liberal, democratic, and secular discussions among members (that is to say, basic Enlightenment era values). In a world where royalty and churches were starting to lose control to the Enlightenment and were pushing back aggressively against "usurpers", groups like the Illuminati had to keep their activities on the downlow. The group attempted to expand its membership by insinuating itself within Freemasonry. By 1785, the Bavarian government grew tired of various hijinks committed by a growing number of secret societies within its borders and banned all such groups. Not wanting to put up a fight, the Illuminati closed up shop. Of course, conspiracy cranks truly believe the group continued on after the 18th century, going super deep state. In recent years, the social media set has come to believe the Illuminati has expanded into the music business and is managing the careers of reality TV stars. Fans pour over music videos and YouTube clips looking for ostensible signs (hand gestures, owl symbols, etc.) their favorite celeb is giving to indicate they're part of the Illuminati. Modern day celebs alleged to be part of the Illuminati include Kayne West, Kim Kardashian, Lady Gaga, Justin Bieber, and Jay-Z. Oh, and apparently Apple is an Illuminati-controlled company. Their required tell? Their corporate logo is an apple with a bite taken out of it. You know, like the forbidden fruit of the Adam and Eve myth? So, yeah, pure Illuminati!

SKULL AND BONES SOCIETY

The Skull and Bones Society grew out of a dispute in 1832 among members of a Yale debating society. Apparently, a handful of the debating societies members didn't like who was getting awards and formed the Skull and Bones. The society also chose alternative monikers including The Order, Order 322, and The Brotherhood of Death. Some things just look better on t-shirts, I suppose. The 322 reference seems to be a nod to 322 BC, which is the year Athenian democracy died and the city state became a plutocracy. It's not clear if the society views that transition as a negative or a positive. By tradition, every spring the society "taps" promising third year students. No more than 15 promising men and women are tapped. They are brought to the group's club house to be initiated, which sounds fun until you realize their club house is pretty much a mausoleum. Legend has it that the society's club house is the final resting place of the bones of Apache leader Geronimo. Conspiracy types are mighty leery of this group as "Bonesmen" seem to live charmed lives after graduating from Yale. The CIA is the career of choice of many Bonesmen. Being smart, going to a top Ivy league school, and having your resume bubble to the top of the stack if the CEO was a fellow Bonesman probably has more to do with it than more mystical claims by conspiracy types.

BOHEMIAN GROVE

The Bohemian Grove is the more popular name for a San Francisco based arts and letters club called the Bohemian Club. It was founded in 1872 with the intent to be a gathering place for artists and musicians. You know, bohemian types. The group quickly discovered the wealthy elite like to hobnob with artists and musicians. The wealthy were more than willing to pay for the privilege.

COUNCIL ON FOREIGN RELATIONS

The Council on Foreign Relations (CFR) was founded way back in 1921. It is a private US think tank that studies issues related to US foreign policy. It counts nearly 5,000 members. Membership is open to only US citizens and green card holders. Members include a large number of political luminaries like Madeleine Albright, George H. W. Bush, Jimmy Carter, Bill and Hillary Clinton, and Condoleezza Rice. The group also includes a smattering of Hollywood types and media personalities including Barbara Walters, Lesley Stahl, Warren Beatty, and George Clooney. Conspiracy cranks generally rope the CFR in with the Trilats and the Bilderberg Group. The group certainly does have very strong ties to major political power brokers. That the group early on warned the US not to get involved in a war in Vietnam and was ignored seems to indicate the group's control is not as absolute as cranks believe.

BILDERBERG GROUP

In the 1950s, Western European nations were starting to slide back into totalitarianism. The Bilderberg Group was formed in the 1950s as a response by exiled Polish politician Józef Retinger and Prince Bernhard of the Netherlands. The intent of the group is to maintain strong ties between Europe and the USA and promote free market capitalism. The group took its name from the location of its first meeting at the Bilderberg Hotel in the Netherlands. The group maintains a certain level of secrecy to promote candid discussions among invited participants. People invited to the group's annual conference include politicians, business leaders, academics, and members of the media. Invitee lists have included Bill Clinton (when he was governor of Arkansas), John Kerry, Paul Volcker, Henry Kissinger, Jared Kushner, Margaret Thatcher, Bill Gates, and Jeff Bezos. Suspicion of this group by the crank wing goes back to the 1960s when Phyllis Schlafly fingered them as a shadowy group trying to control the Republican Party. Alex Jones, however, has probably made the group a household name. They figure prominently in many of his conspiracy rants.

THE CLUB OF ROME

Nothing scares American conspiracy cranks like European intellectuals getting together to discuss things like limiting economic growth. Well, that's The Club of Rome. That the group has Rome in their name sets off end times conspiracy cranks doubly so as they believe a precursor to the end of the world is a one world government centered in Rome. The group was founded in 1968 in Italy although it moved to Switzerland in 2008. It was founded to address the problem of how to deal with increasing populations and decreasing natural resources. No one really paid much attention until 1972 when it published its first report *The Limits to Growth*. The book resonated with the times. Pollution, inflation, and starvation in China and India were in the headlines. Since its publication *The Limits to Growth* has sold over 30 million copies. Conspiracy cranks read a lot into The Club of Rome's notion that we should accept there are limits on growth. They read "therefore the solution is one world government, rounding up people into death camps, making abortion mandatory, and forced sterilization." The Club of Rome never actually explicitly calls for a culling of the human population. *The Limits to Growth* urges sustainable development. But, of course, that's just skeevy European intellectual double-talk!

TRILATERAL COMMISSION

The Trilateral Commission was founded in 1973 by David Rockefeller. David Rockefeller's name attached to anything is a warning bell to the conspiracy bent. It is a private group of over 400 members. Membership is by invitation only. The aim of the group is to encourage discussions and the free exchange of ideas between members of three key economic, political, and cultural powerhouses: North America, Western Europe, and Japan. The group is mostly a collection of CEOs, retired government bureaucrats, and a pile of academics no one has ever heard of. A list of "trilats" is notable for a lack of the world's real movers and shakers. However, some big names are currently members or have been members. Notables include Paul Volcker, Jimmy Carter, Jeffery Epstein, Henry Kissinger, Cyrus Vance, and Walter Mondale. Fringe presidential candidate Lyndon LaRouche (died 2019) was a big promoter of the idea the Trilateral Commission was behind the drug trade. In the 1990s Pat Robertson fingered them in his book *The New World Order* as being one of the groups that was going to take over Real Soon Now.

Six Non-Existent Organizations Said to Run the World

THE COMMITTEE OF 300

There seems to be a class of Americans who fear the UK never really gave up control of America after the American Revolutionary War in the 18th century. Wild-eyed types think the British aristocracy is still calling the shots through a committee of 300 powerful individuals. Hence, the name. They also, apparently, like to go by the name The Olympians. The idea that 300 people are in control of the world finds its genesis in a statement made by a German politician named Walther Rathenau. In 1909 he claimed 300 people who all know each other guided the economies of Europe. One can quibble about the exact numbers but it's not a bad summation of the interlinked, cross-national relationships between politicians and royalty in Europe at the time. World War I was, in essence, a war between three men who were first cousins. Rathenau was assassinated in 1922 by a crazy who assumed Rathenau was himself part of the 300. Before World War II, conspiracy nuts assumed the 300 Rathenau spoke of were Jews and the group was headed by the Rothschilds. The players eventually morphed into the British aristocracy and are alleged to be headed by Queen Elizabeth II. British author John Coleman has turned this fictitious group into something of a cottage industry. He was quite active during the 1990s, writing a number of books about this non-existent group and giving a number of public lectures. Coleman goes by the title "Doctor" but he has never offered what institution awarded him a doctorate. He claims to have worked for MI6 where he stumbled upon cliched "above top secret" documents revealing the existence of the Committee of 300. Like his doctorate, he's never offered any documentation he was ever employed by the intelligence agency. About the only verification of his background and claimed acumen in fields ranging from economics to American constitutional law is a long-winded quote he cites on his website from a Professor Carlos St. John Reuterman. It will come as no surprise, no one can find any university employing a prof by that name or any publication with that name on it.

THE ROUND TABLE GROUP

This group is largely made up by conspiracy cranks. There was an actual group called the Round Table. It was a British think tank set up in 1909 to discuss ways to promote free trade between the UK and its former colonies. Its low point was its endorsement of Australia's "whites only" policy. The Round Table was the basis for the Commonwealth of Nations. The Round Table largely became irrelevant by the 1980s as the Commonwealth became an established trade and cultural group. As conspiracy types see it, however, the Round Table is a front for the more cleverly named Round Table Group. This one, cranks allege, was founded in the 19th century by Cecil Rhodes with the aim to create a one world government run by the Brits. Rhodes was not what you'd call the greatest humanitarian of the 19th century. He was the architect of South Africa's brutal apartheid structure. Again, to stress, there is no actual Round Table Group but conspiracy cranks are not entirely wrong about Rhodes' end game. He was, however, never secret about his beliefs, notably that a world run by British institutions was the best of all possible worlds. He also made it no secret he wanted to train a generation of young Americans to become future "philosopher kings" sympathetic to the British crown. These philosopher kings would guide the American colonists to rejoin the UK at some point. To further this, he established the Rhodes Scholarship. To cast suspicions on anyone today who accepted a Rhodes scholarship is to commit the genetic fallacy. Whatever the intent of the scholarship under Rhodes, today the Rhodes Scholarship is largely just a prestigious scholarship that gives a select group of students from all around the world a free ticket to Oxford. Much like Skull and Bones, when you collect up really super smart young people, it's not surprising some really rise to the top of the food chain. Bill Clinton was, for example, a Rhodes Scholar.

MAJESTIC 12

Majestic 12 (aka MJ-12) is supposedly a secret government organization that has presided over the alien question since the days of Harry Truman. Original members included Vannevar Bush, James Forrestal, and General Hoyt S. Vandenberg. Many in the UFO community believe this group was real and continues in some form today. Supposed knowledge of the group comes from a hoaxed document leaked in the 1980s to various UFO researchers. Conveniently, all twelve members were dead when the document was supposedly leaked. The document itself seems to exist to provide evidence that what crashed at Roswell was an alien craft. It rather neatly, some would say too neatly, parallels the Roswell myth. Curiously a reader wondering about all the juicy stuff (documentation about the alien bodies or pictures of the wreckage) is asked to flip to various appendices. Whoever leaked the document conveniently didn't send along those attachments, probably because those would be even harder to fake. There are many reasons to doubt the authenticity of the MJ-12 document. The biggest problem with it is Truman's signature on the document is an exact match for an authentic Truman signature on another document. It's very hard to sign your name twice exactly the same way. In the late 1980s, arch conspiracy nut Milton William Cooper (died 2001) expanded the mythology of MJ-12 in his writings. In Cooper's fantasy world, MJ-12 brokered a deal with the Greys. The Greys needed humans for ghastly medical experiments. MJ-12 allowed the Greys to take a limited number of humans in exchange for a transfer of astounding technology. Cooper claimed many of our technological advances today came from aliens. The Greys, however, were taking more than their agreed upon quota and were subjecting abductees to ghastly medical experiments that made the Nazis, in Cooper's words, "look like Sunday school children". When

John F. Kennedy became president, he discovered the deal and went ballistic. He was going to expose the coverup. Naturally, that threat signed his own death warrant. MJ-12 had him bumped off one sunny day in Dallas.

THE GREAT WHITE BROTHERHOOD

The Great White Brotherhood is supposedly a group of highly enlightened spiritual masters that are guiding human development. The esoteric movement of the late 19th century was quite taken by this idea. Helena P. Blavatsky made it a cornerstone of her Theosophy movement. Blavatsky's fellow Theosophy traveler Charles Leadbeater most likely came up with the name itself. Although the name sounds pretty racist, the "white" part was not meant to refer to their skin color. In Blavatsky's view most of these masters were of Hindu, East Asian, and Middle Eastern heritage.

PRIORY OF SION

The Priory of Sion is supposed to be an ancient order with an illustrious history of grandmasters. The likes of Leonardo da Vinci, Victor Hugo, and Isaac Newton held the job at the heights of their careers. In reality the group was cooked up by a Frenchman named Pierre Plantard (died 2000) back in the 1950s. He fancied himself a descendant of the Merovingian line of kings. The Merovingians were a French royal dynasty (circa 400-700 AD). Modern mythology alleges the Merovingians are descended from the offspring of Jesus and Mary Magdalene. The notion was popularized by the 1982 book *The Holy Blood and the Holy Grail* and then basically launched into orbit with the publication of *The Da Vinci Code*. These books argued Jesus escaped crucifixion, moved to France (wouldn't you?), and founded the Merovingian royal line. Plantard saw himself as being the fulfillment of a prophecy by Nostradamus of a great monarch being installed on the French throne. He wanted people to believe his Priory of Sion was an ancient order, similar to the Templars but with a modern antisemitic twist. In 1967, he planted a set of fake documents in the French national library hoping one day they would be discovered. He also had a friend create a fake ancient parchment mentioning the group. He then had another friend use the parchments as part of a 1967 book about the supposed treasure of Rennes-le-Château. Plantard hit the jackpot in the 1980s when his document drop in the Bibliothèque nationale de France was discovered by Henry Lincoln while researching *The Holy Blood and the Holy Grail*. Lincoln fused his fantasies of the mysterious origins of Rennes-le-Château with Plantard's hoax and concluded the Priory was part of a centuries long secret society charged with guarding the bloodline of Jesus Christ. Plantard's fantasy about the ancient origins of the Priory of Sion were thoroughly debunked shortly after the publication of *The Holy Blood and the Holy Grail.* Although Plantard was pretty crafty in building out his evidence and playing the long game, his big mistake was using text from a modern Latin edition of the Bible in a document that was supposed to be ancient.

Despite being debunked some four decades ago, there are still people today who think the Priory is a real group working in the shadows to bring about the return of the king.

THE BROTHERHOOD OF THE SNAKE

The Brotherhood of the Snake is supposedly *the* original secret society. It was started 3,000 years ago by an ancient alien called Ea. Ea's original intent was to bring humanity advanced technology and liberate us from spiritual bondage. Of course, whenever ancient astronaut types claim aliens brought us technology, they always mean teaching non-Europeans how to use levers to move large stones. It's never knowledge of germs, vaccines, or basic plumbing. Alas, Ea's plan was stymied by its fellow aliens. The alien overseers put a stop to such niceties and ensured humanity would forever remember Ea as a serpent-like evil doer. Snake imagery has been seeded in our mythology (like the snake in the Garden of Eden) to subtly remind us that seeking escape from spiritual bondage is a totally bad thing. The group remains on Earth today, calling themselves the Brotherhood of the Snake, to ensure humanity remains spiritually repressed. Naturally it does this through an octopus-like network of the usual players (the Illuminati, Masons, David Rockefeller, and Obama). Stories about this secret society have a fairly recent genesis. It pretty much comes out of a 1989 book by William Bramley called *The Gods of Eden*. Other authors picked up on Bramley's ancient cult idea and expanded it. Over the years authors have weaved in Zecharia Sitchin's astoundingly wrong reading of Sumerian mythology, David Icke's fantasies about our ruling class is an ancient bloodline of shape-shifting lizards, and, of course, the antisemitic *Protocols*.

Thirty Four People Who You Might Be Surprised Are Freemasons

1. Mustafa Kemal Atatürk (founder of the Republic of Turkey)
2. Paul Bernardo (Canadian serial killer)
3. Silvio Berlusconi (Italian prime minister)
4. Buffalo Bill (bison hunter and showman)
5. Mel Blanc (voice of Bugs Bunny)
6. Simon Bolivar (leader of South America's independence struggle)
7. Daniel Boone (frontiersman)
8. Omar Bradley (World War II general)
9. Roberto Calvi (head of a bank owned by the Vatican)
10. Winston Churchill (British prime minister)
11. Ty Cobb (baseball player)
12. Gordon Cooper (astronaut)
13. Cecil B. DeMille (movie director)
14. John Elway (two-time Super Bowl champion)
15. John Entwistle (musician, member of The Who)

16. Enrico Fermi (physicist, Nobel prize winner)
17. W. C. Fields (juggler)
18. Henry Ford (founder of the Ford Motor company)
19. John Glenn (astronaut)
20. J. Edgar Hoover (head of the FBI)
21. Tim Horton (NHL hockey player, founder of the Tim Hortons donut chain)
22. Jesse Jackson (civil rights activist)
23. Don King (boxing promoter)
24. Harpo Marx (comedian)
25. Wolfgang Amadeus Mozart (composer)
26. James Naismith (inventor of the game of basketball)
27. Shaquille O'Neal (NBA basketball player)
28. Richard Pryor (comedian)
29. Sugar Ray Robinson (boxer)
30. Joseph Smith (founder of Mormonism)
31. Dave Thomas (founder of the Wendy's restaurant chain)
32. Adam Weishaupt (founder of the Illuminati)
33. Steve Wozniak (co-founder of Apple)
34. Brigham Young (Mormon president)

Fifteen Ways to Signal You're a Member of the Illuminati

Cranks who think the Illuminati went "deep state" after actually closing up shop at the end of the 18th century seem to believe modern members are required to signal their membership in various ways. Why a supposed secret society wants to continually reveal its existence in such a way as the dumbest people on the Internet are able to figure out exactly who is a member is a mystery in of itself. But should you want to make your crank neighbor think a multi-generational Illuminati family has just moved in, here are fifteen ways you can signal you are part of a diminishingly exclusive club.

1. Covering one eye with your hand
2. Giving a Nazi salute
3. Hiding one hand like you are Napoleon
4. Making a pyramid sign with both hands
5. Making an okay sign with your hand
6. Making the "V for victory" sign with your fingers
7. Making the Devil horns with your hand
8. Making use of a birdcage image
9. Making use of a butterfly image
10. Making use of a rainbow image
11. Making use of an owl image
12. Putting a hand over your heart

13. Putting an obelisk on your front yard
14. Sporting an eye-in-the-pyramid tattoo
15. Wearing a peace sign

Nine Influential Crank Documents

A lot of Ufology and conspiracy thinking builds on the work of others. These are nine documents that over the years have influenced crank world building.

THE SHAVER MYSTERIES (1945)

Richard Shaver's *The Shaver Mysteries* emerged from the pulp sci-fi era of the 1940s and seized the minds of the nascent crank wing just before the flying saucer era emerged in 1947. *The Shaver Mysteries* are regarded as a foundational text for conspiratorial thinking and the paranoid wing of modern-day Ufology. It is generally agreed Richard Shaver (died 1975) was schizophrenic. A welder by trade, he came to believe his welding gun allowed him to read thoughts and receive messages. In 1943 Shaver sent a missive to the offices of *Amazing Stories* magazine (a leading sci fi magazine of the day). He implored the editors to publish his discovery of a secret proto language called Mantong. It was, possibly, the language spoken by Atlanteans! He explained each letter has a secret meaning and if you apply those secret meanings to each letter in a word you reveal the true meaning of a word. *Amazing Stories*' editor-in-chief Ray Palmer (died 1977) picked up Shaver's letter from the slush pile and was intrigued. He ran Shaver's Mantong dictionary as a letter to the editor. In "ed" comments below the letter, Palmer suggested maybe the writer had truly discovered a lost language and implored the magazine's readers to test it out. This would set a pattern for Palmer's approach to *The Shaver Mysteries.* Shaver never said the stories were true. Maybe they are. Maybe not. You decide. It's was a precursor to *Fargo*'s "This is a true story" introductory text. Many readers wrote in claiming they proved out the Mantong dictionary and Shaver sure seemed to be on to something. Palmer wrote Shaver back and asked for more. Shaver replied with a 10,000-word story about being prisoner in a vast underground world populated by evil, sadistic Deros (short for detrimental robots), reptilians, and a noble race of Nordic-like Teros. The Deros were not actual robots but small humanoids that did horrible evil things, mainly to scantily clad women, in a robot-like fashion. Shaver revealed the Deros were abducting thousands of humans every year to process into meat. Palmer was hooked by Shaver's tale. He pumped up Palmer's original manuscript (titled "A Warning to Mankind") and published a 30,000-word version in the March 1945 issue titled "I Remember Lemuria". The March issue was a huge hit and sold out. Palmer ordered more stories from Shaver. Between 1945 and 1948 three out of four issues ran a story from Shaver. These stories became known as, collectively, *The Shaver Mysteries*. Local clubs formed to discuss the stories. Readers wrote in with their own tales of being held prisoner by the Deros or having found secret elevators to this underground world. There was no end to readers writing in, anxious to confirm aspects of Shaver's universe. Not all readers of *Amazing Stories* were fans of *The Shaver Mysteries* or, at least, Palmer's "you decide if they're true" approach. Complaints from hard sci-fi fans to the magazine's publisher Ziff-Davis resulted in Palmer's departure. Palmer didn't stay quiet. He founded *Fate* magazine, which itself helped merge Ufology with the broader world of the paranormal. Shaver's inward-looking stories fell out of favor in the 1950s as the fantasy prone started to look outward to space and aliens to explain life's mysteries. However, modern Ufology and the wider world of esotericism have incorporated Shaver's ideas of ancient races, lost continents, underground bases, and abductions into their own mythology. In later years, with a greater understanding of mental illness, critics have questioned Palmer's ethics. At a minimum Palmer was encouraging a deeply disturbed man to plumb ever deeper into delusions that clearly terrified him. At its worst, Palmer was taking advantage of Shaver for monetary reasons. He was using Shaver's mad ramblings as a kind of "idea pump" and was the one writing *The Shaver Mysteries* in full.

THE ANARCHIST COOKBOOK (1971)

William Powell (died 2016) was a teenager working as a manager of a bookstore in Greenwich Village in the late 1960s. He met a lot of Vietnam vets in his shop and became horrified at what happened to them. He grew concerned he himself might get sent to die in Vietnam. For some reason he decided the state needed to be burned to the ground through a violent grassroots revolution. Ah, teenagers. Revolutionists needed bombs, guns, booby traps, communication gear, and, I guess, LSD. So, he quit his job and spent two years researching the book's materials. Powell sold the book to Lyle Stuart (died 2006). Stuart was described as a "first amendment warrior" and had a history of publishing books no other publisher would touch. This ranged from the seedy (the autobiography of Linda Lovelace) to the antisemitic (*The Turner Diaries*). One should note at this point Stuart was of Jewish heritage. So, something for the "Sedition" shelf was par for the course. Stuart gave Powell $35,000 for the copyright. Since publication it has sold over 2 million copies. What's interesting is the book itself became something akin to a household god for hippies, punk rockers, and suburban kids who dreamed of being edge lords. Despite seeming to offer step-by-step guides on how to burn, blow up, or shoot people, not many people really ever seemed to follow any of the book's recipes in the commission of any actual crimes. Since its publication five decades ago, there are, at best, two crimes that can be tied to the book. The reality is most of the instructions in the book are not easy to follow. You need real lab equipment. If you can lay your hands on real lab equipment and know how to hook it up, you probably don't need a book written by an angry teen to make something dangerous. You are capable of doing what Powell did: go to a library and look up the source material Powell used. People in the know have found Powell's ability to transcribe material he found in books like the *U.S. Army Field Manual 5-25* was wanting. He managed to leave out steps and ingredients.

Photo Credit: Karl Mamer

As BBSes and Usenet came on the scene, people attempted to type in various pages from Powell's book or simply created their own versions and posted them under the same title. It kicked off a whole genre of online "mayhem manuals". There were rumors (probably put out by well-meaning people trying to keep kids from burning their faces) that the CIA had purposely posted these instructionally flawed mayhem manuals to the Internet to maim or kill potential anarchists or terrorists. But those allegations are without any substance. Powell later became a devout Anglican and tried to get his book pulled from sale. Unfortunately, he sold his copyright and had no legal means. But, hey, who isn't embarrassed by your angry teenage political rants? Of course, not many of our angry BBS rants from the 1980s are available on Amazon, earning someone else money. It does seem Powell tried to make up for his own perceived hand in whoever might have been on the wrong end of one of his booby traps or improvised weapons. He devoted himself later in life to helping educated children with emotional problems and ADHD.

THE GEMSTONE FILE (1975)

The Gemstone File is the shorter name for a document called *The Key to the Gemstone File*. *The Key to the Gemstone File* (heretofore *The Gemstone File*) has been circulating in one format or another since 1975. *The Gemstone File* was initially passed around as an Nth generation photocopy. *Hustler* magazine also published a version. In the 1980s and early 1990s it became a mainstay on BBS g-file sections and then got propagated in all possible forms via the Internet. *The*

Gemstone File outlines a vast conspiracy behind the murder of John F. Kennedy. But it doesn't stop there. Like most grand conspiracies, the conspirators' end game is to control the world. *The Gemstone File* as we know it today was produced by Stephanie Caruana (a writer for *Playgirl* magazine in the 1970s). It is based on some 300+ pages supplied to her by conspiracy enthusiast Mae Brussell (died 1988). The pages themselves were photocopies of handwritten notes by the originator of the conspiracy theory Bruce Porter Roberts (died 1976). Roberts himself is reported to have written over a thousand pages about his discovery of a secret society called MMORDIS (which, no joke, stood for "Moldering Mass Of Rotten Dribbling Infectious Shit"). Roberts was a fan of Brussell's JFK conspiracy themed radio show *World Watchers International* (which aired on KLRB in Stuart, Oklahoma). He gave her several hundred handwritten pages of his MMORDIS expose. The name of this work takes its name, primarily, from Brussell writing the word "Gemstone" on the file folder she used to hold Roberts' notes. The "Gemstone" appellation also comes from a small side plot within the file that is centered on Roberts himself. Roberts claimed he invented a process for creating synthetic rubies which Hughes Aircraft stole from him to make precision lasers. Was Roberts even capable of making synthetic rubies? Not a lot is known of the man. His obit in a San Francisco paper indicates he served during World War II and pursued a degree in physics but dropped out of university before completing his degree. Some that did meet him or read his original writings came away feeling Roberts was schizophrenic. His unedited papers were disjointed and full of expletives. He wrote nonsense letters to Ralph Nader. It became clear to researchers who gained access to Roberts' original writings that Caruana did a great deal of editing to turn hundreds of pages of word salad into a human readable form. Roberts' supposed ability to produce high quality fake rubies does, however, fill in a glaring plot hole. How does a guy who basically never finished university and lived at home with his parents until he died have access to so much secret information? Apparently, he paid off informants with sacks of high-quality synthetic rubies. Simple! *The Gemstone File* is basically a timeline of events, starting with the rise of Aristotle Onassis in 1932 and ends with Gerald Ford pardoning Nixon in 1974. In between those two points, we learn how Onassis made deals with Joseph Kennedy, Big Oil, the Rockefellers, and the Mafia over the decades to influence politics and enrich himself. The basis for the Kennedy assassination is, in essence, JFK cheated Onassis on a deal originally set up by his father to go easy on the Mafia. Onassis had the Mafia kill JFK. The file fingers the famous Mafia boss Jimmy the Weasel as the trigger man who fired the kill shot in Dallas. The idea that the Mafia killed JFK isn't an original theory but *The Gemstone File* offers probably the best summation of the theory. Roberts writes of the supposed Mafia hit and Onassis ending up with Jackie Kennedy: "An old Mafia rule: if someone welches on a deal, kill him and take his gun and his girl: in this case, Jackie and the Pentagon." There's also another bizarre plot where Onassis supposedly kidnapped Howard Hughes, kept him prisoner on his island of Scorpios, and then took control of Hughes' numerous businesses. Hughes is claimed to have died there and was replaced by a lookalike. The Catholic Church also figures into things too. Turns out the Church has known all along Jesus wasn't Jewish but "an Arab". The Romans entered into an early conspiracy with the Church to turn Jesus into a Jew. And away we go.

THE TURNER DIARIES (1978)

The Turner Diaries, although a work of fiction, became the operating Bible of many right-wing neo-Nazi groups and racist "citizen militias" throughout the 1990s. Its influence persists even into today. It was first self-published in 1978 by William Luther Pierce (died 2002). In the 1960s Pierce was a tenured physics professor but left his position for a job in private industry because he thought the on-campus counterculture movement was being controlled by Jews. He joined the John Birch Society but quickly resigned his membership because they didn't hate visible minorities and Jews enough for his tastes. He then joined the American Nazi Party but left that after realizing Nazis in the 1970s had a bad image among pedestrian racists. He founded his own hate group called the National Alliance. Between 1975 and 1978 Pierce started publishing *The Turner Diaries* in serialized form for his National Alliance's newsletter *Attack!*. Upon completion, Pierce collected the chapters into a book and self-published it in 1978. Events in the serialized version were set in the 1980s but the book edition changed the setting to the 1990s. The book, in essence, details how "whites" are suffering from a "white genocide" imposed by the Federal government and Jews. It shows how "patriots" can rise up and strike back at the Feds with trucks laden with fertilizer bombs. The book offers tips on how hate groups can fund their citizen militias through crime. It also suggests hate groups should infiltrate the military and convince white soldiers to start shooting any Black person in a uniform. This way when African Americans start to defend themselves it can be spun as an uprising of violent Black people. Piece's solution to stopping "white genocide" was to kill all visible minorities, race traitors (women in interracial marriages), and Jews. Basically, his book was a template how racists can foment and fund a race war and commit genocide. The work is graphic and unforgiving and sold half a million copies. Oh yeah, it also has a love story. Because violent racists need love too. Some suggest the book was prophetic and described many events that did transpire in the 1990s. Most critics point out neo Nazi groups like The Order (which took its name directly from *The Turner Diaries*) and individuals like Timothy McVeigh and John William King (who murdered African American James Byrd by dragging him to death behind his pickup) were fans of the book and were acting out various scenarios depicted within the book's pages. At least 12 hate crimes and acts of terrorism can be tied directly back to *The Turner Diaries* since the 1990s.

GRUDGE REPORT 13 (1981)

Project Grudge was an early Air Force investigation of UFOs. The project started and ended in 1949. It was succeeded by Project Blue Book. In 1968 the US Airforce released a compendium of twelve declassified reports from both Project Grudge and Project Blue Book. These reports were originally published between 1951 and 1953. They are all pretty dry reads. In 1955, Project Blue Book released its final report, Special Report #14. It did not classify this one but released it soon after publication to the American public. It was a huge 315-page document offering an incredibly thorough statistical analysis of some 3,200 sightings. When the Air Force later declassified its twelve previous reports, UFO people noticed there were twelve status reports numbered 1-12 and then one special report numbered 14. What happened to report 13? The Air Force noticed this too. In its 1968 release of 12 previously classified reports, it noted it could find no 13th report. Instead of accepting the Air Force is occasionally capable of incorrectly numbering things, Ufologists were certain report 13 was out there and it was far too shocking for the American public. Enter William "Bill" English. English claimed to have been a Green Beret Captain in Vietnam and a former POW who had managed a daring escape. Having proved his mettle, he was sent on a secret mission to investigate a B-52 that was attacked by a UFO over Laos. Ignore for a moment we know the service record of every B-52 that rolled off the production line and none was ever lost over Laos. Also ignore for the moment the Air Force didn't rely on the US Army for search and recovery efforts. They had their own well-trained teams. After his service in Vietnam, English claimed to work for an intelligence group based in the UK. In the late 1970s, while in the UK, he was given the fabled 13th "Grudge Report" to analyze. He reported to his superiors that the report was genuine as, among its 600+ pages, it contained photos from his mission in the 1960s to locate the downed B-52. His wild account of Grudge Report 13 first surfaced in 1981 in the form of a mysterious memorandum to several UFO groups and zine publications. He distributed them under the pseudonym "Captain Toulinet". He gave a few interviews to UFO investigators and the UFO zine *Stigmata* (published in issue 14). At least one UFO investigator came away with the impression English was "unstable" and "nuttier than a fruitcake". In the late 1980s, English dropped his pseudonym and started going by his real name. He also got the ear of John Lear (son of the guy who invented the Learjet and something of a 1980s UFO empresario). Lear gave wider distribution to English's recounting of what he supposedly saw in Grudge Report 13. No one will be surprised to read it contained a fantastical account of crashed saucer retrievals, the capture of live aliens, alien autopsies, and evidence aliens were performing ghastly experiments on humans. English seemed really hung up on human mutilation and covered supposed incidents in excruciating detail. Someone typed in Lear's report on English's claims about Grudge Report 13 and uploaded it to a local BBS. From there it got passed around and became a staple of any well stocked BBS g-file section (a section devoted to downloadable text files). English's account of Grudge Report 13 is now part of the standard ET mythology, helped in part by BBSes and then Usenet. Milton William Cooper amplified English's story by lifting large parts for his own work *Behold a Pale Horse*. English should have probably stuck to his "Captain Toulinet" pseudonym. As his fame grew, people started looking into his rather illustrious military career. People found English's version suspect. One can find him listed on a couple "stolen valor" websites. These are websites that expose "phonies" who publicly trade on made up military service. One stolen valor site published what appears to be his official service record. It makes no mention he was a Green Beret. If accurate, it reveals English served in the army between 1971 and 1974. Private First Class was the highest rank he had attained. The closest he got to Southeast Asia was a two-year deployment in Germany. Before anyone got a hold of his service record there were other red flags about his story. He had trouble citing his unit in a fashion familiar to any military veteran. He couldn't cite the tail number of the B-52 he supposedly risked his life to find in Laos. In 2012, English threw his hat in for a run for the Senate. His biography for his candidacy dropped any mention of him being a Green Beret, an officer, or even a military vet. Typically, these are seen as a plus for any candidate. Odd that. English did confess on his campaign page he was a felon. He revealed he was once convicted of aggravated assault. In his defense, he says he had only pulled a gun trying by to stop "a sexual predator who was raping [his] fourteen-year-old daughter".

DULCE PAPERS (1987)

The Dulce Papers are a mythical set of documents supposedly leaked by a government insider to the UFO community in the 1980s. A handful of unnamed Ufologists were entrusted with copies for safekeeping. Those who ostensibly saw the Dulce Papers said they described an underground base near Dulce, New Mexico that is jointly operated by the US military and an "alien nation". It is important to understand that no one has ever produced the actual Dulce Papers. We are told the copies were all mysteriously recalled. Unlike Area 51 which is a real military base that looms large in UFO mythology, there is no actual base near Dulce. It is the product of cranks riffing on the words of a mentally disturbed man named Paul Bennewitz (died 2003). Bennewitz was a military contractor and a pilot. He liked to fly his light plane around US military bases. He was convinced he was protecting the US from an invasion of space aliens. He would photograph UFOs and then dutifully hand his findings over to the Air Force. It is generally believed he managed to capture a photo of the still-classified F-117A Stealth Fighter. The Air Force grew worried Bennewitz might make his photos public and the Soviets would discern the fighter's existence. The Air Force started to blow a lot of disinformation chaff around Bennewitz. It provided Bennewitz with all kinds of crazy misinformation, notably stories about a magical underground realm located just outside of Dulce, New Mexico. The Soviets likely knew there was no base near Dulce. Anyone talking about aliens in an underground facility under a patch of land that clearly had never been excavated would sound crazy to the Soviets and not worth their time. But as it turns out, no story is too crazy for the UFO community. Bennewitz passed on the disinformation he was being fed to his peers. Word of this mysterious base spread. In 1987 John Lear seemed to be the first to put pen to paper about what he heard about the Dulce Base in a snail mail letter. Someone retyped the letter and distributed it electronically via BBS. In that letter Lear revealed the underground base was home to labs where aliens were conducting horrible experiments on human abductees. In 1979 when the military became aware of this, they tried to send in the crack Delta Force to rescue the human captives. They were unsuccessful and were decimated by the alien defenders. In this letter Lear speaks of including a copy of the fabled Dulce Papers. When Lear writes he's including the Dulce Papers, he's actually referring to hand-drawn reproductions of photos included with the supposed leaked document. Here is where it gets immediately stupid. We are to believe the actual Dulce Papers were all recalled but someone had the foresight to make hand-drawn reproductions instead of taking everything down to Kinkos and churning off actual reproductions. Okay. A year later, in 1989, Lear's associate John Grace (who went by the online name "Val Valerian") published these images in a 300+ page tome called *The Matrix: Understanding Aspects of Covert Interaction With Alien Culture, Technology and Planetary Power Structures.* People just generally refer to it as either *The Matrix* or *The Matrix I* as Grace followed it up with seven more Matrix books. Each Matrix volume grew in size. *Matrix II* was 700 pages long. *Matrix III* hit 900 pages. *Matrix IV* topped at over 1,000 pages. These books are mostly assemblages of long, disjointed narratives of imagined UFO technology, government conspiracies, sacred geometry, spiritualism, marriage advice (no joke!) and even one of those infernal "magic eye" stereograms that were so popular in the 1990s. Part of it reads, frankly, like a source book for an 1980s era sci fi RPG, right down to the amateur line art. *Matrix I* is probably best known for including the so-called Dulce Papers. This amounts to a single paragraph summary of the horrors going on under the deserts of Dulce: human/animal hybrids grown in tanks, blood-drinking aliens, and women being impregnated with space babies. It also included the drawings Lear passed on. They're somewhat ghoulish, featuring images of aliens being grown in pods and a drawing of a tank with pieces of pale meat floating in it. I suppose we're left to imagine the "meat" is human. A year later someone going by the name Jason Bishop III got ahold of the crank talking stick and banged out his own report on Dulce for the Paranet BBS. Bishop's report layers on top the hand of the Illuminati and the RAND corporation. He also ups the horror show factor with his allegation there's a level dubbed the "Night-Mare Hall". Bishop quotes an unnamed source as saying "I have seen multi-legged 'humans' that look like half-human/half-octopus. Also, Reptilian-humans, and furry creatures that have hands like humans and cries like a baby, it mimics human words... I frequently encountered humans in cages, usually dazed or drugged, but sometimes they cried and begged for help." After Bishop's drop, there was no end to people coming forward with tales that they had worked at the base or had seen secret materials confirming what previous authors alleged. Of course, each extended the myth, adding additional levels, different alien types, and secret underground rail systems between places like Area 51 and the Denver airport.

THE KRILL FILES (1988)

This one was a real shit show. In November of 1988, John Grace (aka "Val Valerian") uploaded to the Paranet BBS a 14-page document titled "A Situation Report On Our Acquisition of Advanced Technology and Interaction with Alien Cultures". It claimed to be authored by someone named O.H. Krill. Due to technology limits, many early BBSes of the day had to break the document up into four parts. It became known colloquially as "The Krill Files" although technically it was meant to be one single file. The Krill document was authored by Grace and John Lear. It reads like an attempt by Grace and Lear to pull together and make sense of many of the wild claims being distributed via UFO zines and on BBS systems of the day. In essence it was an early attempt at creating a theory that unifies the fields of Ufology and emerging anti-government conspiracy nuttery. The Krill Files assert as fact a number of completely unsubstantiated claims. The US government was in possession of alien craft and even live ETs. Aliens have been mucking around with our evolution as a species. Humans are being subliminally reprogrammed to be Manchurian candidates. Aliens have bases on Earth and have been mutilating both cattle and humans. The US government has generally looked the other way in return for advanced technology. What Ufology takes today as core dogma was new and disturbing to many. Why Grace and Lear chose to publish under a pseudonym is unclear. The Krill Files regurgitated material Lear had published under his own name. It also roped in material from the 1984 Majestic 12 hoax and borrowed from Bill English's earlier super bogus Grudge Report 13. Perhaps Lear didn't want to be accused of plagiarizing himself and others. There's also a phenomenon in Ufology called reification. As people introduce their own wild claims they tend to build on earlier claims. By citing and including these earlier sources, they offer Ufologists the illusion of independent verification. More they get mentioned in other people's work, the more they become concrete pillars of the UFO mythos. It's possible Lear was hoping people would read the Krill files as confirmation of Lear's own claims (which were at the time controversial). In terms of repetition-is-reification, the master was Milton William "Behold a Pale Horse" Cooper. Cooper had a way of taking other people's made-up stories and claiming they were totally authentic because he had actually seen them first back in the 1970s during his service in the Navy. Not satisfied with letting someone else bask in the "glory" of having finally pierced the veil, Cooper would then pile on pages and pages of his own made up shit. When Cooper read the Krill Files, he predictably claimed they were 100% authentic as he had seen them while working in military intelligence. He tried to extend the mythology by claiming Krill was the name of a space alien held by the US government and the O.H. stood for "Original Hostage". Cooper claimed Krill was given to the US government as live collateral for the right to borrow hicks and hayseeds. Lear was no fan of Cooper. When Lear heard Cooper lying about, well, his lie, Lear went public. Lear publicly admitted he and Grace authored the Krill Files and made up the name. Krill was an inside joke. It's a reference to a 1974 UFO documentary called *UFOs: It Has Begun* (hosted by Rod Serling). That movie refers to a space alien named Crylll. The O and H initials were just made up out of thin air. Many in Ufology were upset at being hoodwinked by both Lear and Cooper. Lear and Cooper were not super popular with the "nuts and bolts" Ufologists who haunted Paranet back then. Here was evidence to most that Lear and Cooper were bullshitting all along. Many took Lear's admission that he made up the Krill name to mean he made up the Krill Files in total. Lear and Cooper found themselves driven out of respectable Ufology for trying to pull a fast one on their totally legit community. Cooper, pinned to the wall, eventually jumped into the antigovernment Patriot movement full hog. He later claimed all that UFO stuff he spouted with such conviction and certitude was actually disinformation. Seems his Navy handlers way back in the 1970s wanted to turn him into an unwitting disinformation agent. Cooper tried to then take down the rest of Ufology by claiming the likes of Lear and Grace were not just dupes like him but actual CIA disinformation agents. Today that muckraking has been largely forgotten. Despite Lear's confession Krill was a made-up pseudonym and never pretended to be the name of a space alien in the original file, many modern Ufologists today claim O.H. Krill was indeed an alien held by the US government.

BEHOLD A PALE HORSE (1991)

Behold a Pale Horse is a book written by Milton William Cooper. Cooper emerged in the UFO field in the late 1980s. He claimed he worked in Naval Intelligence in the early 1970s and had access to secret government documents about the alien presence. He proved to be a polarizing figure and was eventually drummed out of the UFO field. Before his death in 2001, he found his constituency among members of the antigovernment Patriot movement. Post death, *Behold a Pale Horse* is enjoying a resurgence among Qanon types. Because, why not? *Behold a Pale Horse* is many things and consequently there's a chapter in the book for any crank of any bent. If you believe in UFOs and think the government is covering it up, there's a chapter for you. If you think Kennedy was not killed by Oswald, there's a chapter for you. What this book is not, however, is well organized. There's no central thesis. Chapters do not build and mesh together to support a final conclusion. You could randomly reorder the book's chapters or delete several and it wouldn't affect anything Cooper was trying to say. It's pre-Internet Pinterest for conspiracy cranks. One chapter is Cooper expanding on articles he posted to various BBSes in the 1980s. The next is a copy of a proposed Federal law he doesn't like and was in fact never passed (you're not actually told that). The next chapter is a transcript of a phone conversation he had with some random person. Cooper pads his book out with his autobiography, fake documents supplied to him by other people, and random factoids that don't seem to be true (did you know the founder of the Bilderberg group had the power to veto the Vatican's choice for Pope?). There are, however, a few chapters in the book that have generally proved influential with people given to adorning their MAGA hat with tinfoil and MUFON buttons. In Cooper's universe the world is controlled by the Illuminati through an octopus network of groups like the fictitious MJ-12 group, George Bush's Zapata Oil company, politicians ostensibly behind the war on drugs, the Trilateral Commission, the Jesuits, and the Knights of Columbus. Yeah, the Knights of Columbus. Those guys in weird sailor hats that host fish fries and collect pop bottles. Although some of the groups seem to work at cross purposes, that's all disinformation. In Cooper's mind (the mind of a Cold War era navy vet), the Illuminati is using all these groups to spread communism across the globe! He also seems to have had a knack for collecting, and then republishing in this book several known hoaxed documents that, of course, Cooper believes are true. *The Protocols of the Elders of Zion* can be found tossed into the middle of his book. So as not to be accused of being an antisemite, Cooper suggested *The Protocols* were really a bit of a deception. Yes, the plan is true but "Jews" is really code for "Illuminati". Many critics believe the inclusion of *The Protocols* was Cooper's way to sell his book to more right-wing racists. Offer two books in one! Lesser known hoaxed documents Cooper sees fit to include are the Secret Treaty of Verona, a copy of a supposed oath members of the Knights of Columbus take (promising to wage total war against Protestants), and a fake leaked document revealing the Club of Rome's goal to reorganize the world into various kingdoms. The Appendices are probably the most fun to read. It's not clear why he called them Appendices when they are just as random in nature as the material he labelled chapters. But he thoughtfully includes seven of them. Probably the best is Appendix B which includes, in essence, a fan letter from someone named Millard. It's part fanboy mail and part Millard's attempt to pour his heart out to Cooper about a woman who walked out of his life. Millard reasons this woman, who seemed just so perfect for him, was actually a government agent. They wanted this basic nobody to know "they could 'get me' at the deepest, most personal and painful level!"

THE CELESTINE PROPHECY (1993)

The Celestine Prophecy is a book by James Redfield. It was published in 1993. Via a weak narrative format, the book reveals the human race is developing a new spiritual consciousness and predicted by the end of the 20th century society would experience a massive spiritual transformation. That didn't happen unless you want to believe Redfield missed by about 20 years and Qanon is the prophesied spiritual transformation. The book pretends this knowledge, and so much more, comes to us from a secret manuscript found in Peru that was, curiously, written by ancient Mayans in Aramaic. Aramaic is, of course, the language Jesus allegedly spoke. Not sure how any of that works. The book claims this secret manuscript offers twelve insights but only reveals nine of them. This was a convenient hook for the best-selling sequel *The Tenth Insight* and follow on works. The book is ultimately a stew of over warmed New Age mumbo jumbo about auras, vibrations, striving to be our best selves, understanding the nature of coincidences, and abandoning traditional religion as part of your personal journey to evolved spiritualism. Of course, the Catholic Church wants to suppress knowledge of the book and tries to kill the protagonist before he can reveal the insights. The book posits the Mayans attained such a level of spiritual enlightenment in 600 BC they vibrated themselves (and any trace of themselves) straight out of Peru. That's why we don't find actual Mayans in Peru today. The New Age movement was certainly in full swing by 1993. However, given the book remained #1 on the best seller lists for nearly two years, it sent a clear message to mainstream media companies that the New Age was bankable. As well, the book's fact-absent claims about the Mayans and their supposed evolved spiritualism easily teed up the New Age community for 2012.

Three Prosocial Conspiracies You Wish Cranks Would Start to Believe in

THE ANTI-VACCINE CONSPIRACY IS POPULATION REDUCTION

The anti-vaccine conspiracy is part of the Illuminati's plan to reduce the world's population to 500 million people. Anti-vaxxers are really shills of the Illuminati paid to convince a large percentage of the population that vaccines are unsafe and opt into their own genocide as they die off of vaccine-preventable diseases. The best way to thumb your nose at the conspiracy is getting your annual flu shot.

PUBLIC HEALTHCARE FEARS ARE STOKED TO KEEP YOU AT THE BACK OF THE LINE

Public healthcare systems are not rationed by price but by a waiting list based on need. A feature of a public healthcare system is a person can't buy his or her way to the front of the line. The rich and powerful are lying to Americans about public healthcare and making the middle classed and poor vote against it thinking they are doing so in their own self-interest. In reality, the rich and powerful fear that if private healthcare is replaced by a public system, the middle classed and poor will get to the front of the line before them.

THERE IS NO PLAN

Conspiracies are floated by the powers that be to keep people from confronting the shocking truth: there is no plan. Politicians make it up as they go along and there are no adults in charge.

Fourteen Real People Who Were Members of the Bavarian Illuminati

Adam Weishaupt's Order of the Illuminati (aka the Bavarian Illuminati) was founded May 1, 1776. The Illuminati initially consisted of Weishaupt and four of his students. The group dissolved itself in 1785 after the Duke of Bavaria issued an edict banning all secret societies. At its peak the group had upwards of 2,500 members. Here are fourteen notable members.

ADAM WEISHAUPT

Adam Weishaupt (died 1830) was a professor of Canon Law at the University of Ingolstadt. Weishaupt was a Freemason until he found membership dues were getting a little too rich for his blood and formed the Illuminati. Many members used codenames. Weishaupt went by the codename "Spartacus".

FRANZ ANTON VON MASSENHAUSEN

Franz Anton von Massenhausen (died 1815) was one of Weishaupt's students. He was a founding member. He went by the codename "Ajax". Massenhausen fell out of favor with the group as his recruiting efforts were judged substandard. He left the group upon graduation from the University of Ingolstadt.

MAX EDLER VON MERZ

Max Edler von Merz (died 1811) was a student of Weishaupt and a founding member. He went by the codename "Tiberius Milesius". Not much is known about Merz beyond his membership in the Illuminati. He seems to have had a position as a General Commissioner in Munich.

ANDREAS SUTOR

Andreas Sutor (died 1822) was a student of Weishaupt and a founding member. He went by the codename "Erasmus Roterdamus". At some point Weishaupt expelled Sutor, believing he was too indolent.

FRANZ XAVER VON ZWACKH

Franz Xaver von Zwackh (died 1843) was for a time the Illuminati's second in command. His codename was "Cato". After serving a number of years in a leadership position, Zwackh grew tired of Weishaupt's dictatorial control of the organization. He resigned in 1784. A few years after leaving the secret society he fashioned himself a career as a diploma.

JOHANN WOLFGANG VON GOETHE

Johann Wolfgang von Goethe (died 1832) was a German poet, novelist, playwright, and occasional scientist. Goethe might be best known to North American readers as the author of the play *Faust*, in which the title character makes a bargain with the Devil and famously signs the contract in his own blood. From Goethe we get the term "Faustian bargain". Goethe was a Freemason but at some point joined the Illuminati. It does not appear like he was a particularly active member. It's suggested as an adviser to Illuminati member Duke Karl August of Saxe-Weimar, Goethe may have just come along for the ride. However, given Goethe's association with a play about making a bargain with Satan, his membership alone is ripe fertilizer for conspiracy types.

Statue of Johann Wolfgang von Goethe in Frankfurt
Photo Credit: Karl Mamer

ERNEST II, DUKE OF SAXE-GOTHA-ALTENBURG

Ernest II (died 1804) was the Duke of Saxe-Gotha-Altenburg. Ernest II was an astronomy buff and founded the Observatory of Gotha which for a time became a major center for astronomical research in Europe. Ernest II joined the Illuminati in 1783. He took the codename "Quntus Severus". After the Illuminati's dissolution, Ernest II granted Weishaupt sanctuary in his duchy so he could avoid Bavaria's persecution of secret society members.

ADOLPH FREIHERR KNIGGE

Adolph Freiherr Knigge (died 1796) joined the Illuminati in 1780. He took the codename "Philo". Knigge was considered the group's top recruiter. After ascending to the highest levels of Freemasonry in his twenties, he started to look around for something else. He was recruited by Weishaupt on the promise the Illuminati had higher grades that revealed even deeper secrets of "the craft". Knigge was told he would be moved through these grades based on his success at recruiting further members. Knigge found numerous other Freemasons like him who had attained all possible degrees and were looking for something more. Many of his recruits were men who were running local lodges who turned their lodges over to the Illuminati power structure. Although Knigge was amazingly successful at recruiting fellow masons with the promise of higher degrees, he'd sometimes get questions he couldn't answer about them. When he pressed Weishaupt to clue him in, Weishaupt brushed him off initially but eventually was forced to admit there were no higher degrees. Knigge grew disillusioned after this revelation and left the group in 1784. Knigge went on to write a book about etiquette. For a time, he was Germany's "Miss Manners" and his name Knigge became synonymous with good manners.

PRINCE CHARLES OF HESSE-KASSEL

Prince Charles of Hesse-Kassel (died 1836) was for a time the governor of the German state of Schleswig-Holstein and eventually became the commander-in-chief of the Norwegian army. Hesse-Kassel seems to have been given to more mystical aspects of Freemasonry and Egyptian esoterica. He was a member of a Freemason order that was modeled after the Knights Templar. Disappointed that Freemasonry was going in a more rational direction, he joined the Illuminati in 1783. He took the codename "Aaron".

JOHANN JOACHIM CHRISTOPH BODE

Johann Joachim Christoph Bode (died 1793) was a publisher and considered an important Enlightenment-era translator. He should not be confused with Johann Elert Bode who suggested what is known as Bode's Law (or the Titius-Bode Law). Bode's Law was based on a spurious observation that each planet's orbit is roughly based on a multiplication of Earth's distance from the sun. J.J. Bode was recruited into the Illuminati by Adolph Freiherr Knigge in 1783. He went by the codename "Aemilius". Bode eventually published his autobiography titled *Biography of Bode*. He tried to write it in the third person, as if it were written by another author, but kept falling back into first person.

KARL VON ECKARTSHAUSEN

Karl von Eckartshausen (died 1803) was an author of early esoteric works, notably a work of Christian mysticism titled *The Cloud upon the Sanctuary*. In this book he develops a notion of a "society of the Elect". Members have been working in the shadows since the beginning of time to bring about an eventual theocratic republic. His book influenced the founders of the Hermetic Order of the Golden Dawn as well as Aleister Crowley. Eckartshausen joined the Illuminati for a brief time but quit when he concluded the group was given more to human reason than mysticism. He went by the codename "Attilius Regulus".

JENS BAGGESEN

Jens Baggesen (died 1826) was, for a time, a successful Danish poet and comedic writer. He attained rapid success by the age of 21. His meteoric rise was matched by a quick and nasty fall when wrote and staged an opera in 1789 called *Holger Danske*. Although today *Holger Danske* is considered a masterpiece, the production was a flop with 18th century Danish audiences. Opera goers, swept up by anti-German sentiments, stayed away in droves when it became known Baggesen was collaborating with Germans. Baggesen pulled it after only six performances. Unappreciated in his home nation, Baggesen left for Germany and joined an attempted Illuminati revival in 1790. He went by the codename "Immanuel". Danish Prince Friedrich Christian von Schleswig-Holstein-Sonderburg-Augustenburg paid Baggesen to tour Europe and talk up a revival of the order.

JOSEPH VON SONNENFELS

Joseph von Sonnenfels (died 1817) was a patron of Mozart and apparently really had it in for harlequin characters. As a literary critic, he zealously crusaded for their removal from Viennese theater. He apparently got his way. Possibly unrelated to his hate-on for harlequins, he was instrumental in getting Austria to abolish the use of torture. He joined the Illuminati in 1782 and headed up their Vienna outpost. His codename was "Fabius".

DANIEL GOTTHILF MOLDENHAWER

Daniel Gotthilf Moldenhawer (died 1823) was the head librarian of the Royal Danish Library. His personal mandate was to add as many books as possible. He didn't always add books via legal means. He apparently raided the libraries of monasteries during his travels through Europe. Moldenhawer joined the Illuminati in 1782. His codename was "Godeskalk".

Twelve Organizations or Corporations that Supposedly Reveal Conspiratorial Connections in Their Logos

It's fairly common for companies to hide a bit of information in their corporate logo. A classic example is the whitespace between the E and X in FedEx forms an arrow. The smile in Amazon's logo goes from A to Z. Baskin Robbins cleverly hides 31 in its logo. Conspiracy cranks seem to devote an inordinate amount of time pouring over corporate logos looking for Satanic, Masonic, and Illuminati symbols.

AMERICAN MEDICAL ASSOCIATION

The American Medical Association's logo is a snake winding itself around a staff. This is known as the Rod of Asclepius. It's frequently confused with Hermes' staff of caduceus. The AMA is already considered a Satanic organization by many in the alt med field. Some find confirmation in the group's logo. The snake winds its way around the rod and represents a stylized 666.

APPLE APP STORE ICON

The Apple App store icon features a stylized letter A (you know, for "App"). The A is formed from what appears to be a ruler, a pen, and a stylus. Cranks claim this icon actually represents the symbol of Freemasonry! So, stop using your computer, I guess.

CBS

The CBS logo is a stylized eye. Why would a TV network choose an eye? The only possible reason is, of course, the eye is meant to represent the Eye of Horus.

CERN LAB

CERN is probably best known for building and running the Large Hadron Collider (LHC). As well, Tim Berners-Lee invented the World Wide Web in 1989 while employed by CERN. There are a number of conspiracies surrounding CERN and its activities. The LHC is claimed to be a machine purpose built to destroy the world or open portals to other dimensions. CERN proudly displays a statue of Shiva on its grounds. This was a gift from India in the 1960s. Cranks read a lot into the Shiva statue. The CERN logo itself is subject to much consternation by conspiracy theorists and fundamentalist Christians who see Satan in every tub of grocery store margarine. Cranks believe the logo represents 666, the Number of the Beast! In reality the logo's looping curvy lines are stylistic representations of particle collisions.

The CERN logo is property of CERN. Used here for critical comment.

CHASE BANK

JPMorgan Chase is one of the largest financial institutions on Earth. It's no big surprise there is at least one conspiracy theory about its logo. According to Rule 286 of the Conspiracy Operations manual, the company is required to reveal its hidden allegiances via its logo. The Chase logo is a blue octagon with a white square in the middle. Why was that white square left in the middle? So, you can draw a Nazi swastika in it. Yup. That's it. That's the claim.

Illustration: Karl Mamer

DALLAS COWBOY CHEERLEADERS

The Dallas Cowboy cheerleaders were America's first professional cheerleaders. For many, they represent everything super wonderful and patriotic about America. As it turns out, these cheerleaders are leading the cheer for the Illuminati. Did you ever notice their uniforms are adorned with five pointed stars? Where do you see similar stars? On Chinese tanks! Is it more than a little suspect the five-pointed star is also used by Freemasons? Are cheerleaders urging crowds to love each other? No, they're prancing about, tempting people with their bodies, and urging the crowd to hate and kill their neighbor!

WALT DISNEY

Boy, one could write a whole other book about the things supposedly hidden in Disney films. But did you realize Walt Disney himself was a Satanist? No? Good. You're sane. However, that doesn't stop loons from seeing 666 in the W, T, and P in Walt's iconic signature.

Illustration: Karl Mamer

GOOGLE CHROME LOGO

Have you ever stopped to really look at the Google Chrome logo and realize it cleverly hides the Number of the Beast? No? That's because your brain operates normally. However, cranks believe the red, yellow, and green colors spinning off the central blue hub represent a stylized depiction of 666. We're also told, knowingly, the logo is a representation of the "divine king sign". What they won't tell you is there is zero historical context for this term and this sign. But, okay, sure. What is the divine king sign? It is the OK sign people make with their fingers but in wheel format. Apparently, the OK sign, with three fingers up, forms 666 and it's a way of telling others you're a Satanist or member of the Illuminati. The OK sign is also used by extreme Christian white power types. So, I'm not sure how they resolve that they're actually promoting Satanism.

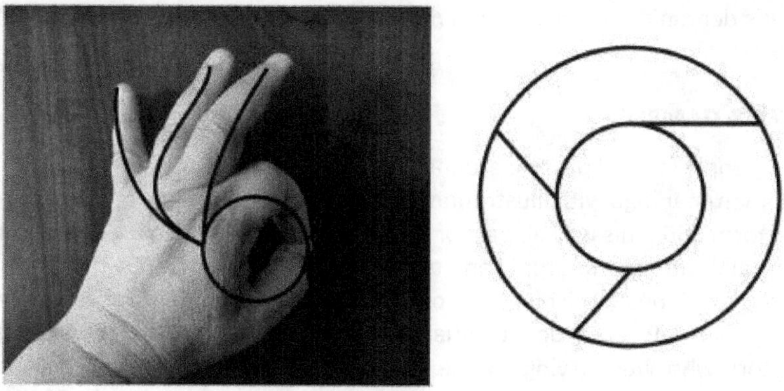

The "divine king sign" (666) and its stylized representation.
Illustration and Photo Credit: Karl Mamer

GMAIL ICON

Google isn't quite done exposing users to occult symbols. Its Gmail icon, which incorporates an M in an envelope image, is actually a representation of a Masonic apron.

MONSTER ENERGY DRINK

Cans of Monster Energy have what looks like a rip caused by three sharp talons. That's what they want you to think. Those who have taken the red pill see each rip representing the sixth letter in the Hebrew alphabet.

Photo Credit: Karl Mamer

The sixth letter (ו) also has the numeric value of six. So, each beverage can reads "666"! It's demonic. Just in case you missed that, every can urges you to "Unleash the beast!"

PROCTER & GAMBLE

In 1981, people started passing around Nth generation photocopies of P&G's man-in-the-moon/star logo with illustrations how you can supposedly connect the logo's stars to form 666. This was all part of the 1980s Satanic panic. P&G's initial stance was basically you can't fix stupid and tried to ignore the rumors. By the 1990s, P&G had enough of loons tarnishing the company's good name. P&G retired its original logo and then lawyered up. It initiated a libel suit against a group of Amway distributors who were trying to use the rumor to bring more suckers into the Amway fold. The case took more than a decade to resolve but in 2007 a jury awarded P&G $19.25 million in a libel suit against four Amway distributors who tried to tie P&G to Satanism.

STARBUCKS

Starbucks seems to go out of its way to wind up Christians every Christmas with its holiday season cup design. It's never quite Jesus-y enough for some. When Christians are not fretting over Starbucks' failure to pump the Christian take on the solstice, they're raging about the company's logo itself. Turns out the logo is replete with Satanic and Illuminati symbolism. For example, in the logo used until 2011, you can connect the three stars in the logo to form a pyramid. A clear Illuminati symbol! If you rotate the logo 180 degrees, the siren image resembles the horned head of Baphomet! Some make a connection between Starbucks' logo and a secret society called The Ordo Baphometis. Admittedly the logo looks awfully familiar. Did Starbucks base its logo on an ancient order of Satan worshipers? Not likely. The Ordo Baphometis seems to be a recent creation of a German esoteric author named Walter Jantschik. He self-published the group's degrees and rituals online back in 2014. It's likely a modern creation and possible the creator took inspiration from the Starbucks logo

Twenty Conspiracies by Country

AUSTRALIA: DISAPPEARANCE OF HAROLD HOLT

Harold Holt was the Prime Minister of Australia in the late 1960s. While out for a swim in the ocean in 1967, he disappeared. His body was never found and it is generally believed he simply drowned. At least one friend accompanying him on his swim saw him enter waters the other swimmers judged unsafe. There is not much of a mystery here. However, various conspiracy theories have emerged over the years. Some believe Holt was really assassinated by the CIA because he was about to pull Australian troops out of Vietnam. Others think he was actually a spy for China and he was "exfiltrated" by a Chinese submarine crew.

BRAZIL: AVRIL LAVIGNE DIED AND WAS REPLACED BY A LOOK ALIKE

In 2011, a Brazilian blog called *Avril Está Morta* (which translates to "Avril is Dead") floated the theory Avril Lavigne committed suicide some time around 2002. Her record company replaced her with a look alike named "Melissa". Apparently, Melissa was a body double the record company employed as a decoy to lure away crazed fans and the paparazzi. The conspiracy theory jumped to the Northern Hemisphere in 2015 when BuzzFeed tweeted about the blog. Lavigne does, however, assure her fans she is very much alive and does not answer to the name Melissa. Ever.

CANADA: THE AVRO ARROW CONSPIRACY

In the 1950s, Canada's Avro aircraft maker was developing a successor to its CF-100 "Canuck" fighter. This new aircraft was a highly advanced delta-wing interceptor called the Arrow. The plane was intended to fly at Mach 2 and shoot down Soviet bombers coming over the Arctic Circle. Shortly after the first Arrow rolled off the assembly line for testing, the Canadian government of the day abruptly canceled the project. The official explanation for the cancellation is the program was expensive, Canada was unable to sign up foreign buyers, and Sputnik made it clear expensive interceptors provided no defense against ICBMs. A certain cult has developed over the Arrow in Canada that centers on a conspiracy theory the US feared Canada would eclipse America's aerospace industry. The US government applied political pressure on the Canadian government to cease all work on domestic fighter aircraft development.

The Avro CF-100 Canuck is the only fighter designed and mass produced in Canada.
Photo Credit: Orval Mamer

DEMOCRATIC REPUBLIC OF THE CONGO: EBOLA CONSPIRACY

Efforts in 2019 to fight an Ebola outbreak in the Democratic Republic of Congo were stymied when gangs of armed men destroyed medical facilities and murder medical workers under the belief Ebola was a conspiracy. Beliefs ranged from the medical workers themselves were the cause of the outbreak to the outbreak was seeded in the population to delay elections.

FRANCE: MOLIÈRE AUTHORSHIP QUESTION

Molière (died 1673) was a playwright. The French consider him one of their greatest writers. He is France's Shakespeare. In 1919 French writer Pierre Louÿs (died 1925) published a series of articles questioning whether or not Molière was capable of authoring the corpus of work attributed to him. Louÿs argued Molière was not known by contemporaries to be a writer and was actually better known as a director for a comedy troupe. How could one of France's greatest writers not be recognized by his contemporaries? Louÿs concluded French playwright Pierre

Corneille actually ghost wrote a majority of Molière's most famous works. Modern day textual analysis of the writings of Molière and Corneille, however, shows distinct writing styles and makes Louÿs' theory unlikely.

GREECE: THE MACHINATIONS OF THE EPSILON TEAM

In 1977, George Lefkofrydis believed he had deciphered the works of Aristotle and found a hidden message from the Greek philosopher. What was the hidden message? Turns out Aristotle was a space alien from a solar system called Mu. The knowledge of Aristotle's true nature has been preserved over the centuries by an unnamed secret society. Lefkofrydis published these findings in a book titled *Spaceship Epsilon: Aristotle's Organon: The Researcher*. Since the book's publication, other writers have expanded on Lefkofrydis' conspiracy theory. Athenian journalist Ioannis Fourakis added to this mythology in a 1989 book titled *Messages of Delphi*. Fourakis claimed Lefkofrydis' secret society called itself the Epsilon Team. According to Fourakis it wasn't just Aristotle who was a space alien but the Greeks themselves are descendants from space aliens sent to Earth as part of a cosmic war against the Jews. Sigh. In many ways, the Epsilon Team conspiracy is Greece's version of Qanon, although it predates that particular band of idiots.

HUNGARY: THE SUPPRESSED HISTORY OF THE HUNGARY LANGUAGE

The Hungarian language is a bit of an odd duck. It's not a Romance language. It's not Slavic. It's not a Germanic language. Hungarian is a Uralic language. It's the only one you'll find in central Europe. As the name implies, Uralic languages trace their origins to the Urals. There are a number of Hungarians that reject this Russian origin, however. The actual origin of Hungarian has been suppressed by historians and linguists. Some hold a belief the language and people don't trace themselves back to the Urals but trace right back to ancient Sumer. In the view of some, Hungarian has far too many similarities to Sumerian to be mere coincidence. This is, however, a case of cherry picking. There is no modern language today that bears any resemblance to the grammar and syntax of Sumerian. Others posit Hungarian traces its roots back to Biblical times. Some see similarities between names in the Bible and Hungarian names. For example, some Hungarians see similarities between "Jesus" and the Hungarian name "Jász". Jesus, therefore, was not actually Jewish but Hungarian.

INDIA: LOVE JIHAD

The so-called Love Jihad is a conspiracy theory that Muslim men are purposely trying to seduce and/or marry Hindu and Christian women in India under false pretenses. Their goal is to convert them to Islam. The Love Jihad conspiracy theory arose in 2009 and was initially spread by right-wing Hindu organizations. The conspiracy was given legitimacy when it was adopted by the Bharatiya Janata Party (one of India's major parties) as a key election issue. The Love Jihad conspiracy caused an Indian court in 2017 to annul the marriage of an Indian woman to a Muslim man against the woman's wishes.

JAPAN: THE FINAL RESTING PLACE OF JESUS

Shingō village in northern Japan is supposedly home to the tomb of Jesus Christ. How Jesus arrived in Japan is a little complicated but the story goes, like a lot of 20-somethings, Jesus spent some time in Japan. He returned to Judea, did the Jesus stuff we know from the Bible, but managed to avoid being crucified when his brother Isukiri took his place on the cross. Jesus made his way back to Japan, married a nice Japanese woman, and became a rice farmer. He had several children and lived to the ripe old age of 106. This story comes to us from a young Shinto priest named Takeuchi Kyomaro (died 1965) who claimed he had uncovered an ancient set of manuscripts that told a secret history of Japan. The story of Jesus in Japan comes to us from these manuscripts. There are many problems with Takeuchi's claim. One of the more obvious ones is that these supposedly ancient documents mention cities like New York, Boston, and Johannesburg.

KOREA: THE K WAS IMPOSED BY JAPAN

You will find a number of ardent Koreans and fans of Korea who spell the country's name with a C. So Corea. There are those who argue the original English spelling of Korea was the C variant. The K variant was later imposed by Imperial Japan sometime before their occupation of the peninsula between 1910 and 1945. The reason for that is Japan didn't want the alphabetically superior Corea to enter the stadium before Japan in the parade of nations at the 1908 Olympics in London. Yeah, petty. There is zero evidence the Japanese did this, however. Arguing against the conspiracy theory, linguistic studies indicate the K variant was well in use by the end of the 19th century. As well, it's doubtful English-speaking nations like the UK and the USA would bow to Japanese pressure at the start of the 20th century to redo their maps to match Japan's preferred spelling.

LUXEMBOURG: THE BOMBERS AFFAIR

Between 1984 and 1986, Luxembourg experienced a number of mysterious bomb attacks. The perpetrators were never identified but their chosen targets were public buildings and infrastructure sites. No one was killed during the attacks and it appears the bombers went out of their way to avoid casualties. Starting in 2014, Luxembourgian prosecutors zeroed in on half a dozen members of the *Brigade mobile de la Gendarmerie* (basically, an elite SWAT-type police unit). The prosecution's theory is the members conducted a series of false flag bombings in hope of getting more funding for their unit from the government. A popular theory today is the bombings were actually committed by a paramilitary stay-behind group. These stay-behind groups were formed by NATO in many Western European nations, including Luxembourg. Should the Soviets steam roll over Western Europe, these units were charged with conducting acts of terrorism and assassination against Soviet occupying forces. The existence of these groups was a well-kept secret until the 1990s when a scandal in Italy revealed their existence. These groups were highly irregular paramilitary units that tended to be staffed by right-wing thuggish types with alleged connections to the fabled Italian P2 Masonic order. Conspiracy theorists believe these clandestine stay-behind groups sought to create conditions throughout Western Europe that were ripe for a coup. Prince Jean, Grand Duke of Luxembourg, somehow became implicated as well. Around the times of the bombings, Prince Jean renounced his right of succession to the Luxembourgian throne. The Grand Duke never revealed his motivation but some saw the timing of his renunciation as suspicious. Luxembourgian prosecutors took the allegations seriously enough that they interviewed Prince Jean in regard to the bombings. The Prince had convincing alibis for many of the dates in question.

MEXICO: THE OSWALD CONNECTION

Some JFK conspiracy theorists view Oswald's visit to Mexico City a month before the Kennedy assassination as evidence he conspired with Mexican communist to bump off the president at the behest of Fidel Castro (died 2016). The evidence for that is the testimony of Mexican reporter Óscar Contreras Lartigue (died 2016). Contreras claimed he was part of a pro-Castro group while studying law in 1963. He professed to a number of American investigators over the years he had met Oswald and had knowledge of the plot. He gave few concrete details, however. He always begged off, hinting his life was in danger if he said too much. In 2020, Gonzaolo Soltero, a professor at the university Contreras claimed to have attended, tried to verify his past. Soltero found Contreras was not actually a student at the time but was a gossip columnist for a local paper. Contreras' writing indicates he was attending a wedding some 300 miles away from Mexico City at the time Oswald paid a visit.

NEW ZEALAND: THE ALL BLACKS WERE POISONED AT THE 1995 RUGBY WORLD CUP

The All Blacks is New Zealand's national rugby team. They're the Dallas Cowboys, the Yankees, and the LA Lakers all rolled into one. They are an enduring symbol of New Zealand and a huge source of pride. In 1995, the All Blacks were in the finals of the Rugby World Cup in Johannesburg. Two days before their game many of the team members experienced diarrhea and vomiting. The team lost to South Africa. Soon after the loss rumors emerged the team had been poisoned. The team's coach posited that a waitress named Suzie was to blame. Suzie mysteriously disappeared shortly before the final. The coach went as far as to hire a private investigator to track down the waitress. The PI came up empty. In 2001, Rory Steyn published his memoirs of working as Nelson Mandela's bodyguard. In his book, he alleged the All Blacks were definitely poisoned. He claims a betting syndicate was behind the poisoning. Ed Griffiths, who was the South African team's CEO at the time, suggested the All Blacks have no one to blame but themselves. He reported seeing team members overindulge in seafood the day before the game.

POLAND: SMOLENSK AIR DISASTER

On April 10, 2010 a plane carrying Polish president Lech Kaczyński and several high-ranking members of the military crashed near the Russian city of Smolensk. Kaczyński's intent was to visit Russia to honor Polish soldiers killed during the Katyn massacre. Both Russian and Polish officials concluded the crash was due to pilot error. The pilots tried to land in thick fog and their aircraft was well below the normal approach altitude. The plane hit a tree during the landing attempt and crashed. It has been suggested the Polish president ordered the pilots to attempt the landing instead of diverting to another airport. After Kaczyński's death his twin brother Jarosław Kaczyński jumped into politics and campaigned on the notion his brother's death wasn't a result of pilot error. He was assassinated by the Russians! Kaczyński parlayed the tragedy into a win for his Law and Justice party in Poland's 2015 election. Upon forming the government, Kaczyński reopened an investigation into the plane crash. Claims emerged of bomb traces being found on the wreckage. Part of the investigation involved exhuming the remains of almost every victim and threats to prosecute the trip organizers, which includes Donald Tusk. Tusk was prime minister at the time of the crash and is currently the president of the European Council.

ROMANIA: MOUNTAIN SPHINX

Photo Credit: Source Wikipedia. Released to the public domain by Raluca Stoica

There's a natural rock formation in Romania's Bucegi National Park that from certain angles resembles the Great Sphinx of Giza. It's on a list of Romania's Seven Natural Wonders. There are a number of conspiracy theories involving historians and geologists who have suppressed evidence the formation is not natural. It is, in fact, proof of ancient contacts between Romanians and Egyptians. Others claim there is a cave under the rock formation. The cave is protected by a force field, preventing people from entering. Should someone manage to figure a way in, legend has it they would find a room created by aliens to store time crystals that preserve the totality of human knowledge and history. Naturally, this has been kept secret by an agreement between the Romanian government and the CIA.

SWEDEN: THE ASSASSINATION OF OLOF PALME

Olof Palme was Sweden's Prime Minister in the mid-1980s. Sweden is a country notorious for its low crime rate and safety. Prime Minister Palme generally avoided the use of bodyguards and went about life as a normal citizen. Unfortunately, an assassin took advantage of that and shot Palme dead as he emerged from a movie house with his wife. His murder was never solved, although Swedish police eventually identified Stig Engström (died 2000) as the most likely suspect. Various conspiracy theories have been proposed. Some people believe Palme was assassinated by a pro-Apartheid group after he gave a keynote address to an anti-Apartheid group. Another theory is that he was murdered by international arms dealers after he discovered a shady deal made by Bofors to sell arms to India. In 1990 American Richard Brenneke (died 2015) started claiming he worked for the CIA and had knowledge Italy's Masonic P2 lodge murdered Palme. The CIA disputes Brenneke's allegations he worked for the agency. Brenneke's credibility was highly questionable. He was indicted for perjury in the late 1980s. He had previously asserted without evidence he was a central figure in the October Surprise and the Mena scandal.

TURKEY: THE SECRET ARTICLES OF THE TREATY OF LAUSANNE

The 1923 Treaty of Lausanne was signed by the allied powers of World War I and recognized the borders of Turkey. There is a belief this treaty was accompanied by secret articles that forbids Turkey from exploiting its natural resources and claiming northern Iraq (and its oil) as its own. These secret articles have kept Turkey from emerging as a first world power. However, as the conspiracy theory goes, these secret articles will expire in 2023 and allow Turkey to seize northern Iraq's oil fields and everyone will be rich!

THE UNITED KINGDOM: THE DEATH OF PRINCESS DIANA

Probably one of the best known and most enduring conspiracies associated with the UK is the belief the UK government and the Royal Family murdered Princess Diana. Both French and British investigations ruled her 1998 death was an accident, caused by her driver being drunk and the lack of the passengers using seatbelts. The sole survivor, bodyguard Trevor Rees-Jones, was the only person wearing his seatbelt. Conspiracy theorists suggest Diana was murdered for two reasons: this would allow Prince Charles to remarry without having to renounce his claim on the throne and fears Diana would marry her Muslim boyfriend Dodi Fayed and bear a Muslim half-brother/sister to future king Prince William. Evidence for the conspiracy is thin. Some allege her driver was in the pay of MI6. Given the driver wasn't himself wearing a seatbelt, it's not clear how MI6 convinced him to die in a crash along with the princess. The MI6 theory was further reinforced by Richard Tomlison, a former MI6 officer. Tomlinson was dismissed from the agency in 1995 following a probationary period. His termination was allegedly due to being a poor fit as a spy. In 2008, he came forward with claims Princess Diana's driver was indeed an MI6 agent and he had knowledge that MI6 was spying on Diana before her death. He added her death eerily mirrored a plan he saw for a proposed assassination attempt on Serbian president Slobodan Milošević. Conspiracy theorists make a lot of hay out of a finding that there were no less than 14 CCTV cameras along Diana's final route and yet not one managed to record the car crash. A British investigation looked into potential CCTV camera footage but found the alleged cameras were actually security cameras pointing at the entrances to buildings.

VIETNAM: THERE ARE STILL MIA AMERICAN POWS BEING HELD

During and after the Vietnam War America did not have an accurate accounting of how many service personnel were captured by the North Vietnamese. Estimates ranged from 600 (based on the best intelligence) to 1,500 (a politically motivated estimate provided by the Nixon government). Following the signing of the Paris Peace Accords, 591 POWs were returned to America. The 1,500 estimate heavily promoted by Nixon along with the US military MIA figure of 2,477 led many to believe the Vietnamese kept hundreds of POWs back. An MIA designation does not necessarily mean the US military has no clue of a soldier's fate. It simply means the US was never able to recover a body. For example, 436 Vietnam-era pilots shot down over the sea will forever be designated as MIA because their remains are non-recoverable. The modern myth of American POWs being held as slaves in Vietnam got its start with the testimony of Private First Class Robert R. Garwood. Garwood was court martialed by the US military for collaboration with the enemy. Garwood sought immunity by claiming he had knowledge of POWs still being held. Films like Chuck Norris' 1978 film *Good Guys Wear Black* and 1985's *Rambo* catapulted the conspiracy theory into the public consciousness like nothing else. Various marginal types seized the spotlight. Probably the most famous was Bo Gritz who led his own private expeditions to Southeast Asia to find POWs. Gritz was financed by H. Ross Perot (died 2019). In 1980, Perot started throwing money around, funding a number of cranks. In one case, Perot was prepared to pay UK businessman Robin Gregson $4.2 million for a videotape that Gregson alleged showed evidence POWs were being held as slaves in Vietnam. There was just one problem. Gregson was at the time sitting in a Singapore jail. Perot paid his $100,000 bail to spring him. Right before Perot was to sign the contract to acquire the tape, Gregson informed Perot he had burned it. Convenient. Congress eventually addressed the issue with 1991's Senate Select POW/MIA Committee headed by John Kerry and John McCain. Their committee concluded there was no compelling evidence "that any American remains alive in captivity in Southeast Asia".

ZAMBIA: REJECTS GMO FOOD DONATIONS

In 2002 Zambia was on the verge of a famine. Their grain crops were wiped out by a drought. The government rejected 35,000 tons of needed food aid because much of the donated grains were GMO products and they were "poison". The Zambian president proclaimed his citizens were not going to be "guinea pigs" despite the food aid being mostly GMO corn that North Americans have been eating safely since the 1990s.

Eight Well-Worn Conspiracy Tropes

CONTRACTUALLY OBLIGATED "TELLS"

The secret societies that are alleged to rule the world don't seem to do a very good job at keeping themselves secret. They reveal their presence and plans via easily decoded symbols and code words. It is almost as if they are contractually required to make it easy for the yokels to discover their existence and understand every detail of their plan.

CRISIS ACTORS

Emergency training will employ crisis actors. These are people that play injured and distressed victims. In 2012, following the Sandy Hook school shooting, conspiracy theorists adopted it as a trope. Real victims of things like mass shootings were actually crisis actors paid to be part of a staged false flag operation.

THE END GAME IS IMMINENT

Conspiracy authors are always warning we're 2 minutes to midnight. The evil plans of the conspirators are in motion and time is short. Alex Jones has traded on this trope since the 1990s. So far his dire predictions the New World Order was on the verge of taking over have failed time and time again.

A FAILED PREDICTION IS EVIDENCE OF THE CONSPIRACY

What happens when conspiracy prophecy fails and the imminent plans do not materialize? Almost never does a conspiracy theorist admit he or she was wrong. "Well, there is no conspiracy." That the conspiracy didn't happen is recast as evidence for how truly powerful the conspirators are. They are able to just disappear without a trace after their plans were revealed by conspiracy theorists!

FALSE FLAG

A false flag operation is typically some kind attack or atrocity committed by one group with the intent to cast blame on another group. A false flag is usually then a pretext for action that was previously considered politically untenable. There are clear historical examples of governments employing false flag operations. The Japanese staged the Mukden Incident as a pretext for the invasion of Manchuria in the 1930s. Conspiracy theorists will always point to the Reichstag fire as a false flag operation launched by Hitler to clamp down on his communist political opponents. However, there's considerable debate among legitimate historians as to whether the Reichstag fire was a false flag operation or an actual act of terrorism by a communist activist. In the mind of modern conspiracy theories, there is almost no tragedy or act of terrorism on domestic soil that isn't a false flag operation by the American government with the aim to seize guns and put patriots in concentration camps.

THE CONSPIRATORS, THE SHEEPLE, AND THE WOKE

Conspiracies have three main players. There are the conspirators. They are incredibly evil, can command vast resources, have access to technology a generation beyond what is known, but make really fundamental errors that reveal their plans time and time again. There is also the woke. They have pierced the veil. Their self-appointed duty is to fight the conspiracy at every turn and inform the world. Alas, there are also the sheeple. These are people who unknowingly do the bidding of the conspiracy.

NO EVIDENCE IS EVIDENCE OF THEIR POWER

Wild-eyed conspiracy theories generally have one thing in common: a complete lack of positive evidence. This is not troubling to conspiracy theorists. It is only proof of how powerful and how deep the conspiracy goes! During the Satanic panic of the 1980s, many people who gave positive evidence against claims of Satanic abuse immediately became suspected of being part of the Satanic cover up.

THERE ARE NO COINCIDENCES

Life can be random at times. This is a fact we all have to deal with. Conspiracy theorists have a very hard time accepting this. They will go to extremes to create a narrative that ties together what any other right-thinking human being would see as mere coincidence.

POLITICS, LAW, & CRIME

Fourteen Fictitious or Misunderstood Treaties or Laws

AGENDA 21

Agenda 21 is a non-binding plan developed by the United Nations to help countries around the world build out a framework for economic growth that is sustainable and environmentally friendly. It was developed during the 1992 Earth Summit in Rio. Many conspiracy cranks think Agenda 21 is a road map to world domination and population reduction by the New World Order and its stooges within the US government. Opposition was stoked in the early days by Tea Party members. They warned American support for Agenda 21 would lead to a loss of American sovereignty. The unhinged opposition that grew out of Tea Party whinging gave us the wonderful concept of "eco-totalitarian regimes" forcing everyone to use bikes to get to work. Opposition to Agenda 21 quickly became a platform of the Republican Party and many red states started proposing laws that would strip public financing from any groups that tried to implement any facet of Agenda 21 at the local level. Since Agenda 21 basically summarized a lot of environmentally friendly projects that groups have been doing for years, like planting trees, these laws actually threatened to cut off the funding of rather benign service groups like the Boy Scouts.

CODEX ALIMENTARIUS

Codex Alimentarius means "Book of Food" in Latin. It's a set of labelling and food safety standards published by the UN. The ultimate goal of this set of standards is to protect public health. The codex has been around since the early 1960s. Almost every domestic industry that wants to trade with the rest of the world needs to meet the standards of the country it wants to trade with. So sane people generally believe if everyone agrees upon one standard, it's better for all. Industries can comply with international standards voluntarily or a nation may mandate it. In the USA a food maker's application of Codex standards is purely voluntary. The FDA sits on various Codex committees but Codex recommendations do not get written into US law. In 2005, with the growing popularity of vitamins and supplements, the Codex released guidelines for those products. As we've seen over the years, what's on the label of an expensive supplement is not always what's in the bottle. American health nuts and people who profit greatly from selling mostly useless supplements started to freak out when they found out the Codex was adding these products to its purview. To whip up fear in their gullible customers and get them to apply pressure on their Congress members, the supplement industry began to make a series of wild-eyed claims. Those included conspiracies about one world government, international control over America's food supply, the complete ban of vitamins, mandated eugenics, and requirements that all food has to be irradiated.

What had been a sleepy, entirely boring international dialog the average person had little motivation to look into suddenly became billed as a super-secret organization long hidden from the American public that's been controlling the world food supply for generations!

THE CONSTITUTION OF THE NEWSTATES OF AMERICA

The Center for the Study of Democratic Institutions (CSDI) was a think tank set up in 1959 using a $15 million grant from the Ford Foundation. Its formation was a response to McCarthyism. Its intended goal was to conceive of ways to protect constitutional liberties through multidisciplinary dialogs. The group hosted talks and basically paid people to write up policy papers. One of its members was a former FDR economist named Rexford Tugwell. One of Tugwell's many pursuits was thinking of different constitutional models. In the late 1940s, Tugwell worked on a "Word Constitution" as a way to curtail the proliferation of nuclear weapons. Under the auspices of the CSDI in the late 1960s, he drafted a constitution for the "Newstates of America". The "Newstates" appellation comes from his plan that the USA should be reorganized into a series of new states, each with no less than 5% of the nation's population. Although Tugwell's constitution was published more than half a century ago and never went anywhere, to this day conspiracy types make a lot of hay over it. To them it's a smoking gun exhibit that the New World Order has a new constitution ready to swap into place. They get pretty worked up over the constitution's mention of the United Nations and a lack of a Bill of Rights. Of course, the original US constitution had no such Bill of Rights either. Milton William Cooper ran a copy of it in full in his book *Behold a Pale Horse*. Cooper seemed disturbed over the CSDI's ostensible ties to the Rockefellers.

THE EXTRA-TERRESTRIAL EXPOSURE LAW

Many UFO cranks point to 1969's Extra-Terrestrial Exposure Law as proof positive the US government knows we're being visited by aliens. They assert the law gives the government the authority to round up citizens in the event of an alien takeover. Cranks that cite this law are typically unaware the law was revoked in 1991. The year of passage (1969) should also be another tip off. The law was meant to give the government the legal authority to quarantine the Apollo astronauts after their return from the moon, as well as other NASA workers who might have accidentally been exposed to Moon dust.

GREADA TREATY

The Greada Treaty is the made-up name of a treaty President Dwight Eisenhower supposedly signed with Grey space aliens on February 20, 1954. Although no one has ever produced a copy of the treaty, Ufologists are very certain of the date the treaty was signed. On February 20, 1954 Eisenhower was known to have been on vacation and paid an unplanned visit to a dentist after he chipped a tooth eating chicken wings. In UFO mythology this unplanned dentist visit was a cover story for a treaty signing between Eisenhower and space aliens at Holloman AFB. The treaty itself allowed space aliens to abduct a number of human beings every year and in return aliens would supply America with increasingly advanced technology. This mythical treaty was popularized in the 1990s by the works of Milton William Cooper and John Lear. Although Cooper later disavowed his space alien claims, the idea Eisenhower signed a treaty with space aliens was too good to let go of by the wider UFO community. The treaty seems to have gotten an actual name in the mid-1990s. The earliest mentions of the treaty name show up in 1995 in talks given by UFO lecturer Philip Schneider (died 1996). Schneider is probably best known for his vivid stories about working at the fictional Dulce Base and his involvement in the so-called Dulce War. The Dulce War was an imagined battle between Grey aliens and their human coworkers.

THE JEWISH TAX ON FOOD

In the late 1990s, Canada's IRS (the Canada Revenue Agency) started to get bogus refund claims for the so-called "Jewish tax on food". What now? Many food products, if you look closely at the label, have a symbol that indicates the product is kosher.

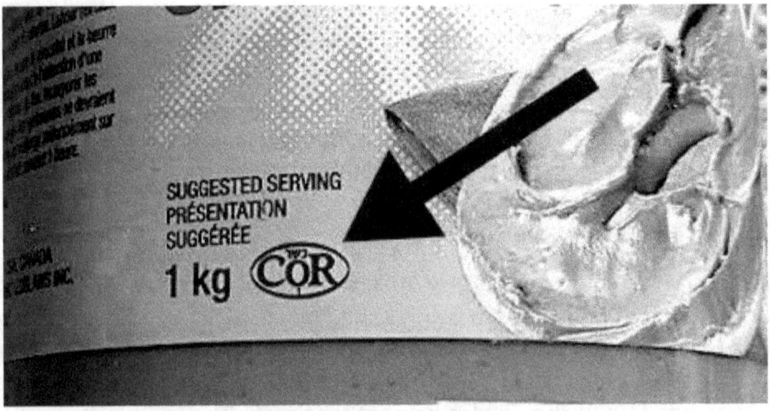

Example of a kosher certification symbol used by the Kashruth Council of Canada.
Photo Credit: Karl Mamer

Many companies apply for such certification. It helps them sell more product, not only to Jewish people but Muslims, Seventh-Day Adventists, lactose intolerant, and a growing segment of the population that thinks a kosher designation means the food is safer. Antisemitic conspiracy nuts claim the cost of attaining and maintaining a kosher designation is incredibly expensive, adds 10% a year to the family food budget, and funds paid to Rabbis get funneled directly to the Rothschilds and Zionist Occupational Government types. The reality is the cost to have a bottle of Coke certified as Kosher adds a fraction of a fraction of a fraction of a cent. The cost is so low, a family's yearly food budget wouldn't change by even a single penny if food makers no longer sought certification. The small fee the Coca-Cola corporation or Nabisco pays to a certification organization simply goes to support the work of the certifying organization. The funds are not funneled back to Israel.

THE LOGAN ACT

Whenever an American citizen speaks with or negotiates in some fashion with a foreign leader, there are people always quick to assert the person is in clear violation of something called the Logan Act. The act is real. It was passed in 1799 and criminalizes American citizens, unless authorized, negotiating with foreign governments "in relation to any disputes or controversies with the United States, or to defeat the measures of the United States". However, the law does not forbid American citizens from conducting any kind of negotiation with any kind foreign government. The intent of the law is to keep American citizens from muddying the diplomatic waters over actual disputes America is having with a potentially hostile power.

NAFTA SUPERHIGHWAY

North America's Corridor Coalition (NASCO) is a small non-governmental organization that advocates for bringing efficiency to trucking (and other modes of ground shipping) across North America. A large number of goods flow between Canada, the USA, and Mexico via a limited set of north-south highways. NASCO's goal is to encourage governments to expand and maximize this inter-continental highway system. Although not a government agency, it does get a small bit of funding from the Federal government. Somehow, in 2007, a NASCO map of proposed improvements to a series of highways used heavily to move goods between countries was copied and relabeled the "NAFTA Superhighway blueprint" by conspiracy cranks. It was billed as proof the government's end goal is a North American superstate. It was claimed the highway would be a quarter mile wide and divide America in two, destroying the local environment along its route as well as local economies. State legislators started passing reactionary laws against state funds being used to support this imagined highway. Of course, no one could ever point out who in government was actually planning this or directing Congress to cough up the money. No one could point to a source demonstrating the highlighted map's route was an official blueprint for constructing a quarter mile wide highway. The reality was NASCO's map simply referred to suggestions of where existing highways could be improved to handle increased international trade brought on by NAFTA. The idea was to eliminate certain obvious choke points and make improvements to existing roads (like removing traffic lights along the corridor).

NATIONAL ECONOMIC SECURITY AND RECOVERY ACT

It's not rare to encounter an engineer who thinks engineering principles can be used to fix the economy. You have known inputs and outputs. So, just math it. You know? In the late 1980s, engineering consultant (drink!) Harvey Barnard (died 2005) released sample legislation he called the National Economic Security and Recovery Act (NESARA). He felt income tax should be replaced by a national sales tax and the US should return to the gold standard. In his view, this would lead to a more stable economy and 0% inflation. He was so confident politicians would see the wisdom of his plan that he sent a copy of his act to every member of Congress and fully expected them to pass it within a week. They did not. He eventually published it on the Internet in the late 1990s and set up a foundation to promote his ideas. Around the same time a group of followers of Ramtha channeler J.Z. Knight were elbow deep in a failing pyramid scheme called the Omega Trust. In 2000, the group had just witnessed the head of the scam arrested after he failed to pay members their promised returns ($50 for every $1 invested!). With Omega Trust's founder languishing in jail, investor and J.Z. Knight adherent Shaini Cadence Goodwin (died 2010) stepped up as the spokesperson for the scam. She told nervous investors to hold tight. Keep the faith as she had uncovered a remarkable secret law that would guarantee their investment and pay them their promised fortune. That law? Yup, NESARA. See, NESARA wasn't just some rando dude's attempt at tax-reform LARPing. It was actual legislation secretly passed by Clinton in the final days of his administration and had the backing of an overwhelming majority of Supreme Court justices. In fact, the time was now to borrow heavily and invest more in the Omega Trust. You couldn't lose because at the core of Barnard's secretly enacted NESARA was debt forgiveness. NESARA offered many things but debt forgiveness wasn't an actual feature of the proposed act. Goodwin told skeptics who were unable to find any paper trail establishing the act was passed that all the records were kept in the World Trade Center before it was destroyed in 2001 by nefarious forces working against NESARA. Convenient. Goodwin was able to keep thousands of Omega Trust investors on the hook for years until her death in 2010. After that, it got even weirder. Several New Age and UFO gurus worked NESARA into their mythologies of space brothers on the verge of mass landings to bring a new era of peace and prosperity and the mass arrests of most of the world's bankers.

NORTH AMERICAN MONETARY UNION

In 1999 Canadian economics professor Herbert Grubel published a report for the right-wing Fraser Institute. In it he advised North America should adopt a common currency, similar to the Euro. He dubbed it the Amero. The notion of the Amero never went beyond one white paper. It's very unlikely America would ever give up the greenback as the U.S. dollar is already the world's reserve currency. Any realistic monetary union would simply be Canada and Mexico adopting the American dollar.

Jerome Corsi, famous for "Swiftboating" presidential candidate John Kerry, seized upon this sleepy Canadian proposal and used it to stoke fears of a looming socialist superstate. An odd framing given the proposal emerged from the Fraser Institute, which is an extreme right-wing Canadian group vigorously opposed to the social state. Fears of the impending monetary union were further reinforced when coin designer Daniel Carr released a set of fantasy coins modeled after what he envisioned the Amero currency would look like. Carr is a legit coin designer, having designed actual circulating American coins. His motivation for releasing images of his fantasy coins is a bit murky. He suggests he wanted to generate dialog on the issue and leaves it at that. Images of the coins started getting passed around. They were stripped of their original provenance and re-dubbed as images of leaked coinage the US mint was prepping for the looming monetary union.

THE SECRET FREE TRADE AGREEMENT

Before NAFTA roped in Mexico, Canada and the USA signed the Free Trade Agreement (FTA) in 1988. In the early 1990s, Shelley Ann Clark, who worked as a secretary for one of the Canadian negotiators, started blitzing the media with claims the FTA as signed into law by Canada and the USA wasn't the real trade deal. A secret trade deal was signed and hidden in a bunker outside of Ottawa. And not just any bunker, mind you. This bunker, for reasons never explained, had to be exactly 16 miles outside of Ottawa. Part of the secret FTA was a nefarious implementation plan that called for fomenting social and economic unrest in Canada. The real, secret FTA wasn't about selling more stuff to the USA but a plan to sell Canada as a nation to America by 2005. Oddly that never happened. No one who believed in this secret FTA could ever explain how a secret trade deal, not enacted by legislation, and seen by no one, could have any legal force.

THE SECRET TREATY OF VERONA

The great European powers met in Verona in 1822 to settle some issues left hanging after the close of the Napoleonic Wars. A year after the Conference of Verona, the UK newspaper *Morning Chronicle* published the so-called "Secret Treaty of Verona". The treaty was supposedly a pledge by the great powers of Europe to resist democratic rule, suppress freedom of the press, and preserve the divine right of kings. Although the treaty was quickly debunked by other newspapers, it arguably helped contribute to the establishment of the Monroe Doctrine. Here was proof positive that the Great European Powers were going to use any means necessary to snuff out democracy wherever it might arise, even in the Americas! Conspiracy cranks have been quoting it and reprinting it in full for nearly two centuries. It is typically used as evidence that the designs of the New World Order are both real and have deep historical roots among the elite ruling class.

THE SECURITY AND PROSPERITY PARTNERSHIP OF NORTH AMERICA

The Security and Prosperity Partnership of North America (SPP) was a working group set up by the George W. Bush administration in 2005 to discuss ways Canada, the USA, and Mexico could increase cooperation in terms of security and economic issues. The discussions wrapped up in 2009. Cable news disgrace Lou Dobbs began warning viewers the SPP was really a plan to merge Canada, the USA, and Mexico into an EU-style North American Union. American nationalists piled on with wild-eyed stories that the SPP was a plan to bring millions of illegal Mexican laborers into the country and steal jobs. North of the border, Canadian cranks joined in: The SPP was really a plan to sell all of Canada's water to the USA and force Canada to adopt America's low labor standards.

US-MOROCCO TREATY OF 1787

The US signed a treaty of friendship with Morocco in 1786. The treaty, in essence, opened Morocco's ports to US ships. The treaty has some historical significance as this was one of the first official recognitions of the USA as an independent nation. It is also America's longest lasting unbroken treaty. There's a sovereign-citizen group who follow something they term "Moorish American Science Temple" or "Moorish Law". They believe there is a 1787 treaty that guarantees Moorish-Americans are exempt from American laws. No such treaty exists. They are, however, likely referring to the 1786 treaty which was ratified in 1787. Their claims of immunity from US prosecution or having to pay taxes stems from the treaty's agreement on how to handle actual Moroccan citizens or property found aboard seized enemy ships. In essence, the citizens and their property "shall immediately be set at Liberty...". This in no way absolves Moroccan-American citizens or even Moroccan citizens doing work in the USA from having to pay taxes. It doesn't place them above the law.

Eleven Examples of Woo Codified in Law

SELLING A HAUNTED HOUSE?

Many states have "stigmatized property" disclosure laws. If someone was murdered in the house or the house was used for criminal activity, the seller is required to disclose that information. Failing to do so can result in a court reversing the sale. A few states in the USA have weighed in on a house's purported status as being haunted. New York state takes the position that if you have worked to create a reputation that your house is haunted, you need to reveal this to the buyer. Just next door, Jersey takes a more relaxed attitude. The seller doesn't have to disclose an alleged haunting unless the buyer asks.

IT'S ILLEGAL TO KILL BIGFOOT

Skamania County in Washington State has had a law on its books since 1969 that makes it illegal to kill "a nocturnal primate mammal variously described as an ape-like creature or a sub-species of *Homo Sapiens*". Although it may sound silly to make it illegal to kill an animal that has never been proven to exist, it does have the knock-on effect of keeping Bigfoot hunters from shooting humans and playing the "I thought I was shooting a Bigfoot" card.

IT'S LEGAL TO WANDER AROUND NEW HAMPSHIRE STATE PARKS DRESSED AS BIGFOOT

In 2009, Jonathan Doyle ascended New Hampshire's Mount Mondnock and donned a Bigfoot costume. As court documents suggest, he was trying to film himself for a hoax. He was told by park rangers he needed a special-use permit. If he didn't have a permit, he'd need to leave. Doyle complied but took the state to court. The case got all the way up to the state's supreme court. The judges ruled requiring a permit would violate a person's free speech right. They noted wedding parties that wanted to photograph themselves in whacky costumes required no such permit.

IT'S ILLEGAL TO REINCARNATE WITHOUT A LICENSE

In a move to control Tibetan religious practices, the Chinese government has required since 2007 that reincarnated lamas must file a reincarnation application and have it approved by the government before they can claim to be a reincarnated soul.

LICENSED TO WOO

A number of states and cities require fortune tellers to have a license. For example, San Francisco requires people engaging in fortune telling to purchase a license and undergo a background check. The city has a pretty expansive definition of what fortune telling encompasses. Among other things, the city views fortune telling as reading of the future or the past via clairvoyance, tarot cards, and coffee grounds. Yeah. Coffee grounds. The license doesn't expect you to actually be able to do any of this. It also covers people merely pretending to tell fortunes. Supporters of licensing woo contend it helps cut down on fraud. Critics argue licensing woo practitioners allows them to suggest they're government approved.

STATES THAT LET NATUROPATHS PRACTICE AS PRIMARY CARE PROVIDERS

Many American states have a shortage of primary care physicians. Some states like New Mexico and Oregon have thrown all good sense to the wind and allow naturopaths to act as primary care providers. So-called Naturopathic Doctors (NDs) when licensed to be primary care providers can do things like prescribe drugs, send people for X-rays, and perform minor surgical procedures. These are the same people who also believe science-based medicine is nothing more than poisoning (with drugs NDs are now allowed to prescribe), burning (sending them for X-rays NDs are now allowed to requisition), and slashing (surgeries NDs are now allowed to perform).

STATES THAT LEGISLATED A FAKE DISEASE INTO REALITY

Lyme disease is not pleasant but it can usually be beaten with antibiotics. There are people with nonspecific symptoms (like pain and fatigue) who have come to believe (usually under the influence of a quack doctor) they're suffering from "chronic Lyme disease" and demand doctors prescribe them long courses of medically unnecessary antibiotics. Although chronic Lyme disease is not a disease any legitimate medical society has ever recognized, so-called sufferers have successfully lobbied states like New York, Connecticut, and half-a-dozen others to enact laws that recognize the rights of so-called "Lyme literate" doctors to treat chronic Lyme disease. Insurance companies are also required to pay for the unnecessary medication.

HOMEOPATHY IS PROTECTED BY LAW

Since 1938, the FDA has exempted homeopathic "treatments" from most regulations that govern the legit pharmaceutical industry. Makers of homeopathic nostrums do not have to submit new drug applications to the FDA, they're exempt from laws that safeguard the way drugs are manufactured, and they don't have to demonstrate their products contain what they advertise.

YOU CAN KILL YOUR CHILD FOR JESUS

At least six US states allow parents to ignore well established and efficacious medical treatments for their child and instead try to cure their child through prayer. If the child dies, the parents cannot legally be charged with neglect or manslaughter. A 1998 article in the journal *Pediatrics* ("Child Fatalities from Religion-motivated Medical Neglect") found that between 1975 and 1995 "One hundred forty fatalities were from conditions for which survival rates with medical care would have exceeded 90%."

SATANIC IMMUNITY

In 1971, Gerald Mayo filed a lawsuit against Satan for causing him misery and ultimately depriving him of his constitutional rights. A judge threw out the case, noting that even if the US Marshals Service could haul Satan into a courtroom, Satan was a foreign prince and would probably be protected by the doctrine of sovereign immunity.

CREATIONISM IN THE CLASSROOM

Many US states have tried and failed to get creationism taught in the classroom, either alongside evolution or in place of. Strategies like equal time (teaching creation "science" alongside evolution) and Intelligent Design have spectacularly failed court tests. Several states have, in the last few years, adopted the strategy of getting religious myth into the science class under the guise of academic freedom laws. These allow teachers to teach the "strengths and weaknesses" of any scientific theory taught in the classroom. Of course, by "any" they mean evolution and global warming. Fortunately, most of these attempts to pass these stealth creation science laws have died in committee. So far only Tennessee and Louisiana have been successful.

Nine Reasons Why Income Tax Is Legal and Constitutional and You Really Do Have to Pay Your Taxes

Photo Credit: Karl Mamer

Unless you're Swedish, you probably don't like paying taxes. However, most of us take the view that some reasonable amount of taxation is necessary to pay for some reasonable amount of government services. There are some that go to extreme lengths to argue (sometimes before a judge) that they don't have to pay any income tax because it's actually unconstitutional. The folly of this should be readily apparent. If it were so obvious income taxes violated the constitution, some really smart lawyers would have argued this decades ago and won. That a group of good lawyers and constitutional scholars have never tried to organize a class action lawsuit by, say, high income earners and win a vastly rich settlement, would indicate people who understand the constitution also understand income tax is legal. But there are always people who think Google makes them smarter than Harvard-trained lawyers. Here are nine arguments so-called tax protestors, or better tax deniers, use to claim you don't have to pay your income tax. Most of the arguments below are so well known that they've been declared by courts as *prima facie* evidence the tax denier is engaging in a frivolous lawsuit. As well, the IRS can legally impose a $5,000 fine for anyone even trying to avoid taxes using one of these "theories".

PROBLEMS WITH THE SIXTEENTH AMENDMENT

The Sixteenth Amendment to the US constitution grants the federal government the right to levy an income tax. It's pretty clear: "The Congress shall have power to lay and collect taxes on incomes, from whatever source derived...". It does not matter if you're a person or a corporation or how you make your income. It does not matter if you're a citizen or a non-citizen. It doesn't even matter if you're earning your money on the other side of the world. Congress can tax any kind of income in any manner it sees fit. The amendment was ratified in 1913. Deniers usually try to attack this broad and crystal-clear power given to Congress by arguing the amendment was not properly ratified back in 1913. This argument was put forward by William Benson and Martin Beckman in a 1985 book called *The Law That Never Was: The Fraud of the 16th Amendment and Personal Income Tax*. Their argument hinges on evidence the states didn't ratify the exact same amendment. Back in the early 20th century, you couldn't just email identical Word docs. When states sent back their paperwork confirming ratification, the text of the amendment they ratified varied in minor ways like wording, punctuation, and capitalization. Congress did notice these small differences before certifying the ratification but judged them to be inconsequential. No state ratified something substantively different. A couple idiots have tried to argue the claims of Benson and Beckman before judges. Benson and Beckman unfortunately did not clue the plaintiffs in that this problem was common to most previous attempts to pass constitutional amendments. It was expected that scribes and typists would make small mistakes. Many amendments have passed despite such problems. Courts have ultimately and consistently ruled (e.g., Miller v. United States 868 F.2d 236, 241) that Congress duly considered these differences, judged them minor, and accepted the ratifications as valid. Once Congress ratifies something, the decision is beyond the jurisdiction of the courts.

INCOME TAX VIOLATES THE FIRST AMENDMENT

The First Amendment is always a go-to amendment for law cranks. They view it as a get-out-of-jail-free card that lets them ignore even the simplest law if they believe it limits their freedom of speech or practice of religion. Tax deniers are no different. They reason since churches don't pay taxes because of the First Amendment, they don't have to pay taxes if it violates their sincerely held religious belief! Their first error is assuming churches don't pay tax because of a First Amendment right. No. Religious institutions are viewed as 501(c)(3) nonprofits and exempt from certain taxes by the tax code. There is no inherent right in the First Amendment to prevent the government from taxing a church. Further churches that qualify for 501(c)(3) status need to meet certain tests. For example, does the church conduct regular religious services, have an actual congregation, or a distinct religious history? Further, priests and ministers still have to pay income tax on their incomes. Courts have also ruled you can't simply choose to ignore basic laws if one claims they're in violation of some imagined moral precept.

INCOME TAX VIOLATES THE FIFTH AMENDMENT

Here tax dodgers argue filing a return or providing financial information is analogous to incriminating themselves. On the face of it, it seems kind of dumb. How is reporting factual information that cannot possibly get you in legal trouble self-incrimination? The Supreme Court dealt with this back in 1927. The court's decision amusingly notes a citizen can't get out of filling in government forms by drawing "a conjurer's circle around the whole matter by his own declaration that to write any word upon the government blank would bring him into danger of the law."

INCOME TAX VIOLATES THE THIRTEENTH AMENDMENT

Okay so if the Fifth Amendment provides no defense, keep looking. Hey, maybe the Thirteenth Amendment offers hope. The Thirteenth Amendment prevents involuntary servitude. In other words, the Thirteenth Amendment outlaws slavery. Tax deniers argue income tax is slavery. In essence they try to assert a false equivalency between compelled compliance with a law and involuntary servitude. Morons who have tried this rarely even warrant much comment on the claim by a judge. Courts find such assertions frivolous and not worth the court's time to even comment.

THE IRS IS NOT A LEGAL GOVERNMENT AGENCY

Another approach to take is accepting income tax itself is legal but the IRS has no actual authority to collect taxes because they're a private corporation and not an agency of the US government. Tax deniers maintain the IRS was never established by "positive law". Sounds so legalish. That's just a fancy way of saying no act of Congress specifically created the agency. True. Their error is then assuming the IRS has no powers or right to collect taxes on behalf of the government because of this. Not true. The agency was created by the Treasury. The Secretary of the Treasury has the authority to collect taxes and has the authority to create any agency he/she sees fit to do that work.

THE 1040 TAX FORM DOES NOT COMPLY WITH THE PAPERWORK REDUCTION ACT

Next time you look at your 1040, look at the top right. There's a number there: OMB No. 1545-0074. It has been on tax forms since 1981. OMB stands for the "Office of Budgetary Management". That office's job is to review government forms sent out to individuals and businesses and ensure they're collecting the least amount of information possible. If the OMB approves, the document gets an OMB number. So while a 1040 sure seems like it has a lot of stuff to fill out, the OMB has determined the IRS is collecting exactly what it needs and not one bit more. This all comes from a 1980 law called the Paperwork Reduction Act (PRA). If a form is not in compliance with the PRA, the law stipulates the individual can refuse to fill out the form and not be penalized. Here tax deniers thought they found a loophole. They claim the tax package the IRS sends you is not in compliance with the PRA. Since it's not in compliance, you don't have to fill it out. The tax denier attack on the 1040 rests on two beliefs. One, while the 1040 does have an OMB control number, the instruction booklet that comes with the 1040 does not. Two, tax deniers assert the PRA requires OMB control numbers to expire after three years yet the 1040 has had the same control number since 1981. Courts were not impressed by either of these two arguments. They've ruled instruction books do not require their own control numbers as they're not forms collecting information. Second, courts have pointed out the numbers themselves don't get retired after three years. Forms are not sports jerseys. The law simply requires agencies have their documents reviewed every three years. If they still pass muster, they can keep the same control number.

ACTUALLY, ONLY FOREIGNERS HAVE TO PAY INCOME TAX

Tax deniers have somehow read the broad powers granted Congress to tax incomes and decided it actually only applies to foreigners earning income within the USA. This is based on a misreading of Internal Revenue Code Section 861. Nonresident foreigners sometimes earn money within the USA. It is meant to clearly define the scope of their tax obligations. When should they be taxed? How much of their income should be taxed? When should they also be taxed on their "world income" like US citizens? This section of the code answers those questions. Tax deniers seem to think this section of the code is meant to define what taxable income means in total. "This money I get. Is it income? Well, what's income? Oh, hey, Section 861 seems to define income as money earned by nonresidents." Courts see this so often they refer to it as "The 861 Argument." Actor Wesley Snipes famously tried to use the 861 Argument to avoid paying taxes on several million dollars of income.

I GET PAID IN FEDERAL RESERVE NOTES NOT ACTUAL MONEY

Another dodge is to argue your compensation is not taxable because it doesn't fall under the definition of money. For example, the labor you perform uploading videos of your cats is paid back by Facebook likes. That's not money and the government can't tax Facebook likes you've earned. Tax deniers go out of their way to explain that the "Federal Reserve Note" (that is, a normal dollar bill) does not constitute actual money. The courts have thrown out several arguments by tax dodgers that since money is debt (a common claim of gold standard cranks) you can't tax debt. Having $10 in your pocket is not a debt. It's quite the opposite. They also misread Article 1, Section 10 of the Constitution. This section prevents the individual states from issuing their own paper money. For example, California can't print off "Dude Bucks" and use it to pay off debts. At best, they can issue gold and silver coins to pay debts. Tax deniers don't understand this section limits the power of individual states and read it as indicating the whole of the United States limits legal tender to gold or silver coins or paper money backed by such. Since the pieces of paper they earned are not backed by gold or silver per Article 1, Section 10, a dollar bill earned is akin to a Facebook like. In reality, the Constitution grants Congress broad powers to declare what is legal tender. Money, not backed by gold or silver, has been declared legal tender.

ACTUALLY, I'M NOT A US CITIZEN OR EVEN A PERSON

Although some tax deniers argue only noncitizens have to pay taxes, there's another batch that takes the exact opposite stand. If you're not a citizen of the USA, you don't have to pay taxes (you do). Of course, tax deniers who claim they're not citizens don't actually go all the way and formally renounce their citizenship. They simply take the position they're actually *just* citizens of their local state (e.g., Michigan) or they're a state unto themselves (as in a member of the Sovereign Citizen movement). Until you formally give up your citizenship, if you were born on US soil, the Fourteenth Amendment automatically makes you a citizen and subject to American law. Another tactic is to claim only the District of Columbia, federal territories like Guam, and indigenous American reservations are subject to income tax. Alas, the Sixteenth Amendment is clear that Congress has the power to collect taxes "among the several States" and the tax code does not excuse you from taxes even if you don't think you're currently living in the USA. The final strategy of the tax denier is simply to argue they're not a person as defined by the Income Tax Act. If one is unsure of the meaning of "person", the tax code defines that. A person is an individual. Tax deniers try to define themselves as something else entirely like a private sector worker or believe if the court can't offer a legal definition of ever more basic terms ("person", "individual") then the law is invalid. They don't realize judges are called judges because they get to judge when basic terms apply to the individual in front of them.

Nine Pseudolegal Claims by Sovereign Citizens

Sovereign citizens are people who live by a weird pseudolegal code that, at its core, proposes they are not citizens of the nation written on their passport and therefore are not subject to most laws. Typically, these are laws that require a person to pay income tax or child support. The sovereign citizen movement is quite large in the USA. The Southern Poverty Law Center estimates at least 300,000 people think of themselves as sovereign citizens. In places outside of the USA the movement goes by names like Freemen on the Land or followers of Moorish Law.

THE COMMON LAW WAS SECRETLY REPLACED BY ADMIRALTY LAW

Sovereign citizens will argue in court that a judge has no authority over them because courts today are based on admiralty law and not the common law. Their jurisdiction is only valid on water and since their court is clearly on land they're limited as to what kind of cases they can try. They believe courts based on common law were secretly replaced either during the American Civil War or when the USA went off the gold standard in 1933. Evidence? Sovereign citizens will assert the gold fringe around the flag in the courtroom is evidence the court is an admiralty court. No sovereign citizen has ever had his case thrown out making this argument.

YOU HAVE TWO IDENTITIES

Sovereign citizens read a lot into court documents that use all caps to represent your name. They insist this represents your corporate identity or "strawman". The real you, the physical tangible human being, is represented by your name written not in all caps. So the real you is John William Smith and the corporate strawman version of you is JOHN WILLIAM SMITH. Makes sense? Yeah, no. You can imagine how judges deal with this claim the first time a defendant uses it in court. The claim boils down to "You're trying to get child support out of the corporate version of me, not the real version of me." Any question the judge poses to the defendant is met with a circular argument that goes something like "which version of me are you addressing?" It's like asking a person what the definition of "is" is. Sovereign citizens will also go to great lengths to hash up their real name for some reason with odd forms of punctuation. So John William Smith will represent his "real" identity as Smith-William:John.

YOUR BIRTH CERTIFICATE IS YOUR PASSBOOK TO A SECRET BANK ACCOUNT

Sovereign citizens believe your birth certificate and/or your social security number really represent a bank account or bond that was opened by the government in your name upon birth. The amount of money in your secret bank account varies with the telling. It ranges from $650,000 on the low end to $20 million on the high end. The imagined amount is always more than enough to cover any debt the sovereign citizen is hauled into court to pay. The defendant will then urge the judge to authorize the financial obligation to be paid out of this secret account. One might naturally wonder why your government opened a secret bank account under your name when it appears to have lots of its own money and the ability to simply create it as needed. Proponents of this pseudolegal theory will typically argue after the government went off the gold standard it had nothing of value to back foreign loans. So, the government offered its citizens as collateral. As a citizen, you're given a "value" (sometimes alleged to be based on your lifetime earning potential) and so-called fiat currency is deposited in a secret account in your name. Sovereign citizen "gurus" will sell "redemption" kits (some costing upwards of $1,500) to the easily duped. These kits promise that after mailing nonsense documents to the IRS or the Treasury filled with Latin gibberish written in red ink, a person can start issuing their own bills of exchange or direct the IRS to take unpaid taxes out of their secret bank account.

THE RIGHT TO FOIST OBLIGATIONS

Sovereign citizens will try to intimidate low-level authorities like traffic cops or meter readers by issuing them a contract full of legalese and a fee schedule with rates that are both prominent and ruinous. For example, their fee schedule will demand a generous hourly rate for their time, fees for use of their copyrighted name if written on a traffic ticket, tolls that must be paid for trespass on their property, mileage reimbursements if they have to drive to a court, and so on. Oh yeah, these fees are all to be paid in gold or silver coinage, not "fiat currency". They believe merely presenting the fee schedule creates an obligation. Courts have universally rejected the legality of these kinds of one-sided contracts. Silence regarding a foisted contract is not consent.

A RIGHT TO TRAVEL

Sovereign citizens generally refuse to pay for license plates and driver's licenses. In their view, this infringes on their unrestricted "right to travel". Sovereign citizens, if pulled over by the police for not having a valid license plate, will go to great pains to claim they're not driving somewhere but traveling. To drive is to suggest they're doing the work of a driver (in the sense of a truck driver). They will freely admit using roads in pursuit of commerce is subject to government control. However, despite clearly being behind a wheel and driving a car, they will claim they're not engaged in actual work for hire or commerce on the road and therefore their right to travel is sacrosanct and cannot be restricted in any way.

CRIMES MUST HAVE AN INJURED PARTY

Sovereign citizens who find themselves in court for failure to pay income tax or unpaid speeding tickets will make the argument that all crimes, by definition, must have a human victim. They believe all they have to do to avoid paying up is simply demand the court produces a human victim. When the court cannot, they'll walk scot-free. But, of course, it doesn't quite work like that.

THE AUTHORITY OF BLACK'S LAW DICTIONARY IS ON PAR WITH THE CONSTITUTION AND BIBLE

Black's Law Dictionary is a legal dictionary used by many lawyers in the USA when preparing legal cases. Lawyers use it the way people should use any well-written Wikipedia article. It's a good first place to start. It gives you a good overview, it sets you off in the right direction, but it's not the final authority. On anything. Sovereign citizens, on the other hand, seem to think anything they look up in *Black's Law Dictionary* is legally binding law and encompasses the totality of the law. So, if a real law's definition of some term seems not to align with how *Black's Law Dictionary* defines the term, sovereign citizens will argue the law is invalid and can be ignored. Because dictionary. Sovereign citizens seem to prefer the second edition of the dictionary (published in 1910) but don't seem to realize it's woefully out of date. This is odd given sovereign citizens are big champions of the common law. They probably quote from the second edition as that version is out of copyright and in the public domain. They don't actually have to pay for it.

THE JUDGE CAN'T SIT IN JUDGMENT UNLESS HE/SHE SHOWS PROOF OF AN OATH

Obviously judges just don't show up one day and preside over a courtroom. Judges are appointed or elected and typically have to swear some kind of oath. There's a paper trail. Sovereign citizens are under the belief it's the duty of any judge to produce that paper trail upon demand. If the judge can't or refuses to, the judge is legally required to recuse him or herself. The reality is in most jurisdictions the law presumes the judge has legal authority to preside and if you feel he or she doesn't, you need to lawyer up and prove it. The burden is not on the judge to establish his or her qualifications before each and every defendant.

CRIMINAL CHARGES ARE COMMERCIAL OFFERS

The Uniform Commercial Code (UCC) is a set of basic laws and clear definitions that govern all commercial transactions in the United States. Sovereign citizens seem to think the U in UCC stands for "Universal" and believe the UCC governs all matters of law (including criminal and civil law). In their view, UCC became the law of the land when the USA went off the gold standard. Yeah, I don't understand the logic either. The upshot is any criminal charge or court summons is necessarily a commercial offer governed by some section of the UCC, and open to negotiation. Sovereign citizens are convinced one can simply write "accepted for value" on a court summons. That then authorizes the court to simply take money out of the sovereign citizen's secret government account in lieu of jailing the person.

Five Lawsuits Against Skeptics

DILWORTH V. DUDLEY

Like many professors, Underwood Dudley got a lot of letters from cranks. Being a math professor, he got a lot of letters from math cranks. He collected them in something he called his crank file. In a 1986 article for his university magazine, he characterized math cranks as people who "just get carried away with their numbers. It's a hobby that's gotten out of control. Most of them are essentially normal in every way, except when it comes to this obsession with numbers." In 1992, Dudley decided to turn his file into a book about math cranks. In one chapter he took William Dilworth to task for his odd obsession with Cantor's proof that some infinities are larger than other infinities. Cantor cranks tend to find the notion of different infinities unsettling and spend a lot of time trying to prove that Cantor's well-established proof is wrong. A few years after the book was published, the book seems to have made it into the hands of Dilworth. Dilworth didn't like being called a crank and sued Dudley in 1995. Dilworth lost. A judge ruled calling someone a crank, while insulting, didn't meet the test of defamation. The judge agreed with Dudley's 1986 definition of a math crank. To call someone a crank just meant you were saying you thought they were obsessive about a certain topic and you felt they were probably wasting their time.

GELLER V. RANDI/CSICOP/PROMETHEUS BOOKS

Self-proclaimed psychic Uri Geller, famous for bending spoons, has sued James Randi, CSICOP, fellow skeptical book publisher Prometheus books at least three times. In 1991, Geller sued Randi and CSICOP for $15 million. Geller objected to a statement by Randi that Geller had tricked scientists with a simple magic trick of "the kind used on the back of cereal boxes...". The court dismissed Geller's 1991 suit. Geller tried again in 1992, suing Randi and Prometheus Books for an inaccurate passage in Randi's *The Magic of Uri Geller*. In an earlier edition (eventually corrected in subsequent publications), Randi wrote that Geller had once been arrested in Israel for claiming psychic powers. Geller wasn't actually arrested but sued by a person who attended his show and was upset he didn't see a demonstration of real psychic powers. Randi admitted to the judge Geller was correct and he had made an unintended factual error. The court agreed the statement was without malice and ruled in Randi's favor. Geller followed that with another 1992 lawsuit alleging he was libeled by two other CSICOP publications (books by the late Vic Stenger and Paul Kurtz). Both authors repeated Randi's claim about Geller's arrest in their books. This suit was thrown out as Geller's lawyers had filed after the statute of limitations date.

HUBER V. HERMES

Brit Hermes studied naturopathy at Bastyr University in Washington state. She got her ND in 2011. She found standards in the field low. After witnessing what she characterized as illegal practices, she gave up being an ND and moved to Germany to study real medicine. While working on her degree, she wrote a blog called *Naturopathic Diaries* where she argued NDs should not be able to call themselves doctors, should not be able to prescribe actual controlled drugs, and they should not be treated as the equivalent of primary care physicians. In one blog post, Hermes criticized a clinic run by Colleen Huber and her claims about successfully treating cancer patients with a special diet. Huber objected and, in 2017, she sued Hermes in Germany. Skeptics raised $80K for Hermes' legal defense. In 2019 a German court ruled Hermes was well within her right to criticize Huber as she was clearly expressing an opinion.

IRVING V. PENGUIN BOOKS LTD

David Irving wrote some popular historical books on World War II in the 1960s and 1970s. Bringing history to the masses is typically a good thing but academic historians became increasingly uncomfortable as Irving started to slide into historical revisionism. In 1993, author Deborah Lipstadt called out Irving as a Holocaust denier in her book *Denying the Holocaust*. Irving sued Lipstadt and her publisher Penguin Books for libel. He sued in the UK where defending a libel suit is both difficult and expensive. Irving, for reasons not clear, decided to act as his own lawyer in the lawsuit. Despite the high bar imposed on Lipstadt and her publisher, they won. They managed to prove Irving was, in the opinion of the judge, "an active Holocaust denier". Irving also proved the old adage "being your own lawyer means you have a fool for a client".

TOBINICK V. NOVELLA

Dr. Steven Novella is the host of *The Skeptics' Guide to the Universe* podcast and a proponent of science-based medicine. In a 2013 blog post, Novella criticized Dr. Edward Tobinick for running a clinic (which Tobinick billed as an "institute") that was using the drug Enbrel for a wide variety of off label uses including Alzheimer's. Novella warned his readers Tobinick's claims had many of the features typical of a quack practitioner: fronting a group with an impressive name, suggesting he/she can treat a wide range of seemingly unrelated diseases, and he/she is being unfairly persecuted by the medical establishment. Tobinick's response was to marshal his evidence and publish his results from double-blind, placebo-controlled studies. No. Actually, he sued. Because Novella spends part of his time treating Alzheimer's patients, he sued Novella for contravening something called the Lanham Act. The Lanham Act (passed 1947) outlaws unfair competition notably through the use of false commercial speech. See, since Novella was apparently in competition for Alzheimer's patients, he was unfairly using his blog to direct Tobinick's own

Alzheimer's patients to his practice at Yale. Crafty! Given Tobinick had clinics in Florida and California and Novella was way up in Connecticut, it didn't seem to make sense Novella was in competition for clients thousands of miles away. But, hey, tell it to a judge. Fans of the *Skeptics' Guide* podcast rallied to the cause and raised huge amounts of money for Novella's defense. Novella tried to get the lawsuit dismissed based on anti-SLAPP regulations. The judge agreed to dismiss the libel claims but did not immediately deny Tobinick's petition that Novella's blog post constituted an unfair competitive practice. Tobinick argued since Novella's blog runs ads and sells merch, therefore Novella's opinions constituted commercial speech. Fortunately, the judge could see the commercial activity of Novella's blog was in support of the education mission and not — like a lot of crank sites — the other way around. In the words of the judge, Novella's blog posts don't constitute commercial speech as "they do not propose a commercial transaction." (Novella never said in essence anything like "his clinic sucks, bring your dollars to my clinic!") The court threw out Tobinick's claim, ruling no reasonable jury was likely to agree Novella's blogging was commercial in nature.

Dr. Steven Novella at The Amazing Meeting 8 in Las Vegas.
Photo Credit: Karl Mamer

Seventeen Paranormal Proponents Who Ended up in Court

KEITH BARNES

For reasons not entirely clear, Keith Barnes liked to hunt Bigfoot in Arkansas while wearing a uniform of his own design, complete with an official looking badge and a Bigfoot Field Research Organization pin. He'd tell people he was a cryptozoologist with the Interior Department. That job does not exist and no one was fooled. People started calling the police on this oddball. The police launched an investigation and in the course of their investigation found evidence Barnes was allegedly in possession of child pornography (a crime that could have found him in jail for ten years). He was arrested April 24, 2018 and it appears in November 2019 he took a plea deal. He avoided spending a decade in the Klink and instead served a couple months in jail for video voyeurism.

SYLIVA BROWNE

Psychic Sylvia Browne (died 2013) didn't see this one coming: a conviction for fraud. In 1992, a California court found that Browne sold securities under false pretense. She convinced people to buy into a gold mining venture she had "feelings" about. Browne received one-year probation after paying restitution to the victims.

RYAN BUELL

Ryan Buell was the host of an A&E TV show called *Paranormal State*. It ran for six seasons between 2007 and 2011. In 2016 Buell sold a bunch of tickets to something he called his "Conversations with the Dead Tour". It would feature himself and psychic kiddie wrangler Chip Coffey. The tour never happened. Many people who bought tickets reportedly never got refunds. In 2016 Buell was arrested for the theft of leased property. It appears he rented a Hertz car in Pennsylvania and failed to return it. He spent 30 days in jail until he made bail.

FRANK COLLIN

In the 1980s Frank Collin styled himself as a New Age author and proponent of hyperdiffusionism. He went by the pen name Frank Joseph. Before Collin wrote a dozen New Age and pseudoarchaeology books under the Frank Joseph pen name, he was a leading figure in the American neo-Nazi movement of the 1970s. His attempts to march neo-Nazi dickheads through the streets of Skokie, Illinois ended up in the Supreme Court. In 1979 the court ruled in *The National Socialist Party of America v. Village of Skokie* that cities could not ban such parades on First Amendment grounds. Around this time Collin also saw the insides of a very different court when he was convicted for child molestation. His fellow Nazis found pictures he had taken of young boys in his desk at his Nazi party HQ and turned him over to the cops.

MILTON WILLIAM COOPER

Milton William Cooper (died 2001) probably did more than anyone to merge Ufology with End Times/New World Order paranoia. At some point in his career as Crank-in-Chief, he abandoned the whole UFO schtick and became a gun-toting patriot. He also stopped paying his income tax because he thought income tax was unconstitutional (it ain't). The government dutifully issued an arrest warrant. Cooper warned he would not be taken alive. On November 5, 2001, Arizona cops came to haul him off to court. Cooper shot one deputy in the head (the officer lived). Arizona's finest returned the favor and blew him away in a hail of police bullets.

RICK DYER

Rick Dyer, the man who gave us at least two Bigfoot-costume-in-a-chest-of-ice "hoaxes", was arrested in 2011 for allegedly committing fraud over eBay. Alleged victims, ranging from Texas to Canada, claimed Dyer took money for goods he sold on eBay but either never delivered the items or delivered them in poor condition.

STEVEN FEDER AND PETER STOTZ

Cousins Steven Feder and Peter Stotz ran the Psychic Readers Network (a psychic phone hotline) in the 1990s. Their most famous psychic went by the name of Miss Cleo. Miss Cleo's real name was Youree Dell Harris (died 2016). Despite promoting herself in televised commercials as some sort of Jamaican psychic priestess, Harris was, in fact, born in America and affected a Jamaican accent as part of her character. Very few callers ever got Miss Cleo on the phone. Feder and Stotz reportedly subcontracted calls out to a stable of 1,000 psychic readers. In 2001, several states and the FCC sued Feder and Stotz for deceptive advertising and billing practices. In 2002 the cousins pleaded no contest to felony fraud. The pair were required to pay a fine as well as pay back callers they had allegedly overbilled.

JOHN FORD

John Ford is the founder of the Long Island UFO Network. People who knew him before his legal troubles described him as a lonely man given to extreme gullibility. Despite having earned a Master's degree in Public Administration, the best Ford was able to do in life was find a job as a security guard at his local Suffolk County courthouse. He eventually left the job on a disability. He devoted himself full time to his Long Island UFO Network. His mission in life seemed to be trying to expose a number of wild-eyed conspiracies about UFO crash cover ups happening in and around Long Island. Ford was upset that local governments and his fellow Republican party members were not rallying to his cause. Ford hatched a plot to basically take over the Suffolk County government. To do that, politicians in his way needed to go. His plan, according to a conversation taped by police investigators, was to poison three county officials with radium. Ford and two other co-conspirators were arrested. Ford was eventually found not fit to stand trial. He was remanded to a psychiatric hospital and eventually released in 2019.

KENT HOVIND

Kent Hovind bills himself as "Doctor Dino". He's a young earth creationist and was a prolific debater on the subject throughout much of the 1990s and early 2000s. He's responsible for introducing dozens of "proofs" the earth is only 6,000 years old. These get repeated to this day *ad nauseum* in religious tracts and by online creationists. In debates, against people with real PhDs, he insisted on being called Doctor as well. His doctorate appears to be from an unaccredited institution called Patriot University. Many regard PU as a diploma mill. He created a backyard theme park in Pensacola, Florida called Dinosaur Adventure Land that supposedly taught kids about young Earth creationism and humans living at the same time with dinosaurs. This park got him in hot water with his local level of government in 2002. He was charged for failing to obey county zoning regulations. Turns out you just can't open a theme park in your backyard without a permit. Hovind ignored his local government's directives and he was then charged with contempt of court. His backyard theme park was locked up by court order. Far more serious than ignoring the need to get permits was Hovind's attempt to avoid paying taxes on his income, which was estimated at over $1 million a year from merchandise sales and speaker fees. In 2006, Hovind was charged with 58 counts related to his tax avoidance activities. He found himself in court. That didn't go well. His lawyer failed to call any witnesses in Hovind's defense. It took a jury only three hours of deliberation to find him guilty on all charges. He was sentenced to ten years in jail. Hovind left prison in 2010 and moved to Alabama. He has reopened Dinosaur Adventure Land in an Alabama town called Repton.

JAMES HYDRICK

James Hydrick made his name in the early 1980s as a psychic with astounding telekinetic powers. He could move, oddly enough, only very light items like a balanced pencil or a page of a phone book with nothing more than the power of his mind and some careful puffs of air. He appeared twice on an early 1980s reality TV show called *That's Incredible*. He told the hosts he learned to hone his amazing ability from a Chinese master. As it turned out, he had learned the trick in prison. If TV shows giving him publicity had stopped to do a background check, they would have discovered Hydrick had spent time in prison for kidnapping and torture. To skeptics, Hydrick is probably most famous for being exposed by James Randi. Randi quickly figured out his trick of blowing air. After being exposed, Hydrick descended into deeper depravity. Following a 1989 appearance on the *Sally Jesse Raphael* show he was arrested by police on an outstanding warrant related to his molestation of several runaway boys in California. He was sentenced to 17 years. After serving his sentence it was determined his pedophilia presented a danger to society and he was remanded to a state mental hospital where he resides today.

PETER HURKOS

Peter Hurkos (died 1988) claimed to be a psychic. He billed himself as "the man with the radar brain". He also fancied himself a psychic detective. He tried to "help" the police in the investigation of the Manson family and the Boston Strangler. In the case of the Boston Strangler he identified the wrong person. Hurkos tried to convince police investigators he had a "radar brain" by offering details of the crime only the police knew. Turns out he got some of this information by posing as a law enforcement officer and harassing witnesses. His trick caught up with him in 1964. Hurkos was trying to find undisclosed details about the kidnapping of Frank Sinatra Jr. He started to brag to a gas station attendant he was an FBI agent. The attendant seemed dubious of Hurkos's claim. As proof, Hurkos showed the attendant he had a rifle with a telescopic sight in his car. After Hurkos left the attendant immediately called the police. The police caught up with Hurkos in New York and arrested him. A jury found him guilty of impersonating a federal agent. He was fined $1,000.

RADOVAN KARADŽIĆ

Radovan Karadžić is a psychiatrist who became a Serbian ultra-nationalist during the 1990s civil war in Yugoslavia. He led Bosnian Serb forces and was responsible for ordering the Srebrenica genocide in 1995. To call this man an asshole would be an insult to assholes. After the civil war, he was indicted by an international tribunal for war crimes but managed to evade capture. While trying to lie low, he grew a white David Letterman beard and practiced as a naturopath in Belgrade and, allegedly, in Vienna. Apparently, he actually believed in the efficacy of "energy medicine". It wasn't a convenient cover. Hiding out in the open didn't work for him forever. He was arrested in 2008 and in 2016 was found guilty of genocide and crimes against humanity. He was given 40 years in the Klink. He appealed his sentence and lost. The court saw fit to up the punishment to life.

STEFAN LANKA

Germany's Stefan Lanka seems to believe viruses like measles and HIV do not exist. He offered a €100,000 prize for anyone who could prove the existence of the measles virus. That measles nearly vanished from the face of the earth after a vaccine was introduced didn't seem compelling enough to him. Measles, you see, is a psychosomatic illness. Dr. David Bardens, also of Germany, took up the challenge and submitted six peer-reviewed papers demonstrating the reality of the measles virus. Lanka naturally waved off the evidence. Bardens took Lanka to court. The court ruled Bardens had met Lanka's burden of proof and ordered Lanka to pay up. Lanka appealed and noted the Bardens had actually failed to heed the full rules. Lanka wanted ONE scientific paper. Bardens had offered more than one. The award was overturned on appeal.

BOB LAZAR

Since 1989 Bob Lazar has been making a large number of unverifiable claims about helping reverse engineer alien technology. His claims about his education (like advanced university degrees) and employment history (work as a physicist for Nellis and Los Alamos) seem at odds with public records. A year after coming forward as some kind of UFO scientist, Lazar was arrested in 1990 for his alleged involvement with a prostitution ring. Lazar pleaded guilty to a lesser charge of pandering and was given a suspended sentence. In 2006 Lazar was charged with violating an environmental law when his company United Nuclear allegedly shipped hazardous chemicals across state lines.

STAN ROMANEK

Stan Romanek is a UFO experiencer with a lot of rather hard to believe stories. He claims he was abducted by aliens and woke up in Betty Hill's nightgown after being returned to Earth. He also alleged he had an alien implant in his leg that he was going to show the world. The implant mysteriously disappeared before anyone could test it. That was not the least of his credibility problems. In 2018 Romanek was found guilty of possession of child pornography. He was sentenced to two years in a halfway house and was put on a list of registered sex offender. His name will remain on that list until at least 2027.

THOMAS SAM

Thomas Sam is an Australian naturopathic lecturer. When his four-month-old baby daughter started suffering from eczema, Sam and his wife decided to treat it with homeopathy instead of actual effective medicine. The eczema got worse until their child died of an infection in 2009. The parents were arrested and convicted of manslaughter by a court in New South Wales. In the words of the judge who put Sam away for ten years "Gloria was subjected to significant pain over an extended period of time and the omission of the offenders to seek proper assistance for her may be characterized accurately as cruelty."

WENDELLE STEVENS

Wendelle Stevens (died 2010) liked to portray himself as a UFO expert. That he was a vocal proponent of Billy Meier's numerous photos of UFOs indistinguishable from trash can lids dangling from a string indicates he was no expert. What he was, however, was a convicted child molester. Stevens was jailed in 1983 (released in 1988) after pleading guilty to numerous counts of child molestation, furnishing obscene materials to a minor, and possession of films or pictures of a minor involved in a sex act. A real charmer.

Sixteen Examples of Questionable Forensic Methods and Police Practices

DNA has become the gold standard for evidence in recent decades. Groups like The Innocence Project have been using it to help exonerate people wrongly convicted of crimes. What has become apparent is many of these people were convicted based on a wide variety of forensic methods and police practices that fall apart under scientific scrutiny. Here are 16 techniques with dubious underpinnings.

ARSON ANALYSIS

In 2004, Cameron Willingham was executed for the alleged 1991 murder of his own children by arson. He was largely convicted based on the findings of Texas fire investigators that the blaze at Willingham's home was purposely started by the use of accelerants. Willingham had no known motive for murdering his children. Subsequent independent reviews of the scene and forensics determined the conclusions were based on folklore and dubious science. Texas revised its forensic standards in 1992 but fire pattern analysis not based on rigorous scientific validation is still being used today.

BITE MARKS

Killers and rapists sometimes bite their victims and leave bite marks. It would seem reasonable then you can match marks left on a victim to the teeth of a suspect. Bite mark evidence has been used in the courts since the 19th century. However, since the 1990s, it's become increasingly obvious using the pattern left by a bite is an unreliable way to identify a perpetrator. One 1999 study found a 63% false positive rate. The problem is skin is not a good medium for preserving accurate dental impressions. Controlled studies have never shown a pattern of good inter-rater reliability for bite mark analysis. Despite the unreliability of bite mark evidence, the Innocence Project reports at least 26 people have been wrongly convicted based largely on such evidence.

BLOOD SPATTER ANALYSIS

Blood spatter analysis is an attempt by forensic examiners to better understand how a crime was committed based on blood stains at a crime scene. Forensic examiners believe based on the shape of certain stains, they can determine the angle and kind of impact that may have caused a fatality. Blood spatter analysis can be used by the prosecution to dispute the defense's theory of a crime. For example, a defendant trying to render aid to a suicide victim might pick up blood stains that a biased examiner concludes is evidence the defendant partook in the murder of the individual. A 2009 study by the National Academy of Sciences concluded blood spatter analysis is a highly subjective endeavor and not based on science. Blood is assumed to behave like water. It doesn't. Blood is assumed to behave in a similar fashion without regard to the surface it lands on. Blood will bounce off certain surfaces giving the examiner the impression the stain is a result of blood exiting the body with greater velocity than it really did. Given blood spatter analysis is usually undertaken by the prosecution, the analyst is subject to bias.

BODY LANGUAGE

Since 9/11, a bit of a cottage industry has emerged trying to teach police officers, airport security guards, and customs officials how to read a person's body language to detect deception. Facial expressions or body postures are ostensibly reliable "tells" when a person is lying. For example, it is claimed people will blink more, glance up to the right, or press their lips together when fibbing. In at least two studies, police officers relying on such assumptions performed worse when trying to detect lies. One 2003 meta-analysis of 11 studies of the efficacy of body language in detecting lying found, at best, people trained in various recognition techniques have only a slightly better chance than the average person at detecting a fib.

CONFESSIONS

A suspect's confession is usually the single most damning piece of evidence that can be used in a court of law. However, one study found over 300 people have been exonerated of crimes they did not commit, yet still confessed to. Some people confess simply to get out of the interrogation room, assuming evidence will clear things up soon enough. A typical method for getting people to confess to crimes they've not committed is by using leading questions and then demonstrating to the accused they know too much about the crime to be innocent. Young people and the mentally disadvantaged have a significantly higher rate of making false confessions.

EYEWITNESS TESTIMONY

Faulty eyewitness testimony is the leading cause of innocent people being convicted. The reality is most witnesses don't see an entire crime play out before their eyes. They see a piece of it. Research about how memories are formed strongly suggests the brain fills gaps in with its own narrative. Likewise, as we recall memories, our brain fills in forgotten parts to preserve a cogent narrative. A witness to a crime is more likely to fill in those gaps with information they've learned about the crime or the police theory of the crime. Assumptions that memories made under highly stressful events, such as being a victim or witnessing a violent crime, are more accurate have long dominated and highly influenced juries. Who of us would think we wouldn't recall the face of a person holding us up at gunpoint? However, research indicates high stress situations impede the reliability of the memory. We are more likely to pay attention to the object that threatens us (like a gun) than the person that threatens us.

FACILITATED COMMUNICATION

Facilitated Communication is a pseudoscientific technique where a so-called facilitator moves the hand of a non-verbal client on a keyboard or some kind of pointing device and supposedly helps unlock the client's "inner voice". Objective research time and time again demonstrates the facilitator is doing the communication. The facilitator can never unlock words or images shown only to the client and not to the facilitator. It's important to understand that a large number of these clients are severely disabled people who have never had a formal education or demonstrated any kind of cognitive ability to recognize words or letters. But, magically, in the hands of a facilitator they compose poetry, write-university level papers, and report sexual abuse. It's the latter that found Facilitated Communication being used in at least 60 courtrooms as evidence. Fortunately, most cases have favored the accused after researchers have been able to clearly demonstrate to courts the claimed evidence is no more reliable than evidence obtained from a Ouija board. Unfortunately, most people accused by these wanton cranks get dragged through the mud for months, temporarily lose custody of loved ones, and see their careers destroyed by the taint of the accusations.

FINGERPRINT ANALYSIS

It is widely held, although not established by any rigorous scientific study, that no two people have the same set of prints. It would seem then if a print collected at a crime scene matches that of an accused individual, it's powerful evidence the person was at the scene of the crime. What is not well known to the average person is the act of matching a set of prints is a very subjective endeavor. There's no agreement on what constitutes a match. Standards differ depending upon legal jurisdiction. As well, recent studies show poor inter-rater reliability. What one examiner calls a match does not reliably predict what another examiner calls a match. Over the last few years it has become increasingly obvious the use of fingerprints as evidence is not backed by objective scientific validation. Recent attempts to validate fingerprint analysis scientifically have discovered false positives are as high as 1 in 18. Multiple studies have shown the examiner can be easily influenced by external suggestions. A lack of blinding can be a problem as well as pressure on the examiner to aid in the conviction.

HAIR ANALYSIS

Hair analysis is based on the idea that hair left at the scene of a crime can be matched up with the hair of a suspect using microscopic comparisons. The idea gained wide popularity as a forensic tool in the 1970s. Skepticism began to emerge soon after. By the mid-1980s the FBI had concluded hair analysis had no objective basis and was purely subjective. It was impossible to use hair left at the scene of a crime to match with a suspect. Hair analysis, however, does have some valid use in the detection of drug use.

HYPNOSIS

Proponents of "forensic hypnosis" claim witnesses put under hypnosis can recall events with greater clarity or even recall so-called suppressed memories. Research indicates hypnotists end up creating false memories instead of recovering them. The use of hypnosis by police became fashionable in 1976 after the media reported it was used during the Chowchilla kidnapping case to help an eyewitness remember a license plate number. By the late 1970s, psychological researchers began to cast doubt on forensic hypnosis. It became increasingly clear the hypnotist could easily distort memories. So called hypnotic regression therapy grew in popularity in the 1990s. And with it came a tidal wave of patients alleging they were sexually abused by close family members or tortured by shadowy Satanic cults operating out of daycares. The work of Elizabeth Loftus did a lot to establish that the brain doesn't simply record memory like a VCR but memory is highly malleable and can easily be influenced by suggestive techniques employed by the hypnotist. Despite the science showing forensic hypnosis is junk and numerous innocent people being jailed on the basis of such evidence, its use is still technically legal in the USA. The Canadian Supreme Court, however, has banned it.

POLICE LINEUPS

We've seen this scene in hundreds of movies. A witness is ushered into a room with a one-way mirror. A group of men are walked into the room and the witness now must choose. It is in that forced choice that bias rules the day. If the witness is not instructed that none of the people in the lineup is the possible perpetrator and answering "none of them" is a perfectly acceptable choice, the chances the witness simply picks a person who looks most guilty due to biases about what a criminal should look like increases. Likewise, the witness's choice can be influenced by cues given by police in the room who know who the suspect is in the lineup.

POLYGRAPHS

Polygraphs are usually what people think of when they think of a lie detector. It's a machine that graphs a number of physiological responses to questions. A supposedly trained polygraph operator can read derivations from baseline measurements of things like blood pressure, heart rate, and skin conductivity. Proponents claim they can detect lies 90% of the time. However, there's a great body of scientific evidence that indicates polygraphs can't detect lies. Polygraphs can be confounded by various easy-to-master techniques like raising baseline readings by recalling scary or distressing memories during neutral questions. Many countries ban the use of polygraphs as evidence. The US Supreme Court determined there was no consensus about the accuracy of polygraphs and suggested a coin flip was as accurate. Oddly, the court left the decision of admissibility of polygraph evidence up to individual jurisdictions.

STATEMENT ANALYSIS

Statement analysis (sometimes known as "Scientific Content Analysis" or SCAN) is the supposed ability to analyze information provided by a suspect and find markers that indicate deception. Markers of what indicates deception is bewildering, inconsistent, and not founded on any objective scientific study. For example, it's claimed if a suspect doesn't use the word "I" when describing his or her activities, he or she is being deceptive. A man lying about murdering his wife will not use "my" to describe his wife. An employee accused of theft might not use typical work terminology when providing an alibi. It's really that dumb. Despite the complete pseudoscientific nature of statement analysis, over 400 police departments around the world use it as an investigative technique.

TORTURE

A bomb is about to go off. The bomber is in custody but won't reveal the location of the bomb or which wire to cut. Time is of the essence, so the daring police detective beats the answer out of the bomber and the day is saved. It makes for high drama but reality has demonstrated time and time again torture is more likely to produce false information. The US army's own field manual on torture suggests torture is unreliable and "can induce the source to say what he thinks the interrogator wants to hear". Torture of captured Taliban soldiers in Afghanistan led to false information that claimed Saddam Hussein was supporting the al-Qaeda movement.

TRAINED OBSERVERS ARE SUPERIOR WITNESSES

There's a belief that people in certain professions like law enforcement, journalism, and aviation have superior observational abilities and their eyewitness testimony is nearly unimpeachable. Studies indicate, however, so-called "trained observers" are generally no better at recalling details of a witnessed crime than the general population.

VOICEPRINT ANALYSIS

Some forensic experts believe voices are like fingerprints. No two are the same. A properly trained examiner can supposedly match a known voice spectrograph to an unknown one. Despite voiceprint analysis being used on high profile crimes like the search for the Yorkshire Ripper and Osama bin Laden, there has only been one large scale study of the accuracy of voiceprint analysis and that was back in 1972. Michigan State researcher Oscar Tosi tested 30 researchers comparing the voices of 250 male students. It is notable no female voices were used for validation purposes. Tosi concluded voice analysis was accurate when both samples are recorded under identical conditions. Comparing a mysterious voice left on an answering machine to a suspect's voice in a sound booth makes for a poor comparison. Likewise, accuracy significantly drops if the known and unknown samples are recorded at different times. Accuracy drops significantly if the difference is greater than one month. Both the FBI and the National Academy of Science have cast doubt on the use of voiceprints. There's no agreement on the admissibility or inadmissibility of voiceprint analysis in courts.

VOICE STRESS ANALYSIS

It is claimed that by analyzing a voice recording, you can detect stress. If you have stress you have deception. Proponents of this form of lie detection suggest low-frequency content of a voice, not normally picked up by the human ear, can indicate the emotional state of the person speaking. There is no evidence any system designed to measure stress through the voice can actually measure stress. Even if it could, as we know with attempts to measure stress via polygraphs, it is not a reliable indicator of deception. Multiple reviews of voice stress analysis have concluded it's useless. Despite this, a few judges in the USA have allowed voice stress analysis in court.

Eleven Odd Fringe Political Parties

ANTI-MASONIC PARTY

The Anti-Masonic party was formed in 1828. It was America's first third party. It became a political force when concerns about Freemasonry spread across the United States in response to the murder of William Morgan, a prominent critic of Freemasonry. The party believed Freemason elites in the government had murdered Morgan and covered it up. For a party of cranks, it did quite well in elections between 1828 and 1838. The party managed on average to get about 10 members elected to the House of Representatives over six elections. President John Quincy Adams was briefly a member of the party and ran under its banner for the Massachusetts governor's race in 1833.

BOILING FROG PARTY

The Boiling Frog Party is an environmentalist party that may not be much bigger than its 2020 presidential candidate Gary Wing and his running mate David Olszta. The party's motto is "It's not easy being green." Wing's party takes its name from the myth that a frog will allow itself to be boiled to death if you gradually increase the temperature of the water it's sitting in. Wing is clear, however, this is not true but it is still a useful metaphor for the slow destruction of the environment through global warming. In 2016, Wing got a lot of press when he ran for the US Senate on the platform people should stop having kids.

MARIJUANA PARTY

The Marijuana Party is active in the US and Canada. A version operated in the UK under the name Cannabis Is Safer Than Alcohol. In Canada's province of Quebec, it operates under the name Bloc Pot (a play on Quebec's Bloc Québécois party). As the name implies, these parties run on a platform of legalizing pot. The Marijuana Party did as well, or better, than one might expect of a fringe party in Canada's 2000 federal election. It managed nearly 2% of the vote. As Canada moved towards officially legalizing pot, the party's activities have declined. In the USA, the party is mostly active in state elections. It doesn't field a presidential candidate but usually endorses the Libertarian candidate.

NATURAL LAW PARTY

The Natural Law Party is a political party based on the tenets of Transcendental Meditation (TM). It was founded in 1992. The party is active in the USA, the UK, Canada, and about eight other nations. Magician Doug Henning (died 2000) ran as a Natural Law candidate in the 1993 Canadian federal election. He came in 6th place and got 1.59% of the vote. The party advocates for the wider practice of TM techniques and "yogic flying" as a way to combat war and crime. Although the party's candidates are generally not vote getters in most nations, the party did manage to get one candidate elected in India and one in Croatia.

NOVEMBER 9TH SOCIETY

The November 9th Society is a British neo-Nazi party. It was formed in 1977. Its name celebrates the beginning of *Kristallnacht* in 1938. Charming. The group's platform includes banning non-white immigration, suppressing gay rights, and promoting Holocaust denial.

THE PIRATE PARTY

The Pirate Party is a transnational political movement that supports free speech, the reform of copyright and patent laws, and technologies that foster forms of direct democracy.

Photo Credit: Karl Mamer

The party was founded in Sweden in 2006 and has spread to many countries across Europe. The Pirate Party has managed to get at least five members elected to the

European Parliament. In 2016, the party won 10 of 63 seats in the Icelandic elections.

RHINOCEROS PARTY

The Rhinoceros Party (or *Parti Rhinocéros* in French) was a satirical Canadian political party. The party was founded in 1963 and folded in 1993 (it has since been reborn as the "neorhino party"). The party's name was inspired by a rhinoceros in a Brazilian zoo that received 100,000 protest votes in the 1958 Sao Paulo civic election. If the rhino had been a qualified human candidate, it would have won the election by a landslide. The Rhinoceros Party had a history of offering impossible yet clever campaign promises. It promised, if elected, to repeal the law of gravity. The party would require car makers to make cars more energy efficient by forcing manufacturers to install larger wheels on the back. The theory here is all cars would be constantly driving downhill. As a make-work project and a way to unify Canada, a Rhino government would build a transcontinental bike path that sloped down in both directions so "Canadians could coast from coast to coast". Although the party never won a seat in Canadian parliament, it came dangerously close in a few ridings.

THE SOCIAL CREDIT PARTY

The Social Credit Party got its start during Canada's depression era. The party's name comes from the Social Credit economic theory. Social Credit suggests if people don't have enough money, print more of it. There are a few problems with that idea. The party quickly became convinced a cabal of Jewish bankers was behind most of the world's problems. One elected party member read parts of *The Protocols of the Elders of Zion* from the floor of parliament. This is now officially enshrined in Canada's permanent *Hansard*. Thanks! Though the party officially reputed antisemitism after World War II, it never really shook its antisemitic, conspiracy driven roots. For example, Canadian Holocaust denier James Keegstra was a long-time member and federal candidate. Social Credit was a bit of a force in Canadian politics during the Great Depression and following the war years but sputtered out by the 1970s. It still has a strong following among conspiracy bent Quebec Catholics. It publishes a magazine called *Michael* (aka the *Michael Journal*) which features bonkers articles about one world government, population reduction, Masonic conspiracies, and the Antichrist.

TECHNOCRACY

Technocracy is a political movement that saw its heyday during the Great Depression. It was founded by engineer (drink!) Henry Scott (died 1970). The movement sought to replace politicians and business leaders with scientists and engineers. It was banned in Canada at the start of World War II out of fears it was an anti-war party but it was allowed back into the political discourse when the government realized the group was all about maximizing war production.

Technocracy's plan was to merge North America into a single "technate". Technocracy posited since scientists and engineers already knew how to balance and transfer energy efficiently, it was logical then to think of a whole continent and all its people (that is, a technate) as a single energy system. Naturally, most of the energy was being provided by the workers. By using scientific and engineering principles (like the Laws of Thermodynamics), work could be maximized. This would produce an abundance of just about everything. The price system would be eliminated and replaced by energy certificates and something they called Energy Accounting. Scientific principles would be used to determine how to efficiently distribute the abundance. Part of that plan involved a canal system that would crisscross the continent and drastically lower transportation costs. Now if you think Technocracy's plan was to turn lowly workers into slaves, their plan was anything but. They wanted to replace the calendar with their own version which was centered around a four-day work week. However, this new calendar eliminated actual weekends and 5 PM "Miller time". Rotating shifts would ensure some part of the population was working 24/7. A three-day weekend, no matter how you sliced it, indeed sounded like paradise.

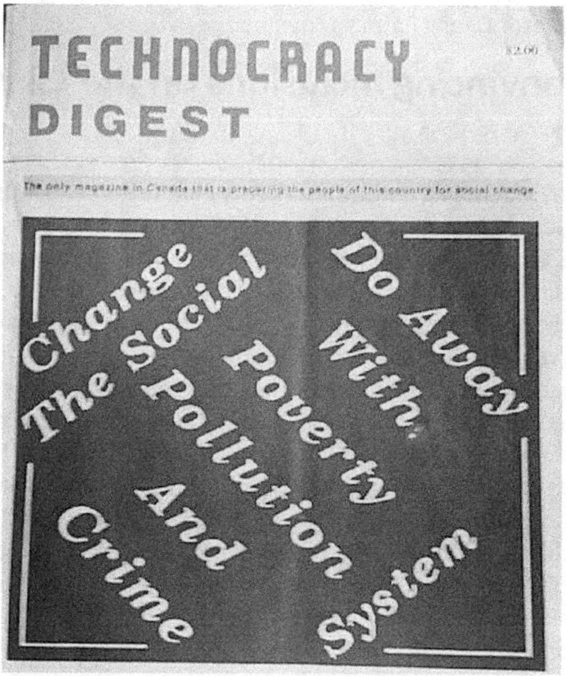

A Technocracy magazine published in Canada in the 1980s.

TRANSHUMANIST PARTY

The Transhumanist Party is a US federal party that believes America needs to increase funding for life extension research. It opposes research into nanotechnology and AI. The party was founded in 2014. The party was not on any presidential ballots in the 2020 election but the party claims they got at least 51 write-in votes.

YOUTH INTERNATIONAL PARTY

The Youth International Party, better known as the Yippies, was an American counterculture party active in the 1960s and early 1970s. In 1968 it nominated a pig known as Pigasus as its candidate for the 1968 election. James Randi would later use the term Pigasus as the name of his annual woo award. However, the name was unrelated to the 1968 Yippie presidential candidate. In 1976 the party ran "Nobody for President". This resulted in clever campaign slogans like "Nobody is perfect" and "Nobody should have too much power". Yippies pioneered the idea of embarrassing politicians and influential figures by shoving a pie in their face. People pied by the Yippies included Phyllis Schlafly (the woman who killed the ERA and gave birth to the founder of Conservapedia), *National Review* publisher William F. Buckley, and Senator Daniel Patrick Moynihan.

Nine Unconvincing Inclusions on the Clinton Body Count List

The Clinton Body Count is a conspiracy theory that has been kicking around since the days of Usenet. Cranks allege Bill and Hillary Clinton have orchestrated a campaign of assassinations. They wantonly kill those who apparently know too much about their alleged crimes (Whitewater, Clinton's infidelities, drug trafficking via Arkansas's Mena airport, Pizzagate, and so on). By way of evidence, cranks repost impressively long lists of people supposedly known to the Clintons who have died under "mysterious circumstances". The idea is no one could be associated with so many people who have died untimely deaths. Therefore, they murdered them. It should be obvious to anyone with two functioning brain cells that two people who have had the Clintons' wide-ranging political careers (governor, President, Secretary of State, presidential candidate) associate with a huge pool of people, many in jobs that are either dangerous (military, law enforcement) or stressful (increasing the likelihood suicide or death by disease). In this list we examine some of the unfortunate people who have passed away and been "memorialized" for having some connection with the Clintons.

HERSCHEL FRIDAY

Herschel Friday was once short-listed by Nixon for the Supreme Court. In the late 1980s, Friday became a Clinton benefactor when then Governor Clinton passed a tax package favorable to Friday. The Clinton Body List asserts Friday died in 1994 after his plane mysteriously exploded. His plane did not mysteriously explode. Friday, who was 72, died when he lost control of his plane during a difficult landing at a small airport in the dark and during inclement weather.

STANLEY HEARD

Stanley Heard was a chiropractor and had treated members of the Clinton family. Clinton's health care taskforce asked Heard to put together a position paper on the economics of using back crackers as primary care providers. Heard died after attending a Clinton briefing on a proposed health plan in September of 1993. Heard and an associate rented a poorly maintained plane and it crashed in flames when one of the engines came apart. The accident, while tragic, was not suspicious. The company that rented out the plane was found to have ignored a large number of FAA airworthiness directives. Versions of the list give no indication why Clinton wanted to kill his back cracker. We're left to suppose being so close to the Clintons he knew too much of something or another.

STANLEY HUGGINS

Stanley Huggins was a lawyer investigating a savings and loan linked to the Clinton Whitewater Controversy. Versions of the Clinton Body Count list claim he died of suicide. He actually died of pneumonia. No sign of foul play was detected.

FLORENCE MARTIN

Somehow the Clinton Body Count list has turned a murdered 69-year old convenience store worker into a CIA accountant bumped off by the Clintons because she apparently had information on Barry Seal. There is no mystery about her killer. DNA found at the scene matched the suspected killer, a former construction worker.

GENE MATTHEWS

Gene Matthews was an Arkansas sheriff. We're told he was likely bumped off because he was investigating Clinton-approved drug deals going on at Mena Airport. Matthews died while trying to arrest Gordon Kahl, a so-called tax protestor and sovereign citizen crank. Matthews and a team of federal law enforcement officials were attempting to arrest Kahl for a parole violation. Kahl fatally wounded Matthews as he attempted to enter Kahl's home. Matthews managed to return fire and killed Kahl. Matthews' fellow federal officers also returned fire with shotguns and accidentally hit Matthews in the thigh. In the mind of Clinton Body Count conspiracists, this constituted an assassination by the Reagan government to protect super liberal Clinton. There's no indication Matthews was ever working on an investigation of the Clintons and activities in and around Mena Airport.

MARY MOHANE

Clinton Body Count lists frequently include someone named Mary Mohane. They claim she was a White House intern shot in a coffee shop. She was ostensibly murdered before she could testify about sexual harassment. There are many problems with this. One there is no Mary Mohane who was a White House intern. Who they are probably referring to, however, is Mary Mahoney. She, along with several hundred other young people, had worked as an intern at the White House. Mahoney was gunned down in a coffee shop although it was a coffee shop she was managing at the time. Mahoney along with two of her other coworkers were murdered by Derek Havord Cooper in the commission of a robbery. Mahoney wasn't willing to give up the day's receipts without a fight. While trying to get Cooper's gun, Cooper shot and killed her and then killed Mahoney's other two coworkers. He then fled without taking any money. There is zero evidence she was ever about to be part of an investigation into sexual harassment.

BILL SHELTON AND KATHY FERGUSON

Bill Shelton was the boyfriend (sometimes described as the fiancé) of Kathy Ferguson. Ferguson herself was the former wife of an Arkansas state trooper accused by Paula Jones of aiding then Governor Bill Clinton in sexual harassment. About four years after Ferguson and her husband divorced, Ferguson entered into a topsy-turvy relationship with Shelton. Each accused the other of cheating. Ferguson killed herself, leaving a suicide note describing problems with Shelton. About a year later, a guilt-stricken Shelton killed himself over Ferguson's grave. He left a note explaining his motives. Although it might seem odd a Clinton co-defendant would die under mysterious circumstances, it's a bit of a haul to suggest the ex-wife, four years after the fact and then her new boyfriend were murdered by the Clintons.

PAUL TULLY

Paul Tully was a fairly high ranking official in the Democratic National Committee. In 1992 he was found dead in his hotel room. The Clinton Body Count asserts no cause of death was ever determined and alleges he was part of a Clinton "damage control squad". Therefore, he knew too much. It claims, ominously, an autopsy was never performed. Contrary to the list's claims, an autopsy was performed and found Tully had died from a massive heart attack. He was known to be a chain smoker, a heavy drinker, and not in the greatest of physical shape.

JOHN WILSON

The Clinton Body Count alleges John Wilson, who died in 1993, had ties to the Whitewater controversy. The list also claims Wilson died of a "suspicious" suicide. It's hard to understand why Wilson is on this list. He had no demonstrable connection to the Clintons or Whitewater. His suicide was anything but suspicious. He was known to have battled depression for a long time and had tried to kill himself on multiple occasions.

Seventeen Politicians from the Woo Wing

Scary political clown found in Osaka's America-mura.
Photo Credit: Karl Mamer

GREG ABBOTT

Texas Governor Greg Abbott seemed to believe a US military exercise in his state, codenamed Jade Helm 15, was in preparation for a war against the people of Texas. Rumors flew that mothballed Walmart stores in the state were being used to stockpile weapons for Chinese troops. In response to an entirely routine military training, Abbott ordered the Texas State Guard to be deployed so Texans could be assured "their safety, constitutional rights, private property rights and civil liberties will not be infringed."

MICHELLE BACHMANN

Michelle Bachmann was a Congressperson from Minnesota between 2007 and 2015. Over the years, she's expressed support for a wide range of bizarre conspiracies. In 2009 she suggested the government had a plan to use the census to round people up and put them in concentration camps. During the Obama administration she told followers Hillary Clinton was working as a spy for the Muslim Brotherhood. She signaled herself a global warming denier when in 2009 she asserted there's never been a single study that shows carbon dioxide is a harmful gas. She's suggested Democratic administrations are linked to swine flu outbreaks. Without evidence, she alleged an Obama state visit to India cost the taxpayers $200 million a day and involved commandeering hundreds of rooms at a five-star hotel.

BEN CARSON

Ben Carson, by all accounts, is a brilliant surgeon. He was also Trump's HUD secretary. He is not, however, well versed in history or the historicity of the Bible he apparently puts much faith in. While stumping for the 2016 Republican nomination he claimed the great pyramids were built by Joseph as grain storehouses. The story of Joseph is, however, set in Egypt's Middle Kingdom period, which is about five centuries after the Giza pyramids were built.

HAIM ESHED

Haim Eshed was a general in the Israeli military and the head of Israel's space program until 2011. In 2020, he gave an interview to an Israeli newspaper stating he believes the US government has been in contact with space aliens for years and had signed a secret agreement with a Galactic space federation. The current head of Israel's space program suggested Eshed "went too far" with his claims and denied knowledge of any human-alien space treaty. In a book about his life, published in 2020, Eshed made additional claims like space aliens prevented a nuclear disaster during the Bay of Pigs invasion.

BARRY GOLDWATER

Barry Goldwater (died 1987) was a US senator and former GOP presidential candidate. He had an intense interest in UFOs. At one point he attempted to get permission to visit a building at Wright-Patterson reputed to contain alien saucers or alien bodies. He was denied. Shortly before his death, he gave an interview to Larry King where he revealed he believed the government was conducting highly secret UFO investigations.

GARY GOODYEAR

Gary Goodyear was Canada's Minister for Science and Technology from 2008 to 2013. He gutted funding for basic scientific research and ordered Canada's National Research Council to focus less on what he derisively called "blue sky" projects and more on science with business applications. Goodyear is a chiropractor by training which made him an odd choice for a science role. Goodyear appeared to be a creationist when he was asked by a reporter if he believed in evolution. His reply was "I am a Christian, and I don't think anybody asking a question about my religion is appropriate."

MARJORIE TAYLOR GREENE

Marjorie Taylor Greene won a House seat in the 2020 election. Her social media posts suggest she is a Qanon supporter and believes in Pizzagate. She has argued the Sandy Hook shootings were a false flag and California forest fires could be caused by "solar space generators" funded by the Rothschild family. She has also offered support for the impossible claim George Soros was a Nazi.

PAUL HELLYER

Paul Hellyer (died 2021) was Canada's Minister of National Defense in the late 1960s. He was probably best known to Canadians for unifying the three branches of the Canadian military (Army, Air Force, Navy) into a single Canadian Forces command structure. He was a Liberal politician for a number of years. In 1997 he formed his own fringe party called the Canadian Action Party. Hellyer was also a darling of the UFO "exopolitics" movement. In 2005, Hellyer started demanding world governments disclose what they *really* know about UFOs. He believed Earth is being visited by at least four different alien species.

JAMES KEEGSTRA

James Keegstra (died 2014) was the former mayor of Eckville, Alberta and a perennial candidate for a Canadian fringe party called the Social Credit Party. He is one of the few people in Canada ever to be successfully convicted for hate speech. After his mayoral term, he became a high school teacher and began teaching his kids the Holocaust was a fraud and Jews were treacherous child killers.

ROBERT F. KENNEDY JR.

Robert F. Kennedy Jr. is the son of the late Robert F. Kennedy. Kennedy has never held political office but uses his family political legacy to insert himself in a wide variety of crank causes. He's one of the leading proponents of vaccine misinformation. He's recently branched out into promoting various conspiracy theories about the Covid pandemic and 5G wireless technology. Kennedy also doubts the facts of the assassination of his father and his uncle. He doubts the lone gunman theory and he does not believe Sirhan Sirhan was responsible for the murder of his father.

LYNDON LAROUCHE

Before Donald Trump ran for office, Lyndon LaRouche (died 2019) was probably the best-known example of a conspiracy monger seeking office. LaRouche was famous for being on every presidential ticket between 1976 and 2004. He typically ran as a candidate for a fringe party. LaRouche was a proponent of a number of financial conspiracies. He was an advocate for returning to the gold standard. He believed the British were behind most international conspiracies. He claimed there was a secret network of British financial organizations that fomented wars to protect their interests. He believed Queen Elizabeth II was the world's foremost drug trafficker. In 1989, LaRouche was sentenced to 15 years in jail for tax fraud and taking more than $30 million from his followers under false pretenses. While in jail he still ran for president in 1992. He received 26,344 votes.

JOE MANCHIK

Joe Manchik was the Green candidate for Ohio's 12th Congressional district in 2016 and 2018. Although the GOP candidate won in both elections, Manchik won enough votes to give the Democratic candidate concern that Manchik might be a spoiler. Nothing odd here, really, for a Green candidate. What made Manchik actually odd was during his campaign he revealed he believed he was a descendant of space aliens from the Pleiades. He did assure voters, however, he was himself not an illegal alien.

ELIZABETH MAY

Elizabeth May stepped down in 2021 as the leader of Canada's Green Party. While leader, she warned followers about the supposed harmful effects of Wi-Fi. In 2011, from her Wi-Fi enabled phone, she tweeted she was glad she didn't have Wi-Fi at home. The irony seemed lost on her. She also opposed things like smart meters. In 2014 May submitted a petition to Canada's parliament calling on the government to open an investigation into the terrorist attacks on 9/11. When asked if she was truly insane and believed in 9/11 conspiracies, she suggested she didn't agree with the contents but felt it was her obligation to present every batshit petition cranks submit to her.

ROY MOORE

Roy Moore was Chief Justice for Alabama's Supreme Court and famously lost a safe Republican Senate seat in 2017 when allegations were raised that he was a sexual creep. Moore was also one of the leaders of the Obama Birther Conspiracy. Moore believes the massacre of children at Sandy Hook was divine retribution. Moore is linked to groups that seek to rewrite the history of the American Civil War.

RON PAUL

Ron Paul is a retired federal politician and failed presidential candidate. He is noted for having a, umm, ardent fanbase of people with a Libertarian bent who think returning to the gold standard will bring untold prosperity. In 1978, Paul's think tank the Foundation for Rational Economics and Education began publishing a series of newsletters like *The Ron Paul Investment Letter*, *The Ron Paul Survival Report*, and *The Ron Paul Political Report*. The papers were filled with conspiratorial, racist, and homophobic content. You can find lines in Paul publications like "opinion polls consistently show only about 5% of blacks have sensible political opinions" and "if you have ever been robbed by a black teen-aged male, you know how unbelievably fleet-footed they can be". Charming. No article was published under a byline. Paul denies he wrote any of the controversial material although he has defended the content.

JOHN PODESTA

John Podesta was Bill Clinton's chief of staff from 1998 to 2001. He also ran Hillary Clinton's 2016 presidential campaign. Podesta is a big supporter of the so-called "Disclosure Project" which seeks to compel the government to release everything it knows about UFOs. Podesta is rather tight lipped on his own beliefs about aliens, but he is a wee bit too interested in the aims of the disclosure crowd to be anything other than a believer to some extent.

JESSE VENTURA

Jesse Ventura was a professional wrestler and a one-term governor of Minnesota. Between 2009 and 2013 he hosted a reality TV show called *Conspiracy Theory with Jesse Ventura*. Ventura has espoused support for conspiracy theories involving the Kennedy assassinations and the events surrounding 9/11.

NE WIN

Ne Win (died 2002) was a Burmese dictator who ran the country between 1962 and 1988. He pursued an isolationist economic policy that bankrupted the nation. Based on economic advice from astrologers, he decided the economy's problem was that none of the currency was evenly divisible by nine. Win recalled larger banknotes and had them replaced by notes with denominations like 45 and 90. The brilliance of this economic policy is founded on the fact each note's face value was divisible by nine and the digits themselves added up to nine. See? No one did. He stepped down a year later after a violent student uprising.

Twenty Five Political Leaders Who Are Reputed to be Shape-Shifting Space Lizards

David Icke was an un-notable soccer player for an English football team in the early 1970s. He had to quit playing early in his career for health reasons and, in the 1980s, found work with the BBC as a sports announcer. He developed some minor celebrity status at the BBC until he was let go for refusing to pay the Thatcher government's newly introduced Poll Tax. In the 1990s he entered what was known as his "turquoise" phase. He wore a turquoise-colored tracksuit and told anyone who would listen he was the son of the god of the Bible. In 1999 he published his seminal work *The Biggest Secret* which revealed his belief that the world is run by a cabal of alien lizards from a planet in the Draco constellation. These lizards take human form. Most prominent politicians and members of royal families are, according to Icke, space lizards. Below is a list of politicians who, over the years, Icke has alleged are shape-shifting space lizards. They are likely not space lizards.

1. Al Gore
2. Bill Clinton
3. Dalton McGuinty (former premier of Ontario)
4. David Rockefeller
5. Dick Cheney
6. Franklin Roosevelt
7. George H. W. Bush
8. George W. Bush
9. Gerald Ford
10. Henry Kissinger
11. Hillary Clinton
12. Adolf Hitler
13. Jeb Bush
14. Jimmy Carter
15. John F. Kennedy
16. Lyndon B. Johnson
17. Mikhail Gorbachev
18. Queen Elizabeth II (plus every member of the royal family)
19. Richard Nixon
20. Ronald Reagan
21. Stalin
22. Ted Heath
23. Ted Kennedy
24. Tony Blair
25. Winston Churchill

Ten Extraordinarily Dumb Conservapedia Pages

Conservapedia was started by homeschool advocate Andrew Schlafly in 2006. He perceived the reality-based agenda of Wikipedia had a liberal bias. "Liberal" seems to be defined by the site as views inconsistent with fundamentalist Christian theology. Here are some amazingly dumb pages found on that site.

ATHEISM AND OBESITY

This article seems mostly to exist to call prominent atheists "fat". We get pictures of P.Z. Myers, the late Perry DeAngelis (former co-host of *The Skeptics' Guide to the Universe* podcast), George R. R. Martin, and about 70 more luminaries. By way of contrast we get an illustration of Jesus and Chuck Norris. We're told Jesus kept the pounds off because he probably ate a healthy Mediterranean diet.

BLACK HOLE

Conservapedia categorizes this page under "Liberal Pseudoscience" along with "Dark Matter" and "WormHole". For reasons not clear, the Black Hole article assures us the concept of the worm hole has never caught on with the public "due to its unattractive terminology". The article doesn't make it clear how black holes amount to pseudoscience. The article has a brief section that reminds readers scientists have never directly observed a black hole but goes on to list all the ways black holes have been observed and detected. The article admits at the very bottom, in parentheses, that many black holes have been found.

CARL SAGAN

The Conservapedia article doesn't waste much time suggesting Sagan was a pot smoking liberal and his pot smoking likely inspired "pseudoscientific fantasies that he foisted on his gullible public". What were those? The only one the article lists is evolution is a fact. For reasons not clear, the article about Sagan then immediately diverges into a reminder that Thomas Huxley was an agnostic and agnostics find gay sex, abortion, and shacking up morally permissible. If you can get past this screed, the article does quickly summarize Sagan's actual scientific contributions (e.g., theorizing the surface of Venus was hot and Europa may have an ocean of liquid water below its icy exterior).

DUNGEONS AND DRAGONS

You'd think Conservapedia would be all over this one with righteous Satanic panic rage. Oddly, no. It's a fairly concise history of the game and even offers a defense of Christians playing the game. The article is notably absent of pictures. Looking at the article's Talk tab it seems there were pictures but they were removed by other editors upset they displayed occult imagery. For a while the article's lead image featured a player in a t-shirt bearing a pentagram.

MOON

The Wikipedia version of this page begins with a section on the etymology of the name. The Conservapedia's page skips this as it suggests pagan origins of the term and dares to link the reader to a list of lunar deities. Conservapedia jumps right into a list of long debunked reasons the moon supports young Earth creationism.

PROFESSOR VALUES

It's probably no surprise a site created by a homeschooling advocate has a page intended to scare parents into what awaits little Johnny when/if he gets into a real college with real professors. This page lays it out: Indoctrination, intolerance, safe spaces, crime, bizarre behavior, and bestiality (the ultimate liberal value, apparently).

PROJECT STEVE

Project Steve is a satirical response to creationists who created a list of so-called scientists that dissent from Darwin. Conservapedia seems to recognize the satirical nature of Project Steve. However, should readers be gob smacked by the number of Steves who have signed onto Project Steve, the article takes pains to point out lists such as these are illegitimate. Science isn't dependent on scientists voting in favor of a theory. Of course, no mention is made of similar lists like lists of scientists who dissent from Darwin or climate change.

SECOND LAW OF THERMODYNAMICS

Conservapedia comes right out and asserts the Second Law of Thermodynamics is "the tendency towards disorder in the absence of intelligent intervention". By way of support, the article quotes Isaac Asimov analogizing how your room gets messy if you don't keep on top of it. The article can't really get around systems with net energy gains (oh, like the earth being supplied with energy from the sun) can result in increased order. In those cases, the article takes pains to remind the reader that can only happen with intelligent intervention. This is one of those articles where it's helpful to read the Talk section. It does seem this is an article Schlafly has taken a special interest in, reversing additions made by some authors trying to bring reality to the article.

VIDEO GAME

By the third sentence of this article Conservapedia starts getting into how video games are responsible for violence and moral decline. It suggests kids can break video game addictions by taking up chess. About the only positive thing Conservapedia can muster for video games is they can, somehow, lead liberals to "virtual conservative viewpoints".

WIKIPEDIA ON BESTIALITY

Here we have two things Conservapedia seems obsessed about in one convenient article. The article takes issue with Wikipedia's article on zoophilia having a section titled "Arguments for Zoophilia". There is no mention the article also has a section on "Arguments against bestiality". It's almost as if the authors of the article don't understand that reporting what people say about a given topic is not an endorsement *of* the topic. Should you want to know more about bestiality, the article provides links to 14 other articles raging about roughly the same topic

Eleven Odd or Disturbing Manifestos

A manifesto is defined as a public declaration of policy and aims. People on the fringes of society have long seen it as a way to get their demands before the public eye. Below are some notable manifestos released by outsiders, kooks, cranks, and other odd balls over the years.

2083: A EUROPEAN DECLARATION OF INDEPENDENCE

First Sentence: "Historical revisionism is the attempt to change commonly held ideas about the past."

Final Sentence: "The conclusion is undeniable."

Before Norwegian Anders Breivik went on a killing spree in 2011, he distributed his 1,500-page manifesto titled *2083: A European Declaration of Independence*. It's largely a cut-and-paste job from other sources Breivik found on the web. "2083" in the title is a reference to Breivik's belief Europe will be rocked by a civil war which will be concluded by that year. It is also the year Breivik imagined Europe will start deporting all Muslims and executing traitors (feminists, people who support multiculturalism, and people he judged as politically correct).

THE CONSCIENCE OF A HACKER

First Sentence: "Another one got caught today, It's all over the papers."

Final Sentence: "You may stop this individual, but you can't stop us all... after all, we're all alike."

The Conscience of a Hacker was published in the *Phrack* ezine in 1986. It is colloquially referred to as the "Hacker Manifesto". Although it should not be confused with *A Hacker Manifesto* (published 2004). *The Conscience of a Hacker* is not long. It's just over 500 words long. Hackers are known for their economic approach to using resources, be they computing power or words. *The Conscience of a Hacker* was published under the pseudonym "The Mentor". It is known today the true author was former Legion of Doom member Loyd Blankenship. Blankenship penned the short manifesto after he was arrested for hacking. His manifesto tried to define hackers as those driven by curiosity and in search of knowledge. Hackers are not driven by evil intent. Hackers are outsiders, who reject knowledge "spoon-fed... at school when we hungered for steak." They are not the criminals the media portrays.

A DECLARATION OF THE INDEPENDENCE OF CYBERSPACE

First Sentence: "Governments of the Industrial World, you weary giants of flesh and steel, I come from Cyberspace, the new home of Mind."

Final Sentence: "May it be more humane and fair than the world your governments have made before."

A Declaration of the Independence of Cyberspace was written by John Perry Barlow in 1996. Barlow founded the Electronic Frontier Freedom. It rings in at about 850 words. It urges governments to stay out of cyberspace. He argues if government is by the consent of the governed, no one in cyberspace is asking to be governed. The problems lawmakers think need to be addressed in cyberspace don't exist. It was an optimistic manifesto for its time but, alas, governments around the world either didn't read it or rejected it.

INDUSTRIAL SOCIETY AND ITS FUTURE

First Sentence: "The Industrial Revolution and its consequences have been a disaster for the human race."

Final Sentence: "This is a significant question to which historians ought to give their attention."

The *Industrial Society and Its Future* manifesto is probably better known as the *Unabomber Manifesto*. It's a 35,000-word work by Ted Kaczynski. It was published in 1995 by both *The New York Times* and *The Washington Post.* At the time, the Unabomber's identity was unknown and both papers published it hoping someone would recognize the writing style. Sure enough, Kaczynski's brother did and turned him in. *Industrial Society and Its Future* posits technology has led to a destruction of small human communities and has forced humans to live in big, unfriendly cities. Those who oppose the march of progress are actively prevented from engaging in dissent by ever-growing police powers and counter programming in the media. For a time during the go-go 1990s, tech workers slaving away to build the new Internet economy treated the *Unabomber Manifesto* as something of a cultural totem.

THE INVASION OF THE HUMAN RACE

First Sentence: "The man who has appeared on KNBC for the last 3 years is not my biological father"

Final Sentence: "I ask for a Congressional Investigation and Federal protection for my family, and those involved."

Gary Stollman is the son of a former KNBC on-air personality. He used that connection in 1987 to gain access to KNBC's studios. He wandered onto the set of a live airing of *Fight Back* and pointed a gun at David Horowitz (died 2019). Stollman handed Horowitz an 800-word statement written on a yellow paper pad and demanded he read it on air. The station went to black while Horowitz read his statement. Stollman was arrested but only spent a few weeks in detention. He was ultimately sentenced to three-years probation. His original manifesto, the one he handed Horowitz, was a screed about his treatment in various mental facilities, a nefarious CIA plot to cause people vertigo with special light bulbs installed in subway stations, and how his family was replaced by clones created by aliens. In the early-to-mid 1990s, Stollman discovered the Internet's Usenet system and began propagating his delusions. He refashioned his original 800-word statement into a 25,000-word manifesto titled "The Invasion of the Human Race". The online version is broken up in a highly incoherent fashion. It's composed of multiple posts strewn across different Usenet groups but always identified by his signature "INVASION!!!" subject line. In his online version, he expanded on CIA skullduggery and the organization's persecution of him. He claims the CIA had planted fake students in his university classes to keep an eye on him. Eventually God and Jesus came to his rescue. Oh, and Carrie Fisher (Princess Leah) wanted to date him and began stalking him. Stollman, who is in his late 60s today, was still posting on Usenet about the INVASION!!! as late as 2010.

LEFT WING MANIFESTO BY THE LEFT WING NATIONAL COUNCIL

First Sentence: "The world is in crisis."

Final Sentence: "The Communist International calls the proletariat of the world to the final struggle!"

Technically this 8,000-word document is called just the Left Wing Manifesto. However, two different socialist organizations released manifestos of the same name in 1919. To distinguish, historians typically appended the authoring group's name. The other Left Wing Manifesto was released by the Left Wing Section of the Socialist Party. But not the entire Left Wing Section of the Socialist Party, just the Greater New York branch (or "local"). Confused? What's important to understand is the version by the Left Wing National Council was the centerpiece of the landmark Gitlow v. New York Supreme Court case. The manifesto was published by Benjamin Gitlow in a newspaper called *The Revolutionary Age*. He was arrested and convicted for his violation of New York's Criminal Anarchy Law. His conviction was overturned by the Supreme Court in 1925. The ruling established that the Fourteenth Amendment guarantees states themselves are bound by provisions of the First Amendment, notably the freedom of speech and the freedom of the press.

MEIN KAMPF

First Sentence: "On April 1st, 1924, I began to serve my sentence of detention in the Fortress of Landsberg am Lech, following the verdict of the Munich People's Court of that time."

Final Sentence: "The adherents of our Movements must always remember this, whenever they may have misgivings lest the greatness of the sacrifices demanded of them may not be justified by the possibilities of success."

Mein Kampf is, of course, Hitler's well-known manifesto published in 1925. It weighs in at over 200,000 words. *Mein Kamp*f outlines Hitler's belief in a Jewish conspiracy to control the world. Before Hitler became chancellor of Germany, he earned the equivalent of over $7 million from its sale. By the end of World War II, there were over 10 million copies in print. *Mein Kampf* was issued to every newly married couple in Germany and every soldier fighting on the front lines.

MY TWISTED WORLD: THE STORY OF ELLIOT RODGER

First Sentence: "Humanity... All of my suffering on this world has been at the hands of humanity, particularly women."

Final Sentence: "Finally, at long last, I can show the world my true worth."

Elliot Rodger (died 2014) was an early Incel and one of the first to take out his lack of success with women by murdering six people. Before he went on his killing spree in 2014, he emailed what became known as the *Elliot Rodger Manifesto* to several acquaintances. His manifesto, over 20,000 words, is part autobiography, part suicide note, and part laundry list of all the supposed injustices done to him by women and men who were more successful at dating. Rodger died by his own hands after his rampage. Since 2014, he has become a hero to the Incel community and has influenced other men to commit similar acts of mass murders.

RUSSELL–EINSTEIN MANIFESTO

First Sentence: "In the tragic situation which confronts humanity, we feel that scientists should assemble in conference to appraise the perils that have arisen as a result of the development of weapons of mass destruction, and to discuss a resolution in the spirit of the appended draft."

Final Sentence: "In view of the fact that in any future world war nuclear weapons will certainly be employed, and that such weapons threaten the continued existence of mankind, we urge the governments of the world to realize, and to acknowledge publicly, that their purpose cannot be furthered by a world war, and we urge them, consequently, to find peaceful means for the settlement of all matters of dispute between them."

The Russell–Einstein Manifesto was issued by Bertrand Russell in 1955. It was co-signed by Albert Einstein. The manifesto urged world leaders to reconsider their pursuit of nuclear weapons and dedicated themselves to resolving conflict in a peaceful manner. It's a good example of a sensible manifesto ignored by the people in charge, despite being backed by people of the day who were very hard to ignore. The manifesto did act as a kick-off to the ongoing Pugwash Conference.

SCUM (SOCIETY FOR CUTTING UP MEN) MANIFESTO

First Sentence: "Life in this society being, at best, an utter bore and no aspect of society being at all relevant to women, there remains to civic-minded, responsible, thrill-seeking females only to overthrow the government, eliminate the money system, institute complete automation and destroy the male sex."

Final Sentence: "Men who are rational, however, won't kick or struggle or raise a distressing fuss, but will just sit back, relax, enjoy the show and ride the waves to their demise."

Valerie Solanas (died 1988) was probably best known for her attempt to murder Andy Warhol in 1968. A year before she shot Warhol, she published the *SCUM Manifesto*. The *SCUM Manifesto* posits the primary thing wrong with men is they want to be women. It mirrors Freud's "penis envy" with male "pussy envy". Unable to become women, men just go out of their way to turn the world into a horrible place to live. The only solution to this man problem is to eliminate them. Once men are gone, women will live in a utopia without disease and a want for money.

WHO IS KYLE ODOM?

First Sentence: "Born and raised in North Idaho."

Final Sentence: "They keep track of every 'wild' human on the planet and manage us like animals in a zoo. Our 'freedom' is a carefully crafted illusion."

Kyle Odom shot Tim Remington, an Idaho pastor, in 2016 and then jumped a flight to Washington D.C. He was arrested by the Secret Service after he was caught throwing USB drives over the fence of the White House. Before he attacked the pastor (who lived and eventually got elected to Congress), he sent a 5,000-word manifesto to members of the media containing an assurance he wasn't crazy. His manifesto detailed how an intelligent race of hypersexual Martian lizards, millions of years more advanced than humanity, had ruined his life. For those seeking actual proof, he urged people to ask President Obama about the reality of Martians living deep under the Martian surface and the moon. He knew everything. Odom believed these Martians were responsible for writing the Bible as a way to control humans and satisfy their sexual desires. His manifesto personally called out Pastor Remington as being a shape-shifting Martian lizard man. Odom, obviously distraught by paranoid delusions of being persecuted, sought counseling from Remington but seemed to be put off when the pastor only offered him pat advice about submitting to Jesus. Odom's manifesto suggests he shot Remington and had planned to shoot more people to draw attention to the reality of these space lizards.

Eleven Poorly or Misleadingly Named Groups and Societies

"The Catholic League is just one divorced guy in his garage." — Penn Jillette

Many fringe groups try to puff up their importance or scope with names that sound more official than they are. Others take names that obscure their real aims. Here are eleven excellent examples.

CHILDREN'S HEALTH DEFENSE

Children's Health Defense is not a charity that helps poor children get needed medical care. It is an anti-vaccination lobby group set up by Robert F. Kennedy. Originally it was called the World Mercury Project until 2018. The group alleges, based on junk science, there's a pandemic of autism, ADHD, food allergies, and autoimmune diseases among children and the cause is exposure to things like vaccines, fluoridated water, and Wi-Fi.

DISCOVERY INSTITUTE

The Discovery Institute is probably best known as a think tank that promotes creationism under the refashioned term "Intelligent Design". It was founded in 1990. It is probably best associated with the legal thumping Intelligent Design received following the 2005 *Kitzmiller v. Dover Area School District* ruling.

GLOBAL ENERGY BALANCE NETWORK

The Global Energy Balance Network was founded in 2014 as a Coca-Cola front group. The group positioned itself as a science-based organization looking into the causes of obesity. Remarkably, the group promoted the idea obesity had nothing to do with junk food but a lack of exercise. The group was dissolved about a year after its founding.

INSTITUTE FOR HISTORICAL REVIEW

The Institute for Historical Review was founded in 1978 by a former member of the far-right British National Front. It publishes numerous materials that deny the reality of the Holocaust. It tries to portray itself as a scholarly organization that merely seeks to take a sober second look at events purported to have happened during the war years.

Entrance to the Auschwitz death camp.
Photo Credit: Karl Mamer

Yeah. In pursuit of trying to appear like a legitimate historical society, it published the non-peer reviewed *Journal of Historical Review* up until 2002. In it you'll find articles like "How Many Jews Died in the German Concentration Camps" and "Is the Diary of Anne Frank genuine?" David Irving was a regular speaker at the group's conferences throughout the 1980s and 1990s.

NATIONAL ASSOCIATION FOR RESEARCH & THERAPY OF HOMOSEXUALITY

The National Association for Research & Therapy of Homosexuality is a group that promotes so-called conversion therapy and tries to argue being gay is a mental disorder that needs to be cured. Conversion therapy has been characterized as rank pseudoscience, banned in many nations, and denounced by nearly every major medical association. The group made headlines in 2010 when its so-called scientific adviser was found to be touring Europe with a "rent boy".

NATIONAL COUNCIL FOR GEOCOSMIC RESEARCH

This group sounds like a scientific body of astronomers and cosmologists but in reality it's a group of astrologers. The group was founded in 1971. According to the organization's website, the group's name was chosen "to avoid the popular misunderstanding of astrology and to set the tone for a serious and scholarly approach to the study of correspondences between life on earth (geo) and the cycles of the cosmos." An examination of the group's current board is light on members with deep experience in traditional disciplines associated with Earth and space sciences. Board members include a director with a BA in English, a director with an MA in Psychology, and a treasurer with HR experience.

NATURAL PHILOSOPHY ALLIANCE

The Natural Philosophy Alliance conjures up images of bearded men sitting around discussing epistemology and Hegel. The group is anything but. It was founded in 1994 by John Chappell (died 2002) to oppose Einstein's Theory of Relativity. A bit late in the game. The group could only muster about 40 members. The group seems to have folded after holding exactly one conference in 2014. Former members respawned the group as the John Chappell Natural Philosophy Society. They maintain a website and message board where they discuss "problems" with relativity.

NEW ENGLAND ANTIQUITIES RESEARCH ASSOCIATION

The New England Antiquities Research Association isn't, as the name suggests, a group of trained archaeologists. The group, founded in 1964, promotes hyperdiffusionism and pseudohistorical ideas like ancient Celts settled the Americas before Columbus.

PHYSICIANS AND SURGEONS FOR SCIENTIFIC INTEGRITY

Physicians and Surgeons for Scientific Integrity (PSSI) was founded in 2006. Hey, doctors all about responsibility! Sounds great. Nope. It is really a group of medical doctors that oppose evolution. It was co-founded by Geoffrey Simmons, a doctor of internal medicine. Simmons is also a fellow of the Discovery Institute. In 2006 the group held a highly publicized "Doctors Doubting Darwin" rally at the Sun Dome on the grounds of the University of South Florida. The group does not seem active today.

ASSOCIATION OF AMERICAN PHYSICIANS AND SURGEONS

The Association of American Physicians and Surgeons was founded in 1943 to fight a bill that would provide government-funded healthcare. Since then it has been using its vague, encompassing name (and a journal it publishes) to lend scientific respectability to crank claims. The group's position is, for example, HIV does not cause AIDS, climate change is not caused by human activity, there's a link between vaccines and autism, and masks are not effective to help prevent the spread of Covid. Membership includes Joseph Mercola, Ron Paul, and Rand Paul.

INTERNATIONAL LYME AND ASSOCIATED DISEASES SOCIETY

The International Lyme and Associated Diseases Society promotes a chronic variant of Lyme disease. There is no evidence Lyme disease persists after successful treatment with antibiotics. The group advocates that people with a set of nonspecific symptoms like pain and fatigue are likely suffering from so-called chronic Lyme disease and require long-term antibiotic treatment. The weight of scientific evidence indicates the disease doesn't exist and an extended course of antibiotics is both unnecessary and medically dangerous.

Five Politically Incorrect Facets of the Paranormal

ANCIENT ASTRONAUTS PSEUDOSCIENCE IS RACIST

People who promote ancient astronaut pseudoscience follow a standard template: I don't know how these ancients did something or knew something about nature, therefore aliens helped them! As many have pointed out, very few of the greatest accomplishments by Western Europeans are claimed to be the work of aliens. No one, for example, claims the Notre Dame cathedral in Paris was built by space aliens. However, the accomplishments of Black and brown people are frequently fodder for this kind of sad speculation.

CRYPTOZOOLOGY UNETHICALLY REPURPOSES INDIGENOUS LEGENDS AND ARTIFACTS

Indigenous people have rich, complex beliefs about their relation to nature and the divine. Cryptozoology cherry picks these beliefs and repurposes their meaning. For example, ceremonial masks used by indigenous peoples in North America are prominently displayed in many Bigfoot museums in North America. There they are reinterpreted as a people's unsophisticated attempts at understanding a flesh-and-blood creature they've apparently been living side by side with for thousands of years. Sacred stories that indigenous people should be communicating and interpreting to the public are shoved aside by cryptozoologists who seize the mic and alter these stories to fit their own narrative.

SEND IN THE EXORCIST NOT A THERAPIST

There are many mentally disturbed people who not only don't get the help that they need but get the help they don't need. Reality TV's popularization of the notion that homes are being haunted by demons drives more and more people to call in an exorcist instead of a competent mental health professional. This can have deadly consequences. There have been dozens of deaths due to exorcisms over the years. One of the most high-profile cases, which formed the basis for the 2005 film *The Exorcism of Emily Rose*, was in 1976. A young German woman named Anneliese Michel had been diagnosed with temporal lobe epilepsy and depression. She was put through an exorcism that killed her.

POLTERGEISTS AND TROUBLED PEOPLE

Poltergeists are supposed to be noisy ghosts. They knock and throw things around. A mythology has developed around this pseudoscience that poltergeists are associated with boys or girls usually in the tween to teen range. Digging into many of these stories, it becomes clear the young person at the center of the case is troubled in some way. The cases of Julio Vasquez (the Miami Poltergeist) and Tina Resch (the Columbus Poltergeist) are classic examples of troubled people at the center of high-profile poltergeist cases. In the late 1960s, Vasquez was a worker at a warehouse stocking china and glass souvenirs. Whenever he was around, breakages shot up. So obviously it was a poltergeist. Vasquez was investigated by famed parapsychologist William Roll (died 2012). Roll determined Vasquez was a very disturbed young man living with a stepmother who wanted him out of the house. Vasquez admitted to Roll he felt happy when things around him broke. Instead of hooking up Vasquez with a therapist, Roll coined a term: Recurrent Spontaneous Psychokinesis. Vasquez reportedly got married, settled down, and had a good outcome. The case of Tina Resch is a sadder story. In 1984, following a viewing of 1982's *Poltergeist* movie, Resch started to emulate what she saw. Again Roll was called in, along with a huge retinue of reporters and TV cameras. Resch did a pretty good job of tossing items around the house while reporters were not looking. However eventually she was caught on tape picking up a lamp and throwing it. After the poltergeist explanation evaporated, it became obvious Resch was a foster child in desperate need of attention. A couple years after the spotlight was off, Resch was tossed out by her foster parents. Resch jumped into an abusive marriage and had a daughter. After escaping her marriage, Roll took her in. Instead of getting her counseling, he treated her as a parapsychology research subject. Sadly, Resch's problems resulted in her beating her infant daughter to death. In 1994 she was sentenced to life in jail.

UFOLOGY HAS A RACISM PROBLEM

White supremacists as of late have been allying themselves with the UFO world. Ufology's obsession with the notion ancient aliens had to help Black and brown people build their civilizations is a perfect "in" for right-wing white nationalists. MUFON's Pennsylvania State Director John Ventre in 2018 posted a racist rant on his Facebook, taking Netflix to task for supposedly promoting white genocide. In his words, "Everything this world is was created by Europeans and Americans. F'ing blacks didn't even have a calendar, a wheel or a numbering system until the Brits showed up." One would think MUFON would bounce him. However, MUFON Executive Director Jan Harzan simply dismissed Ventre's words as the kind of stuff that happens on social media. MUFON Director of Communications Roger Marsh likewise failed to address Ventre's statements and instead suggested calling Ventre a "white supremacist" was uncalled for. The worst one could maybe attribute to him is he was a lousy editor and liked to serve visitors "cheap food at his mansion".

VISIT OUR
Refreshment
Center
DURING THE INTERMISSION
...OR ANY TIME
CULTURE

Thirteen Tropes About Science, Scientists, and Skeptics Found on TV, in Movies, and in Culture

BAFFLED SCIENTISTS

Any new or surprising scientific finding is always described as leaving scientists baffled, scrambling for answers, or scratching their heads. Usually, if the article goes on long enough, you'll find scientists are actually more delighted by the new findings and probably have some reasonable hypotheses they're chomping at the bit to test.

BRAVE, MAVERICK ROGUE SCIENTISTS ALWAYS SAVE THE DAY

There's the notion that stunning breakthroughs in the world of medicine and engineering come from lone scientists who are mocked by their colleagues for their non-conformist ideas. Modern science is rarely advanced by lone scientists working on the fringes of mainstream research. A quick survey of the Nobel science prizes shows in the last 50 years, only 17% of Nobel prizes were won by a single individual.

DISCOVERIES ALWAYS REQUIRE RE-WRITING THE TEXTBOOKS

Reporters love to suggest a new discovery invariably means scientists have to re-write their textbooks. For example, if evidence is found that agriculture got an earlier start than 10,000 BC, articles will claim scientists are going to need to re-write their textbooks! Reporters usually fail to grasp a single study doesn't necessarily mean whole theories need to be tossed out. Likewise, scientists are usually careful in their language. For example, they may say the best evidence suggests agriculture started no later than 10,000 BC but that doesn't mean scientists think agriculture started no earlier than 10,000 BC.

NEUTERED SCIENTISTS

Scientists are so deep into their research they have, in the words of Short Round, "No time for love, Doctor Jones". Scientists, like anyone else, have libidos. Many don't realize what Einstein did with his Nobel prize money. He was pretty sure a Nobel prize was in the not too distant future. So he cut a deal with his wife. When he won, he'd give her the money and she'd give him a divorce so he could take up with his cousin/lover Elsa.

PUT A PHYSICIST IN CHARGE!

Physics is hard. Really hard. You need to be really smart to be a good physicist. Somehow there's the assumption that physicists are these kinds of omni-disciplinary scientists that can understand and interpret a wide body of knowledge.

Need someone to debate a creationist? Send in a smart physicist! Need someone to run your country or business? Give the job to a physicist!

SCIENTISTS AND SKEPTICS ARE CLOSED MINDED

Scientists and skeptics refuse to entertain new ideas. Science is just about having a series of "we were right all along" parties. This trope probably comes from a misunderstanding of the null hypothesis. One assumes your hypothesis is incorrect until you can mount sufficient evidence for it. Many woo proponents tend to think they're the first ones with a brilliant new idea yet don't realize the idea was rigorously tested generations ago.

SCIENTISTS ARE GOING TO KILL US ALL

Scientists are, like the dwarves of Moria, always delving too deep into life's mysteries. They never ponder the consequences. Eventually they're going to release some kind of Balrog upon humanity. Death will come in the form of a zombie plague, an artificial black hole that sucks the earth into a singularity, Ice-9, or some GMO that wipes out all our crops. The reality is researchers operate under pretty strict ethical rules set out by institutional review boards, particularly when it comes to experimenting on humans and aspects that affect human health.

SCIENTISTS ARE EMINENTLY RATIONAL

Yes, you need to be rational when following the scientific method. But that doesn't imply it rules other aspects of your life. Sir Isaac Newton's contributions to science and math are beyond compare. At the same time, he generated a tremendous body of work about the occult, alchemy, and sacred geometry. Newton viewed himself as handpicked by his god as one of the few people in history able to correctly interpret scripture. He also had some pretty nutty ideas about the end of the world. Newton, for example, tried to calculate when the world would end and determined it would be no earlier than 2060.

SCIENTISTS ARE THE PARAGON OF VIRTUE

Many times scientists are depicted in TV and movies as people with unimpeachable morals. Obviously, scientists lie, cheat, steal, and willingly aid in genocide.

SKEPTICS ARE AFRAID OF X

Skeptics are sometimes portrayed as being zealous guardians of a certain paradigm, like there is no other life in the universe or science can answer all possible questions. The possibility that aliens are visiting the earth or Bigfoot lives in the forests of the Pacific Northwest is terrifying to skeptics.

SKEPTICS DON'T BELIEVE IN ANYTHING INCLUDING LOVE

There's the notion that being a skeptic means you have to be skeptical of everything. A common retort to skeptics asking true believers for evidence to back their claim is "well, do you need scientific proof your child loves you?" Some also confuse being skeptical with being a denialist. The official story must be assumed to be wrong.

SKEPTICS ARE ATHEISTS

It's certainly true that most skeptics are also atheists. While they tend to be fellow travelers and hang out at many of the same conferences, one can still be a skeptic and have religious faith. Such skeptics tend to put religious belief into a category of untestable claims and therefore not an appropriate area to bring to bear the tools of skepticism.

SKEPTICS ARE REALLY JUST CYNICS

Penn Jillette once suggested a good demarcation between a skeptic and a cynic. A cynic questions a person's motivations for belief. A skeptic questions the person's evidence for that belief. It's also a pretty good "check digit" to keep you from falling into cynicism.

Ten Movies and TV Shows Ruined by Later Scientific Advances

ANTI-VAXXERS HAVE KEPT MEASLES ALIVE AND WELL INTO THE 24TH CENTURY?

In a 1989 episode of *Star Trek: The Next Generation* called "The Long Ladder", Worf comes down with a disease common to Klingon children. Dr. Pulaski, attempting some rare bedside manner, likens it to a common human childhood disease called measles. Measles was certainly a thing in the mid-to-late 1980s. Stepped up vaccination efforts in the 1980s helped eliminate it from the USA by the year 2000. Anti-vaxxers have caused a resurgence in recent years, however. One would imagine measles would be eliminated, somehow, by the 24th century. Although it is possible it is still around in the 24th century due to the continued efforts of anti-vaxxers over the centuries.

CRACK IN THE WORLD QUICKLY SHOWED ITS CRACKS

The film *Crack in the World*, released in 1965, was, for its time, an excellent science fiction/disaster movie. The plot revolves around scientists trying to drill through Earth's crust to access the magma underneath to generate geothermal energy. The movie posits the earth's crust is one solid shell. Unfortunately, the shell is a little too solid so scientists attempt to blast through it with a nuclear bomb. This creates a crack that threatens to split the world in two and boil away the oceans. One problem though, the movie was released just as scientists had gathered in 1965 at the Royal Society of London to put the finishing touches on plate tectonics which established the earth wasn't one solid shell but plates that coasted atop the asthenosphere. One cannot crack that which already is cracked into about seven different pieces.

DISNEY'S DEMISE OF THE DINOSAURS

Disney's *Fantasia* (1941) has a segment, scored to Stravinsky's "Rites of Spring", where a global drought caused by a superheated Earth is suggested as the cause of the extinction of the dinosaurs. It wasn't until 1980 that evidence emerged that the dinosaurs were wiped out by a process started when a comet or meteor smacked into the earth right around the Yucatan Peninsula. This has become the dominant theory today.

FERMAT'S LOST THEOREM

A 1989 episode of *Star Trek: The Next Generation* ("The Royale") notes that Fermat's Last Theorem was still unproven in the 24th century. Fermat's Last Theorem was actually proven six years later in 1995. Perhaps the proof was lost over the centuries?

PILTDOWN PLOT HOLE

The rather schlocky 1953 horror movie *The Neanderthal Man* has one scene where a scientist tries to illustrate human evolution with a chart. On the chart of our lineage is Piltdown Man. A few months after the movie was released, the team of Kenneth Page Oakley, Sir Wilfrid Edward Le Gros Clark, and Joseph Weiner were able to demonstrate conclusively that Piltdown Man was a forgery.

PLANET OF THE APES SUGGESTS CHIMPS ARE PEACEFUL

Both the 1968 movie and the novel upon which it was based suggest chimps are rather peaceful primates whereas apes are much more brutal. Even Jane Goodall thought this until 1974 when she documented what became known as the "Gombe Chimpanzee War". Studies since then have confirmed war is not uncommon among chimpanzee societies.

TUBE TVS ARE THE FUTURE

Most science fiction movies and TV shows rarely show flat panel displays until flat panel displays became common in reality. An interesting exception is *2001: A Space Odyssey*. Kubrick's 1968 film has a number of scenes with the astronauts interacting with what appears to be flat panel tablet computers no different from iPads of today.

SPIELBERG'S PLUCKED DINOSAURS

Today we understand most species of dinosaurs had feathers. However, that wasn't well established when *Jurassic Park* was released in 1993. Three years later in 1996 the first non-avian dinosaur fossil with feathers was discovered (a species known as a *Sinosauropteryx*). Since then no *Jurassic Park* sequel has ever depicted feathered dinosaurs. *Jurassic World* (released in 2015) did attempt a retcon: the cloned dinosaurs didn't have feathers because they weren't built from complete dino DNA. They lacked the DNA that would have given them plumage.

STAR TREK II FAILS TO PREDICT LASIK SURGERY

In a scene early in *Star Trek II* McCoy informs Kirk he can't treat his vision problems with a drug called Retinax V, so his only alternative is for Kirk to use a pair of vintage reading glasses. *Star Trek II* was released in 1982. A decade later, in 1992, LASIK eye surgery was approved by the FDA. One today is left to wonder why Bones didn't offer Kirk LASIK as another possible alternative.

YOU CAN'T SPIN IT THAT WAY

A 1994 episode of *Deep Space Nine* ("Rivals") asserted neutrinos have two possible spin directions. Half spin clockwise. Half spin counterclockwise. Part of the plot hinges on an anomaly where 80% of neutrinos detected on the station are spinning clockwise. It was discovered several years after the episode broadcast that all neutrons spin in the same direction.

Four Cursed Movies

There have been a few movies over the decades that are claimed to be cursed. Similar to the Mothman Death List (or the Clinton Body Count), it's not difficult to find a few films over the decades that have clusters of seemingly tragic deaths among the large pool of people who worked on a particular film.

THE OMEN (1976)

The Omen was pitched to producer Harvey Bernhard by an associate. The associate cautioned to not actually make the film as a movie about the Antichrist would surely be stopped by the Devil himself. Believers in this curse point to several instances of bad luck. For example, Gregory Peck's son committed suicide shortly before the film was to be made. The hotel the film's executive producer was staying at in London was bombed by the IRA. A trainer at a zoo where a scene was filmed was mauled by a tiger a day after the movie's crew filmed a scene (the scene involved baboons).

***POLTERGEIST* (1982)**

When many people think of a cursed movie, Steven Spielberg's 1982 *Poltergeist* might be the one that comes to mind. Shortly after the movie was released, actress Dominique Dunne (who played the film's irascible teenage daughter) was murdered by her stalkerish ex-boyfriend. Heather O'Rourke, who played the little girl abducted by the poltergeist, followed in death a few years later. She died at the age of 12 due to a congenital bowel defect. No other deaths have really been connected to the film. Spielberg is still quite alive and successful. The actors who played other major roles (JoBeth Williams and Craig T. Nelson) are still quite alive (as of this writing).

***ATUK* (1988, NEVER FINISHED)**

Atuk is an unfinished comedy movie. It is based on Mordecai Richler's 1963 novel *The Incomparable Atuk*. The movie is a fish-out-of-water story about an Inuit poet named Atuk trying to make his way in the big city. The property has kicked around development hell since the early 1970s. John Belushi was originally tapped to play Atuk although he died of a drug overdose. Production managed to start in 1988 with Sam Kinison in the lead role. Kinison had creative differences and exited the production shortly after it began. The film was scrapped and Kinison was killed by a drunk driver in 1992. John Candy, Chris Farley, and Phil Hartman were reportedly interested in seeing the film completed. Each of those actors died in the prime of their careers.

***THE CROW* (1994)**

Bruce Lee's son Brandon Lee was killed during the making of *The Crow*. Some claim the son died of the same curse that took his father. Others point to additional tragedies that befell this film. For example, the set was destroyed by a hurricane. A prop truck caught on fire. A worker was electrocuted and injured on set when he backed his truck up into some high-tension power lines. Brandon Lee died when a prop gun was fired at him in a scene. No one noticed the prop gun still had a round lodged in the barrel before being loaded with a blank. Critics note any tragedy or accident on the set is better explained by the fact the film was made on the cheap and corners were cut.

Eleven Woo Documentaries by Sunn Classic Pictures

Sunn Classic Pictures was a Utah-based film company formed in 1971 by a Mormon. It was known for creating a number of paranormal-themed documentary features. Sunn Classic brilliantly filled two market niches Hollywood had largely abandoned in the 1970s. G-rated films were not getting regularly made. Hollywood favored darker, more violent movies featuring anti-hero types. Hollywood was also ignoring a large potential audience of working-class families that had given up going to the local movie theater. To fill these niches, Sunn pioneered and merged two forms later rediscovered by cable TV: Reality is cheap to film and working-class people love shows about cryptids, UFOs, and paranormal topics with a religious bent. Sunn did very well for itself. It would turn out a movie for $500,000, rent hundreds of movie theaters across North America, and advertise the hell out of the film. Sunn pictures were typically promoted as "coming to your town for a limited time engagement". By renting entire theaters Sunn was able to end-around distribution channels that did not favor G-rated films. This process was known as "four walling" because you temporarily owned the four walls of the theater. Although Sunn took on 100% of the risk, it allowed Sunn to keep 100% of the box office. The strategy never failed. Its films grossed tens of millions of dollars, with 100% of the gross going straight back to Sunn. A large number of Sunn paranormal documentaries released in the 1970s still inform crank mythology today.

THE MYSTERIOUS MONSTERS (1975)

The Mysterious Monsters grossed over $10 million at the box office. It was actually one of 1975's top 20 performers. It out grossed two Disney re-issues (*Snow White* and *Bambi*). The movie was narrated by Peter Graves (who was well known to audiences then as the lead in the TV series *Mission: Impossible*). It looked at the evidence for Bigfoot, yeti, and the Loch Ness monster. It is probably one of the first times the general public got a look at the Patterson-Gimlin footage of a supposed female Bigfoot. You can find versions of it floating around on the Internet including YouTube. The film's best evidence and excuses why cryptozoologists can never manage to offer better evidence (say, a body) are still largely the same lines of evidence and special pleading modern day cryptozoologists trot out today. A modern-day skeptic can only come away from viewing this film with the impression the field of cryptozoology has remained remarkably unchanged over the last five decades.

THE OUTER SPACE CONNECTION (1975)

The Outer Space Connection is the final film in a trilogy of ancient astronaut documentaries produced by Alan Landsburg and narrated by Rod Serling. The first two (*In Search of Ancient Astronauts* and *In Search of Ancient Mysteries*) were made-for-TV specials that aired in 1973. Sunn picked up the third for a four wall distribution. This film is notable for very little except for it being the last major production Serling ever worked on before his death in 1975 of a heart attack (Serling smoked 2-3 packs a day). These Landsburg/Serling ancient astronaut films are considered "pilot" episodes of the *In Search Of* series, a show that launched thousands of 1970s kids into the world of skepticism.

IN SEARCH OF NOAH'S ARK **(1976)**

In Search of Noah's Ark was the 6th highest grossing film of 1976. It took in nearly $24 million. Not bad for a production budget of half a million. It was based on a book of the same name by author David Balsiger (died 2017). It was narrated by Brad Crandall, a one-time New York radio personality from the late 1960s. Following the success of *In Search of Noah's Ark*, Crandall became the go-to VO artist for Sunn documentaries. The movie made popular the notion that at about 14,000 feet, on the side of Mount Ararat, there's a long, rectangular boat sticking out of a glacier. If only explorers could get to it! Alas. That movie suggests politics of the day and dangerous Kurd tribesmen prevent serious expeditions. The movie pads its 90-minute run time with tales of 19th century expeditions to Ararat that all reported finding the ark. Interspersed are reenactments of Noah and his family, including one overlong scene that features Noah's wife trying to teach a chimp how to make bread.

THE AMAZING WORLD OF PSYCHIC PHENOMENA **(1976)**

The Amazing World of Psychic Phenomena was hosted by Raymond Burr (died 1993). Burr was best known for portraying Perry Mason on television. The movie offers a grab bag investigation into the world of the paranormal. The movie covers everything from astral projection to psychic detectives. It also kicks off Sunn pictures' fascination with evidence for life-after-death. The film, to give itself an air of legitimacy, starts with an interview of astronaut Edgar Mitchell (died 2016). Mitchell does what Mitchell did best: offering generalizations about supernatural phenomena but never actual specifics. It offers a look at a young Uri Geller doing spoon bending.

THE LINCOLN CONSPIRACY **(1977)**

The Lincoln Conspiracy is a weird attempt to absolve the South for the assassination of Abraham Lincoln. Historians tell us John Wilkes Booth was part of a conspiracy of Confederate sympathizers to kill Lincoln and three key government officials responsible for the defeat of the Confederacy. Booth's assassination of Lincoln is the final historical cherry stain on top of a despicable pro-slavery legacy. But what if Lincoln wasn't, in fact, assassinated by a conspiracy of Confederate sympathizers but was actually murdered by Northerners upset Lincoln planned to go easy on the South? Well, that's the hypothesis of this movie. Sunn promoted this movie as being based on long suppressed documents that suggest Lincoln's Secretary of War led a group of Northern senators to plan and execute the assassination. Booth was not actually killed but his death was faked. A patsy named James W. Boyd was killed instead and his body was passed off as Booth's. This is the first Sunn film to break from Sunn's documentary style in favor of docudrama. A number of reenactments of supposed conversations are used to reveal the nature of the supposed Northern conspiracy. Brad Crandall returns as narrator and strings together the reenactments in voice over. The acting is deplorably bad in all the reenactments. Fred Grandy (*Love Boat*'s Gopher) has a small role in this turkey. Despite absolving the South for the death of a well-regarded president, the film only did a fraction of what *In Search of Noah's Ark* did the previous year. It took in just north of $5 million. Still not bad a return on a half-a-million-dollar investment.

BEYOND AND BACK (1978)

Sunn teased near death experiences a bit in 1976's *The Amazing World of Psychic Phenomena*. *Beyond and Back* takes on the subject in full. The movie opens with a story about a psychiatrist named Eleanor Stevens who started researching the forbidden topic of near-death experiences. IMDB makes it clear Stevens is played by an actress and it's not clear if there ever existed a real thanatologist by that name. No searches turn up any reference to such a researcher. In fact, none of the people depicted as representing some of the best cases of near-death experiences (NDEs) exist outside of this documentary. It does seem, however, the movie changed the names of people involved in some NDE cases. For example, the movie recounts an NDE experienced by George Ritchie although the movie changes his name to "Paul Kelley". Thanatologist Eleanor Stevens might be a stand-in for NDE researcher Raymond Moody who published a book about NDEs a few years before this movie was made. It does mention a few historical people like Duncan MacDougall (died 1920). MacDougall was behind the famous "21 grams" experiment. This is the notion the human body loses 21 grams after death which represents the weight of a soul leaving the body. This may well be the general public's first introduction to this NDE rubber ducky. *Beyond and Back* also recycles some footage of Uri Geller from *The Amazing World of Psychic Phenomena* in an attempt to establish that there exists some kind of psychic energy that can live beyond death. Roger Ebert had this film on his list of his most hated films ever. However, it did pretty good box office, raking in $23.7 million.

THE BERMUDA TRIANGLE (1979)

This movie was based on the 1974 Charles Berlitz book of the same name. By 1979 Sunn seems to have fallen into a predictable pattern of stringing together a number of dramatizations of supposed real events with Brad Crandall as the voice of authority. This movie features, of course, a retelling of Flight 19 (US Navy torpedo bomber pilots lost during a training mission in 1945). Just when you're about to conclude this movie was a simple retelling of ships and planes supposedly lost at sea, the movie ropes in the legend of Atlantis, Edgar Cayce, the Bimini "road", and (for good measure) the Philadelphia Experiment. The movie offers a lot of science-y explanations as to why the area has an unusually high rate of accidents (it doesn't, of course). Crandall asserts unnamed physicists have suggested energy vortexes or portals into extra dimensions are to blame. Oh, yeah, and UFOs. *The Bermuda Triangle* was another big payday for Sunn, earning them $22 million.

BEYOND DEATH'S DOOR **(1979)**

This movie is an unremarkable follow up to 1978's *Beyond and Back*. The movie dispenses with the documentary approach altogether. It uses a docudrama framing structure to explicate the features of NDEs. We watch as a trio of ED docs become convinced there is life after death after witnessing patients experience a succession of NDEs. In one hilarious scene a skier is dragged into the boiling pits of hell with her ski goggles and toque still firmly on. It's not at all clear if any of the NDE dramatizations are based on actual cases. Most likely not. Maybe the greatest irony in the film is the disconnect between Sunn's mission statement to turn out Christian family fare and their hiring of director Henning Schellerup. Before contracting with Sunn, Schellerup made a number of racy R-rated films intended to be "enjoyed" by single men in cars at drive-in movie theaters.

IN SEARCH OF HISTORIC JESUS **(1979)**

Henning Schellerup, who was previously involved in the creation of such family films as *Mama's Dirty Girls* and the X-rated *Trader Hornee*, returns as director for this documentary about Jesus's historicity. Spoiler: Sunn thinks he's super historical. Similar to previous Sunn Classic documentaries, much of the explication is done via reenactments that are universally overlong, poorly acted, and lacking any actual point. Jesus resembles actor Zach Galifianakis with extra unkempt hair and beard. Anyone viewing this movie hoping to see a more traditional documentary about newly discovered archeological evidence or interviews with Biblical scholars will be sorely disappointed. One of three scientists featured in this movie is actually an actor. There is no indication he's representing the words or research of an actual scientist. In terms of newly discovered texts, the movie goes off the Christian script. It dips into theosophy's *Aquarian Gospel* (written in the 20th century) in an attempt to fill in Jesus's missing years as a boy and teen. We're introduced to the idea Jesus traveled variously to India and Britain. And, not surprising given Sunn Classic Pictures is based in Utah, it attempts to shore up Mormon claims that Jesus visited the Americas. The final act delves into the rare access the Catholic church gave scientists in 1978 to the Shroud of Turin. The film introduces the public for the first time to most of the pro-shroud talking points that continue today.

ENCOUNTER WITH DISASTER **(1979)**

The disaster film genre of the 1970s was pretty played out by 1979. Sunn gambled there was still a little gas left in the tank, especially if your disaster film had what no other disaster film of the 1970s had: footage of real people dying. This film pretty much plays like a proto-*Faces of Death*. We are treated to footage of people being killed at an airshow, people dying in car crashes, and people falling to their death. They cover the 84 lives wiped out at Le Mans in 1955 and of course the Hindenburg conflagration. They also run footage of people desperately (and unsuccessfully) trying to escape the Joelma skyscraper fire in 1974,. The Joelma segment is particularly graphic, reminiscent of events we saw played out on 9/11. We see people leaping from the burning building and their bent bodies after hitting the pavement. After about 90 minutes of man's inability to survive disasters, both natural and man-made, the movie curiously ends on a hopeful note. It offers the story of a Pan Am passenger plane (*Sovereign of the Skies*) that had to ditch in the ocean in 1956. All passengers and crew survived.

THE INCREDIBLE DISCOVERY OF NOAH'S ARK (1993)

In 1993 Sunn produced a made-for-TV sequel for CBS called *The Incredible Discovery of Noah's Ark* with Balsiger returning as a writer for the show. It was hosted by Darren McGavin, famous for playing Kolchak in *The Night Stalker* series. *The Incredible Discovery of Noah's Ark* is probably best remembered for falling hook-line-and-sinker for a Randi-like sting run by Gerald Larue. Larue (died 2014) was a professor of Biblical history and archaeology. Larue had his friend George Jammal appear on the show as an explorer with "definitive" proof the ark was real. Jammal was able to convince Sunn he had a piece of the ark acquired as part of a rogue two-man expedition to Ararat. After the show aired, Larue revealed the nature of the hoax. The piece of the ark was just a modern railroad tie that had been baked with some teriyaki sauce to make it look old and dried up. The show's producers accused Larue of trying to slander Jammal. They simply couldn't believe they were hoaxed, despite doing nothing to verify Jammal's story or performing a C-14 test on the wood.

Eleven Horror Movies Supposedly Based on a True Story

THE EXORCIST (1973)

The novel that spawned the movie was based on a number of exorcisms performed on a 14-year-old boy in the late 1940s. The boy is only known by the pseudonym "Roland Doe". The exorcisms supposedly took place in a Jesuit hospital in St. Louis, Missouri. His family seemed to have some deeply held occult beliefs. For example, his aunt was a spiritualist who gave Doe a Ouija board. One of the priests attending the exorcism made audio recordings and *The Exorcist* author William Peter Blatty (died 2017) used those recordings as well as other accounts as research. Many of the elements of the Doe exorcism found their way into the book and movie: cursing at the priests, urinating, spitting at the priests, and the like.

THE HILLS HAVE EYES (1977)

Wes Craven's *The Hills Have Eyes* tells the tale of a family lost in the Nevada desert. They're stalked by a family of cannibals. The film is roughly based on a 15th century

Scottish legend of the Bean family. The Beans lived in caves by day and emerged at night to rob, murder, and eventually eat travelers. Historians doubt the authenticity of this legend although it's suggested the legend emerged from a 15th century famine that actually forced some people into cannibalism.

THE ENTITY (1982)

The Entity is a 1982 movie based on a 1978 book of the same name. Both are based on Doris Bither (died 1999) who claimed in 1974 she was sexually assaulted by a trio of male ghosts. Bither befriended a pair of parapsychologists from UCLA and told them her story. Bither's story was tragic but ghosts were probably not the cause. Bither was a product of a traumatic childhood and had a substance abuse problem. Her relationship with her four children was strained and the family lived in a condemned home. Alas, the parapsychologists "helping" decided her real problem was poltergeists.

POLTERGEIST (1982)

Spielberg's 1982 blockbuster *Poltergeist* was based on a supposed poltergeist haunting from 1958. It was known as the Seaford Poltergeist case. A family with two teenagers started to report bottle lids popping off mysteriously and items being knocked off of shelves. Skeptics who looked at the case noted every so-called unexplained incident happened when the teenage son was present and always near the son's presence. A psychologist who interviewed the family discovered the son quite loathed his father. It's not unbelievable the son was trying to wind up the dad.

A NIGHTMARE ON ELM STREET (1984)

In the late 1970s, a number of Hmong males who had been resettled by the USA after the Vietnam War were dying in their sleep. They were generally considered healthy. Many reports indicated the men had bad dreams or night terrors before their death. It became dubbed the "Asian Death Syndrome". Filmmaker Wes Craven used the news reports as the basis for his film *A Nightmare on Elm Street*.

DEAD RINGERS (1988)

Dead Ringers is a horror movie about twin gynecologists. I will say no more other than it's a David Cronenberg film. As bizarre as the concept is, it is partially based on the lives of real-life twin gynecologists Steward and Cyril Marcus. Both became prominent New York lady parts doctors. Their careers spiraled down as both became addicted to drugs. Both brothers ended up dying within days of each other from drug overdoses.

THE MOTHMAN PROPHECIES (2002)

The Mothman Prophecies was based on a 1975 book of the same name by John Keel. The book, in essence, relates Keel's investigation into reports of sightings of the Mothman cryptid in Point Pleasant, West Virginia in the late 1960s. The book itself lacks a cogent narrative structure. The movie ditches most of Keel's meanderings into the world of Ufology, men in black, and government conspiracies but retains the book's conclusion that the Mothman is some kind of harbinger of doom.

THE EXORCISM OF EMILY ROSE (2005)

This film is supposed to be based on Anneliese Michel (died 1976). Michel was a German woman with temporal lobe epilepsy. Michel didn't respond well to her medication and the family decided she was actually possessed by the Devil. They got a pair of priests to conduct an exorcism. Actually, they conducted not just one but 67 exorcisms on Michel. She died as a result of the ordeal. The priests were found guilty of negligent homicide. The film version changes the setting to the USA and combines the two priests into one character.

THE HAUNTING IN CONNECTICUT (2009)

The Haunting in Connecticut is based on a 1992 book called *In a Dark Place: The Story of a True Haunting* by Ray Garton. It tells the supposed true story of a family who unknowingly moved into a home once used as a mortuary. Odd things started happening and then, well, they called in Ed and Lorraine Warren. Garton has backed away from his account. He now claims he was hired by the Warrens to simply deliver a good story. This was not an easy task as there were a lot of problems with the family's account. When he discussed with the Warrens that the family's story seemed to not make sense, he alleges the Warrens admitted the family was crazy and he should simply make stuff up.

THE CONJURING (2013)

The Conjuring is based on the lives of self-proclaimed demonologists Ed and Lorraine Warren. The Warrens were terribly unimpressive proponents of the paranormal and evidence they offered skeptics of such was woefully dubious. The film itself isn't based on any one specific incident but is a supposed synthesis of the Warrens' case files.

THE QUIET ONES (2014)

The Quiet Ones is a British horror film about a parapsychology professor who ropes his grad students into an experiment to create a poltergeist. The film is based on the so-called Philip Experiment that was run in 1972 in Toronto. Two parapsychologists and some close associates created a backstory for a fake spirit (which they named Philip) and tried to see if they could conjure it into reality during a seance. Nothing unusual happened despite multiple attempts. However, when the experiments relaxed the control conditions enough to allow trickery (e.g., they turned the lights off) suddenly "Philip" began to "talk" through table rapping. Oddly enough, "Philip" was never able to provide information that wasn't known to the participants.

Seven Examples of Great Media Resources Going to the Woo Side

A&E

A&E is a cable station that began life in the mid-1980s as the Arts & Entertainment network. It was probably best known for its daily one-hour *Biography* series as well as various documentaries and miniseries inspired by classic novels. In 2002 it released a pretty decent adaptation of Ursula K. Le Guine's book *The Lathe of Heaven*. In the early 2000s, A&E turned to reality programming. It started offering paranormal shows like *Ghost Hunters* and *Celebrity Ghost Stories.* Its show *Psychic Kids* is probably its most problematic offering. The show features psychic investigator Chip Coffey who teaches children how to harness their supposed psychic abilities. Critics have pointed out the ethics of adults knowingly or unknowingly deceiving children into thinking they have abilities that have never been shown to exist.

HISTORY (FORMERLY THE HISTORY CHANNEL)

It has become almost mandatory these days to put scare quotes around "History" when writing about the History channel. It's the poster child for something called "channel drift". That is the slow eroding of a cable channel's original format in pursuit of junk programming with broader audience appeal. In the 1990s History ran a lot of World War II documentaries and was often referred to as the "Hitler Channel". In 2010, it scored a huge hit with *Ancient Aliens* and the channel started turning out mystery mongering shows that distort actual history. Particularly galling to fans of actual history was the channel's run of *America Unearthed*. During its three seasons, host Scot Walters gave voice to crackpot theories of American history like Templars stashed the Ark of the Covenant in America and Egyptians buried treasure in the Grand Canyon.

TLC

Way, way back TLC stood for "The Learning Channel". It started in the 1980s as a non-profit network whose motto was "A place for learning minds". It positioned itself as a more technical and niche rival to the Discovery Channel. In 1991 the Discovery Channel purchased the Learning Channel and shortened its name to TLC.

In the late 1990s and early 2000s it moved into reality programming featuring freakish examples of American family life like *Jon & Kate Plus 8, Toddlers & Tiaras, Sister Wives*, and the Duggars.

AMERICAN HEROES CHANNEL

Those who missed the days of History being the Hitler channel found some refuge in the American Heroes Channel (AHC). However the channel has started to slide into some Ufology and conspiracy programming.

NATIONAL GEOGRAPHIC

National Geographic was a once storied magazine that was purchased by climate change denier Rupert Murdoch back in 2015. Although the magazine and its cable network still produce some good content (like 2020's reboot of *Cosmos* and *Is It Real?*), cracks have started showing under Murdoch's ownership. One wonders how in good conscience the company could let a show like *Chasing UFOs* get airtime on its network. Following the sale to Murdoch the magazine started running a number of paranormal- and Christian-themed "special issues". For example, "Strange but True: Secrets of the Supernatural Revealed" (2015), "The Story of Jesus" (2019), "50 Most Influential Figures of the Bible", "Secret Societies" (2020), "Atlas of the Bible" (2021), and "Jesus and the Apostles" (2021).

THE NEW YORK TIMES

The New York Times is arguably one of the most respected sources of journalistic truth on the planet. So, it's kind of sad to see the storied paper apparently setting up a UFO desk and giving numerous column inches to near death experience author Leslie Kean to hype up flying saucers. In the guise of doing hard hitting investigative journalism, in the words of Dr. Steven Novella, Kean "gets there by putting blinders on and ignoring the vast skeptical literature that puts these events into a better light."

THE TRAVEL CHANNEL

The Travel Channel gave us one of the greatest foodie/travel shows ever: *Anthony Bourdain: No Reservations.* Sadly after the Bourdain glory days, the Travel Channel's rating started to slide. In 2018 it turned to paranormal programming. The channel offers no less than eight ghost themed shows (*Ghost Adventures, Ghost Brothers, Ghost Nation*, et al.). Cryptid themed shows include *Into the Unknown* and *Mountain Monsters.*

Eighteen Great Skeptical Podcasts

Skepticality is arguably the first skeptical podcast, beating *The Skeptics' Guide to the Universe* to iTunes by a month. Alas *Skepticality* is no more. The last episode was published in 2019. Don't fret. Here are eighteen skeptical podcasts still releasing new episodes (as of this writing).

AMERICAN HYSTERIA

Host Chelsey Weber-Smith examines urban legends, moral panics, media hype, and conspiracy theories that have plagued the American psyche over the years. Topics include the Satan panic, video game violence, the war on Christmas, and the gay agenda moral panic.

BE REASONABLE

Be Reasonable is a podcast hosted by *The Skeptic* magazine editor Michael Marshall. Each episode features Marshall engaging with a true believer via a friendly, non-confrontational dialog. The goal of *Be Reasonable* is to help skeptics understand what true believers believe and why they believe it. Marshall gently prods the true believer to consider alternative explanations which usually results in the guest sputtering highly improbable babble as to why the rational approach is off the table.

COGNITIVE DISSONANCE

Hosts Tom and Cecil each week take on a wide variety of skeptical, political, and religious topics. The duo do not hide their political leaning which are firmly to the left of the spectrum. The humor can be not always politically correct.

LITTLE ATOMS

Little Atoms is a UK podcast that tends to cover cultural topics of interest to skeptics. The show interviews authors who write works of fiction and nonfiction books that cover humanist, scientific, religious, and skeptical themes.

GEOLOGIC PODCAST

The *Geologic Podcast* is hard to categorize as a show purely about skepticism. Host George Hrab, a high profile skeptic, delivers a weekly podcast full of skits, stories about his life, grammar lessons, music lessons, and skeptical commentary.

IN RESEARCH OF

Hosts Blake Smith (of *Monster Talk* fame) and archaeologist Jeb Card (former host of the *Archaeological Fantasies* podcast) take a sober second look at Leonard Nimoy's influential mystery mongering show *In Search Of*. For many skeptics of a certain age, *In Search Of* was their gateway into skepticism. The hosts are revisiting this show of their youth and slowly rewatching the series in broadcast order. Each episode breaks down a show they watched and offers the needed skeptical commentary that was never included with the show.

OH NO, ROSS AND CARRIE

Each episode features hosts Ross Blocher and Carrie Poppy relating their latest investigation into a paranormal topic or bizarre religious group. Blocher and Poppy jump in feet first, joining cults, learning to be exorcists, trying to follow various alt med dietary regimes, and a whole lot more. Their attempt to become Scientologists is probably some of their best work.

POINT OF INQUIRY

Point of Inquiry began in 2005. It is the Center for Inquiry's official podcast. The podcast has had a number of hosts over the last 15 years, including D. J. Grothe, Karen Stollznow, Indre Viskontas, and James Underdown. The show uses an interview format. It pairs the host with a high-profile guest who offers deeper insight into topical issues of the day. High profile guests have included Oliver Sacks, Neil deGrasse Tyson, Eugenie Scott, and Ann Druyan.

MONSTER TALK

Monster Talk's hosts Blake Smith and Karen Stollznow are self-admitted fans of cryptids and spooky stories. However, the hosts take on the topic from a fair, open-minded, and rational point of view. The podcast has featured interviews with many believers in cryptids.

OUR FAKE HISTORY

Our Fake History is hosted by high school history teacher Sebastian Major. In each episode, Major takes a look at popular historical topics and teases out facts from myth. Of particular note is his three-part series on Helena Blavatsky and the founding of Theosophy.

QANON ANONYMOUS

Each week, hosts Travis View, Julian Feeld, Jake Rockatansky, Annie Kelly, and Liv Agar examine the latest news out of the Q sphere. The show is half hilarious and half terrifying.

THE REALITY CHECK

The Reality Check was started by the Ottawa skeptics in 2008. It's a Canadian podcast similar to *Skeptics' Guide*. A small cast of hosts examine a variety of skeptical topics of the day and occasionally interview guests. *The Reality Check* team usually examines three skeptical topics per show and covers each in depth.

RELIC: THE LOST TREASURE PODCAST

Relic takes a sober, sensible look at legends of lost treasure and a subculture that will expend no end of real treasure to find them.

THE SCIENTIFIC ODYSSEY

The Scientific Odyssey podcast is hosted by physics professor Dr. Chad Davies. In each episode he covers the history of various sciences. Davies, who teaches a course on pseudoscience at Gordon State College in Georgia, infuses a light amount of skepticism and philosophy into many episodes. The show is wonderfully produced and reminiscent of Carl Sagan's *Cosmos*.

SCIENCE VS

In each episode of *Science Vs* host Wendy Zuckerman examines various social anxieties, fads, and misconceptions and determines what the science actually says. She takes on topics like probiotics, Covid, expensive dog food, detoxing, and UFOs.

THE SKEPTICS' GUIDE TO THE UNIVERSE

The Skeptics' Guide to the Universe cracked their mics back in 2005. The team (Dr. Steven Novella, Bob Novella, Jay Novella, Evan Bernstein, and Cara Santa Maria) has been delivering a nearly unbroken chain of weekly shows on a wide range of skeptical topics. The show is probably most famous for Dr. Novella's application of informal logical fallacies to skeptical issues.

SKEPTICS WITH A K

Skeptics with a K is a weekly general skeptical podcast. It's the official podcast of the Merseyside Skeptical Society. Unsure where Merseyside is? Don't ask anyone from Merseyside. They'll tell you it's at the side of the Mersey. It's ably and entertainingly hosted by Mike Hall, Michael Marshall, and Alice Howarth. Main host Hall is a *Doctor Who* fan so you will get a bit more banter about Time Lords and Daleks than you typically get on a skeptical podcast.

SKEPTOID

Skeptoid is a short format podcast on paranormal topics. It was started in 2006. Host Brian Dunning has been accused of infusing a wee bit too much of his Libertarian beliefs into the podcast, although he seems to have listened to such criticism and course corrected over the last number of years. One would be remiss in not pointing out Dunning served 15 months in prison after being convicted of wire fraud. If one gets past the controversies, Dunning usually serves up a well-researched, logically sound analysis of skeptical topics. He's a big believer in trying not to debunk something until he's dug up the primary sources and confirmed if the event ever happened.

SQUARING THE STRANGE

Squaring the Strange features hosts Benjamin Radford, Pascual Romero, and Celestia Ward. Each week they take a deep dive into a skeptical topic like the Minnesota iceman, Bigfoot, psychic detectives, and various urban legends.

TALES FROM THE RABBIT HOLE

Mick West, creator of the *Metabunk* website, interviews skeptics and true believers. True to form, West brings a lot more technical detail to topics like UFOs, 9/11, chemtrails, and flat earth.

Seven Great Skeptical Books

ABOMINABLE SCIENCE! BY DANIEL LOXTON AND DONALD R. PROTHERO

So-called "Bigfoot skepticism" seems to have fallen out of favor in recent years. Many seem to miss so-called Bigfoot skepticism is an excellent gateway into how to think critically and what constitutes good evidence for incredible claims. Daniel Loxton and Donald Prothero demonstrate this brilliantly. *Abominable Science!* is a detailed analysis and take-down of methodologies employed by various cryptozoologists. The book also reveals the odd intersection between cryptozoology and creationism.

DEMON HAUNTED WORLD BY CARL SAGAN

Sagan's 1995 book is considered by many the best all-around book on skepticism. It gave us not only Sagan's 9-point "baloney detection kit" but the great analogy of the invisible dragon in his garage. No matter what test you propose to verify the reality of the dragon, you are always met with an explanation as to why the test won't work on this particular dragon. Sagan sums up brilliantly "Now what's the difference between an invisible, incorporeal, floating dragon who spits heatless fire and no dragon at all?"

FADS AND FALLACIES IN THE NAME OF SCIENCE BY MARTIN GARDNER

Martin Gardner's 1957 survey of all things woo set the template for most general works of skepticism that followed. What's disturbing is a lot of things Gardner debunked in the 1950s are still alive and well today. *Fads and Fallacies* covers things like flat Earth, Dianetics, dowsing, naturopathy, creationism, chiropractors, and ESP.

FLIM-FLAM! PSYCHICS, ESP, UNICORNS, AND OTHER DELUSIONS BY JAMES RANDI

Penn Jillette attributes reading this book to his life-long love for skepticism. Randi's 1980 book lays bare the sloppy controls used by parapsychologists. He details a number of his own tests of claimants for his paranormal prize. Back then the amount was only $10,000 (about $30,000 today adjusted for inflation). It's written in Randi's signature curmudgeonly style. Of von Däniken, Randi comments "the only facts in his four books that I depend on are the page numbers."

NONSENSE ON STILTS: HOW TO TELL SCIENCE FROM BUNK BY MASSIMO PIGLIUCCI

Massimo Pigliucci is one of those rare people who have achieved advanced degrees in science (a PhD in Genetics and a PhD in Biology) and in Philosophy (a PhD in Philosophy of Science). Pigliucci is uniquely qualified, as he demonstrates in *Nonsense on Stilts*, to discuss what some call the demarcation problem. When does science become pseudoscience?

THE SKEPTIC'S DICTIONARY BY ROBERT TODD CARROLL

The late Robert Carroll (died 2016) launched *The Skeptic's Dictionary* as a website back in 1994. As the title suggests, the book is a comprehensive list of woo organized from A to Z. Each entry is a concise overview of a woo topic, the problems with it, and what the science says. It's an excellent desktop reference, a good starting point on most woo topics, and, dare I say, a good bathroom reader!

SCIENTIFICAL AMERICANS BY SHARON HILL

In this book, professional geologist and lover of the paranormal Sharon Hill examines the rise of what she terms Amateur Research and Investigation Groups (ARIGs). You can't turn on a cable channel without finding at least one show devoted to the exploits of an ARIG, be they ghost hunters or trying to find Bigfoot. Not being a working scientist doesn't mean you can't do science. There are plenty of amateur scientists that make valuable contributions to mainstream science, with amateur astronomers being at the top of the list. ARIG paranormal groups, however, have contributed precious little. Hill gets to the bottom of what these groups are doing wrong. They are going through the motions of science (i.e., being "scientifical") without understanding the need for rigor and controls.

Eighteen Woo Celebrities

DAN AYKROYD

Dan Aykroyd has had a lifelong interest in the paranormal and it led him to write the script for *Ghostbusters*. According to Aykroyd, he comes from a long line of mystics starting with his great grandfather. He grew up in a home full of paranormal literature and magazines like *Fate*. Following the success of *Ghostbusters,* Aykroyd went on to host a number of paranormal themed TV shows like *Psi Factor*, *History's Mysteries*, and *Hotel Paranormal*. Since 2007, Aykroyd has been selling a brand of high-end vodka called Crystal Head Vodka. Befitting his paranormal beliefs, it's shaped like a crystal skull. It is advertised as being triple filtered through "Herkimer diamonds". Herkimer diamonds are actually quartz crystals that somewhat resemble diamonds.

Crystal Head Vodka will set you back about $60 in Canada
Photo Credit: Karl Mamer

JIM CAVIEZEL

Jim Caviezel might be best known for playing Jesus in Mel Gibson's *The Passion of the Christ*. In 2021 he started voicing support for a number of unhinged Qanon conspiracies like the notion Democrats are torturing and then murdering children to harvest their adrenochrome.

PRINCE CHARLES

Prince Charles (along with much of the royal family) is deep into homeopathic woo. He's a patron of London's Faculty of Homeopathy. Prince Charles also owns a company called Duchy Originals that sells herbal detox products that have been described as "outright quackery" by Edzard Ernst.

TOM DELONG

Tom DeLong is the founder of the band Blink 182. He's been a lifelong believer in the reality of UFOs, Bigfoot, and various conspiracy theories. Along with long-time Uri Geller advocate Harold Puthoff, DeLong formed a public company called To the Stars Academy of Arts & Sciences. The company seems to be involved in producing paranormal-themed TV shows as well as trying to reverse engineer alien technology. In 2019 the company trumpeted it had in its possession "exotic material samples from UFOs" and it was going to reverse engineer them. Two years on, no new technology has come out of DeLong's company.

ROBERT DE NIRO

Robert De Niro caused some controversy in 2016 when he helped promote Andrew Wakefield's anti-vaccine documentary *Vaxxed*. He initially allowed it to be played at his Tribeca Film Festival although later pulled the screening after significant protest. In 2017 he also joined forces with Robert F. Kennedy Jr. to offer a $100,000 reward for anyone who could offer "a peer-reviewed scientific study demonstrating that thimerosal is safe in the amounts contained in vaccines currently being administered to American children and pregnant women." De Niro claims "I, as a parent of a child who has autism, I'm concerned. And I want to know the truth. I'm not anti-vaccine. I want safe vaccines."

MEL GIBSON

Mel Gibson has been accused of subscribing to antisemitic conspiracy beliefs. In 2006, he was arrested for driving drunk. During the arrest officers reported he angrily asserted "Jews are responsible for all the wars in the world". This may come as no surprise when you examine the beliefs of his father Hutton Gibson (died 2020). The elder Gibson was an open Holocaust denier, a 9/11 Truther, and a traditionalist Catholic who believed Vatican II was "a Masonic plot backed by the Jews". In 2006 Mel Gibson started funding a traditionalist Catholic church in Pennsylvania. Gibson's church adheres to the sedevacantism position. That is, the Roman Catholic church has not sat a valid pope since the death of Pius XII in 1958.

DAVE GROHL

Dave Grohl and other members of his band the Foo Fighters promoted the HIV denial group Alive & Well in the early 2000s. Much to Grohl's credit in recent years he's admitted his HIV denial stance was a product of being young and dumb. They've scrubbed all support for HIV denialists off their social media sites. Today Grohl and the Foo Fighters back reputable HIV groups like Elton John's AIDS Foundation.

KYRIE IRVING

Kyrie Irving is a professional basketball player. In 2016 he helped the Cleveland Cavaliers win the NBA championship. A year later he started asserting the earth is flat and urged people to do their own research. In 2017, he also suggested he also subscribed to a Jim Marrs conspiracy theory that John F. Kennedy was planning to break up the international banking cartel and was killed for it.

SHIRLEY MACLAINE

Actress Shirley MacLaine has appeared in classic films like *Terms of Endearment* and Hitchcock's *The Trouble With Harry*. In 1983, MacLaine published a book titled *Out on a Limb* that detailed her belief in reincarnation, UFOs, and mediums. The book was a huge bestseller and launched the career of many channelers like J.Z. Knight. In the mid-1980s, MacLaine started to claim in a previous life she was the brother of Knight's 35,000-year-old Ramtha. In 2013, MacLaine merged her notions of reincarnation with her newfound fascination with the supposed laws of karma. In her book *What If...?*, she wondered if maybe the victims of the Holocaust had it coming to them. No shit. She asked readers to consider if maybe Holocaust victims were Roman soldiers who had persecuted Christians in a previous life and got their karma balanced by dying in Hitler's gas chambers. It is not clear how that made it past an editor and a publisher.

BILL MAHER

Bill Maher can, at times, be a darling of the political left. Many skeptics on the left have, however, treated him like a hot potato shooting radioactive murder hornets out of his eyes when he started promoting a number of bizarre alt med claims. In 2009 he rather bluntly tweeted "If u get a swine flu shot ur an idiot." Amazingly, the same year he also won the Richard Dawkins Award, which effectively rendered that award meaningless. In 2016, Maher hosted a doctor named Samir Chachoua on his HBO show. Chachoua believes he cured Charlie Sheen of AIDS by injecting him with the encephalitis virus. In 2021, Maher called the media coverage of the Covid pandemic "panic porn".

JENNY MCCARTHY

Jenny McCarthy is the face of vaccine denial. She believes the MMR vaccine caused her child's autism. McCarthy also believes she's a being called an Indigo. This is a pseudoscientific belief that super spiritually advanced people are being born. These Indigo people will help save us all. According to the *Jenny McCarthy Body Count* page, McCarthy's anti-vaccine activism has potentially contributed to the deaths of over 150,000. She's not really helping.

MEHMET OZ

Mehmet Oz is better known as Dr. Oz. Oz hosts a daytime health show. Oz occasionally uses his show to promote pseudoscientific ideas and endorse an endless stream of quack health products. On his show, he has featured homeopaths, anti-GMO advocates, psychic mediums, and doctors promoting "reparative therapy" (the bogus idea you can "cure" gay people). Oz's hard on for woo isn't new. It has been reported that in the 1980s, Oz started letting Reiki practitioners into his surgery room.

RANDY QUAID

Following a charge of burglary in September 2010 (and a host of other legal troubles), Randy Quaid and his wife Evi sought refugee status in Canada. The Quaids claimed a group called the Hollywood Star Whackers had targeted them. The group had already killed Heath Ledger and had poisoned Jeremy Piven by lacing his sushi with mercury. They were next. For some reason. Their request didn't get very far. Evi Quaid was, however, granted Canadian citizenship based on her parents' citizenship. Randy was denied status as his wife's husband.

ALICIA SILVERSTONE

Alicia Silverstone, star of *Clueless*, published a parenting book in 2014. Almost nothing good comes from celebrities giving parenting advice. True to form, in her book she made the remarkable assertion that commercial tampons contain fragrances that have "hormone-upsetting, fertility-knocking phthalates that are snuggling up to your hoo-ha". A hoo-ha is another word for a lady's vagina. She didn't stop there. She joined a long list of celebrity "warrior moms" who promote vaccine hesitancy.

PAT SAJAK

Pat Sajak is the long running host of *The Wheel of Fortune*. In 2014 he outed himself as a climate change denialist when he bizarrely tweeted "I now believe global warming alarmists are unpatriotic racists knowingly misleading for their own ends. Good night."

STEVEN SPIELBERG

Probably no one has done more to advance the Grey alien type than Steven Spielberg. For good reason, in the 1970s Spielberg honestly believed Earth was being visited by aliens. He had originally set out to make a low-budget documentary about UFO sightings but after the success of *Jaws*, Spielberg was given enough financing and creative control by his studio that he could make *Close Encounters of the Third Kind*. Spielberg eventually concluded the ET hypothesis was untenable. He brilliantly realized that given everyone today carries a high-quality camera on their phone, there should be many good quality photos of UFOs. But there aren't.

BEN STEIN

Ben Stein is probably most recognized by people as the droning economics teacher in the movie *Ferris Bueller's Day Off*. Skeptics probably know him better for his support of intelligent design. He cowrote and starred in *Expelled: No Intelligence Allowed* which worked really, really hard to tie evolution to the rise of Nazi Germany.

NAOMI WOLF

Naomi Wolf rocketed to fame as a feminist thought leader with her 1991 book *The Beauty Myth*. She was an advisor to Bill Clinton and Al Gore. Her views on feminism were well regarded for a number of years but she slowly turned towards conspiracy mongering by the 2010s. Her online postings started to warn about looming fascist coups on the verge of taking over America. In 2014 she claimed US troops were being sent to West Africa to bring Ebola back to America to infect the population. In 2019, she published a book on society's repression of gay people. A central thesis of her book was Victorian era courts regularly put to death gay men. In a BBC interview it was pointed out to her she had failed to understand the 19th century legal term "death recorded". Wolf assumed it meant the person was executed, when it really meant the death sentence was commuted or the person was pardoned. Her American publisher canceled the release of her book and pulped printed copies. During the Covid epidemic, Wolf ramped up conspiracy mongering about the virus, 5G, and vaccines. In June of 2021 Twitter pulled her account for various violations of its Terms of Service.

Twenty Two Celebrities Who Are Scientologists

1. Kirstie Alley (actor famous for playing a Vulcan in a *Star Trek* movie)
2. Anne Archer (actor famous for her role in the film *Fatal Attraction*)
3. Sonny Bono (died 1998, singer, Congressman, and former husband of Cher)
4. Nancy Cartwright (the voice of Bart Simpson)
5. Kate Ceberano (actor and granddaughter of L. Ron Hubbard's governess)
6. Tom Constanten (keyboardist for the Grateful Dead between 1968 and 1970)
7. Chick Corea (died 2021, jazz musician)
8. Tom Cruise (actor famous for his role in the film *Risky Business*)
9. Jenna Elfman (actor famous for her lead role in the TV series *Dharma & Greg*)
10. Chloe Fineman (*Saturday Night Live* comedian)
11. Peaches Geldof (died 2014, model and daughter of Bob Geldof)
12. Isaac Hayes (died 2008, a musician famous for writing the musical score for the film *Shaft* and playing the voice of Chef in *South Park*)
13. Vivian Kubrick (the daughter of Stanley Kubrick)
14. Charles Lakes (first Black American gymnast to compete in the Olympics)
15. Juliette Lewis (actor famous for her role in the film *Natural Born Killers*)
16. Elisabeth Moss (actor famous for roles on *Mad Men* and *The Handmaid's Tale*)
17. Michael Peña (actor famous for his lead role in Netflix's *Narcos*)
18. Giovanni Ribisi (actor famous for roles in *Sneaky Pete* and *Lost in Translation*)

19. Michelle Stafford (daytime soap opera actor famous for roles in *The Young and the Restless* and *General Hospital*)
20. John Travolta (actor famous for his role in the *Grease* film adaptation)
21. Greta Van Susteren (news anchor who has appeared on CNN, Fox News, and MSNBC)
22. Edgar Winter (musician famous for performing the song "Free Ride)

Five Musicians Who Are Claimed to Have Faked Their Own Death

Being a rock star is rewarding but, unless you're Keith Richards, it doesn't seem to be a profession that promotes longevity. We've lost numerous musicians to drugs, violence, and simple accidents associated with their frenetic lifestyles. Some people refuse to accept their favorite musician has passed on. It seems every passing of a musician in his/her prime is accompanied by claims the musician faked his/her own death for one contrived reason or another.

MICHAEL JACKSON (DIED JUNE 25, 2009)

It's no secret some celebs are worth more after they die. The theory goes that Jackson was in a debt hole he couldn't climb out of. However, if he simply faked his death, the value of his works would triple overnight and result in a cash windfall that would help him escape his debts. Evidence? Well, there's a silly website called Michael Jackson Sightings that features blobsquatch images of an ostensibly extant Michael Jackson attending weddings, sitting in the backseat of cars, and hanging out in Norway.

ELVIS PRESLEY (DIED AUGUST 16, 1977)

If not the first, the Presley myth is at least the best-known example of a celebrity that apparently faked his own death. The evidence? Well, there are blobsquatch photos of fat, old greasy dudes that sort of look like fat, old Elvis. There's the fact Elvis's middle name on his birth certificate was Aron but his middle name was spelled Aaron on his tombstone. So, that can't be him! In reality, Elvis' middle name was mistakenly spelled Aron on his birth certificate and that became his legal name on various official government documents. However, Presley liked to go by the more Biblical Aaron and that became his stage name. Reasons given for Elvis faking his own death range from him being tired of always being in the spotlight to having to go underground after he supposedly gave damaging testimony about various Mafia types.

KURT COBAIN (DIED APRIL 5, 1994)

The official story is Cobain was high on heroin and committed suicide by shooting himself with a shotgun. An autopsy revealed he had ingested a large amount of the drugs. Some claim he was too high to operate a shotgun. Hence, Cobain, ostensibly a very private person, faked his own death to get away from the pressures of fame. Some allege Cobain didn't want out of the spotlight entirely. He just didn't want to be the poster child for kids of the 1990s so he faked his death and then joined the band Weezer under the name of Rivers Cuomo.

JIM MORRISON (DIED JULY 3, 1971)

Morrison ostensibly died of a drug overdose in a Paris hotel room but since no autopsy was performed, it's anyone's guess. The idea that Morrison faked his own death was given some legs by his bandmate Ray Manzarek who suggested Morrison was certainly rich enough to pay a doctor to write him up a death certificate.

JIMI HENDRIX (DIED SEPTEMBER 18, 1970)

Hendrix is another one of those greats people can't believe is gone. There's a theory that Hendrix really wanted to be an actor. So, he faked his own death and reinvented himself as none other than Morgan Freeman. This half-baked idea is based on a few thin "facts": Freeman and Hendrix look a bit alike, Freeman's acting career started slightly after Hendrix died, and both of them love(ed) music. While it's correct to say Freeman's first credited film role was in 1971's *Who Says I Can't Ride a Rainbow!*, Freeman worked in musical theater in the 1960s and worked on Broadway in 1968.

Six Museums Devoted to Pseudoscience

BIG VALLEY CREATION SCIENCE MUSEUM (BIG VALLEY, ALBERTA)

Canada's version of Ken Ham's Creation Museum was created by an Alberta oil field worker who describes himself as a "grade one dropout". It opened in 2007 at the cost of $280,000. The museum has an exhibit space of 900 square feet, which is about the size of a two-bedroom condo. Exhibits include a supposed fossilized teddy bear (demonstrating fossils are young, I guess), pictures of a supposed carving of a stegosaurus on a Cambodian temple wall, and a rather bizarre scroll that tries to prove the British Royal Family can trace its lineage directly back to Adam and Eve (but can't we all, according to creationists?).

CREATION MUSEUM (PETERSBURG, KENTUCKY)

The Creation Museum is Ken Ham's $27 million monument to young Earth creation pseudoscience. It opened in 2007. It depicts humans and dinosaurs coexisting. One museum display suggests stories of dragons are evidence humans and dinosaurs lived side by side. Weak sauce. The museum also features a planetarium that offers shows on "creationist cosmology", which isn't a real thing. For good measure, it lays the blame for racism and genocide at the feet of people who try to use reason to understand the natural world. Employees must sign a statement of faith if they wish to work there. They must affirm, among other things, the only legitimate marriage is between one man and one woman and no line of evidence can be valid if it contradicts the Bible. Real museums don't care who you marry. For many good reasons, the Creation Museum is not accredited by the American Alliance of Museums.

EXPEDITION: BIGFOOT! THE SASQUATCH MUSEUM (CHERRY LOG, GEORGIA)

Though this museum is pretty far from Bigfoot's reputed home in the Pacific Northwest, at 4,000 square feet, it is the largest Bigfoot museum in the world. It also claims to have the most complete collection of Bigfoot artifacts and lore assembled in one place. It certainly may be the only museum in the world where you can see, live and in person, a 36-inch long Bigfoot turd. The museum opened its doors in 2006. It was partially inspired by the 1972 cult film *The Legend of Boggy Creek*.

INTERNATIONAL UFO MUSEUM AND RESEARCH CENTER (ROSWELL, NEW MEXICO)

It should come as no surprise there's a UFO museum in downtown Roswell. It opened its doors in 1991 in what was a former movie theater. It is largely devoted to the legend of a saucer crash in 1947. It features an alien prop used in the 1994 TV show *Roswell*. Some enterprising individual, maybe trying to recoup the cost of admission, took photos of the prop and then sold them to *Penthouse* magazine. Publisher Bob Guccione (died 2010) trumpeted he had "real" alien autopsy photos but *Roswell*'s producer quickly came forward to point out the photos were an exact match for the prop he had donated to the museum.

LOCH NESS CENTRE AND EXHIBITION CENTRE (DRUMNADROCHIT, SCOTLAND)

This museum was opened in 1999 by Sir Ranulph Fiennes. He was listed in the 1984 edition of the *Guinness Book of World Records* as the world's greatest living explorer. It takes a somewhat non-credulous look at Nessie, attempts to find the monster, and various hoaxes perpetrated over the years. The museum seems to have been largely designed by Adrian Shine. Shine was, for a long time, one of the leading proponents of the monster's existence but the general lack of evidence after decades of searching and the lack of biomass to support a breeding population of sea serpents has led him to conclude sightings are most probably just people's imaginations.

PSYCHIATRY AN INDUSTRY OF DEATH MUSEUM (HOLLYWOOD, CALIFORNIA)

Psychiatry isn't without its problems. But that's not quite the pseudoscience behind this museum. Run by the Citizens Commission on Human Rights, widely understood to be a Scientology front, this museum seems to exist to sow unfounded fear about psychiatry. The museum tries to tie psychiatry to a large number of historical tragedies, including the Holocaust, Apartheid, and 9/11. Apparently, 9/11 was carried out at the behest of Big Psychiatry to boost the sale of antidepressants. Its opening in 2005 was a major outing for Scientology's celebrity royalty. The likes of Anne Archer, Jenna Elfman, and Lisa Marie Presley turned up for the ribbon cutting.

Twenty Seven Skeptical Quotes from Dana Scully of *The X-Files*

Ufology was really going the way of bell bottoms in the 1990s when the Fox network brought it back via their show *The X-Files*. FBI agent Dana Scully was the voice of skepticism, although it was more like a Ufologist's idea of a skeptic. Once in a while, the show's writers did manage to knock a good skeptical line out of the park for Scully.

ON UFOS

"Just because I can't explain it doesn't mean I'm going to believe they're UFOs."

"Mulder, why would alien beings travel light years to Earth in order to play 'doctor' on cattle?"

"Extraterrestrial visitors from beyond who, apparently, have nothing better to do than buzz one mountain over and over again for 700 years."

"The truth is out there, but so are lies."

"Abduction lore has become so prevalent in our society that you can ask someone to imagine what it would be like to be abducted and they'd concoct an identical scenario. "

"How do we know that he's not just another crackpot whose encyclopedic knowledge of extraterrestrial life isn't derived exclusively from reruns of *Star Trek*?"

ON BELIEF IN THE AFTERLIFE

"I'd settle for a life in this one."

ON CONSPIRACY CRANKS

"Did you see the way they answered the telephone? They probably think that every call that they get is monitored and they're followed wherever they go. It's a form of self-delusion. It makes them think that what they're doing is important..."

"And we thought we could rest easy with the fall of the Soviet Union."

"There is a lot of money to be made in scaring people."

ON CRYPTOZOOLOGY

"Do you recall what Barnum said about suckers?"

"You saw something, I'll give you that but I'm not about to go in and sell it. Not when it's nothing more than a sighting in a dark alley."

"Anxieties taking the form of a hideous monster for whom the most horrific human attributes can be ascribed. What we can't possibly imagine ourselves capable of we can blame on the ogre, on the hunchback, on the lowly half-breed."

"Mulder, this is just a dopey hoax."

ON THE NEED FOR EVIDENCE

"Please explain to me the scientific nature of the whammy."

"Maybe you're just seeing what you wanted to see."

"Because it's easier to believe the lie. Isn't it?"

ON FAITH

"Maybe true faith really is a form of insanity."

"I don't have time for your convenient ignorance."

"The whole idea of a benevolent entity fits perfectly with what I'm saying. That a spirit would materialize or return for no other purpose than to show itself is silly and ridiculous."

"You already believe, Mulder. What difference would it make? I mean, what would proof change for you?"

ON USING HYPNOSIS AS EVIDENCE

"I know that it has its therapeutic value, but it has never been proven to enhance memory. In fact, it actually worsens it since people in that state are prone to confabulation."

ON THE SATANIC PANIC

"I don't suppose there have been any actual reports of stolen infants, or mass graves being uncovered anywhere in town, or that you found an altar, or any other evidence of a black mass?"

"...the FBI's study also found that in most cases, like the McWarren preschool trial, witnesses were often prompted in their statements by rumors of stories that were being circulated and that there was in fact nothing to support them."

"Well, possession is a common claim by criminals who have dissociative disorders. It's how they distance themselves from their actions."

ON USING ANCIENT LEGENDS TO VALIDATE THE REALITY OF A PSEUDOSCIENTIFIC CLAIM

"I'm not sure that modern reconstruction of ancient fairy tales lends any more credence."

Five Famous People Claimed to Be Aliens or a Supernatural Creature

DAVID BOWIE

This one is obvious but it seemed to take a long time for people to put the pieces together. A guy who sang a lot about being from space and urging us to love aliens might actually have been from space. Nina Simone commented once that Bowie was "not human". The idea seemed to gain some traction when an issue of *Fortean Times* ran a mysterious classified ad from someone offering proof that Bowie was an alien. What is the proof? Well, aside from the obvious attempt to hide his alienness in the open, Bowie had a fear of flying. Was this because he had a fear of flying or was it because he was from such an advanced civilization that modern jet planes seemed downright rickety to him?

NICOLAS CAGE

An antique dealer stumbled on a Civil War era photo of a Confederate POW who bore a striking resemblance to Nicolas Cage. His conclusion was he was looking at an actual photo of Nicolas Cage from 150+ years ago and therefore Cage must be an immortal vampire. The rumor reached Cage himself who commented on a 2012 episode of *David Letterman* that it could not be a photo of a vampire as everyone knows you can't take a photo of a vampire. An excellent point which led people to conclude Cage was not a vampire after all but was, in fact, a time-traveling space alien.

QUEEN ELIZABETH II

David Icke seems to think humanity is controlled by a multi-generational cult of shape-shifting space lizards. Keep in mind David Icke also claimed at one time he was the chosen son of God and was required to wear a turquoise tracksuit. The Queen (plus her kids, plus about every politician who wields any level of power greater than that of a small-town mayor) is one of them. The only way dear mumsy can keep her form is by drinking human blood.

KEANU REEVES

It has been suggested Keanu Reeves is an immortal being. Not only has he not seemed to age much since appearing in *Bill and Ted's Excellent Adventure*, but Internet denizens have noted he bears a striking resemblance to both Charlemagne (died 814) and French actor Paul Mounet (died 1922).

JOHN TESH

Starting around 1996, a Michigan-based group calling itself the National Anti-Tesh Society (NATAS which is "Satan" backwards) began issuing posts on Usenet groups alleging John Tesh was an alien scout sent by an other worldly organization called Echelon. The group eventually started engaging in live protests outside of his concerts to raise world awareness about the "Tesh threat". What is the evidence Tesh is an alien? Well, he seems to write music at an inhuman pace, he enters the stage by descending from above (like a UFO), and some people think his head isn't quite human shaped. Most people assume NATAS was a parody group, like the Church of the SubGenius. If true, not everyone got the joke. Rumors of Tesh being an alien have dogged him for years.

THE AMAZING MEETING
8
JULY 8-11 2010
SOUTH POINT CASINO, LAS VEGAS
SKEPTICISM

Thirteen Quotes by Skeptics and Scientists

ISAAC ASIMOV

"The most exciting phrase to hear in science, the one that heralds the most discoveries, is not 'Eureka' but 'That's funny'."

ARTHUR C. CLARKE

"A faith that cannot survive collision with the truth is not worth many regrets."

MARIE CURIE

"I was taught that the way of progress was neither swift nor easy."

ALBERT EINSTEIN

"Everything should be made as simple as possible, but not simpler."

PHIL PLAIT

"Don't be a dick." (From a speech where Plait argued that skeptics should not come across as jerks when trying to deal with cranks.)

Phil Plait delivering his "don't be a dick" talk at TAM 8.
Photo Credit: Karl Mamer

ELIZABETH LOFTUS

"To be cautious, one should not take high confidence as any absolute guarantee of anything."

ALAN MOORE

"The truth is, that it is not the Jewish banking conspiracy, or the grey aliens, or the twelve-foot reptiloids from another dimension that are in control, the truth is far more frightening; no-one is in control, the world is rudderless."

STEVEN NOVELLA

"Ignorance is a low-energy state. It takes constant vigilance and work to climb out of it."

KARL POPPER

"Whenever a theory appears to you as the only possible one, take this as a sign that you have neither understood the theory nor the problem which it was intended to solve."

JAMES RANDI

"Uri Geller may have psychic powers by means of which he can bend spoons; if so, he appears to be doing it the hard way."

CARL SAGAN

"If we are not able to ask skeptical questions, to interrogate those who tell us that something is true, to be skeptical of those in authority, then, we are up for grabs for the next charlatan (political or religious) who comes rambling along."

EUGENIE SCOTT

"I never say that evolution is a fact. Evolution is a theory. It's much more important than a fact, because theories explain things."

CAROL TAVRIS

"In the horrifying calculus of self-deception, the greater the pain we inflict on others, the greater the need to justify it to maintain our feelings of decency and self-worth."

Eight Quotes Misattributed to Skeptics, Scientists, and Fellow Travelers

CHARLES DARWIN

"How I wish I had not expressed my theory of evolution as I have done."

Claimed to be Darwin's deathbed words. Debunked by his daughter who was present at Darwin's passing.

ALBERT EINSTEIN

"Evil is the result of what happens when man does not have God's love present in his heart."

This quote seems to have been invented out of whole cloth. It surfaced in various emails circulating in 1999. People love making up Einstein quotes that ostensibly lend support to Christian beliefs, Astrology, and other forms of woo.

HARLAN ELLISON

"If you make people think they're thinking, they'll love you; but if you really make them think, they'll hate you."

The line is from a 1958 book on personal finance.

MURRAY GELL-MANN

"That which is not forbidden is mandatory."

This has become a cliché about quantum physics. It seems to have originated in T.H. White's *The Once and Future King*.

PENN JILLETTE

"Luck is probability taken personally."

Jillette believes the quote originates with statistician Chip Denman.

JAMES RANDI

"Everyone who believes in telekinesis, raise my hand."

Randi used it in a 2002 newsletter once but says he did not coin it. It seems to have been coined by a Washington D.C. radio DJ in 1992.

CARL SAGAN

"If it can be destroyed by the truth, it deserves to be destroyed by the truth. "

The quote actually originates with fantasy author P. C. Hodgell.

NEIL DEGRASSE TYSON

"In this moment, I am euphoric. Not because of any phony god's blessing. But because, I am enlightened by my intelligence. "

The quote originated as a meme on Reddit.

Twenty Skeptical Neologisms and Phrases

BLOBSQUATCH

According to cryptozoological author Loren Coleman this term was coined on a Bigfoot forum by Vito Quaranta. Originally meant to refer to the bad photographic skills of Bigfoot hunters, it has come to refer to any kind of blurry photo offered to support a claim of the existence of some entity (a sasquatch, a UFO, a ghost, the Loch Ness monster, etc.).

DARK BEER ARGUMENT

I hope the gentle reader will allow me to exert my privilege as the author to enter this neologism of my own creation. Full disclosure: it's used by almost no one besides the author. But it should be. It's a check digit all should use when getting into an argument and trying to bring to bear the tools of skepticism. Take a step back. Is this argument ultimately over a matter of taste? If it is, it is a Dark Beer Argument. Some people love dark beer (i.e., stouts and porters). Some people hate dark beer. There is no constellation of facts and logic that can make a person who doesn't enjoy dark beer suddenly decide to love dark beer. Hence, is what you're arguing over really a matter of personal taste, opinion, or some highly abstract moral consideration? If so, you're engaged in a Dark Beer Argument and you should probably stop trying to use the tools of skepticism.

GALILEO GAMBIT

When a person's pseudoscientific claims are not wholly embraced by the establishment because of a complete lack of evidence, typically the person will liken him/herself to Galileo in some way. The unspoken train of thought goes "he too was laughed at, therefore I'm like him and I'm probably right too". The best rejoinder is frequently given as "Sure they laughed at Galileo but that also laughed at Bozo the Clown."

GISH GALLOP

Named for creationist Duane Gish's debating style. In debates on evolution, Gish would offer a long stream of inaccuracies, weak arguments, and made up "facts". It left the evolution expert having to unpack a series of bad arguments and explain in detail why each was wrong. The evolution proponent would quickly discover it's easier to make stuff up than it is to explain why any single claim is wrong. The evolution proponent would quickly run out of time. The audience would be left feeling like the evolution side was unable to rebut many or most of Gish's arguments. Today it can refer to any pseudoscientific claim that piles on a huge number of inaccuracies and evidence-free assertions. Any episode of the History channel's *Ancient Aliens* is a good example of the Gish Gallop. To debunk any one 40-minute episode, it could take several hours of careful explanation as to why each and every claim is wrong.

GODWIN'S LAW

Coined by attorney Mike Godwin, the law states "as an online discussion grows longer, the probability of a comparison involving Nazis or Hitler approaches one". Godwin's Law is usually used to imply once the Nazi allegations start flying, the debate has reached either a natural end or an end to its usefulness for informative discourse.

INVISIBLE DRAGON

From Carl Sagan's *Demon Haunted World*. This is a claim for which there can never be falsifying evidence or the absence of any expected evidence is explained away by an all new ability or phenomenon.

JEAN DIXON EFFECT

Named after psychic Jean Dixon, it's a tendency to remember the hits and forget the misses. The term was coined by mathematician John Allen Paulos.

JUST ASKING QUESTIONS

Conspiracy nuts will throw out all manner of unsupported assertion. When called on it, they will typically suggest they're "just asking questions". Sometimes this tactic is referred to as "JAQing off".

NOBEL DISEASE

Many have noticed that a few Nobel prize winners suddenly switch gears and endorse or engage in pseudoscientific research. The canonical example would be two-time Nobel prize winner Linus Pauling. Pauling spent the final years of his career publishing on Vitamin C quackery. Robert Carroll, author of *The Skeptic's*

Dictionary, suggests the term was coined by blogger David Gorski (who also goes by the online name "Orac").

PELTS AND PAWS

Used to refer to people who investigate cryptids (like Bigfoot) from a purely naturalistic point of view. They do not imbue the cryptid with any kind of supernatural origin or paranormal abilities. The term seems to have been coined by Peter Rogerson in a May 2005 issue of a UK Fortean zine called *Magonia*. Archaeologist Jeb Card (co-host of the *In Research Of* podcast and author of the book *Spooky Archaeology*) has popularized it. The term is meant to parallel Ufology's use of the term "nuts and bolts", which implies an investigative approach to the UFO phenomenon that does not imbue aliens with supernatural or paranormal qualities.

POE

A reference to Poe's Law. Poe's law states that unless a parody of a religious idea is well signaled, it is ultimately indistinguishable from a real religious belief. So to call something a Poe is to caution it might be a clever work of parody. Before spreading a link or story as an example of how dumb the other side can be, first determine if it is, in fact, a Poe.

REAL SOON NOW

A way to sarcastically signal one is doubtful the promised evidence or free energy device will see the light of day. It comes from science fiction fandom and was popularized for many years by sci fi author Jerry Pournelle(died 2017) in his *BYTE* magazine column.

SCIENTIFICAL

Coined by skeptical blogger Sharon Hill. It is also the title of her book *Scientifical Americans*. Scientifical implies paranormal researchers who are trying to do science or think they're doing science but failing to follow or understand the scientific method. Ghosts hunters are a great example. They run around old houses taking measurements with thermometers and EM detectors but have no real idea why they're doing it or what to do with the data.

SCRIPTID

Coined by Blake Smith of the *Monster Talk* podcast. Sometimes written as "scryptid". It's a cryptid that clearly has a movie, TV show, or some work of fiction as its origin point and has morphed into an allegedly real creature over the years.

THE STUPID, IT BURNS

Coined by skeptical blogger David Gorski (aka, Orac). It's an indication the argument or idea is so stupid, it causes the reader's brain a burning pain just reading it.

TRUE BELIEVER

A catch all term to describe people who accept "woo" uncritically. True believers also tend to be the most vocal or intractable supporters of a particular flavor of woo.

TURTLES ALL THE WAY DOWN

This phrase alludes to the notion that a given claim is a never-ending regression of special pleading or attempts to dodge requests for a clear definition of the proposition. Basically, you'll never get to the bottom of it. It's a reference to an anecdote that a flat Earther once claimed the flat Earth resides on the back of a turtle. When challenged as to what the turtle stands upon, the flat Earther retorted "another turtle". When then asked what that second turtle stands on, the flat Earther shot back "very clever, it's turtles all the way down."

UNSINKABLE RUBBER DUCKY

Coined by James "the Amazing" Randi. It's a metaphor for the sad reality that no paranormal hokum, no matter how old or how many times it gets debunked over the decades or centuries, ever truly goes away. It might fall out of fashion for a time but ultimately gets rediscovered by a new generation that is unaware of years of extant criticism.

WEIRD WASHING

The action of removing paranormal or spiritual content from stories involving cryptids or space aliens. Usually this is done by "pelts and paws" cryptozoologists or "nuts and bolts" Ufologists who want a body of eyewitness evidence to be taken seriously. This term was coined by Timothy Renner of the *Strange Familiars* podcast.

WOO

A shortened version of woo-woo. It has become a catch-all noun for any kind of unsupported pseudoscientific position or religious claim.

James Randi's Seven Greatest Debunkings

URI GELLER UNABLE TO PERFORM ON CARSON (1973)

Uri Geller was scheduled to appear on the *Tonight Show with Johnny Carson* in 1973. Geller was at the start of his growing fame. He was going to demonstrate his astounding powers in front of Carson. Carson himself was an amateur magician and a friend of Randi. Randi helped Carson develop some pre-show protocols, including not letting Geller bring his own props and ensuring the objects Geller was going to manipulate were locked away from not only Geller but any stagehand. What ensued was twenty minutes of extremely painful television to watch. It's clear Geller knew he had no avenue to bring to bear his normal abilities. He spent half his allotted time waving his hands over various cans, trying to detect water. He tried to bend a spoon that couldn't be gaffed. After about sixteen minutes he started whining he was feeling pressure to perform. An amusing exchange went like this:

Geller: I don't feel strong. I'm feeling being pressed.

Carson: I'm not trying to pressure you.

Geller: You're telling me to try this and try that.

Carson: I thought that was the idea.

Geller reminded Carson he had been tested under controlled conditions and what was going on that night was not a controlled test. When Carson asked Geller what constituted a controlled test, Geller struggled to answer. He finally suggested nothing was under a bell jar and controlled conditions had cameras running and scientists watching. Geller might have been unaware he was in front of a camera.

$40,000 DOWSING TEST (1980)

Dowsing is a belief a person can psychically sense a hidden substance like water or gold. A dowser usually employs some kind of rod or rods that respond to the ideomotor effect and supposedly point to the location of the hidden substance. In 1980 Randi and partner Dick Smith tested ten of Australia's top dowsers. A prize of $40,000 was on offer for any dowser that could meet or beat their own self-proclaimed accuracy rate. Many consider this one of the most definitive tests of dowsing ever conducted. The setup was elaborate. A field was plowed and ten parallel water pipes were buried. One pipe was randomly selected to have the water turned on. The other nine did not. The subjects all agreed the setup was valid and a fair test of their ability. As a control against any accusation that Randi cheated, the dowsers witnessed every step of the construction. No objections were raised at any point. Before the true test, water was run through one pipe and dowsers were able to demonstrate their powers worked on that pipe in an unblinded fashion. This way they could not explain away their failure because of interference from clouds or sunlight. Dowsers had a 1 in 10 chance of picking the right pipe during the blinded trial. The subjects confidently predicted between an 80% and 100% accuracy rate. In another test, another group of dowsers who claimed the ability to find precious metals were asked to find gold and brass bars hidden in one of ten boxes. Results were no dowser was able to find the brass bar in 26 attempts (0 out of 10). The gold bar was found only four times in thirty six attempts (about 1 in 10). Water dowsers were only able to find the water ten times out of 50 tries (1 in 5). Technically, the dowsers beat the 1 in 10 chance when finding water, however the dowsers were required to achieve a hit rate that met or exceeded their predicted accuracy rate (which averaged about 92%).

JAMES HYDRICK BLOWS IT (1981)

In 1980, following a demonstration of his ability to psychically move a delicately balanced pencil on *That's Incredible!* (an early reality show), James Hydrick became a brief sensation. He was featured on various talk shows. His oeuvre never extended beyond being able to move items delicately balanced. Randi quickly figured out he was using his breath to move these objects (a pencil, a dollar bill on the head of a pin, and pages of a phonebook) with puffs of air. Hydrick had perfected this technique to the point where he could seem to be turning his head away from the object just before it moved, thereby answering critics who suggested he was blowing on the object. In 1981 Bob Barker had Hydrick on his TV show *That's My Line* with Randi to test his abilities to move pages of a phonebook. Hydrick first went through his normal routine of moving a pencil. Next Barker brought out a phonebook and Randi sprinkled light foam peanuts around it. Obviously, any puff of air would move the peanuts as readily as a couple pages of a phonebook. (For those who are not old enough to remember phonebooks, pages were made from notoriously thin tissue). Amazingly, Hydrick's powers abandoned him. Hydrick tried to claim the foam peanuts were causing static electricity that countered his psychic abilities. Hydrick failed spectacularly.

PROJECT ALPHA (1981)

James Smith McDonnell (died 1980) was the chairman of McDonnell Douglas and founder of the McDonnell Aircraft Corporation. He was also a believer in paranormal phenomena. Shortly before he died, he gave St. Louis' Washington University a large endowment to set up the McDonnell Laboratory for Psychical Research. Physicist Peter Phillips, who was interested in spoon bending, agreed to run what became known as the "MacLab". This was huge news. Parapsychologists had long complained funding was limiting their research. Finally, they had the financial backing to truly science this! Randi wrote a letter to Philips suggesting important protocols. Among these was a key protocol: the subjects themselves should not be allowed to change the controls. That one seems obvious, but you'd be surprised. Randi also passed on his customary advice that people researching phenomena that attract a broad range of tricksters should have a magician in the lab. Phillips rebuffed Randi's offer. Randi then launched Project Alpha along with two young magicians by the names of Mike Edwards and Steven "Banachek" Shaw. Edwards and Shaw got themselves selected as test subjects by demonstrating uncanny abilities to bend spoons with their minds... I mean hands. Edwards and Shaw went on to exploit every lax parameter Randi had warned Phillips to tighten up.

Banachek at The Amazing Meeting.
Photo Credit: Karl Mamer

They were also given explicit instructions if they were ever asked if they were just using tricks to immediately answer yes and that they were sent by Randi. In 1981, Philips announced he was ready to publish preliminary results. Randi sent another letter to Phillips to suggest he take a hard look at the results and ask himself if they could be a result of trickery. To help prod Philips, Randi leaked stories he had sent in a couple plants. Word got back to Philips and he tightened up some controls. Edwards and Shaw had limited success hoodwinking Philips and his cast of researchers after that. After three years of this, Randi decided it was time for the big reveal. In 1983, Randi held a press conference with Edwards and Shaw joining him. The MacLab continued along for a couple more years and then closed up shop in 1985 having run through its original grant.

PETER POPOFF EXPOSE ON JOHNNY CARSON (1986)

Peter Popoff was a big deal in the 1980s. At his religious crusades, he had an uncanny ability to call out specific details about people in the audience. He seemed to know their names, addresses, and specific illnesses they came to be healed of.

Randi and a small team of skeptics noticed Popoff wore what appeared to be a hearing aid while doing his bit. For a man who claimed he could channel his god's power to heal intractable diseases like cancer it was odd he couldn't heal himself of a slight hearing defect. It was pretty obvious what Popoff's trick was. Before the show, desperate people hoping Popoff would heal them filled out prayer cards with details about themselves. Plants in the pre-show crowd also circulated, chatted people up, and took notes. During show time Popoff's wife read these details into a microphone and transmitted them to Popoff's earpiece. Randi and his team brought a scanner to a show and captured the wife's audio. In May of 1986, Randi went on *The Tonight Show* and showed a video of Popoff with his wife's audio track synced. Popoff initially alleged the video was a fake but later admitted it was true. The radio thing was just a little helper he used occasionally. Randi's efforts initially drove Popoff off the air. Donations dwindled and his ministry was forced to declare bankruptcy. Popoff stayed down for about a decade but emerged again in 1998, buying time on cable networks, and selling things like Miracle Spring Water and anointing oil. He also led the charge on the emerging prosperity gospel. He would mail cheap trinkets to followers and request they send him "seed" checks written out in amounts equal to Bible verses. For example, if you wanted the blessing promised in *Psalm 34:19*, you'd write him a check for $34.19. People really fall for this stuff.

RANDI REVEALS THE TRICKS OF PSYCHIC SURGEONS (1986)

Peter Sellers and Andy Kaufman had three things in common. They were comedic geniuses out of step with their times. They died young. They both sought out the help of psychic surgeons instead of science-based medical care. From a young age, Sellers suffered from a heart condition that led to a heart attack at age 38. His doctors recommended conventional surgery if he wanted to prolong his life. He opted instead to have "blockages" removed by psychic surgeons in the Philippines. He dropped dead in 1980 at the age of 54. Andy Kaufman was diagnosed with lung cancer in 1983. He attempted to treat his cancer by going to the Philippines and submitting himself to the same kind of tricksters that did absolutely nothing for Sellers. He died at age 35. Randi followed up his expose of Peter Popoff on *The Tonight Show* by exposing the secrets of psychic healers. To millions of viewers, he demonstrated how a huckster with a few well practiced moves and a palmed bag of blood and chicken guts could make it appear the psychic surgeon's hand was entering a person's body and removing tumors and arterial blockages. NBC's switchboards lit up after Randi's demonstration. Not, as one would expect, by people outraged at the blood and gore being broadcast but by people wondering how they could book themselves a session with an actual psychic surgeon.

THE CARLOS HOAX (1988)

Charlatans made a lot of bucks out of psychic channeling in the 1980s. Actress Shirley MacLaine made it the central topic of daytime talk shows by promoting J.Z. Knight's fantasy she could channel the spirit of a Lemurian warrior who supposedly lived over 35,000 years ago. Channeling's central gimmick is the channeler pretends to go into a trance, talks in a funny voice, and tells lonely people what they want to hear. Despite being able to see through the eyes of ancient people, channelers have never been able to produce useful information for archaeologists. It would just take one channeler to show how to translate a lost language like Linear A to convince the world. By the late 1980s, channeling was starting to catch buzz in Australia. Randi, Argentinian performance artist Deyvi Orangel Peña Arteaga (who is probably better known as Jose Alvarez and Randi's future husband), and an Australian investigative journalism show named *60 Minutes* hatched a plan to demonstrate the media's growing gullibility over this trending phenomenon. Alvarez promoted himself as a channeler who was in contact with a 35,000-year-old spirit named Carlos. Alvarez blitzed the media with press packets loaded with false claims easily checked. For example, Alvarez claimed he had played packed houses at theaters in the USA that simply didn't exist. He was touring with something called the Atlantis Crystal which was just a glob of tar Randi found in a parking lot. The media questioned nothing. It ate it up. Similar to Randi's Project Alpha protocols, if any reporter ever once asked Alvarez if he was faking, Alvarez was to answer honestly and tell them James Randi sent him. A free performance at the Sydney Opera House drew a capacity crowd. Randi, Alvarez, and the *60 Minutes* show eventually revealed the nature of the hoax. Unfortunately, as these things go, some people refused to believe they were fooled, preferring to believe "they" got to Alvarez and forced him to make a false confession.

Fourteen Founding Members of CSICOP

1. Paul Kurtz (died 2012, Professor Emeritus of Philosophy at the State University of New York at Buffalo)
2. Marcello Truzzi (died 2003, sociology professor at at New College of Florida)
3. George O. Abell (died 1983, astronomer at UCLA)
4. James Randi (died 2020, magician)
5. Martin Gardner (died 2010, science writer)
6. Isaac Asimov (died 1992, science fiction writer)
7. Bart Bok (died 1983, astronomer at University of Arizona)
8. Brand Blanshard (died 1987, philosopher)
9. L. Sprague de Camp (died 2000, author of books debunking pseudoscience)
10. Sidney Hook (died 1989, philosopher at the Hoover Institution and professor emeritus at New York University)

11. Ernest Nage (died 1985, philosopher at Columbia)
12. Willard Van Orman Quine (died 2000, philosopher at Harvard)
13. B. F. Skinner (died 1990, psychologist at Harvard)
14. Carl Sagan (died 1996, astronomer at Cornell)

Ten Notable Winners of James Randi's Pigasus Award

Every year, between 1979 and 2012, James Randi gave the Pigasus Award to various individuals and groups that promoted mindless woo. For a number of years, the award was called the Uri. A nod to Uri Geller, no doubt. The award's name later changed to the Pigasus, possibly owing to Geller's reputation for being litigious. The award consists of a plaque featuring a flying pig. It is unknown if any award winners ever showed up at Randi's offices to accept the prize.

WILLIAM A. TILLER (1979)

William A. Tiller won one of the first Pigasus Awards back in 1979. Tiller is a retired materials scientist. In his retirement he pursued psychic research. He was featured in 2004's *What the Bleep Do We Know!?*. He also contributed to *The Healing Power of Water*, a book by water crank Masaru Emoto (died 2014). Back in 1979 Randi took him to task for suggesting although the evidence of psychics is poor and people promoting it have credibility problems there's so much substandard evidence it should be taken seriously anyway.

DOROTHY ALLISON (1980)

Dorothy Allison (died 1999) was another in a long line of self-proclaimed psychic detectives that tried to associate themselves with high-profile investigations and make vague assertions of helping the police. Randi gave her a Pigasus in 1980 for her non-help with the Atlanta child murders. Her contribution to the police investigation was giving them 42 names of possible suspects. In addition to providing a list of what were probably just random names, she boasted her mere presence in the city had a powerful psychic effect on the murderer and he wouldn't be able to kill any of "her little angels" as long as she was in the city.

THAT'S INCREDIBLE (1980)

Early reality show *That's Incredible* helped promote the career of James Hydrick. Hydrick claimed he had the ability to move objects with his mind. Curiously they were very light objects which Randi demonstrated were being moved by Hydrick simply blowing on the objects. Randi took the show to task for promoting a simple magic trick as a psychic ability.

KNBC (1981)

Randi took this LA TV station to task for airing a hoaxed video by so-called psychic Tamara Rand. The day after the Reagan assassination attempt, Rand and small-time Vegas talk show host Dick Maurice (died 1989) filmed what was purported to be an interview that took place some two months before Reagan was shot. For reasons unclear, Rand and Maurice assumed the camera operator and other technicians in the studio would remain silent after they discovered their work was being used to dupe the American public. They did not remain silent. The hoax quickly unraveled.

ROBERT BIGELOW (1996)

A couple decades before the Skinwalker Ranch mythology entered the popular conscience Randi was already criticizing the lunacy. Randi awarded Robert Bigelow the prize for purchasing the ranch as well as funding alien abduction cranks John Mack and Bud Hopkins. According to Randi, Bigelow purchased the ranch for $230,000. In 2016, Bigelow flipped it for $4.5 million. That's a nearly 2,000% return on his investment. Woo sells.

BILL MAHER (2000)

Skeptics tend to skew left. Many in the early 2000s were enamored with talk show host Bill Maher's political views. However, many quickly recoiled in horror when they discovered Maher had a strong woo side. Randi called him out early for using his *Politically Incorrect* TV show to help promote a number of psychics who claim they can speak to the dead.

COLIN A. ROSS (2008)

Colin A. Ross is a psychiatrist. He believes he can beam energy out of his eyes, enough energy that he can make a speaker beep. There was some discussion of him attempting Randi's Million Dollar Challenge, but nothing came of it. When Ross is not busy trying to make speakers beep with the power of his eyes, he seems to be a proponent of the reality of Satanic ritual abuse and multiple personality disorders.

KEVIN TRUDEAU (2008)

Randi gave Kevin Trudeau a Pigasus for selling junk weight loss advice and disease cures "they don't want you to know about" despite fines and sanctions by the SEC and FTC. Trudeau's cavalier attitude towards various court orders to cease his schemes caught up with him. In 2014 he was found in contempt of court and sentenced to 10 years in prison.

CVS/PHARMACY (2010)

In 2010 Randi took CVS/Pharmacy to task for happily selling homeopathic placebos alongside actual medication. To this day CVS/Pharmacy still sells products like Oscillococcinum.

SYRACUSE UNIVERSITY (2012)

It's sad when an actual university has to be singled out for supporting woo. It's even sadder when that woo is Facilitated Communication. Facilitated Communication is the claim that a person trained in certain techniques can help severely mentally disabled people communicate by helping them type. The problem is blinded tests reveal the "facilitator" is doing the communicating, not the disabled person. The scientific debate on this ended a couple decades ago. Randi took Syracuse to task for continuing to promote and train Ouija board operators with a science-y name.

Eleven Winners of the Bent Spoon Award

Every year since 1982, the Australian Skeptics society has been giving the Bent Spoon Award to "the perpetrator of the most preposterous piece of paranormal or pseudo-scientific piffle". Here are eleven notable winners.

TOM WARDS (1982)

Tom Wards won the inaugural Bent Spoon Award for his poor prognostication abilities as a psychic. If you took Wards' word for it, his success rate was 85%. However, an investigation of 130 predictions he issued revealed he only got about 8% right. Wards' career took a downturn after an investigative TV show revealed Wards was giving identical psychic readings to each and every customer.

PETER BROCK (1986)

Peter Brock (died 2006) was an Australian race car driver. In 1986 he won a Bent Spoon Award for promoting and selling an attachment he claimed harnessed orgone energy and could improve a car's performance. His "energy polarizer" (which he sold for nearly $500 a pop) was a slab of resin that held two magnets and some crystals. Brock didn't actually invent the device himself. It was invented by a Melbourne chiropractor.

THE DEPARTMENT OF THE ATTORNEY-GENERAL OF AUSTRALIA (1994)

In 1994, the Attorney-General's department changed its rules for sick leave. Previously staff members had to present a doctor's note as part of their application. In 1994, the AG also started accepting notes from naturopaths, herbalists, iridologists, and chiropractors.

MARLO MORGAN (1996)

Marlo Morgan is an American chiropractor. In 1990 she self-published a book called *Mutant Message Downunder*. It purported to describe her kidnapping by a tribe of Aborigines who referred to themselves as the "Real People Tribe". They forced her to complete a four-month walkabout in Australia's outback. During their shared walkabout, they used tea tree oil to cure her of various ailments, taught her to levitate, and charged her with sharing their wisdom with the rest of humanity. Though her book was self-published it sold well and was picked up by HarperCollins. It was on *The New York Times'* best-seller list for 31 weeks. An Aboriginal spokesperson described the book as a "fabricated fantasy of deceit". In 1996, a group of Aboriginals actually flew to New York to protest after hearing rumors that her book was going to be made into a movie.

VIERA SCHEIBNER (1997)

Viera Scheibner won the 1997 Bent Spoon Award for promoting anti-vaccine pseudoscience. Scheibner is a retired geologist. Geologists don't get a lot of training in medicine or virology. At one-point Scheibner claimed the symptoms of shaken baby syndrome weren't caused by, you know, tired and angry parents shaking their crying baby but by vaccinations.

MIKE WILLESEE (1999)

You don't get many people who win both a major (pro) skeptical award and then a Bent Spoon Award. Such is the case, however, with Mike Willesee (died 2019). For much of Willesee's career he was a hard-hitting investigative journalist and presenter on Australian TV. Willesee covered many topics like dowsing, creationism, and cryptozoology from a skeptical point of view. He helped expose 1986's Bent Spoon Award winner Peter Brock and his bogus energy polarizer. For his efforts the Committee of Skeptical Inquiry (CSICOP) gave Willesee its Responsibility in Journalism award. Willesee, however, took a turn in the 1990s. He became a devout Catholic and started reporting on religious topics. In 1998 he made a short documentary titled *Signs From God* that promoted the reality of stigmata. He was awarded a Bent Spoon for these efforts.

JOHN CHRISTIE AND LOU BRITS (2001)

John Christie and Lou Brits claimed they invented a free energy device they called the Lutec Free Energy Generator. It was claimed to produce 30 times more energy than put into the device. Christie and Brits were no garage inventors who fooled themselves playing around with magnets. The marketing of their impossible device was extremely sophisticated, including getting a patent in Australia. Unlike the USA which has a blanket prohibition on accepting patent applications for free energy machines, Australia has no such rule.

MERYL DOREY (2009)

Meryl Dorey is the founder of the Australian Vaccination Network. If that sounds like a group that promotes vaccines, you'd be sadly mistaken. The Australian Skeptics gave her a Bent Spoon for her tireless efforts to unnecessarily worry parents about vaccines. Her group, at one point, tastelessly compared being vaccinated to rape.

HOMEOPATHY PLUS! (2012)

Homeopathy Plus! is an Australian company offering homeopathic services. It won a Bent Spoon for its history of making highly unsubstantiated claims about the efficacy of homeopathy, particularly regarding vaccine-preventable childhood diseases. In 2014 the company demanded the World Health Organization allow homeopaths into West Africa to help treat Ebola victims. In 2015 the company was fined $115,000 for selling homeopathic nostrums and promoting them as effective alternatives to vaccines.

BRIAN MARTIN (2016)

Brian Martin, a humanities professor at the University of Wollongong, won a Bent Spoon for approving anti-vaccination activist Judith Wilyman's PhD thesis "A critical analysis of the Australian government's rationale for its vaccination policy". People with actual medical degrees questioned whether Martin was qualified to assess Wilyman's paper on the science of vaccines. Martin himself was a supporter of anti-vaccine activism. In 2014 *The Australian* newspaper revealed Martin had been "a former paid member of the anti-vaccine Australian Vaccination Network and previously supervised a student, Michael Primero, who has links to Andrew Wakefield, who sparked a worldwide scare by linking the measles, mumps and Rubella vaccine to autism".

PETE EVANS (2015 AND 2020)

Pete Evans is an Australian celebrity chef. He is also the only person to win two Bent Spoon Awards. In 2015 he won for his public stance against fluoridation and his promotion of Joseph Mercola. In 2020 he won a second for promoting a device called the BioCharger NG. The Australian Medical Association described it as little more than a "fancy light machine". Evans claimed the device was programmed with a "recipe" based on the Covid virus and could be used to treat the disease.

Eight Pro Skeptical Awards

Skeptic organizations just don't give mocking awards to woo proponents. They also honor hardworking skeptics. Here is a list of major awards offered by skeptic groups and fellow travelers.

AMERICAN HUMANIST ASSOCIATION HUMANIST OF THE YEAR (1953-PRESENT)

The AHA's Humanist of the Year award goes to the person that year that "has made a significant contribution to the improvement of the human condition". Notable winners over the decades include Margaret Sanger (promoter of the birth control pill), B.F. Skinner, Carl Sagan, Isaac Asimov, P. Z. Myers, and Bill Nye. Lawrence Krauss was awarded one in 2015 but it was revoked in 2018 following allegations of sexual misconduct.

ROBERT P. BALLES PRIZE (2005-PRESENT)

The Robert P. Balles Prize is given out by the Committee for Skeptical Inquiry. It is given to the creator of a published work that best exemplifies healthy skepticism, logical analysis, or empirical science. The prize comes with a $2,500 check. Past winners have included Joe Nickell, Steven Novella, and science journalist Julia Belluz.

BRAVO AWARD (1995-PRESENT)

The Bravo Award is given out by the New Zealand skeptics to "media professionals and those with a high public profile who have provided food for thought, critical analysis and important information on topics of relevance to our interests."

EMPEROR HAS NO CLOTHES AWARD (1999-2019)

The Emperor Has No Clothes Award was given out by the Freedom from Religion Foundation for the individual that year who "told it like it is" on the subject of religion. It does not appear an award was given out for 2020 or 2021, so this award may be defunct. Past winners have included Janeane Garafalo (2001), Ron Reagan (2004 and 2009), Julia Sweeney (2006), Sean Carroll (2014), and Adam Savage (2018).

IIG AWARD (2007-2012)

For a brief time, the fabled Independent Investigations Group awarded an annual "Iggie" for a media production that had "the most positive presentation of scientific, rational, and critical thinking" during the year. Winners included *The Simpsons* episode "The Monkey Suit" (2007), the SyFy network's *Eureka* (2010), Phil Plait's Discovery Channel show *Bad Universe* (2011), and Snopes.com (2012).

THE PHILIP J. KLASS AWARD (2006-2016)

The Philip J. Klass award was given out by the National Capital Area Skeptics in honor of "outstanding contributions in critical thinking and scientific understanding". Past winners included James Randi (2007), Penn & Teller (2012), and Phil Plait (2013).

OCKHAM (2012-PRESENT)

The Okham is given out by the UK's *The Skeptic* magazine. Since 2012 it has been awarding the Ockham (a Lucite trophy depicting a razor) to individuals or groups that have "done outstanding work to promote critical thinking within the last 12 months". Past winners have included Professor Edzard Ernst and skeptical blogger Hayley Stevens.

SKEPTIC OF THE YEAR (1996-PRESENT)

Skeptic of the Year is awarded by the Australian Skeptics. It's awarded to someone who has been particularly active in the movement for the year.

Ten Cash Prizes for Demonstrating Paranormal Ability

In 2015 James Randi retired his Million Dollar Paranormal Challenge. There are, however, still skeptical organizations offering lucrative prizes for demonstrating the reality of some paranormal phenomenon.

1. $11,550 - Swedish Humanist Association
2. $12,000 - North Texas Skeptics
3. $16,096 - Harry Houdini Prize (Russia)
4. $16,535 - Association for Skeptical Enquiry (UK)
5. $30,368 - SKEPP's Sisyphus Prize (Belgium)
6. $60,737 - Alfredo Barrago's Bet (Italy)
7. $70,791 - Puzzling World's Psychic Challenge (New Zealand)
8. $75,820 - Australian Skeptics (Australia)
9. $157,913 - Offered by TV presenter Sima Nan (China)
10. $250,000 - Center for Inquiry Investigations Group

All figures are from Wikipedia and in US dollars.

Eight Skeptical Podcasts that Won or Were Finalists for a Parsec Award

Between 2006 and 2018 the Parsec Awards were presented each year at DragonCon. The award recognized excellence in science fiction and science fiction adjacent podcasting. A number of skeptical podcasts have, over the years won a Parsec or were runners up.

1. *Skepticality* (2007, won Best Speculative Fiction News Podcast)
2. *Geologic Podcast* (2007, finalist Best Audio Production)
3. *365 Days of Astronomy* (2009, won Best Speculative Fiction Infotainment Podcast)
4. *Skeptoid* (2010, won Best Fact Behind the Fiction Podcast)

5. *The Skeptics' Guide to the Universe* (2011, finalist Best Fact Behind the Fiction Podcast)
6. *Monster Talk* (2012, won Best Fact Behind the Fiction Podcast)
7. *Token Skeptic* (2013, finalist Best Fact Behind the Fiction Podcast)
8. *Astronomy Cast* (2018, finalist Best Fact Behind the Fiction Podcast)

The 9-5 Life of Twenty Eight Well Known Skeptics

Being a skeptic, while rewarding in many intangible ways, rarely pays the bills. Here is a list of well-known skeptics and what they do outside of skeptical activism to keep the lights on and put food on the table.

DEREK BARTHOLOMAUS

Derek Bartholomaus has been a long-time thorn in the side of UFO Contactee Billy Meier. Bartholomaus' site billymeierufocase.com doggedly deconstructs how Meier's UFO evidence can easily and exactly be duplicated with common household objects. Bartholomaus is also the creator of the Jenny McCarthy Body Count page. The page tracks the number of vaccine preventable injuries and deaths that have occurred since the actress went on her anti-vax crusade. As a day job, Bartholomaus works in the TV and film industry as a post-production producer.

ROBERT BARTHOLOMEW

Robert Bartholomew is an author of a number of skeptical books about UFOs, cryptids, and mass hysteria. He's a Fellow of the Committee for Skeptical Inquiry. Professionally, Bartholomew works as a medical sociologist and does a lot of research into how society applies medical labels to unpopular but otherwise normal human behaviors. He also works as an instructor at New Zealand's Mission Heights Junior College and teaches History at Auckland's Botany College.

EVAN BERNSTEIN

Evan Bernstein is co-host of *The Skeptics' Guide to the Universe* podcast. He has an accounting practice and prepares taxes. Bernstein is famous for largely disappearing from the podcast during tax season.

KENNY BIDDLE

Kenny Biddle has developed a reputation as a buster of ghost tech. In the pages of the *Skeptical Inquirer* magazine and on his own YouTube channel he examines whether or not ghost hunting equipment (like EMF detectors) can actually do what the ghost hunters claim they can do. He's rather uniquely qualified for this as his day job is working as an aviation mechanic and electronics technician.

SUSAN BLACKMORE

Susan Blackmore holds a PhD in Parapsychology. In the 1980s, she did a lot of research into psychokinesis and out-of-body experiences. By the late 1980s she began to find the paranormal hypothesis an unsatisfying explanation for the phenomena she was investigating. She has since become a leader in the skeptical field and has expanded a lot on Richard Dawkins' idea of memes. She gave a TED talk on memes in 2008. Since 2009 she has been a visiting professor at the University of Plymouth.

ROSS BLOCHER

Ross Blocher is co-host of the *Oh No, Ross and Carrie* podcast. Blocher has worked for Disney for a number of years in various animation and managerial roles.

DEREK COLANDUNO

Derek Colanduno is an early skeptical podcasting pioneer. He started co-hosting the *Skepticality* podcast in 2006. He is, by trade, a software engineer for an environmental engineering firm. He also works as a graphic designer as a side hustle.

PERRY DEANGELIS

Perry DeAngelis was a co-host of *The Skeptics' Guide to the Universe* podcast until passing away in 2007. He helped cofound the Connecticut Skeptical Society which morphed into the New England Skeptical Society which then gave birth to *The Skeptics' Guide to the Universe* podcast. For a time DeAngelis was co-owner of a game shop. He also worked with his father managing properties.

REED ESAU

Reed Esau founded the Skepticamp movement in 2007. Skepticamps are intended to be day-long fluid conferences that encourage audiences to interact much more with scheduled speakers. Esau is a software architect and he based Skepticamp on the "unconference" format popular with software developers. At an Amazing Meeting he realized there was a niche for local skeptical conferences that were less hierarchical and "celebrity skeptic" driven.

TIM FARLEY

Tim Farley created the website *What's the Harm*, a site that tracks deaths and injuries caused by alt med. When not working on his website, he's a computer security analyst.

JULIA GALEF

Julia Galef was the long-time co-host of the *Rational Speaking* podcast. She has a BA in statistics from Columbia University. She co-founded the Center for Applied Rationality. It is a non-profit that offers workshops where participants learn how to make more rational decisions.

PAMELA GAY

Pamela Gay has been co-hosting the *Astronomy Cast* podcast since 2006. Gay has a PhD in Astronomy. She works as a senior scientist for the Planetary Science Institute.

HARRIET HALL

Harriet Hall (aka "The SkepDoc") writes prolifically about pseudoscience and questionable medical practices. Many of her articles have appeared in recent years in *Skeptical Inquirer*, for which she is a contributing editor. She is also a weekly contributor to the *Science-Based Medicine* blog. She has a regular column in *O, The Oprah Magazine* titled "The Health Inspector". Hall received her MD from the University of Washington and did her internship in the Air Force (the second woman ever to do so). During a long career as an Air Force physician, she held various positions from flight surgeon to Director of Base Medical Services and did everything from delivering babies to taking the controls of a B-52. She retired with the rank of colonel. She recently published *Women Aren't Supposed to Fly: The Memoirs of a Female Flight Surgeon.*

SARA HEAD

Sara Head founded the *Archaeological Fantasies* podcast and blog of the same name. Head was one of the first people to do a podcast that took on the world of pseudoarchaeology on a regular basis. She works as a cultural resource management archaeologist. These are the people who make sure construction projects don't dig up and destroy historical objects.

SHARON HILL

Sharon Hill works as a professional geologist and policy expert for the Pennsylvania government. For a long time, she ran the *Doubtful News* blog. *Doubtful News* was a wonderful clearing house for woo topics that cropped up in the media. Hill addressed them in depth with a sober second look. In 2017 she published the book *Scientifical Americans: The Culture of Amateur Paranormal Researchers*. In recent years she's backed off of organized skepticism. She's found the "skeptic" label laden with negative connotations and limits her own outreach attempts.

ROBYNN MCCARTHY

Robynn McCarthy (better known as "Swoopy") was an early skeptical podcasting pioneer. She started co-hosting the *Skepticality* podcast in 2006. She is currently in medical school and helps pay the bills by working as an audio editor and graphic designer.

SEBASTIAN MAJOR

Sebastian Major is host of the *Our Fake History* podcast. When he's not debunking historical myths on his podcast, he works as a History teacher in Canada.

BOB NOVELLA

Robert "Bob" Novella is co-host of *The Skeptics' Guide to the Universe* podcast. He works in IT as a Systems Specialist. Every October he helps design and run a "Forest of Fear" attraction.

STEVEN NOVELLA

Dr. Steven Novella is generally recognized as the driving force behind *The Skeptics' Guide to the Universe* podcast. When he's not working on the show, Novella works as a clinical neurologist and professor at Yale.

PASCUAL ROMERO

Pascual Romero is the co-host of the *Squaring the Strange* podcast. He's a heavy metal guitar player and vocalist. He has been in bands like Pathology and In This Moment. He has worked in movies and television, including productions like *Heroes* and *3:10 to Yuma*. HIs day job is as a communications engineer in New Mexico.

EUGENIE SCOTT

Eugenie Scott has a PhD in biological anthropology and served as the Executive Director of the National Center for Science Education. The NCSE site offers many great resources for people battling creationists online. Scott and her team were advisors to the *Kitzmiller* "intelligent design" trial. Scott has appeared on a number of TV shows and podcasts, always calmly and knowledgeably advocating for evolution. Scott is probably one of the best advocates for science and reason who has ever lived.

ABBIE SMITH

Abbie Smith is a scientist who researches HIV and endogenous retroviruses (ERVs). For a number of years, as a grad student, she blogged under the name "ERV". She took on many skeptical topics including HIV and vaccine denial. As a guest on various podcasts, she's gone toe-to-toe with a number of cranks. One of Smith's greatest achievements was tackling and putting an end to a bit of pseudoscience involving chronic fatigue syndrome (CFS). A researcher was making a lot of hay that she had detected a mouse virus called XMRV in the samples of people claiming to suffer from CFS. Smith demonstrated her research was flawed and likely a product of lab contamination. After graduating, Smith curtailed her blogging and skeptical activities.

BLAKE SMITH

Blake Smith is co-host of the *Monster Talk* podcast as well as the *In Research Of* podcast. Smith served in the US navy for a number of years in the early 1990s. Smith earns a bit of side coin writing for *Fortean Times, Skeptical Inquirer, Skeptic Magazine*, and *Weird Tales*. When not podcasting, Smith is a software developer who specializes in automating workflows and making software a little less monstrous by improving user experiences.

Photo Credit: Blake Smith

HAYLEY STEVENS

Hayley Stevens is a UK blogger and was the host of a podcast called *The Spooktater* which ran from 2018 to 2020. She also co-hosted the *Righteous Indignation* podcast which ran from 2009 to 2012. Hayley began life as a ghost hunter but became more skeptical over time. She's currently working to complete a BSc in Honors Psychology and occasionally writes for the UK's *The Skeptic* magazine. She distanced herself from organized skepticism over the last few years due to toxic attitudes among her skeptical brethren.

Photo Credit: Hayley Stevens

KAREN STOLLZNOW

Karen Stollznow is an author of a number of skeptical books as well as host of podcasts such as *Point of Inquiry* and *Monster Talk*. She's written several skeptical books, notably *God Bless America: Strange and Unusual Religious Beliefs and Practices in the United States* and *Language Myths, Mysteries and Magic*. She holds a PhD in Linguistics. When she's not podcasting or writing skeptical books, Stollznow works as a teacher and researcher. She is currently doing research for the Griffith Centre for Social and Cultural Research.

DAVE THOMAS

Dave Thomas was the co-host of *Science Watch*, a skeptical radio show that ran for a number of years on Albuquerque's KABQ radio station. The weekly shows were distributed as a podcast. He's been president of the New Mexicans for Science and Reason since 1998. The group has done remarkable work challenging creationism and other forms of pseudoscience in the state of New Mexico. He's been a major critic of the Aztec UFO hoax. Thomas graduated from the New Mexico Institute of Mining and Technology. He has a Bachelor's degree in Physics and a Master's degree in Math. For a day job, Thomas works as scientist/programmer for a company that makes instruments that monitor seismic activity.

BRIAN THOMPSON

Brian Thompson is host of the very meta investigative journalism parody podcast *Whatever Happened to Pizza at McDonald's*. He is the former host of the skeptically themed podcast *Amateur Scientist*. Thompson also works as a comedy writer and voice-over artist. Thompson used to be a high school teacher and field coordinator for the James Randi foundation.

CELESTIA WARD

Celestia Ward is a co-host and content producer of the *Squaring the Strange* podcast. She is one of the heads of Two Heads Studio, a Vegas-based business specializing in caricature art. You'll find her skeptical writings in *Skeptical Inquirer* and on CFI's blog.

Twelve Skeptics Who Made It Work

For most skeptics, skeptical activism is "merely" an all-consuming hobby. There are, however, a surprisingly large number of skeptics who have managed to parlay their love for skepticism into a full-time career.

BRIAN DUNNING

Brian Dunning is host of the *Skeptoid* podcast. Skeptical podcasting and related skeptical activities seem to be his current full-time job. Previous to his life as a skeptical podcaster, Dunning was CTO at a dot com startup called Buylink and worked as an editor for technical trade publications.

SUSAN GERBIC

Susan Gerbic is the founder of the It's About Time non-profit and head of the organization's Guerrilla Skepticism on Wikipedia project. The project's aim is to improve the scientific and factual content on Wikipedia pages, notably pages on Wikipedia that tend to get overrun with woo. She also busts high profile grief vampires who claim to be clairvoyant mediums who can talk to your dead loved ones. She has a BA in Behavioral Studies from California State University. Before skeptical activism became her life's pursuit, she worked as a baby photographer and manager of a JCPenney portrait studio for 34 years.

GEORGE HRAB

George Hrab is a musician, part of a band called The Philadelphia Funk Authority, and host of the long-running *Geologic* podcast, which blends skepticism, music theory, grammar, and comedy into one rather delightful and rollicking weekly podcast. Hrab seems to support himself via his podcast and sales of his music.

MICHAEL MARSHALL

Michael Marshall is a co-host of the UK podcast *Skeptics with a K*. He helped found the Merseyside Skeptics Society. Marshall has been a tireless critic of homeopathy. In 2010 he organized the 10:23 Campaign. Marshall and other skeptics imbibed overdoses of homeopathic remedies to demonstrate their ineffectiveness. Marshall is on the board of the Good Thinking Society. That skeptical group's efforts led to the UK government defunding homeopathy. Marshall is currently the editor-in-chief of *The Skeptic*, the UK's oldest skeptical magazine. Before becoming a full-time skeptical activist, Marshall got a BA in English and worked in marketing and web design.

JOE NICKELL

Joe Nickell has had a wide range of careers. He's been a carnival pitchman, a magician, and a private detective. He seems to mostly support himself as a paranormal investigator for hire, an author (he's written about 30 books), and a frequent "token skeptic" on a wide range of paranormal TV shows.

JAY NOVELLA

Jay Novella is co-host of *The Skeptics' Guide to the Universe* podcast. These days he is fully employed as COO for the production wing of the podcast. Previously he was a software engineer.

CARRIE POPPY

Carrie Poppy is co-host of the *Oh No, Ross and Carrie* podcast. She has worked as a sketch comedy performer, director of communications for the James Randi foundation, and a project coordinator for an animal welfare organization. She has a Master of Science degree in Journalism. Poppy seems to support herself via her podcast and work as an independent journalist.

BENJAMIN RADFORD

Benjamin Radford is a skeptical author and podcaster. He is the Deputy Editor of *Skeptical Inquirer* magazine. He writes for a wide variety of publications on skeptical topics and appears on many TV shows as the voice of reason. Radford has a Master's degree in Education.

CARA SANTA MARIA

Cara Santa Maria is a co-host of *The Skeptics' Guide to the Universe*. She has a rich and varied career as a science communicator, working as a correspondent for networks like Al Jazeera America, CNN, and Netflix.

JAMES UNDERDOWN

James Underdown is the founder of the Independent Investigations Group, now the Center for Inquiry Investigations Group (CFIIG). Since 2000, the CFIIG has done a lot of work "in the field" debunking a wide variety of wild claims, and currently offers a $250,000 prize for anyone who can prove paranormal ability under test conditions. This is the largest active paranormal challenge in the world. With the aid of Olympian Dominique Dawes, the CFIIG tested and exposed the "power balance bracelet" as nothing but expensive placebo jewelry. Underdown has a BA in English. He has worked as a carpenter and standup comedian. Today he is the executive director of The Center for Inquiry Los Angeles, a position he's held since 1999.

REBECCA WATSON

Rebecca Watson is the founder of the Skepchick movement. The group is dedicated to promoting critical thinking to women around the world. She was a former co-host of *The Skeptics' Guide to the Universe* and *Little Atoms* podcasts. She holds a Bachelor of Science degree in communications. She was working as a magician when an encounter with James Randi led her to skeptical activism. Watson appears to be a full-time skeptical activist and advocate for women's rights. She has an asteroid named after her (153289 Rebeccawatson).

MICK WEST

Mick West is the creator of the popular *Metabunk* website. He's the author of the book *Escaping the Rabbit Hole: How to Debunk Conspiracy Theories Using Facts, Logic, and Respect*. He's the go-to guy on so-called "chemtrails" as well as a wide range of conspiracy theories and pseudoscience topics that make the news. Skepticism seems to be his retirement project. He was a video game developer and co-founded Neversoft Entertainment in 1994. He later sold the company to Activision.

General Image Credits

Cover image: Karl Mamer

Chapter images: Karl Mamer

Acknowledgement of Sources

The author would like to acknowledge the use of the following resources as major sources in the writing of this book:

Abominable Science! by Daniel Loxton and Donald R. Prothero
Archaeological Fantasies podcast (Sara Head)
Bad Archaeology website
Bad UFOs blog (Robert Shaeffer)
Doubtful News website via archive.org (Sharon Hill)
Exposing PseudoAstronomy podcast (Stuart Robbins)
Good Math/Bad Math blog (Josh Gemmell)
Material prepared by guests of the author's *Conspiracy Skeptic* podcast
Metabunk blog (Mick West)
Jason Colavito's blog
Monster Talk podcast (Blake Smith, Karen Stollznow) and material prepared by guests
Quackwatch website (Center for Inquiry)
Randi.org (James Randi)
Rationalwiki website
Saucer Life podcast (Aaron Gulyas)
Science-Based Medicine blog (Dr. Steven Novella et al.)
Skeptic magazine website
Skeptoid website and podcast (Brian Dunning)
Snopes.com
The Skeptic's Dictionary website (Robert Todd Carroll)
Tetrapod Zoology blog (Darren Naish)
Various online materials authored by Benjamin Radford, Joe Nickell, and James Oberg
What's the Harm? website (Tim Farley)
Wikipedia

About the Author

Praise "Bob"!
Photo Credit: Karl Mamer

Karl Mamer is a Canadian but don't hold that against him. Karl has a BA in something almost not worth mentioning from a Canadian university almost not worth mentioning. He graduated with a GPA definitely not worth mentioning. He's a technical writer by trade and he has been a skeptic since the BBS days of the 1980s. Karl is the long-time host of the *Conspiracy Skeptic* podcast. Karl was the first person in Canada to write a regular column about the Internet and net culture for a major Canadian daily paper. His "Cyberspace" column appeared every other Sunday in the pages of the *Toronto Sun* between 1995 and 2000. He also authored four nonfiction guides to the Internet which were published by Reed Canada. Of note was his 1996 guide *The Truth Is in Here: Conspiracies, Mysteries, Superstitions, Kooks, and Cults on the Internet*. Karl has lived in Montreal, Seattle, Seoul, and Windsor, Ontario. He currently lives in a sub-suburb of Toronto with his super talented wife and his even more talented daughter.

P.S. He offers apologies to the many excellent engineers in the world for the running "drink!" gag in this book.

Index

B

C

D

E

F

G

H

I

J

K

L

M

N

O

P

S

T

U

V

W

X

Y

Z

www.ingramcontent.com/pod-product-compliance
Lightning Source LLC
LaVergne TN
LVHW010710061126
851857LV00014B/1116